Lecture Notes in Physics

The Lecture Notes in Physics

The series Lecture Notes in Physics (LNP), founded in 1969, reports new developments in physics research and teaching – quickly and informally, but with a high quality and the explicit aim to summarize and communicate current knowledge in an accessible way. Books published in this series are conceived as bridging material between advanced graduate textbooks and the forefront of research to serve the following purposes:

• to be a compact and modern up-to-date source of reference on a well-defined topic;

• to serve as an accessible introduction to the field to postgraduate students and nonspecialist researchers from related areas;

• to be a source of advanced teaching material for specialized seminars, courses and schools.

Both monographs and multi-author volumes will be considered for publication. Edited volumes should, however, consist of a very limited number of contributions only. Proceedings will not be considered for LNP.

Volumes published in LNP are disseminated both in print and in electronic formats, the electronic archive is available at springerlink.com. The series content is indexed, abstracted and referenced by many abstracting and information services, bibliographic networks, subscription agencies, library networks, and consortia.

Proposals should be sent to a member of the Editorial Board, or directly to the managing editor at Springer:

Dr. Christian Caron
Springer Heidelberg
Physics Editorial Department I
Tiergartenstrasse 17
69121 Heidelberg/Germany
christian.caron@springer.com

Kolumban Hutter
Alfons A. F. van de Ven
Ana Ursescu

Electromagnetic Field Matter Interactions in Thermoelastic Solids and Viscous Fluids

 Springer

Authors

Kolumban Hutter
Bergstrasse 5
8044 Zürich
Switzerland
E-mail: hutter@vaw.baug.ethz.ch

Dr. Ana Ursescu
Hügelstrasse 21
64283 Darmstadt
Germany
E-mail: ana_ursescu@yahoo.com

Dr. Alfons A. F. van de Ven
Department of Mathematics
and Computer Sciences
Technical University Eindhoven
P.O. Box 513
5600 MB Eindhoven
The Netherlands
E-mail: a.a.f.v.d.ven@tue.nl

K. Hutter et al. *Electromagnetic Field Matter Interactions in Thermoelastic Solids and Viscous Fluids*, Lect. Notes Phys. 710 (Springer, Berlin Heidelberg 2006), DOI 10.1007/b 11803263

ISSN 0075-8450
ISBN 978-3-642-07208-6   e-ISBN 978-3-540-37240-0

Springer is a part of Springer Science+Business Media
springer.com
© Springer-Verlag Berlin Heidelberg 2006
Softcover reprint of the hardcover 1st edition 2006

Cover design: WMXDesign GmbH, Heidelberg

Preface

This book is the second and substantially enlarged edition of the Springer Lecture Notes No. 88: *Field Matter Interactions in Thermoelastic Solids*, which appeared in 1978 and is out of print for about two decades. Since then, the basic issues addressed by the authors in that book have little changed: Because of the nonunique definition of the electromagnetic field quantities in ponderable bodies, constitutive postulates, e.g., for the stress tensor and other field quantities, must adequately be postulated, if two theories aiming to describe the same physical phenomenon yield for the same physical problem the same values of the observables. In the Lecture Notes, equivalence relations were established, which guarantee such equivalences for thermoelastic bodies, but no applications of the complex theory were given. Nevertheless, there was a continuous demand for the book, which was fulfilled by producing photocopies.

In the meantime, however, the authors continued to work on continuum problems of electromagneto-mechanical interactions, in which the theoretical models or simplifications were applied to practical problems. A.A.F. Van de Ven (AV) worked with students and postdoctoral fellows for more than two decades on problems of magnetoelastic instabilities, i.e., buckling of ferromagnetic and superconducting beams, plates and more complex structures, and on magnetoelastic vibrations of the same type of structures. In the latter problems, it is specifically the eigenfrequencies that need to be determined, inclusive of their dependence on the applied magnetic field or electric current. K. Hutter (KH) who was not involved with electrodynamics for 10 years, took up electromagnetic continua again about 15 years ago and concentrated on applications in fluids, as well as electrorheology. In this field, equivalence of formulations is equally a question of electromagneto-mechanical interactions. Here, the central theme is the postulation of adequate constitutive relations, which achieve the electrorheological effect, namely the transition from low viscous fluid behaviour to very high viscous response when the electric field is switched on. The application to plane Poiseuille flow of a theory was developed in a Ph.D. dissertation by W. Eckart, to 2D pipe flow with various arrangements of electrodes along the walls of the bottom and the lid of the 2D channel, which was made by Ana Ursescu (AU). Ursescu is joining us as the third author of this book.

This then outlines the content of the present book. In its first part it contains the material of the first edition and is due to AV and KH. Only small changes have been made to this text. Few adjustments were necessary because of the additions that were made. A few supplementing references are given to account for the recent literature. In the applications, the new chapter on *Magnetoelastic (In)stability and Vibrations* is due to AV and the chapter on *Electro-rheological Fluids* has been drafted by AU along with KH. The entire text has been screened for consistency and homogenization by AV and KH.

The increasing interest in electromagnetic problems in the last decade of the last century manifested itself in the appearance of a new journal in 1990, *International Journal of Applied Electromagnetics in Materials* (IJAEM), founded by K. Miya from Tokyo, and Richard K.T. Hsieh from Stockholm. The first plans for this new journal were made in 1986, during a IUTAM-Symposium in Tokyo alongwith Miya, Hsieh, Gerard Maugin, Francis Moon, Junji Tani, and one of the authors, AV. The birth of the new journal was accompanied by a series of ISEM symposia (International Symposia on Applied Electromagnetics in Materials, under the chairmanship of Miya) from 1988 until the present time. The 12th ISEM took place in 2005 in Salzburg, Austria.

One proviso to the style of the book should be mentioned. Reading the text is not easy. When developing results we are brief and we often outline the steps as to how a result is reached but do not present any details. Thus, the reader is expected to perform the in-between steps, or perhaps even consult the literature. Such an approach is almost unavoidable in electromagneto-mechanical interactions. The computations that are in principle not difficult, but generally involved and long, cannot be presented in detail as it would make the book twice as voluminous, and most likely rather boring over long stretches. We regard this as an acceptable compromise.

We wish to thank our sponsors and many of our friends in this field who have directly or indirectly contributed to this project. AV thanks the Technische Universiteit Eindhoven for their general support to the research on electromagnetoelastic interactions over many years (from 1975 onwards) and he especially thanks, many students for their contributions in particular M. Couwenberg, P. Rongen, P. Smits, and P. van Lieshout, who essentially supported the research on magnetoelastic instabilities. Moreover, he highly appreciates the cooperation and the many fruitful discussions in this area with A.O. J. Tani, Y. Shindo, K. Miya, Shu-Ang Zhou, B. Marusewski, and J.P. Nowacki. KH thanks the Darmstadt University of Technology and the Deutsche Forschungsgemeinschaft (German Research Foundation) for nearly 20 years of support. He also acknowledges the support of W. Eckart and AU in particular for their Ph.D. and postdoctoral fellowships in which the Electrorheological work presented in this book was created. He also thanks Professor K.R. Rajagopal and M. Růžička for their interest in our work on electrorheology.

We also thank E. Vasilieva and A. Maurer for compiling the text of the first edition and helping with the editing of the entire manuscript. Finally our thanks go to Dr. Ch. Caron from the Springer Verlag for his interest in this book and his willingness to publish it.

Darmstadt, Eindhoven   *K. Hutter*
Spring 2006   *A. Ursescu*
                                                            A.A.F. Van de Ven

Preface to the First Edition

The last two decades have witnessed a giant impetus in the formulation of electrodynamics of moving media, commencing with the development of the most simple static theory of dielectrics at large elastic deformations, proceeding further to more and more complex interaction models of polarizable and magnetizable bodies of such complexity as to include magnetic dissipation, spin-spin interaction and so on and, finally, reaching such magistral synthesis as to embrace a great variety of physical effects in a relativistically correct formulation. Unfortunately, the literature being so immense and the methods of approach being so diverse, the newcomer to the subject, who may initially be fascinated by the beauty, breadth and elegance of the formulation may soon be discouraged by his inability to identify two theories as the same, because they look entirely different in their formulation, but are suggested to be the same through the description of the physical situations they apply to. With this tractate we aim to provide the reader with the basic concepts of such a comparison. Our intention is a limited one, as we do not treat the most general theory possible, but restrict ourselves to non-relativistic formulations and to theories, which may be termed deformable, polarizable and magnetizable thermoelastic solids. Our question throughout this monograph is basically; what are the existing theories of field-matter interactions; are these theories equivalent, and if so; what are the conditions for this equivalence? We are not the first ones to be concerned with such fundamental ideas. Indeed, it was W.F. BROWN, who raised the question of non-uniqueness of the formulation of quasistatic theories of magnetoelastic interactions, and within the complexity of his theory, he could also resolve it. PENFIELD and HAUS, on the other hand, were fundamentally concerned with the question how electromagnetic body force had to be properly selected. This led them to collect their findings and to compare the various theories in an excellent monograph, in which they rightly state that equivalence of different formulations of electrodynamics of deformable continua cannot be established without resort to the constitutive theory, but at last, they dismissed the proper answer, as their treatment is incomplete in this regard. For this reason the entire matter was re-investigated in the doctoral dissertation of one of us (K. HUTTER), but this work was soon found unsatisfactory and incomplete in certain points, although the basic structure of the equivalence proof as given in Chap. 3 of this tractate, was essentially already outlined there. Moreover, HUTTER was still not able to compare certain magnetoelastic interaction theories so that what he attempted remained a torso anyhow.

The difficulties were overcome by VAN DE VEN in a series of letters, commencing in fall of 1975, in which we discussed various subtleties of magnetoelastic interactions that had evolved from each of our own work. The correspondence was so fruitful that we soon decided to summarize our efforts in a joint publication. It became this monograph, although this was not our initial intention. Yet, after we realized that a proper treatment required a presentation at considerable length, we decided to be a little broader than is possible in a research report and to write a monograph, which would be suitable at least as a basis for an advanced course in continuum mechanics and electrodynamics (graduate level in the US). We believe that with this text this goal has been achieved. We must at the same time, however, warn the reader not to take this tractate as a basis to learn continuum mechanics and/or electrodynamics from the start. The fundamentals of these subjects are assumed to be known.

Our acknowledgements must start with mentioning Profs. J.B. ALBLAS (Technological University Eindhoven) and Y.H. PAO (Cornell University). They were the ones who initiated our interest in the subject of magnetoelastic interactions. While performing the research for this booklet and during our preparation of the various draughts we were supported by our institutions, the Federal Institute of Technology, Zürich and the Technological University, Eindhoven, and were, furthermore, encouraged by Prof. J.B. ALBLAS, Eindhoven, Prof. D. VISCHER, Zürich, Prof. I. MÜLLER, Paderborn, Prof. H. PARKUS, Vienna, Dr. Ph. BOULANGER, Brussels and Dr. A. PRECHTL, Vienna. The support and criticism provided by them, directly or indirectly, were extremely helpful. We are grateful to these people not only for their keen insight and willingness to discuss the issues with us, but also for their encouragement in general.

During the initial stage and again towards the end of the write-up of the final draught of this monograph K. HUTTER was financially supported in parts by the Technological University, Eindhoven, to spend a total of a two months period (September 1976 and April 1978) at its Mathematics Department. Without the hospitality and the keen friendship of the faculty and staff members of this department and especially of Prof. J.B. ALBLAS and his group, the work compiled in these notes would barely have been finished so timely. The burden of typing the manuscript was taken by Mrs. WOLFS-VAN DEN HURK. It was her duty to transform our hand-written draughts into miraculously looking typed sheets of over 200 pages. Her effort, of course, is gratefully acknowledged.

Eindhoven and Zürich   K. Hutter
in the summer of 1978   A.A.F. Van de Ven

Contents

Part I Equivalence of Different Electromagnetic Formulations
in Thermoelastic Solids

List of Symbols

A_{ij}	Matrix of (anisotropic) heat conduction
$\boldsymbol{A}$	Matrix of thermal and electrical conduction
$\boldsymbol{A}, A_\alpha$	Vectorial electromagnetic potential
a	Large semi-axis of elliptical cross-section
$\boldsymbol{B}, B_i$	Magnetic induction field, magnetic flux density
$\left.\begin{array}{l}\boldsymbol{B}^M, B_i^M\\ \boldsymbol{B}^L, B_i^L\\ \boldsymbol{B}^S, B_i^S\end{array}\right\}$	Magnetic induction in the MAXWELL-, LORENTZ- and statistical formulations
$\boldsymbol{B}^a, B_i^a$	Auxiliary magnetic induction field
$\overset{\circ}{\boldsymbol{B}}, \overset{\circ}{B}_i$	Magnetic induction in the rest frame
$\overset{\circ}{\boldsymbol{B}}{}^a, \overset{\circ}{B}{}_i^a$	Auxiliary magnetic induction in the rest frame
$\mathbb{B}, \mathbb{B}_\alpha$	LAGRANGEan magnetic induction
$\mathbf{B}, \mathbf{B}_\alpha$	LAGRANGEan magnetic induction, used in constitutive models
$\boldsymbol{B} = \overset{\circ}{\boldsymbol{B}} + \mathcal{O}(\dot{x}^2/c^2),\ B_i = \overset{\circ}{B}_i + \mathcal{O}(\dot{x}^2/c^2)$	
$b^{(e)}_{\alpha\beta\gamma\delta}$	Electrostrictive constants
$b^{(m)}_{\alpha\beta\gamma\delta}$	Magnetostrictive constants
$\mathbf{b}, \mathbf{b}_\alpha$	Perturbed LAGRANGEan magnetic induction
$\boldsymbol{B}, B_{ij}$	Left CAUCHY–GREEN deformation tensor
BS	BIOT–SAVARD method
$\boldsymbol{B}_0$	Magnetic induction in the rigid-body state
b	Small semi-axis of elliptical cross-section, mean diameter of two concentric tori
C	Speed of light, thermal constant
C_w	Specific heat
$\boldsymbol{C}, C_{\alpha\beta}$	Right CAUCHY–GREEN deformation tensor
$\boldsymbol{C}^{-1}, C_{\alpha\beta}^{-1}$	Inverse of right CAUCHY–GREEN deformation tensor
$c_{\alpha\beta\gamma\delta}, c_{ijkl}$	Elastic constants, elasticities
CM	Combined method
c	Speed of light in vacuo
$\mathbb{C}$	Symbol for independent constitutive quantity
$d\boldsymbol{A}, dA_\alpha$	LAGRANGEan vectorial surface element

$d\boldsymbol{a}, da_i$	EULERian vectorial surface element
$d\boldsymbol{l}, dl_i$	EULERian vectorial length element
dv, dV	EULERian (LAGRANGEan) volume element
$\boldsymbol{D}, D_i$	Dielectric displacement
$\boldsymbol{D}^M, \boldsymbol{D}_i^M$	MAXWELLian dielectric displacement
$\boldsymbol{D}^a, D_i^a$	Auxiliary dielectric displacement
$\overset{\circ}{\boldsymbol{D}}, \overset{\circ}{D}_i$	Dielectric displacement in the rest frame
$\overset{\circ}{\boldsymbol{D}}^a, \overset{\circ}{D}_i^a$	Auxiliary dielectric displacement in the rest frame
$\mathcal{D}_i = \overset{\circ}{\mathcal{D}}_i + \mathcal{O}(\dot{x}^2/c^2)$	
$\mathbb{D}, \mathbb{D}_\alpha$	LAGRANGEan dielectric displacement
$\mathbb{D}^a, \mathbb{D}_\alpha^a$	Auxiliary LAGRANGEan dielectric displacement
$\mathbf{d}, \mathrm{d}_\alpha$	Perturbed LAGRANGEan dielectric displacement
$\mathbf{d}^a, \mathrm{d}_\alpha^a$	Perturbed LAGRANGEan auxiliary dielectric displacement
D	Plate rigidity
d	Distance of two concentric tori
d_{ij}	Stretching, rate of strain tensor
D_0	Regularization parameter for the constitutive relation of the extra stress tensor
$\boldsymbol{E}, E_{\alpha\beta}$	LAGRANGEan strain tensor, GREEN strain tensor
$\boldsymbol{e}, e_{ij}$	EULERian strain tensor, FINGER strain tensor
$\mathbf{e}, \mathrm{e}_{\alpha\beta}$	Perturbed GREEN strain tensor
$\boldsymbol{E}, E_i$	Electric field strength
$\left.\begin{array}{l} \boldsymbol{E}^C, E_i^C \\ \boldsymbol{E}^M, E_i^M \\ \boldsymbol{E}^L, E_i^L \\ \boldsymbol{E}^S, E_i^S \end{array}\right\}$	Electric field strength in the CHU-, MAXWELL-, LORENTZ- and statistical formulations
$\boldsymbol{\mathcal{E}}, \mathcal{E}_i$	Electromotive intensity
$\overset{\circ}{\mathcal{E}}$	Electric field strength in the rest frame
$\mathbb{E}, \mathbb{E}_\alpha$	LAGRANGEan electric field strength,- electromotive intensity
e_{ijk}	Three-dimensional permutation tensor
e^{ABCD}	Four-dimensional perturbation tensor
$\mathbf{e}, \mathrm{e}_\alpha$	Perturbed LAGRANGEan electromotive intensity
E	YOUNG's modulus
$\boldsymbol{e}_i (i = 1, 2, 3)$	Unit vectors
e_{ij}	Infinitesimal strain tensor
E_0	Scale for electric field
$\boldsymbol{F}, F_i$	Specific body force, BIOT–SAVARD force
$\boldsymbol{F}^0$	BIOT–SAVARD force in the rigid-body state
$\mathbf{f}$	BIOT–SAVARD force in the perturbed state
$\boldsymbol{F}^e, F_i^e$	Electromagnetic body force
$\boldsymbol{F}^{\mathrm{ext}}, F_i^{\mathrm{ext}}$	External body force

$\boldsymbol{F}^{\text{Lorentz}}$	LORENTZ force
$\boldsymbol{\mathcal{F}}^e, \mathcal{F}^e_\alpha$	Total electric force due to surface tractions of the MAXWELL stress tensor just outside the body
$\boldsymbol{\mathcal{F}}^{\text{ext}}, \mathcal{F}^{\text{ext}}_\alpha$	Total force on a body due to external body force and surface traction distributions
$\boldsymbol{\mathcal{F}}$	Deformation gradient
$\left.\begin{array}{l} F_{\Delta p}(Q) \\ F \\ \mathrm{f}_1, \mathrm{f}_2, \mathrm{f}_3 \end{array}\right\}$	Quantities measuring the electrorheological effect
$\boldsymbol{G}, G_\alpha$	LAGRANGEan electromagnetic momentum density
$\boldsymbol{g}, g_i$	EULERian electromagnetic momentum density
G	Elastic shear modulus
$\boldsymbol{G}, G_{\alpha\beta}$	LAGRANGEan strain tensor (GREEN strain tensor)
G	Shear modulus
G_I	Intermediate state
G_Y	Bending moment in a ring
$\boldsymbol{H}, H_i$	Magnetic field strength
$\left.\begin{array}{l} \boldsymbol{H}^C, H^C_i \\ \boldsymbol{H}^M, H^M_i \end{array}\right\}$	Magnetic field strength in the CHU- and MAXWELL formulations
$\boldsymbol{H}^a, H^a_i$	Auxiliary magnetic field strength
$\boldsymbol{\mathcal{H}}^a, \mathcal{H}^a_i$	Auxiliary effective magnetic field strength
$\overset{\circ}{\boldsymbol{\mathcal{H}}}, \overset{\circ}{\mathcal{H}}_i$	Magnetic field strength in the rest frame
$\mathbb{H}, \mathbb{H}_\alpha$	LAGRANGEan effective magnetic field
$\mathbb{H}^a, \mathbb{H}^a_\alpha$	Auxiliary LAGRANGEan effective magnetic field strength
$\mathbf{H}, \mathsf{H}_\alpha$	LAGRANGEan magnetic field-strength, used in the constitutive relations
H^{AB}	Magnetic field-electric displacement four-tensor
$\mathbf{h}, \mathsf{h}_\alpha$	Perturbed LAGRANGEan effective magnetic field
$\mathbf{h}^a, \mathsf{h}^a_\alpha$	Perturbed LAGRANGEan auxiliary effective magnetic field
h	Gap width
I	Thermodynamic potential in determining the internal energy from electromagnetic integrability conditions
I_{AB}	Moment of inertia of the body relative to its center of mass in the rigid-body state
I_y	Moment of inertia of an area
J	JACOBIan determinant, determinant of the deformation gradient
J	Thermodynamic potential, determining the contribution of the deformation to the internal energy
$\mathbf{j}, \mathsf{j}_\alpha$	Perturbed LAGRANGEan electric current field
$\left.\begin{array}{l} \mathrm{j}^1_{\alpha\beta\gamma} \\ \mathrm{j}^2_{\alpha\beta}, \mathrm{j}^3_{\alpha\beta}, \mathrm{j}^5_{\alpha\beta} \\ \mathrm{j}^4_\alpha \end{array}\right\}$	Coefficients arising in the constitutive relation for the LAGRANGEan conductive electric current

J	Homogeneous quadratic functional of the perturbation fields of which δJ is the second variation of the action functional L
J_0, J_1	BESSEL functions of zeroth and first order
$\left.\begin{array}{l} \boldsymbol{J}, \boldsymbol{J}^{\mathrm{e}}, \boldsymbol{J}^{\mathrm{m}} \\ J_i, J_i^{\mathrm{e}}, J_i^{\mathrm{m}} \end{array}\right\}$	Non-conductive current density,-electric, -magnetic current density
$\boldsymbol{\mathcal{J}}, \mathcal{J}_i$	Conductive current density
$\left.\begin{array}{l} \boldsymbol{\mathcal{J}}^{\mathrm{e}}, \mathcal{J}_i^{\mathrm{e}}, \\ \boldsymbol{\mathcal{J}}^{\mathrm{m}}, \boldsymbol{\mathcal{J}}_i^{\mathrm{m}}, \end{array}\right\}$	Electric, magnetic current densities
$\overset{\circ}{\boldsymbol{\mathcal{J}}}, \overset{\circ}{\mathcal{J}}_i$	Conductive current in the rest frame
$\left.\begin{array}{l} \mathbb{J}, \mathbb{J}_\alpha \\ \mathbb{J}^{\mathrm{e}}, \mathbb{J}_\alpha^{\mathrm{e}} \\ \mathbb{J}^{\mathrm{m}}, \mathbb{J}_\alpha^{\mathrm{m}} \end{array}\right\}$	LAGRANGEan conductive current density LAGRANGEan electric (magnetic) conductive current density
$\mathscr{J}$	Common factor of J
$\mathcal{K}$	Magnetic part of J, $\mathcal{K} = \mathcal{B}_0^2 K, \mathcal{K} = I_0^2 K$
k	Longitudinal pressure gradient
L	Characteristic length
$L^{(e)}$	Thermoelectric constant
$L^{(m)}$	Thermomagnetic constant
$\boldsymbol{L}$	Action integral
$\mathcal{L}, \mathcal{L}_i$	LAGRANGE densities
$\boldsymbol{L}, L_i$	Specific body couple,
$\boldsymbol{L}, L_{ij}$	Dual tensor of the body couple vector
$\boldsymbol{L}^{\mathrm{ext}}, L_i^{\mathrm{ext}}$	Dual tensor of the external body couple vector
$\boldsymbol{\mathcal{L}}^{\mathrm{e}}, \mathcal{L}_\alpha^{\mathrm{e}}$	Total electromagnetic couple relative to the center of gravity in the rigid-body state due to the MAXWELL stress tensor just outside the body
$\boldsymbol{\mathcal{L}}^{\mathrm{ext}}, \mathcal{L}_\alpha^{\mathrm{ext}}$	Total external couple due to body couples, body forces and surface tractions
$\left.\begin{array}{l} \boldsymbol{M}^C, M_i^C \\ \boldsymbol{M}^M, M_i^M \\ \boldsymbol{M}^L, M_i^L \\ \boldsymbol{M}^S, M_i^S \end{array}\right\}$	Magnetization density in the CHU-, MAXWELL- , LORENTZ- and statistical formulations
$\left.\begin{array}{l} \boldsymbol{\mathcal{M}}, \mathcal{M}_i \\ \mathcal{M}^M, \mathcal{M}_i^M \end{array}\right\}$	Magnetization density, – in the MAXWELL formulation in the rest frame
$\mathbb{M}, \mathbb{M}_\alpha$	LAGRANGEan magnetization density
$\mathbf{M}, \mathsf{M}_\alpha$	LAGRANGEan magnetization density per unit mass
m	Viscosity scale for the power law constitutive relation for the extra stress tensor

$\left.\begin{array}{l}\mathbf{m}, \mathrm{m}_\alpha \\ \mathbf{m}^C, \mathrm{m}_\alpha^C \\ \mathbf{m}^L, \mathrm{m}_\alpha^L\end{array}\right\}$ Perturbed magnetization, – in the CHU- and LORENTZ formulations

$\left.\begin{array}{l}\mathrm{m}_{\alpha\beta\gamma}^1 \\ \mathrm{m}_{\alpha\beta}^2, \mathrm{m}_{\alpha\beta}^3, \mathrm{m}_{\alpha\beta}^5 \\ \mathrm{m}_{\alpha\beta}^4\end{array}\right\}$ Coefficients arising in the constitutive relation for the perturbed magnetization

M_{rr} Radial bending moment in a ring-like structure
$\mathbb{M}a$ MASON number
$\boldsymbol{n}, n_i$ EULERian unit normal vector
$\boldsymbol{N}, N_\alpha$ LAGRANGEan unit normal vector
n, n_0, n_∞ Exponents in the power law of the extra stress tensor
N_1, N_2 Normal stress differences
$n_{AB}(A, B =$ Coefficients in the parameterization of η_{gen}
$1, 2, 3)$

$\boldsymbol{O}, O_{ij}$ Orthogonal 3×3 matrix
$\boldsymbol{P}, P_i$ Polarization density

$\left.\begin{array}{l}\boldsymbol{P}^C, P_i^C \\ P^M, P_i^M \\ P^S, P_i^S\end{array}\right\}$ Polarization density in the CHU-, MAXWELL- and statistical formulations

$\mathbb{P}, \mathbb{P}_\alpha$ LAGRANGEan polarization density
$\mathbf{P}, \mathsf{P}_\alpha$ LAGRANGEan polarization density per unit mass
$\mathbf{p}, \mathrm{p}_\alpha$ Perturbed LAGRANGEan polarization density

$\left.\begin{array}{l}\mathrm{p}_{\alpha\beta\gamma}^1 \\ \mathrm{p}_{\alpha\beta}^2, \mathrm{p}_{\alpha\beta}^3 \\ \mathrm{p}_\alpha^4\end{array}\right\}$ Coefficients arising in the constitutive relation of the perturbed polarization

p Pressure
q Transverse load of a beam or plate
q Perturbed LAGRANGEan charge density
$\mathbf{q}, \mathsf{q}_\alpha$ Perturbed LAGRANGEan energy flux
$\boldsymbol{q}, q_i$ Energy (heat) flux vector
$\boldsymbol{q}^S, q_i^S$ Energy (heat) flux vector in the statistical formulation
$\boldsymbol{Q}, Q_\alpha$ LAGRANGEan energy (heat) flux vector
$\mathcal{Q}$ Electric charge density, surface charge density
$Q, Q^{\mathrm{e}}, Q^{\mathrm{m}}$ Charge density, electric-, magnetic-
$\overset{\circ}{Q}$ Charge density in the rest frame
$\mathbb{Q}, \mathbb{Q}^{\mathrm{e}}, \mathbb{Q}^{\mathrm{m}}$ LAGRANGEan charge density, electric-, magnetic-.

$\left.\begin{array}{l}\mathrm{q}_{\alpha\beta\gamma}^1 \\ \mathrm{q}_{\alpha\beta}^2, \mathrm{q}_{\alpha\beta}^3, \mathrm{q}_{\alpha\beta}^5 \\ \mathrm{q}_\alpha^4\end{array}\right\}$ Coefficients arising in the constitutive equation for the perturbed LAGRANGEan energy flux

Q Quantity in the evaluation of I_0
$Q_j(j = 1, .., n)$ Contribution to the magnetic transverse load of beams and plates

Q	Discharge
r	Energy supply density
r^{e}	Electromagnetic energy supply density
r^{ext}	External energy supply density
$R_{\alpha\beta}$	Orthogonal matrix, describing the rotation of a rigid-body
R	Radius of a circular cross-section
$\boldsymbol{r}_1, r_2$	Position vectors on wires, $-1,2$
$\mathbb{R}e$	REYNOLDS number
s	Entropy supply density
s	Perturbation entropy density
$\left.\begin{array}{l} s^1_{\alpha\beta} \\ s^2_\alpha, s^3_\alpha \\ s^4 \end{array}\right\}$	Coefficients arising in the constitutive relation for the perturbed entropy density
$\mathscr{S}$	Set of equations characterizing the equilibrium state of a magnetoelastic system subject to an external field
$\mathscr{S}^1$	Analogous set as $\mathscr{S}$ describing the perturbed state
$\mathscr{S}^0$	Analogous set as $\mathscr{S}$ describing the rigid-body state
S_i	Element of $\mathscr{S}$
S^0_i	Element of $\mathscr{S}^0$
s_i	Element of $\mathscr{S}^1$
T, T^0	Stress tensor, - in rigid-body state
T^1	Perturbed stress tensor
T_y	Traction in the y-direction along the boundary of the cross-section
T	Kinetic energy
$\boldsymbol{T}, T_{i\alpha}$	First PIOLA–KIRCHHOFF stress tensor
$T^M_{i\alpha}$	First PIOLA–KIRCHHOFF–MAXWELL stress tensor
$\boldsymbol{T}^p, T^p_{\alpha\beta}$	Second PIOLA–KIRCHHOFF stress tensor
$\boldsymbol{t}_i (i = 1,2)$	Unit vector tangential to the i-th infinitely thin wire
t^e_{ij}	Extra CAUCHY stress tensor
$\boldsymbol{t}, t_{ij}$	CAUCHY stress tensor
$\boldsymbol{t}^M, t^M_{ij}$	CAUCHY–MAXWELL stress tensor
$\boldsymbol{t}^{(n)}, t^{(n)}_{ij}$	Stress vector, traction
$\boldsymbol{t}, t_{i\alpha}$	Perturbed first PIOLA–KIRCHHOFF stress tensor
$t_{\alpha\beta}\delta_{i\alpha}t_{\alpha\beta}$	
$\left.\begin{array}{l} t^1_{\alpha\beta\gamma\delta} \\ t^2_{\alpha\beta\gamma}, t^3_{\alpha\beta\gamma} \\ t^4_{\alpha\beta} \end{array}\right\}$	Coefficients arising in the constitutive relation of the perturbed first PIOLA–KIRCHHOFF stress tensor
U	Internal energy density
$\bar{U}$	Particle displacement from the initial to the intermediate configuration
$\boldsymbol{u}$	Particle displacement from the intermediate to the final state

U	HELMHOLTZ free energy, internal energy
u	Elliptical coordinate
$\boldsymbol{V}, V_i$	Velocity, arising in the LORENTZ transformation
V	Transverse displacement in beam and plate bending
v	Elliptical coordinate
VM	Variational method
V_0	Scale for material velocity
V	Electric potential
v_1	Longitudinal velocity
v_2	Transverse velocity
W_N	Speed of propagation
w_n	Speed of displacement
W^M	MAXWELL modified internal energy density
W	Bending energy of beams or plates, part of the elastic energy independent of $\boldsymbol{B}$
$w^{(m)}(z) = u^{(m)}(z) + iv^{(m)}(z)$	Complex displacement
$\left. \begin{array}{c} \boldsymbol{x} \\ x_i \end{array} \right\}$	Position vector
$\boldsymbol{X}, X_\alpha$	Particle position in reference configuration
$Z = (D_0^2 + \frac{\dot{\gamma}^2}{2})$	Regularized second stretching invariant in simple shear flow

Greek Symbols

$\alpha_1, \ldots, \alpha_6$	Scalar, parameters in the isotropic representation of the extra stress tensor
$\alpha_{AB}(A = 1, \ldots, 6; B = 1, 0)$	Scalar, parameters in the isotropic representation of the extra stress tensor
$\hat{\beta}_i(i = 1, 2)$	Parameters in the constitutive equation for the shear stress
β	Thickness to width ratio
β_1, β_2	Parameters arising in the CASSON-like constitutive function
$\begin{array}{c} \beta_{\alpha\beta} \\ \beta_{ij} \end{array}$	Coupling matrix of thermal and electrical conduction
γ	Entropy production density
$\dot{\gamma}$	Shear rate in simple shear flow
$\gamma = 0.5772$	EULER number
δ	Regularization parameter in the constitutive relation for stress
δJ	Second variation of the action integral L
δL	First variation of the action integral
$\delta_{\alpha\beta}, \delta_{ij}$	KRONECKER delta
ε	Effective permittivity
$\varepsilon = \pi a/(4l)$	Slenderness parameter

ε	Regularization parameter in the constitutive relation for stress
ε_0	Dielectric constant in vacuo, electric permittivity of the free space
$\tilde{\varepsilon}, \bar{\varepsilon}$	LEGENDRE transformed internal energy
$\varepsilon_{\alpha\beta\gamma}^{(e)}$	Piezoelectric constants
$\varepsilon_{\alpha\beta\gamma}^{(m)}$	Piezomagnetic constants
$\boldsymbol{\xi}, \xi_\alpha$	Particle position in intermediate state
$\boldsymbol{\Xi}, \Xi_\alpha$	Center of mass in the rigid-body state
ξ	Parameter to regularize the power law term of the extra stress tensor at zero electric field
Ξ	General physical quantity
$\boldsymbol{\Xi}$	Displacement from the initial to the intermediate state
η	Entropy density
η^{AB}	Magnetic field strength-electric displacement four-tensor
η_0, η_∞	Viscosities in the bi-viscosity models
η_{gen}	Generalized viscosity
Θ	Absolute temperature
θ	Perturbed temperature
$\kappa_{ij}, \kappa_{\alpha\beta}$	EULERian, LAGRANGEan matrix of heat conductivity
$\kappa = -\gamma - \ln\big((1+\beta)/2\big)$	
$\Lambda = 1 + (2\beta\mu\varepsilon^2\kappa)^{-1}$	
Λ_{AB}	Matrix of extended LORENTZ group
λ	LAMÉ constant, eigenvalue of a matrix
$\lambda_\alpha^{(e)}$	Pyroelectric constant
$\lambda_\alpha^{(m)}$	Pyromagnetic constant
μ_0	Magnetic permeability in free space
μ	Magnetic permeability
μ_{AB}	Magnetization four-tensor
ν	Thermoelastic constant
$\nu_{\alpha\beta}$	Thermoelastic coupling tensor
ν	POISSON number
π_α, π_i	Electromagnetic energy flux vector
π^{AB}	(Magnetization)-polarization four-tensor
ρ	Mass density in present configuration
ρ_0	Mass density in reference configuration
Σ, σ	Singular surface
σ^A	Electric current, electric charge density four-vector
$\sigma_{\alpha\beta}$	Matrix of electric conductivity
$\sigma_A (A = 1, 2, 3)$	Effective electric conductivities
σ_{ABC}	Effective electric conductivity constants
Φ	Physical variable (unspecified)

Φ, Φ_i	Entropy flux (EULERian)
Φ, Φ_α	LAGRANGEan entropy flux, electromagnetic scalar potential
$\Phi^{(0)}$	Magnetic potential in the rigid-body state
Φ_j	General flux quantity
φ_{AB}	Electric field-magnetic induction four-tensor
φ	Perturbed magnetic potential
$_2\chi^{(m)}$	Second-order magnetic susceptibility
$_2\chi^{(e)}$	Second-order electric susceptibility
$_4\chi^{(e)}$	Fourth-order electric susceptibility
$_4\chi^{(m)}$	Fourth-order magnetic susceptibility
ψ	External perturbed magnetic potential
$\bar{\psi}_{0,1,2}$	Parameters in the constitutive relation for the normal stress components
ψ	HELMHOLTZ free energy density
$\tilde{\psi}$	LEGENDRE transform of ψ
$\bar{\psi}$	LEGENDRE transformed ψ
$\check{\psi}$	LEGENDRE transformed ψ
ψ^{AB}	Covariant four-tensor (e.g. for polarization)
ψ_{AB}	Contravariant four-tensor (e.g. magnetization)
τ^B, τ^C, τ^U	Shear stress of the BINGHAM, CASSON and CASSON-like model
τ_y	Yield stress in the BINGHAM and classical CASSON model
Ω, ω	Electromagnetic energy density
$\boldsymbol{\Omega}, \Omega_\alpha$	Angular velocity in the rigid-body motion
ω	Electromagnetic specific energy density

Other Symbols

$A^\star$	Convective derivative
$\mathcal{O}$	Order symbol
$(\ )^{\boldsymbol{\cdot}} = \dfrac{d}{dt}$	Total or material time derivative
$[\![f]\!] = f^+ - f^-$	Jump of f at a singular surface
$\langle f \rangle = \frac{1}{2}(f^+ + f^-)$	
$f_{(ij)}$	Symmetric part of f_{ij}
$f_{[ij]}$	Skew-symmetric part of f_{ij}
$\overset{*}{f}_i$	Corotational derivative of f_i
f_E	f, evaluated in thermostatic equilibrium
$\bar{\bar{f}}$	$\bar{f}$, evaluated in the intermediate state
$\|\ f\ \|$	Norm of f, usually lim sup $|\ f\ |$
F^0	F, evaluated in the rigid-body state

1 General Introduction

With the boundaries of physical science expanding rapidly, present day engineers must necessarily assimilate information and knowledge of subjects that continue to become more and more complex. In the study of the nature and mechanical behavior of engineering materials, however, their knowledge seldom progresses beyond the level of elementary theory of elasticity and viscous fluids and the basic concepts of the theory of plasticity and viscoelasticity. More esoteric theories of material behavior and the interaction with various fields are generally left out of consideration or soon abolished as being mathematically intractable or economically unjustifiable.

This book is an account on one of the above mentioned more esoteric theories. What we have in mind is the response of deformable bodies (solids or fluids) to electromagnetic fields. Indeed, the interaction of electromagnetism with thermoelastic fields is not only a challenging scientific problem, but it is increasingly attracting also engineers in the materials sciences and in nanotechnology from a fundamental and in the nuclear power and electronic industry from a purely applied point of view.

The response of deformable bodies to electromagnetic fields is difficult to describe for several reasons. First, it is *fundamentally* nontrivial because the formulation of electrodynamics in ponderable bodies is nonunique. In other words, one generally introduces four different electromagnetic vector fields, e.g. the electric field strength E and the magnetic induction B plus two other fields. Some authors work with electric displacement D and magnetic field strength H, others use polarization P and magnetization M instead, but there are several variants how to introduce the latter two quantities. This as such would not make yet electrodynamics of deformable media fundamentally difficult, as only unique transformations from the variables (E, D, B, H) to (P, M) would be needed. The difficulty arises with the definitions of the electromagnetic stress tensor and specific body force in the momentum equation (and in theories with polar structure corresponding definitions of the electromagnetic couple stress and specific body couple). Consequently, for two scientists using different electromagnetic variables and seemingly unmatchable expressions for stress and specific body force, but aiming to describe the same electromagneto-mechanical phenomenon, there arises the question, whether the two formulations can ever yield the same results for the

K. Hutter et al.: *Electromagnetic Field Matter Interaction in Thermoelastic Solids and Viscous Fluids*, Lect. Notes Phys. **710**, 1–6 (2006)
DOI 10.1007/3-540-37240-7_1   © Springer-Verlag Berlin Heidelberg 2006

measurable quantities in the mathematical reproduction of the same physical situation. In Part I of this book we shall give an answer to this difficult problem. The form of the answer will demonstrate that thermodynamic arguments involving the Second Law of Thermodynamics are needed to resolve this fundamental question.

The analysis presented in Part I will be performed on the level of non-relativistic approximation. This is a matter of judicious compromise, as our interest is in technical applications for which mechanical equations are classical and Galilean. Alternatively, the Maxwell equations are Lorentz invariant. To merge the two, an approximation will be introduced that is, in electromagnetic variables and combinations thereof, accurate up to terms of order $O(v^2/c^2)$, which will be omitted. Here, v denotes the modulus of the material velocity and c is the speed of light.

Second, the description of deformable bodies to electromagnetic fields is also difficult, because the formulation of rather simple initial boundary value problems leads, in general, to rather complex solution procedures, i.e., the equations are generally *structurally coupled* and *complex*, i.e. *nonlinear*. When deformations of solid bodies are concerned, then it is often so that the deformations from the initial to the final positions are small. Under such circumstances one may compose the total response to an external magnetic field of an *intermediate configuration* that is known plus a correction from it that is small, so that linearization of the perturbed equations is justified. In many situations it is justified to identify the intermediate configuration with the initial configuration, also called *rigid-body state*. The electromagnetic fields as a response to the external driving fields are then in a first step calculated for the rigid-body state and then corrected for the deformation. In these perturbed equations the perturbed electromagnetic fields are coupled with the mechanical equations for the displacement field. They are nonlinear but on the prerequisite that the displacements and the corrections of the electromagnetic fields due to these displacements are equally small, these equations may be linearized. This makes the analysis much simpler, even though for many problems analytical solutions still cannot be found. We shall demonstrate in the last chapter of Part I how this decomposition can systematically be implemented.

Part II of this book is devoted to two different, but technically important engineering applications. Here, the complexity of the problem formulation may be simply due to the surprising subleties of the electromagneto-mechanical interactions that require adjustment of the novel reader's attitude to the delicacy of the results. We will shortly give an example.

We shall treat two different examples. The first concerns the reaction of beams or plates to an external magnetic field. Early experiments, performed in the 60s of the last century and theories describing the buckling of flat and wide cantilever beams did not yield the same functional relation for the buckling value of the external magnetic field B_0. The reason was found more than

20 years later. On the basis that the ratio "beam thickness to beam width", $h/w \ll 1$, was small, it was concluded that the finiteness of the width of the beam could be ignored and $w \to \infty$ or $h/w = 0$ could be assumed. This is too simple, because for the perturbed magnetic potential, satisfying a 2D HELMHOLTZ equation and not a 2D LAPLACE equation, it is not allowed to approximate the narrow rectangular cross-section by an infinitely wide one (no matter how small the ratio h/w becomes). Once this was recognized and the perturbed magnetic fields were calculated on the basis of the correct geometry of the beam, experimental and theoretical analytic results could be brought into coincidence. An analytical solution to this problem was found for a BERNOULLI beam with elliptical cross-section and arbitrary aspect ratio. Rectangular cross-sections face difficulties because of singularities that develop in corners. One procedure in the determination of the magnetoelastic critical magnetic field is to start from the homogeneous set of linear equations with homogeneous boundary conditions for the perturbed fields and to determine the smallest value for the magnetic field in the rigid-body state that allows for a nontrivial solution. This is a linear eigenvalue problem and the parameter characterizing this critical magnetic field is the eigenvalue of this linear boundary value problem with nontrivial eigenvector. For ferromagnetic materials this linear problem is replaced by a variational formulation with higher-order accuracy. This variational formulation is applied to the magnetoelastic response of two parallel circular rods, to a superconducting torus and sets of two concentric tori and a pair of tori with circular cross sections.

The solution of the linear elliptic partial differential equations requires use of conformal mapping techniques and complex variable theory, which, in these times of immediate application of electronic computation are often considered to be difficult, even though they belong to the classical syllabus of the education in theoretical engineering, physics and applied mathematics.

Another approximation that has been popular in determining the critical current in superconducting structures is the application of the law of BIOT–SAVARD. This law provides a formula for the electromagnetic force in an infinitely thin wire due to a second infinitely thin wire, both carrying an electric current. For parallel wires, finite and infinite helical wires the Law of BIOT–SAVARD allows determination of the buckling current. Application of the variational method or a combined method using the variational formulation and the law of BIOT–SAVARD shows that the BIOT–SAVARD approach can only yield reliable results, provided the distance of the wires or of consecutive tori is large as compared to the diameter of the wire. Interesting and somewhat counterintuitive is the result that the buckling current of an infinitely long helical wire is never approached by the buckling current of a helical wire with $n \to \infty$ coils.

The second example of electro-mechanical interactions is provided in the last chapter where *electrorheological fluids* are treated. These are actually mixtures of an electrically neutral fluid with suspended polarizable particles. Phenomenologically, when the electrical field is switched off the particles, which are of subgrid scale, are randomly distributed and the mixture, treated as a fluid, appears to have small viscosity. Alternatively, when the electric field is switched on, the polarizable particles arrange in columns more or less perpendicularly to the flow and so stiffen the motion against shearing. This becomes manifest as a substantial increase in viscosity almost to the extent that the mixture now behaves much like a solid.

The theoretical description of electrorheological fluids as we employ it is based on a single constituent continuum concept and postulations of the electromagnetic stress tensor, specific body force, energy supply and energy flux. These are derived from the MAXWELL equations in the MAXWELL–MINKOWSKI formulation by deriving the electromagnetic balance laws of momentum and energy. What emerges are e.g. two forms of electromagnetic stresses and corresponding specific electromagnetic forces. By a scaling analysis these are reduced to their nonrelativistic counterparts. The constitutive theory is more complex than the corresponding theory in Part I, because the mechanical stress tensor depends not only on the density, temperature, electric field but also on the rate of strain tensors.

The difficulty in arriving at a theoretical formulation for electrorheological fluids is in a proper three-dimensional formulation of the constitutive relation for the "mechanical" stress tensor and how it matches experiments, mostly in two-dimensional POISEUILLE-type flows such that the Second Law of Thermodynamics is satisfied. This leads to BINGHAM-type and power-law type stress-stretching relations and extensions thereof in which the dependence on the electric field is particularly subtle.

The optimal model is subsequently applied to two-dimensional channel flows of which the walls are alternatively covered by infinitely thin charged electrodes or left uncovered. The electric field that would be uniform, if electrodes were infinitely long, will now be inhomogeneous. In particular at electrode ends square root singularities develop, and if electrodes are not long in comparison to the gap width, the local electric fields will mutually interact. Of interest is, how these inhomogeneities affect the discharge through the two-dimensional channel or how the pressure drop between two positions along the channel axis will depend on the relative electrode lengths and their gap widths. Additional questions are, how finitely thick electrodes will affect the pressure drop, if they confine or enlarge the gap width or yield a converging or diverging channel stretch. Such questions are also asked when electrodes are corrugated or beamed and thus destroy the nearly viscometric flow of the fluid.

This is roughly speaking the content of the book.

After the appearance of the first edition of this book, the interest in electromagnetomechanical problems increased strongly. This was manifested for instance by the start of a new journal *International Journal of Applied*

Electromagnetics in Materials (IJAEM) in 1990, edited by K. MIYA and R.K.T. HSIEH. In 1995, the journal changed its name slightly in *International Journal of Applied Electromagnetics and Mechanics* with K. MIYA as editor-in-chief. Parallel with this journal a series of symposia called ISEM (International Symposium on Applied Electromagnetics in Materials) found place every 2 years, the first being held in 1988 in Tokyo and the up to now last one in 2005 in Salzburg, Austria. The proceedings of these symposia appeared, besides in [189] and [190], as special volumes of IJAEM. For instance, the proceedings of the 2003-symposium in Versailles, France, can be found as a special issue in [191]. During this period, also, several new books were published, of which we here only mention the works of ERINGEN and MAUGIN [70, 71], MAUGIN et al. [137], TIERSTEN [238], and SHU-ANG ZHOU [215, 216].

Besides the two fields of applications dealt with in the Chaps. 7 and 8 of this book, a lot of other fields exists in which electromagneto-mechanical models are applied. Far from aiming to be general, we mention here as examples the following three subjects:

Piezoelectric Devices

This subject is so diverse and knows so many different applications, such as non-destructive testing by means of sonar, ignition lighters, ultrasonic transducers in remote controls, transducers in ink-jet printers and many more, that a complete review is far beyond the scope of this book. Also, the amount of literature appearing yearly on this subject is immense. Therefore, we restrict ourselves here to mentioning only the classical treatments by TIERSTEN [235], IKEDA [101] and TAYLOR [232], and two, more recent ones, by DING and CHEN [58] on 3D problems piezoelasticity, and by QIN [193] on fracture in piezoelectric materials. Finally, KAMLAH [106] gives an account on ferroelectric and ferroelastic piezoceramics.

Cracks in Magnetoelastic or Piezoelectric Plates

The investigation of crack propagation and the (non-destructive) detecting of cracks in magnetoelastic or piezoelectric materials is an important subject in fracture mechanics. One of the authors having a great impetus in this respect is certainly Y. SINDO, who published since 1981, [211], up to now on cracks in magnetoclastic or piezoelectric layers and plates; see e.g. [212] and [213] and, of more recent date, [125] and [214]. Some very recent papers are by HASANYAN and PILIPOSIAN [87], Y. and I. PODIL'CHUCK [183], LI and YANG [119], and WANG and JIANG [262].

Waves in Electromagneto-Elastic Solids

This is a very classical subject that received attention from the first beginnings of the research on electromagneto-elastic interactions by the founding fathers of this theory such as PAO, PARKUS, ERINGEN, MAUGIN and many others. We restrict ourselves to giving only a few more recent references, which consider successively surface waves in magnetoelastic conductors, HHEFNI et al. [88], waves at piezoelectric interfaces, ROMEO, [201, 202] and JANG and ZHOU, [103] and flexure waves in electroelastic plates by LANCIONI and TOMASETTI [117].

2 Basic Concepts

In this chapter we shall lay down the foundations of electrodynamics and continuum mechanics as it will be used in this book. We start with kinematics and then turn to the equations of balance, the MAXWELL equations and the balance laws of mass, momenta, energy and entropy. We introduce the concept of material objectivity in the nonrelativistic approximation and briefly outline constitutive equations of thermoelasticity.

2.1 Kinematics

As is common in continuum mechanics, we regard a body as a three-dimensional manifold embedded in EUCLIDian 3-space. Its elements are called particles. Let $\mathcal{R}_R$ be its reference configuration and $\mathcal{R}_t$ its configuration at time t. Instead of $\mathcal{R}_t$ we shall subsequently write $\mathcal{R}$, and we shall refer to $\mathcal{R}$ as the present configuration of the body. Parts of the body will be denoted by V_R and $\mathcal{V}$, dependent on whether they are referred to the reference configuration and the present configuration, respectively. The boundary of V_R and of $\mathcal{V}$ will be denoted by ∂V_R and $\partial \mathcal{V}$, respectively. The position of a particle in $\mathcal{R}_R$ will be designated by $\boldsymbol{X}(X_\alpha, \alpha = 1, 2, 3)$, whereas that in the present configuration $\mathcal{R}$ is $\boldsymbol{x}$ $(x_i, i = 1, 2, 3)$. A motion of the body is then described by the mapping

$$x_i = \chi_i(X_\alpha, t) , \qquad (i, \alpha = 1, 2, 3) . \tag{2.1.1}$$

We assume χ to be invertible and this is tantamount to assuming that the functional determinant

$$J := \det\left(\partial \chi_i / \partial X_\alpha\right) , \tag{2.1.2}$$

never vanishes and may without loss of generality be assumed to be positive. In the above and throughout this monograph symbolic and Cartesian tensor notation is used. Greek indices refer to the material coordinates $\boldsymbol{X}$, and Latin indices to the spatial coordinates $\boldsymbol{x}$. Summation convention will be used over doubly repeated indices and commas preceding indices indicate differentiations with respect to space variables. If A_{ij} denote the components of a second rank tensor, then

$$A_{(ij)} = \tfrac{1}{2}(A_{ij} + A_{ji}) , \quad A_{[ij]} = \tfrac{1}{2}(A_{ij} - A_{ji}) \tag{2.1.3}$$

K. Hutter et al.: *Electromagnetic Field Matter Interaction in Thermoelastic Solids and Viscous Fluids*, Lect. Notes Phys. **710**, 7–29 (2006)
DOI 10.1007/3-540-37240-7_2

denote its symmetric and skew-symmetric parts such that

$$A_{ij} = A_{(ij)} + A_{[ij]} \, . \tag{2.1.4}$$

Moreover tr $\boldsymbol{A} = A_{ii}$ will denote its trace. The symbol d/dt or the superimposed dot will designate differentiation with respect to time t, holding the particle $\boldsymbol{X}$ fixed, i.e. for $\Phi = \hat{\Phi}(\boldsymbol{X}, t)$

$$\frac{d\Phi}{dt} \equiv \dot{\Phi} := \frac{\partial \hat{\Phi}(\boldsymbol{X}, t)}{\partial t} \, . \tag{2.1.5}$$

Here, $\dot{\Phi}$ is called the material time derivative. Likewise, $\partial/\partial t$ will denote differentiation with respect to time t, holding the spatial position $\boldsymbol{x}$ fixed. Hence, with $\Phi = \tilde{\Phi}(\boldsymbol{x}, t)$,

$$\frac{\partial \Phi}{\partial t} := \frac{\partial \tilde{\Phi}(\boldsymbol{x}, t)}{\partial t} \, , \tag{2.1.6}$$

where $\partial \Phi/\partial t$ is called the local, or partial, time derivative. It is easy to see that (2.1.5) and (2.1.6) may be combined to yield

$$\frac{d\Phi}{dt} = \dot{\Phi} = \frac{\partial \tilde{\Phi}}{\partial t} + \dot{x}_i \frac{\partial \tilde{\Phi}}{\partial x_i} \, . \tag{2.1.7}$$

The local deformation of a body in the neighborhood of a particle $\boldsymbol{X}$ may be characterized to first order by the deformation gradient $F_{i\alpha}$, defined by

$$F_{i\alpha} := \frac{\partial \chi_i(\boldsymbol{X}, t)}{\partial X_\alpha} = x_{i,\alpha} \, . \tag{2.1.8}$$

Its inverse exists and is written as

$$F_{\alpha i}^{-1} = X_{\alpha,i} \, . \tag{2.1.9}$$

Of particular interest are objective combinations of $F_{i\alpha}$. The right (left) CAUCHY-GREEN deformation tensors $C_{\alpha\beta}$, (B_{ij}) are defined by

$$C_{\alpha\beta} := F_{i\alpha} F_{i\beta} \, , \quad B_{ij} = F_{i\alpha} F_{j\alpha} \, . \tag{2.1.10}$$

In applications it is more convenient to use the LAGRANGEan (EULERian) strain tensors, $G_{\alpha\beta}$, (E_{ij}) or the deformation tensors,

$$G_{\alpha\beta} := \tfrac{1}{2}(C_{\alpha\beta} - \delta_{\alpha\beta}) \, , \quad E_{ij} := \tfrac{1}{2}(\delta_{ij} - B_{ij}) \tag{2.1.11}$$

instead. Here,

$$\delta_{\alpha\beta} = \begin{cases} 1 & \alpha = \beta \, , \\ 0 & \text{otherwise} \end{cases} \tag{2.1.12}$$

is the KRONECKER delta.

The velocity of a material particle is the time rate of change of its position. In the LAGRANGEan formulation it takes the form

$$\dot{x}_i = \frac{\partial \chi_i(X_\alpha, t)}{\partial t} = \hat{v}_i(X_\alpha, t) . \tag{2.1.13}$$

The EULER representation of the velocity reads

$$\dot{x}_i = \hat{v}_i(X_\alpha^{-1}(x_n, t), t) = \tilde{v}_i(x_n, t) . \tag{2.1.14}$$

The time rate of change of the velocity of a material element is its *acceleration*. We express this in the EULER representation as

$$\ddot{x}_i = \dot{v}_i = \frac{d\tilde{v}_i(x_n, t)}{dt} = \frac{\partial \tilde{v}_i}{\partial t} + L_{ij}\tilde{v}_j , \tag{2.1.15}$$

in which L_{ij} is the *spatial velocity gradient* defined as

$$L_{ij} = \dot{x}_{i,j} = (\text{grad}\,\dot{\boldsymbol{x}})_{ij} . \tag{2.1.16}$$

Its symmetric, d_{ij}, and skew-symmetric, W_{ij}, parts

$$d_{ij} := L_{(ij)} , \qquad W_{ij} := L_{[ij]} , \tag{2.1.17}$$

are called *stretching* or *rate of strain tensor* and *vorticity* or *spin tensor*, respectively.

Combining (2.1.8) with (2.1.14) we see that

$$\dot{F}_{i\alpha} = (x_{i,\alpha})^{\boldsymbol{\cdot}} = \hat{v}_{i,\alpha}\tilde{v}_{i,j}F_{j\alpha} = L_{ij}F_{j\alpha} ,$$

or after multiplication with $F_{\alpha k}^{-1}$

$$L_{ik} = \dot{F}_{i\alpha}F_{\alpha k}^{-1} . \tag{2.1.18}$$

This formula expresses the spatial velocity gradient in terms of the material deformation gradient.

2.2 Equations of Balance

2.2.1 The Balance Laws of Mechanics

It is the ultimate goal of any thermodynamic theory, be it a theory of a simple fluid or a fairly complicated description of electromechanical interaction phenomena, to calculate within a body the independent fields of this theory as functions of space and time. For the determination of these fields, we need field equations, which are obtained when the balance laws of mechanics and electrodynamics are combined with constitutive equations. Basic ingredients of any theory are thus the balance equations which are discussed below.

We start with the balance laws of mechanics. The integral expressions of the laws of conservation of mass, balance of momentum, moment of momentum and energy in the spatial and material description are well-known. For the former they are

$$\frac{d}{dt}\int_{\mathcal{V}} \rho d\nu = 0 \; , \tag{2.2.1}$$

$$\frac{d}{dt}\int_{\mathcal{V}} \rho \dot{x}_i d\nu = \int_{\partial \mathcal{V}} t_{ij} da_j + \int_{\mathcal{V}} \rho F_i d\nu \; , \tag{2.2.2}$$

$$\frac{d}{dt}\int_{\mathcal{V}} \rho x_{[i}\dot{x}_{j]} d\nu = \int_{\partial \mathcal{V}} x_{[i}t_{j]k} da_k + \int_{\mathcal{V}} \rho(L_{ij} + x_{[i}F_{j]}) d\nu \; , \tag{2.2.3}$$

and

$$\frac{d}{dt}\int_{\mathcal{V}} (\tfrac{1}{2}\rho\dot{x}_i\dot{x}_i + \rho U) d\nu = \int_{\partial \mathcal{V}} (\dot{x}_i t_{ij} - q_j) da_j + \int_{\mathcal{V}} (\rho\dot{x}_i F_i + \rho r) d\nu \; . \tag{2.2.4}$$

Here, ρ is the mass density per unit volume, t_{ij} is the CAUCHY stress tensor, F_i the total body force due to electromagnetic fields and external actions, L_{ij} the dual of the body couple L_k, i.e.

$$L_{ij} = \tfrac{1}{2}e_{ijk}L_k \; , \tag{2.2.5}$$

U is the internal energy per unit mass, q_i the energy flux, consisting of heat flux and non-thermal energy flux, and r the total energy supply, due to the electromagnetic fields and due to heat. Bracketed indices indicate antisymmetric tensors. Note that the stress tensor t_{ij} and the stress vector $t_i^{(n)}$ on a surface element with exterior unit normal $\boldsymbol{n}$ are related by

$$t_i^{(n)} = t_{ij}n_j \; . \tag{2.2.6}$$

Often, in the literature the stress tensor is defined through

$$t_i^{(n)} = \bar{t}_{ji}n_j \; . \tag{2.2.7}$$

This definition implies $\bar{t}_{ij} = t_{ji}$. Moreover, t_{ij} is not symmetric in general. It is assumed that body force, body couple and energy supply can be decomposed into two parts; one is due to the electromagnetic fields, the other is supposed to be externally applied and known from the outset. Hence,

$$\rho F_i = \rho F_i^{\mathrm{e}} + \rho F_i^{\mathrm{ext}} \; ,$$

$$\rho L_{ij} = \rho L_{ij}^{\mathrm{e}} \; , \tag{2.2.8}$$

$$\rho r = \rho r^{\mathrm{e}} + \rho r^{\mathrm{ext}} \; .$$

Here, ρF_i^{e}, ρL_{ij}^{e} and ρr^{e} are thought to be expressed in terms of electromagnetic field quantities, while ρF_i^{ext} and ρr^{ext} are known. We have assumed that there are no externally applied body couples.

For sufficiently smooth fields, the global balance laws (2.2.1)–(2.2.4) assume their local forms

$$\dot{\rho} + \rho\dot{x}_{i,i} = 0 \,,$$

$$\rho\ddot{x}_i = t_{ij,j} + \rho F_i^{\mathrm{e}} + \rho F_i^{\mathrm{ext}} \,,$$

$$t_{[ij]} = \rho L_{[ij]}^{\mathrm{e}} \,,$$

$$\rho\dot{U} = t_{ij}\dot{x}_{i,j} - q_{i,i} + \rho r^{\mathrm{e}} + \rho r^{\mathrm{ext}} \,,$$

(2.2.9)

where use has been made of (2.2.8). Here, and throughout this work by $\dot{x}_{i,j}$ is meant

$$\dot{x}_{i,j} = (\dot{x}_i)_{,j} = F_{\alpha j}^{-1}\dot{F}_{i\alpha} \,. \tag{2.2.10}$$

As follows from (2.2.9)$_1$, the present density ρ is related to ρ_0, the density in the reference configuration, by

$$\rho = \frac{\rho_0}{|J|} \,. \tag{2.2.11}$$

2.2.2 The Maxwell Equations

As is well-known, there are several formulations of electrodynamics, all of which may be derived from different postulations. For reference we refer the reader to PENFIELD and HAUS ([177], Ch. 7). It is not our intention to derive the electromagnetic balance laws from various charge and electric circuit models, because we aim at describing the equivalence of these formulations rather than emphasizing their differences.

We start with the *Conservation of Magnetic Flux.*

Let B_i be the magnetic flux density (or *magnetic induction*) and let $\mathcal{E}_i$ be the effective electric field strength, sometimes also called the *electromotive intensity*. The conservation law of magnetic flux (FARADAY law) may then be expressed as

$$\frac{d}{dt}\int_S B_i \mathrm{d}a_i + \int_{\partial S} \mathcal{E}_i \mathrm{d}l_i = 0 \,. \tag{2.2.12}$$

This relation must hold for any material surface S with boundary ∂S in $\mathbb{R}^3$.

By applying (2.2.12) to a closed surface $S = \partial\mathcal{V}$, one can derive the GAUSS–FARADAY law, which states that

$$\frac{d}{dt}\left(\int_{\partial\mathcal{V}} B_i \mathrm{d}a_i\right) = 0 \quad \Rightarrow \quad \int_{\partial\mathcal{V}} B_i \mathrm{d}a_i = 0 \,, \tag{2.2.13}$$

in which the constant of integration has been set to zero. For sufficiently smooth fields, the balance laws (2.2.12) and (2.2.13) can be written in local form. To this end, we use the divergence, STOKES' and REYNOLDS' transport theorems as follows:

$$\int_{\partial \mathcal{V}} B_i da_i \quad = \int_{\mathcal{V}} B_{i,i} dv \, ,$$

$$\int_{\partial S} \mathcal{E}_i dl_i \quad = \int_S e_{ijk} \mathcal{E}_{i,j} da_i \, , \qquad (2.2.14)$$

$$\frac{d}{dt} \int_S B_i da_i = \int_S \overset{\star}{B}_i \, da_i \, .$$

Substitute these into (2.2.12) and (2.2.13) and localize the emerging equations. This yields

$$e_{ijk} \mathcal{E}_{k,j} + \overset{\star}{B}_i = 0, \quad B_{i,i} = 0 \, . \qquad (2.2.15)$$

In the above

$$e_{ijk} = \begin{cases} 1, \text{ if } i,j,k \text{ is an even pemutation of } 1,2,3 \, , \\ -1, \text{ if } i,j,k \text{ is an odd pemutation of } 1,2,3 \, , \qquad (2.2.16) \\ 0, \text{ else} \, , \end{cases}$$

is the completely skew-symmetric LEVI–CIVITÀ tensor and the convective derivative $\overset{\star}{\psi}_i$ of a vector ψ_i is defined by

$$\overset{\star}{\psi}_i := \frac{\partial \psi_i}{\partial t} + \psi_{i,j} \dot{x}_j + \psi_i \dot{x}_{j,j} - \psi_j \dot{x}_{i,j} \, . \qquad (2.2.17)$$

A proof of (2.2.14)$_3$ is given e.g. in CHADWICK [39] or HUTTER & JÖHNK [100].

The second basic law of electromagnetism is the law of *Conservation of Charge*. Let $\mathcal{Q}$ be the *electric charge density* and $\mathcal{J}_i$ the *conductive current*. The conservation law of electric charges may then be expressed by the global balance law

$$\frac{d}{dt} \int_{\mathcal{V}} \mathcal{Q} dv + \int_{\partial \mathcal{V}} \mathcal{J}_i da_i = 0 \, . \qquad (2.2.18)$$

This equation holds for any material volume $\mathcal{V}$ with boundary $\partial \mathcal{V}$. It is a purely formal matter to introduce a field D_i such that

$$\int_{\mathcal{V}} \mathcal{Q} dv = \int_{\partial \mathcal{V}} D_i da_i \, , \qquad (2.2.19)$$

holds for any material part $\mathcal{V}$ of the body. This introduction of D_i suggests to call it charge potential; however, we shall refer to D_i as the *dielectric displacement*.

It follows from (2.2.18) and (2.2.19) that the conservation of charge is satisfied if a field $\mathcal{H}_i$ exists such that

$$\int_S \mathcal{J}_i \mathrm{d}a_i + \frac{d}{dt} \int_S D_i \mathrm{d}a_i = \int_{\partial S} \mathcal{H}_i \mathrm{d}l_i \qquad (2.2.20)$$

holds for any material surface $\mathcal{S}$ with boundary $\partial \mathcal{S}$. Indeed, if we choose $\mathcal{S} = \partial \mathcal{V}$ in (2.2.20), then $\partial \mathcal{S} = \emptyset$ and (2.2.20) reduces to (2.2.18). Equation (2.2.20) is called AMPÈRE's law. We shall call $\mathcal{H}_i$ the *effective magnetic field strength* or the *magnetomotive intensity*. Note that $\mathcal{H}_i$ and D_i as introduced by (2.2.19) and (2.2.20) are not unique.

Again, assuming sufficient smoothness of the fields, the balance laws (2.2.18), (2.2.19) and (2.2.20) may be brought into local form. From (2.2.19) and (2.2.20) we then obtain

$$D_{i,i} = \mathcal{Q}, \qquad e_{ijk} \mathcal{H}_{k,j} - \overset{\star}{D}_i = \mathcal{J}_i, \qquad (2.2.21)$$

and from (2.2.18)

$$(\mathcal{J}_i + \mathcal{Q} \dot{x}_i)_{,i} + \frac{\partial \mathcal{Q}}{\partial t} = J_{i,i} + \frac{\partial \mathcal{Q}}{\partial t} = 0, \qquad (2.2.22)$$

which is the local conservation law of electric charges. For convenience we have also introduced here the *non-conductive current* J_i defined by

$$J_i := \mathcal{J}_i + \mathcal{Q} \dot{x}_i. \qquad (2.2.23)$$

The equations (2.2.21) are called the inhomogeneous MAXWELL equations, in contrast to the homogeneous ones, (2.2.15), because on the right-hand sides they contain electric charge and current densities.

Note that not all of the above equations are independent. Indeed, the conservation law of charge may be derived from the inhomogeneous MAXWELL equations. In other words, when integrating the MAXWELL equations, the conservation law of electric charges must be fulfilled along with the MAXWELL equations. As a result, at most seven of the above electromagnetic field variables can be considered basic, while for the remaining ones constitutive equations must be established. Of course, this does not mean that one cannot establish constitutive equations for more electromagnetic field variables. Indeed, there is a valid point to treat both $\mathcal{J}_i$ and $\mathcal{Q}$ as dependent constitutive variables. If this is done, however, further integrability conditions must be satisfied.

In the above presentation we have not made any appeal to a specific model, how the moving material body may contribute to the electromagnetic field vectors B_i, $\mathcal{E}_i$, D_i and $\mathcal{H}_i$. We shall not do this here either and refer to the pertinent literature (PENFIELD and HAUS [177], FANO, CHU and ADLER [73], TRUESDELL and TOUPIN [244], PAO [172], ERINGEN and MAUGIN [70] [71], MAUGIN et al. [137], TIERSTEN [238], ZHOU [215, 216] and others). The reader may be puzzled, however, by the occurrence of the convective time derivatives in (2.2.15) and (2.2.21), which contain implicitly the motion. If he so desires, he may formally eliminate the latter by simply defining two new fields through

$$\mathcal{E}_i = E_i + \mathrm{e}_{ijk}\dot{x}_j B_k \,, \qquad \mathcal{H}_i = H_i - \mathrm{e}_{ijk}\dot{x}_j D_k \,. \qquad (2.2.24)$$

Here, E_i and H_i are the well-known MINKOWSKIan electric and magnetic field strengths. We note that relations (2.2.24) are not the only ones through which the motion can be eliminated.

We might further mention that of the four fields B_i, $\mathcal{E}_i$, D_i and $\mathcal{H}_i$ only two are considered basic, while the remaining two will have to be described by constitutive assumptions. In practice, some of the above field quantities are replaced by others, which may allow a better physical insight, but this does not change the fundamental fact that there are two basic vectorial field quantities, while the remaining ones must be expressed in terms of the former. We shall come back to this at a later stage.

2.2.3 Material Description

The balance laws listed in the previous two sections are written in the spatial or EULERian formulation. Corresponding to these equations there is a material or LAGRANGEan description.

Let dv, da_k and dl_k be the volume-, area- and length-increments in the spatial description, while dV, dA_α and dL_α are the corresponding increments in the material description. Then the following well-known identities hold between these increments:

$$dv = |J|dV \,, \qquad da_k = JF_{\alpha k}^{-1}dA_\alpha \,, \qquad dl_k = F_{k\alpha}dL_\alpha \,. \qquad (2.2.25)$$

Here a remark concerning the second equation is in order. Since the normal vector on a closed surface is defined as the outward normal and outward normals given by relation $(2.2.25)_2$ are transformed into inward normals when $\mathrm{sgn}J < 0$, we must replace in $(2.2.25)_2$ J by $|J|$ whenever the surface is closed. Hence, if dA_α and da_k are area-elements of a closed surface, we have

$$da_k = |J|F_{\alpha k}^{-1}dA_\alpha \,. \qquad (2.2.26)$$

Although we may restrict J to positive values, use of $|J|$ instead of J in the following equations is preferred in order to make the respective quantities objective under the full group of (EUCLIDian) transformations, (see Sect. 1.6). With the relations (2.2.25) the mechanical balance laws (2.2.1)–(2.2.4) may be transformed into the following forms

$$\frac{d}{dt} \int_{V_R} \rho_0 dV = 0 \,, \qquad (2.2.27)$$

$$\frac{d}{dt} \int\limits_{V_R} \rho_0 \dot{x}_i dV = \int\limits_{\partial V_R} T_{i\alpha} dA_\alpha + \int\limits_{V_R} \rho_0 F_i dV \; , \qquad (2.2.28)$$

$$\frac{d}{dt} \int\limits_{V_R} \rho_0 x_{[i} \dot{x}_{j]} dV = \int\limits_{\partial V_R} x_{[i} T_{j]\alpha} dA_\alpha + \int\limits_{V_R} \rho_0 (L_{ij} + x_{[i} F_{j]}) dV \; , \qquad (2.2.29)$$

$$\frac{d}{dt} \int\limits_{V_R} \left(\frac{1}{2} \rho \dot{x}_i \dot{x}_i + \rho U \right) dV = \int\limits_{\partial V_R} (\dot{x}_i T_{i\alpha} - Q_\alpha) dA_\alpha + \int\limits_{V_R} (\rho_0 \dot{x}_i F_i + \rho_0 r) dV \; .$$

$$(2.2.30)$$

Here, integration is over reference volume and surface, respectively. Further, $T_{i\alpha}$ is the *first* PIOLA–KIRCHHOFF *stress tensor* and Q_α is the *material energy flux vector*, which are related to t_{ij} and q_i according to

$$
\begin{aligned}
T_{i\alpha} = |J| F_{\alpha j}^{-1} t_{ij} \; , \qquad t_{ij} = |J^{-1}| F_{j\alpha} T_{i\alpha} \; , \\
Q_\alpha = |J| F_{\alpha i}^{-1} q_i \; , \qquad q_i = |J^{-1}| F_{i\alpha} Q_\alpha \; .
\end{aligned}
\qquad (2.2.31)
$$

Assuming for the external source terms decompositions similar to those listed in (2.2.8) and supposing sufficient smoothness of the fields involved, we find that (2.2.28)–(2.2.30) imply

$$\rho_0 \ddot{x}_i = T_{i\alpha,\alpha} + \rho_0 F_i^{\mathrm{e}} + \rho_0 F_i^{\mathrm{ext}} \; ,$$

$$T_{[i\alpha} F_{j]\alpha} = \rho_0 L_{ij} \; , \qquad (2.2.32)$$

$$\rho_0 \dot{U} = T_{i\alpha} \dot{F}_{i\alpha} - Q_{\alpha,\alpha} + \rho_0 r^{\mathrm{e}} + \rho_0 r^{\mathrm{ext}} \; ,$$

whereas (2.2.27) integrates to yield $\rho_0 = \rho_0(\boldsymbol{X})$.

Apart from the stress tensors t_{ij} and $T_{i\alpha}$ we shall occasionally also use the so-called *second* PIOLA-KIRCHHOFF *stress tensor* defined by

$$T_{\alpha\beta}^P := T_{i\alpha} F_{\beta i}^{-1} = |J| t_{ij} F_{\alpha j}^{-1} F_{\beta i}^{-1} \; . \qquad (2.2.33)$$

Notice that $T_{\alpha\beta}^P$ is symmetric if and only if t_{ij} is symmetric. There exists also a material formulation of the electromagnetic balance laws (see [92]). To derive the material counterpart of (2.2.12), (2.2.13) and (2.2.18)–(2.2.20), recall that the integrals occurring in these equations must only be transformed back to the reference configuration. With the aid of (2.2.25) we obtain straightforwardly

$$\frac{d}{dt} \int\limits_{S_R} \mathbb{B}_\alpha dA_\alpha + \int\limits_{\partial S_R} \mathbb{E}_\alpha dL_\alpha = 0 \; , \qquad (2.2.34)$$

$$\int\limits_{\partial V_R} \mathbb{B}_\alpha dA_\alpha = 0 \; , \qquad (2.2.35)$$

$$\int_{\partial V_R} \mathbb{D}_\alpha dA_\alpha = \int_{V_R} \mathbb{Q} dV \,, \tag{2.2.36}$$

$$-\frac{d}{dt}\int_{S_R} \mathbb{D}_\alpha dA_\alpha + \int_{\partial S_R} \mathbb{H}_\alpha dL_\alpha = \int_{S_R} \mathbb{J}_\alpha dA_\alpha \,, \tag{2.2.37}$$

$$\frac{d}{dt}\int_{V_R} \mathbb{Q} dV + \int_{\partial V_R} \mathbb{J}_\alpha dA_\alpha = 0 \,. \tag{2.2.38}$$

Here, hollow quantities are the material counterparts of $B_i, \mathcal{E}_i, D_i, \mathcal{H}_i, \mathcal{J}_i$ and $\mathcal{Q}$ and they are related to these according to the transformation rules

$$
\begin{aligned}
\mathbb{Q} &= \mathcal{Q}|J| \,, & \mathcal{Q} &= |J^{-1}|\mathbb{Q} \,, \\
\mathbb{J}_\alpha &= \mathcal{J}_i|J|F_{\alpha i}^{-1} \,, & \mathcal{J}_i &= |J^{-1}|F_{i\alpha}\mathbb{J}_\alpha \,, \\
\mathbb{D}_\alpha &= D_i|J|F_{\alpha i}^{-1} \,, & D_i &= |J^{-1}|F_{i\alpha}\mathbb{D}_\alpha \,, \\
\mathbb{H}_\alpha &= \mathcal{H}_i F_{i\alpha}\mathrm{sgn}J \,, & \mathcal{H}_i &= F_{\alpha i}^{-1}\mathbb{H}_\alpha \mathrm{sgn}J \,, \\
\mathbb{B}_\alpha &= B_i J F_{\alpha i}^{-1} \,, & B_i &= J^{-1}F_{i\alpha}\mathbb{B}_\alpha \,, \\
\mathbb{E}_\alpha &= \mathcal{E}_i F_{i\alpha} \,, & \mathcal{E}_i &= F_{\alpha i}^{-1}\mathbb{E}_\alpha \,.
\end{aligned}
\tag{2.2.39}
$$

The quantities $\mathbb{E}_\alpha, \mathbb{B}_\alpha$ etc. will be called the material or LAGRANGEan electromagnetic fields. Again, although J may be assumed positive we have written these formulas such that the hollow quantities transform under the EUCLIDian transformation group as scalars (see Sect. 1.6).

For sufficiently smooth fields the global laws (2.2.34)–(2.2.38) can be written in local form. Then they are

$$
\begin{aligned}
e_{\alpha\beta\gamma}\mathbb{E}_{\gamma,\beta} + \dot{\mathbb{B}}_\alpha &= 0 \,, & \mathbb{B}_{\alpha,\alpha} &= 0 \,, \\
e_{\alpha\beta\gamma}\mathbb{H}_{\gamma,\beta} - \dot{\mathbb{D}}_\alpha &= \mathbb{J}_\alpha \,, & \mathbb{D}_{\alpha,\alpha} &= \mathbb{Q} \,, \\
\dot{\mathbb{Q}} + \mathbb{J}_{\alpha,\alpha} &= 0 \,,
\end{aligned}
\tag{2.2.40}
$$

where all differentiations are with respect to the material coordinates. As was mentioned already before, this set of equations is a dependent one since the conservation of charge is already implicitly contained in the MAXWELL equations. As a consequence, at most seven variables are independent, while the remaining ones must be given by constitutive equations.

At the moment, the LAGRANGEan fields are introduced purely formally. That they are of advantage will be demonstrated in Chaps. 4 and 5.

2.3 The Entropy Production Inequality

It is a fact of experience that real physical processes are irreversible. This means that processes cannot, in general, be traversed backward in time, i.e., time reversal does not lead to a physically realizable process. This fact is called the second law of thermodynamics and its mathematical realization is the entropy production inequality. It is based on the assumption that there exists an additive quantity η, called the entropy, which satisfies the balance law

$$\frac{d}{dt} \int_V \rho\eta d\nu + \int_{\partial V} \phi_i da_i - \int_V \rho s d\nu = \int_V \rho\gamma d\nu . \qquad (2.3.1)$$

Here, ϕ_i is *the entropy flux*, s the *entropy supply* and γ the *entropy production*. For sufficiently smooth fields (2.3.1) implies

$$\rho\dot\eta + \phi_{i,i} - \rho s = \rho\gamma , \qquad (2.3.2)$$

and it is the expression of the second law of thermodynamics that

$$\gamma \geq 0 ; \qquad (2.3.3)$$

hence

$$\rho\dot\eta + \phi_{i,i} - \rho s \geq 0 . \qquad (2.3.4)$$

This inequality must hold for all thermodynamic processes, i.e. processes that satisfy the balance laws of thermomechanics and electrodynamics as well as the constitutive relations. We shall set the entropy supply s equal to the external energy supply divided by the absolute temperature Θ, i.e.

$$s = \frac{r^{\text{ext}}}{\Theta} , \qquad (2.3.5)$$

but we shall, in general, not assume that "entropy flux equals heat flux divided by absolute temperature". Thus, (2.3.4) becomes

$$\rho\dot\eta + \phi_{i,i} \geq \frac{\rho r^{\text{ext}}}{\Theta} . \qquad (2.3.6)$$

The material counterpart of (2.3.1) can easily be derived. In its local form, it reads

$$\rho_0\dot\eta + \Phi_{\alpha,\alpha} \geq \frac{\rho_0 r^{\text{ext}}}{\Theta} , \qquad (2.3.7)$$

where

$$\Phi_\alpha = |J|F_{\alpha i}^{-1}\phi_i , \qquad (2.3.8)$$

is the LAGRANGEan entropy flux.

Before we proceed it seems to be worthwhile to justify the approach we take regarding the entropy inequality (2.3.4) and the interpretation we give to the entropy supply and to the entropy flux. Our aim is not the justification of one particular theory against any other one. On the contrary, we shall adopt certain results obtained by using a particular entropy principle in each theory and aim at a comparison of such theories. This comparison should be made on the level of fully developed theories.

2.4 Jump Conditions

Let
$$\Sigma = \hat{\Sigma}(\boldsymbol{X}, t) = \tilde{\sigma}(\boldsymbol{x}, t) = 0 \,, \tag{2.4.1}$$

be a smooth orientable surface, not necessarily material, and let W_N and w_n be its *speed of propagation* and *speed of displacement*, respectively, i.e.

$$W_N := -\frac{\partial \hat{\Sigma}/\partial t}{(\hat{\Sigma}_{,\alpha} \, \hat{\Sigma}_{,\alpha})^{1/2}} \,, \qquad \mathrm{w}_n := -\frac{\partial \tilde{\sigma}/\partial t}{(\tilde{\sigma}_{,i} \, \tilde{\sigma}_{,i})^{1/2}} \,. \tag{2.4.2}$$

Assume further that the electromagnetic field quantities $\mathcal{B}_i, \mathcal{E}_i$ etc., or $\mathbb{B}_\alpha, \mathbb{E}_\alpha$ etc., as well as all mechanical quantities listed in Sect. 1.3.1 or 1.3.3 may suffer finite jumps across the surface Σ. In particular we assume $\chi_i(\boldsymbol{X}, t)$ to be continuous on Σ, but its first derivatives may suffer finite jumps. Hence, it follows that, although

$$\hat{\Sigma}(\boldsymbol{X}, t) = \tilde{\sigma}(\boldsymbol{\chi}(\boldsymbol{X}, t), t) = \hat{\sigma}(\boldsymbol{X}, t) \,,$$

is one and only one surface in V_R, the values for W_N are the same on both sides of the surface only when $\dot{\chi}_i$ and $F_{i\alpha}$ are continuous. Waves in which $\dot{\chi}_i$ and $F_{i\alpha}$ may jump are called *shock waves*.

By applying the global balance laws to a part of the body ($\mathcal{V}$ or V_R) containing the surface of discontinuity, we can derive jump conditions for the fields occurring in these laws. We suppose the methods of derivation to be known and therefore only list the results. The jump conditions obtained thereby will depend on whether one is dealing with the material or the spatial description. In the spatial description, the balance laws (2.2.12), (2.2.13) and (2.2.18)–(2.2.20) reveal that

$$[\![B_i]\!]n_i = 0 \,, \qquad [\![e_{ijk}\mathcal{E}_j n_k + \dot{x}_i B_k n_k - B_i(\dot{x}_k n_k - \mathrm{w}_n)]\!] = 0 \,,$$

$$[\![D_i]\!]n_i = 0 \,, \qquad [\![e_{ijk}\mathcal{H}_j n_k + \dot{x}_i D_k n_k - D_i(\dot{x}_k n_k - \mathrm{w}_n)]\!] = 0 \,, \tag{2.4.3}$$

$$[\![J_i n_i - \mathcal{Q}\mathrm{w}_n]\!] = 0 \,.$$

Here,
$$[\![\Phi]\!] := \Phi^+ - \Phi^- \,, \tag{2.4.4}$$

denotes the jump of Φ across the surface $\tilde{\sigma}(\boldsymbol{x}, t)$, $\Phi^\pm$ are the values of Φ immediately on the positive and negative side of the surface, respectively, and $\boldsymbol{n}$ is the unit normal vector at a point on the singular surface pointing into the positive side of $\tilde{\sigma}(\boldsymbol{x}, t)$. We remark that conditions (2.4.3) tacitly assume that $\mathcal{J}_i$ and $\mathcal{Q}$ are finite on the surface of discontinuity. This is a restriction; e.g. it excludes surface distributions of charge and current.

In the material description the counterparts of (2.4.3) read as follows

$$\llbracket \mathbb{B}_\alpha \rrbracket N_\alpha = 0 \,, \qquad e_{\alpha\beta\gamma} \llbracket \mathbb{E}_\beta \rrbracket N_\gamma + \llbracket \mathbb{B}_\alpha W_N \rrbracket = 0 \,,$$

$$\llbracket \mathbb{D}_\alpha \rrbracket N_\alpha = 0 \,, \qquad e_{\alpha\beta\gamma} \llbracket \mathbb{H}_\beta \rrbracket N_\gamma + \llbracket \mathbb{D}_\alpha W_N \rrbracket = 0 \,, \tag{2.4.5}$$

$$\llbracket \mathbb{J}_\alpha N_\alpha - \mathbb{Q} W_N \rrbracket = 0 \,.$$

If we set $W_N = 0$, or $\mathrm{w}_n = \dot{x}_i^+ n_i = \dot{x}_i^- n_i$, then the surface of discontinuity is material. In that case, the jump conditions may serve as boundary conditions. We like to note here that, in the sense of the definition given above, the boundary of a body in a vacuum is not a material surface of discontinuity because on this boundary:

$$\dot{x}_i^+ = 0 \,, \qquad \text{and} \qquad \dot{x}_i^- n_i \mathrm{w}_n \,. \tag{2.4.6}$$

For the derivation of the jump conditions of momentum and energy, we start with the expressions for the electromagnetic body force and energy supply. We consider it to be known that the electromagnetic momentum and energy supply terms appearing in (2.2.2), (2.2.4), (2.2.28) and (2.2.30) can always be written in the form

$$\rho F_i^{\mathrm{e}} = t_{ij,j}^M + \frac{\partial g_i}{\partial t} \,, \qquad \rho_0 F_i^{\mathrm{e}} = T_{i\alpha,\alpha}^M + \dot{G}_i \,,$$

$$\rho r^{\mathrm{e}} + \rho F_i^{\mathrm{e}} \dot{x}_i = \pi_{i,i} + \frac{\partial \omega}{\partial t} \,, \qquad \rho_0 r^{\mathrm{e}} + \rho_0 F_i^{\mathrm{e}} \dot{x}_i = \Pi_{\alpha,\alpha} + \dot{\Omega} \,. \tag{2.4.7}$$

In the above relations, $T_{i\alpha}^M$ and t_{ij}^M are electromagnetic stress tensors, which sometimes are called MAXWELL *stress tensors*, G_i and g_i are the *electromagnetic momenta* in the material and spatial description, respectively, Π_α and π_i are the material and spatial representation of the *electromagnetic energy flux* and Ω and ω are *electromagnetic energy densities*. All these quantities are expressible in terms of the electromagnetic fields, the motion and their derivatives. Hence, although we assume the electromagnetic fields to suffer at most a finite jump across the singular surface Σ, such an assumption does not hold in general for ρF_i^{e} and ρr^{e} (or $\rho_0 F_i^{\mathrm{e}}$ and $\rho_0 r^{\mathrm{e}}$), which may become unbounded because of the occurrence of the above mentioned gradients. Apparently, for the weak forms of (2.4.7) one must write

$$\int_V \rho F_i^{\mathrm{e}} d\nu = \frac{d}{dt} \int_V \mathrm{g}_i d\nu + \int_{\partial V} (t_{ij}^M - \mathrm{g}_i \dot{x}_j) da_j \,,$$

$$\int_V (\rho r^{\mathrm{e}} + \rho F_i^{\mathrm{e}} \dot{x}_i) d\nu = \frac{d}{dt} \int_V \omega d\nu + \int_{\partial V} (\pi_i - \omega \dot{x}_i) da_i \,, \tag{2.4.8}$$

in the spatial formulation, and

$$\int\limits_{V_R} \rho_0 F_i^{\mathrm{e}} dV = \frac{d}{dt} \int\limits_{V_R} G_i dV + \int\limits_{\partial V_R} T_{i\alpha}^M dA_\alpha \,,$$

$$\int\limits_{V_R} (\rho_0 r^{\mathrm{e}} + \rho_0 F_i^{\mathrm{e}} \dot{x}_i) dV = \frac{d}{dt} \int\limits_{V_R} \Omega dV + \int\limits_{\partial V_R} \Pi_\alpha dA_\alpha \,,$$

(2.4.9)

in the material description, respectively. The fields $G_i, \mathrm{g}_i, T_{i\alpha}^M, t_{ij}^M, \Pi_\alpha, \pi_i, \Omega$ and ω are related to each other by

$$G_i = |J| \mathrm{g}_i \,,$$

$$T_{i\alpha}^M = |J| F_{\alpha j}^{-1} (t_{ij}^M - \mathrm{g}_i \dot{x}_j) \,,$$

$$\Pi_\alpha = |J| F_{\alpha i}^{-1} (\pi_i - \omega \dot{x}_i) \,,$$

$$\Omega = |J| \omega \,.$$

(2.4.10)

As long as all fields are sufficiently smooth, the left-hand and right-hand sides of (2.4.8) and (2.4.9) are equivalent. This is no longer so if ρF_i^{e} and ρr^{e} may be unbounded. Then the left-hand sides of (2.4.8) and (2.4.9) may not be meaningful at all, whereas the expressions on the right-hand sides still make sense. We therefore postulate the expressions on the right-hand sides to be the appropriate global statements.

Substituting $(2.2.8)_{1,3}$ into (2.2.2) and (2.2.4) and using (2.4.8) yields the global form of the balance laws of linear momentum and energy appropriate for the derivation of the jump conditions. Under the assumption that the external sources ρF_i^{ext} and ρr^{ext} remain finite at Σ, these global laws, together with the balance of mass (2.2.1), lead to the following set of jump conditions

$$[\![\rho(\dot{x}_i n_i - \mathrm{w}_n)]\!] = 0 \,,$$

$$[\![(\rho \dot{x}_i - g_i)(\dot{x}_j n_j - \mathrm{w}_n)]\!] - [\![t_{ij} + t_{ij}^M - g_i \dot{x}_j]\!] n_j = 0 \,,$$

$$[\![(\tfrac{1}{2}\rho \dot{x}_i \dot{x}_i + \rho U - \omega)(\dot{x}_j n_j - \mathrm{w}_n)]\!] - [\![\dot{x}_i t_{ij} - q_j + \pi_j - \omega \dot{x}_j]\!] n_j = 0 \,.$$

(2.4.11)

Similarly, in the material description,

$$[\![\rho_0 W_N]\!] = 0 \,,$$

$$[\![(\rho_0 \dot{x}_i - G_i) W_N]\!] - [\![T_{i\alpha} + T_{i\alpha}^M]\!] N_\alpha = 0 \,,$$

(2.4.12)

$$[\![(\tfrac{1}{2}\rho_0 \dot{x}_i \dot{x}_i + \rho_0 U - \Omega) W_N]\!] - [\![\dot{x}_i T_{i\alpha} - Q_\alpha + \Pi_\alpha]\!] N_\alpha = 0 \,.$$

Note that the balance of mass simply implies that ρ_0 may jump only along with a jump of W_N, or on material surfaces, where $W_N = 0$. For $W_N = 0$, $(2.4.11)_{2,3}$ constitute the boundary conditions for the tractions and the energy flux of matter and fields, respectively. On the other hand, if ρ_0 is continuous on Σ, $(2.4.12)_1$ implies continuity of W_N.

2.5 Material Objectivity

Although we consider the principle of material frame indifference to be known in general, we would like to call upon it here, mainly because of the complications resulting from the combination of the mechanical balance laws with those of electrodynamics.

Let x_i be the Cartesian coordinates of a particle as measured by a stationary observer and let x_i^* be the Cartesian coordinates of the same particle as measured by another observer in his frame of reference. We call transformations that relate x_i with x_i^* by a rigid-body motion EUCLIDian transformations. They have the form

$$x_i^* = O_{ij}(t)x_j + b_i(t) , \qquad t^* = t , \qquad (2.5.1)$$

where O_{ij} is an orthogonal (time dependent) matrix and where b_i is an arbitrary (time dependent) vector. A EUCLIDian transformation for which O_{ij} is not time dependent and for which $b_i = -V_i t$ is called a GALILEI *transformation*, and it is well-known that the balance laws of classical non-relativistic mechanics are invariant and frame independent under such transformations. They are, however, frame dependent with respect to EUCLIDian transformations.

Let $a, \boldsymbol{a}$ and $\boldsymbol{A}$ be scalar, vector and second rank tensor quantities over $\mathbb{R}^3$, and assume that two observers in their reference systems $\{O, \boldsymbol{e}_1, \boldsymbol{e}_2, \boldsymbol{e}_3\}$, $\{O^*, \boldsymbol{e}_1^*, \boldsymbol{e}_2^*, \boldsymbol{e}_3^*\}$ measure the components

$$a, a_i, A_{ij} \quad \text{and} \quad a^*, a_i^* A_{ij}^* , \qquad (2.5.2)$$

respectively. Then the starred and the unstarred quantities are related to one another through a transformation that is dictated by the transformation group from the unstarred to the starred reference system. If this transformation has the form

$$a^* = (\det \boldsymbol{O})^p a ,$$

$$a_i^* = (\det \boldsymbol{O})^p O_{ij} a_j , \qquad (2.5.3)$$

$$A_{ij}^* = (\det \boldsymbol{O})^p O_{ik} O_{il} A_{kl} ,$$

with $p = 0$ or $p = 1$, then $a, \boldsymbol{a}, \boldsymbol{A}$ are called an objective scalar, objective vector and objective tensor, respectively. For $p = 0$ ($p = 1$) these quantities are specified as polar (axial) objective scalar, vector or tensor, respectively. For brevity, the attribute "polar" is often omitted for polar field quantities. For instance, $\det \boldsymbol{O}$ is an axial scalar, the curl is an axial vector operator, but the gradient is polar. If $a_i = \dot{x}_i$ is the velocity it is easily seen that under the GALILEI group it is an objective polar vector, but when the group is EUCLIDian the velocity is not an objective vector quantity.

In contrast to classical mechanics, the MAXWELL equations are neither invariant under EUCLIDian nor under GALILEI transformations. The transformation group here is the **extended LORENTZ group**. This group may be explained as follows. Let (x_i, t) denote the position x_i and time t of a particle as measured by one observer and let $(x_i^\star, t^\star)$ be those as measured by another observer, who translates relative to the first observer with constant velocity V_i. A transformation of the form

$$x_i^\star = x_i + V_i \left\{ \frac{x_k V_k}{V^2} \left[\frac{1}{\sqrt{1 - V^2/c^2}} - 1 \right] - \frac{t}{\sqrt{1 - V^2/c^2}} \right\}$$
$$= (x_i - V_i t)(1 + \mathcal{O}(V^2/c^2)) \, ,$$

$$t^\star = \frac{t - \dfrac{x_k V_k}{c^2}}{\sqrt{1 - V^2/c^2}} = \left(t - \frac{1}{c^2} x_k V_k \right) (1 + \mathcal{O}(V^2/c^2)) \, ,$$

(2.5.4)

is then called a *special* LORENTZ *transformation*. Special in this transformation is that the frames x_i and $x_i^\star$ are parallel. If they are also rotated with respect to each other then $x_i^\star$ on the left-hand side of (2.5.4)$_1$ must be replaced by $O_{ji} x_j^\star$, where O_{ij} is a constant orthogonal matrix. The group of these transformations is called the extended LORENTZ group. In four-dimensional notation

$$\mathbf{x}_A^\star = *_{AB} \mathbf{x}_B \, , \qquad (A, B = 1, 2, 3, 4) \, , \tag{2.5.5}$$

where the four vector $\mathbf{x}_{(4)}$ is the ordered quadrupel $(\boldsymbol{x}_{(3)}, t)$ and, to within terms of the order of (V^2/c^2), $*_{AB}$ is given by

$$\Lambda_{AB} = \begin{pmatrix} O_{11} & O_{12} & O_{13} & -O_{1k}V_k \\ O_{21} & O_{22} & O_{23} & -O_{2k}V_k \\ O_{31} & O_{32} & O_{33} & -O_{3k}V_k \\ -V_1/c^2 & -V_2/c^2 & -V_3/c^2 & 1 \end{pmatrix} . \tag{2.5.6}$$

The LORENTZ group is the analogon to the GALILEI group and there is no immediate analogon to the EUCLIDian group.

The principle of material frame indifference (material objectivity) states that the material response ought not to be frame-dependent. But with respect to which transformation group? Obviously, since classical mechanics is invariant under the GALILEI group and the special theory of relativity is so under the LORENTZ group, the material response must be invariant with respect to one of these groups, depending on whether one deals with classical mechanics or with special relativity. We consider it to be known that the principle of material objectivity of classical mechanics as stated by NOLL requests the material response to be invariant under EUCLIDian transformations. There have been objections raised against the general truth of this (see MÜLLER [159]), but this will be of no relevance here, because the results of this monograph (except Chap. 8) will also hold true if we restrict ourselves in those cases to GALILEI transformations.

In order to gain some insight into the transformation properties of the electromagnetic field variables, recall that the fields E_i, B_i, H_i and D_i introduced in (2.2.24) can be written in the form of two skew-symmetric covariant and contravariant four-tensors, namely as

$$
\varphi_{AB} = \begin{pmatrix} 0 & B_3 & -B_2 & E_1 \\ -B_3 & 0 & B_1 & E_2 \\ B_2 & -B_1 & 0 & E_3 \\ -E_1 & -E_2 & -E_3 & 0 \end{pmatrix}, \quad \eta^{AB} = \begin{pmatrix} 0 & H_3 & -H_2 & -D_1 \\ -H_3 & 0 & H_1 & -D_2 \\ H_2 & -H_1 & 0 & -D_3 \\ D_1 & D_2 & D_3 & 0 \end{pmatrix},
$$

$$(2.5.7)$$

and that these tensors transform under general transformations of the form $x^{\star A}(x^B)$ according to

$$
\varphi^{\star}_{AB} = \frac{\partial x^C}{\partial x^{\star A}} \frac{\partial x^D}{\partial x^{\star B}} \varphi_{CD} , \qquad \eta^{\star AB} = \frac{\partial x^{\star A}}{\partial x^C} \frac{\partial x^{\star B}}{\partial x^D} \eta^{CD} . \tag{2.5.8}
$$

With

$$
\sigma^A := (\mathcal{J}_i + \mathcal{Q}\dot{x}_i, \mathcal{Q}) , \tag{2.5.9}
$$

it is then a routine matter to show that the MAXWELL equations (2.2.15) and (2.2.21) are given by

$$
e^{ABCD} \frac{\partial \varphi_{CD}}{\partial x^B} = 0 , \qquad \frac{\partial \eta^{AB}}{\partial x^B} = \sigma^A , \tag{2.5.10}
$$

where

$$
e^{ABCD} = \begin{cases} -1, \text{ if } ABCD \text{ is an even permutation of } 1234 , \\ 1, \text{ if } ABCD \text{ is an odd permutation of } 1234 , \\ 0, \text{ else } , \end{cases} \tag{2.5.11}
$$

is the four-dimensional permutation tensor. The equations (2.5.8) hold for any transformation $x^{\star A} = x^{\star A}(x^B)$. If one chooses EUCLIDian transformations, one can show that

$\mathcal{Q}$	transforms as an objective scalar ,
$\mathcal{E}_i, D_i, \mathcal{J}_i$	transform as objective polar vectors ,
$\mathcal{H}_i, B_i$	transform as objective axial vectors .

$$(2.5.12)$$

For the proof of these statements, the reader may consult Appendix A. In particular, these quantities are objective under GALILEI transformations. In the sequel we shall also introduce other electromagnetic variables, such as polarization and magnetization vectors, some of which are also objective vectors in the above sense.

The MAXWELL equations (2.5.10) would formally be invariant under general transformations of the form $x^{\star A} = x^{\star A}(x^B)$, as can easily be checked by substituting (2.5.8) into (2.5.10), if there would not be a relation of the form

$$\eta^{AB} = \eta^{AB}(\varphi_{CD}) \,, \tag{2.5.13}$$

that is not invariant under the most general transformations. Indeed, a relation (2.5.13) even exists in vacuo, in which case it reads

$$D_i = \varepsilon_0 E_i \,, \qquad H_i = \frac{1}{\mu_0} B_i \,. \tag{2.5.14}$$

These relations are sometimes referred to as the MAXWELL–LORENTZ **aether relations**. Equations (2.5.14) restrict the class for which the MAXWELL equations (2.5.10) are invariant to the extended LORENTZ group. In general, (2.5.13) gives a model for electromagnetic field interactions with matter.

Based on the properties (2.5.12) we may then request as is done classically, that the material response be invariant under the EUCLIDian group. This is an approximation, because the MAXWELL equations can never be rendered LORENTZ invariant this way. On the other hand, dependent on the choice of electromagnetic body force, body couple and energy supply, the balance laws of mechanics may be GALILEI invariant this way. Theories of this nature have been proposed by, amongst others, TOUPIN [241], LIU and MÜLLER [127], PAO and HUTTER [171], ALBLAS [9], VAN DE VEN [249], HUTTER [95], DE GROOT and SUTTORP [53], ERINGEN and MAUGIN [70, 71], TIESTEN [238], MAUGIN et. al. [137], and SHU-ANG ZHOU [216]. For reasons that will become apparent shortly, we shall call these treatments **non-relativistic** theories.

Of course, the above selection of the transformation group is not satisfactory, and it should be replaced by one treating the mechanical and electromagnetic equations alike. Such a description must necessarily be relativistic. Hence, the material response must be invariant under the extended LORENTZ group. Considering motions only that are small relative to the velocity of light, we may then drop all terms containing a factor c^{-2} (this statement depends on the choice of units and in the way we state it here, SI-units are implied). In this way one arrives at MAXWELL equations which are LORENTZ invariant except for terms with a c^{-2}-factor. The same holds true for the constitutive relations and the mechanical equations. Theories obtained in this fashion may also be called non-relativistic, because we may look at their constitutive treatment in the light of EUCLIDian transformations, as we shall soon see.

Finally there are formulations in which only the terms of order $\dot{x}^2/c^2$ are dropped, while terms with a c^{-2}-factor (but no $(\dot{x}^2/c^2)$-factor) are kept. Such theories will be called **semi-relativistic**.

In order to render these ideas a little more precise, consider a stationary frame of Cartesian coordinates and a particle moving with velocity $\dot{x}_i$. An inertial frame, in which the particle is instantaneously at rest, is called the **rest**

frame. Electromagnetic variables as measured by an observer in this frame are called *rest frame values* and they are related to those in the stationary frame by a LORENTZ transformation. Rest frame values will be denoted by a superimposed ring, viz. $\overset{\circ}{E}_i$ etc. Of course, the MAXWELL equations also hold in the rest frame (whereby all variables and operations carry the symbol ∘). Transforming these equations back to the original frame reveals the transformation rules for the variables $B_i, \overset{\circ}{B}_i, \mathcal{E}_i, \overset{\circ}{\mathcal{E}}_i$ etc. To within the semi-relativistic approximation the transformation rules are

$$\overset{\circ}{\mathcal{B}}_i = \mathcal{B}_i + \mathcal{O}(\dot{x}^2/c^2)\,, \qquad \mathcal{B}_i = B_i - \frac{1}{c^2}e_{ijk}\dot{x}_j\mathcal{E}_k\,,$$

$$\overset{\circ}{\mathcal{E}}_i = \mathcal{E}_i + \mathcal{O}(\dot{x}^2/c^2)\,,$$

$$\overset{\circ}{\mathcal{D}}_i = \mathcal{D}_i + \mathcal{O}(\dot{x}^2/c^2)\,, \qquad \mathcal{D}_i = D_i + \frac{1}{c^2}e_{ijk}\dot{x}_j\mathcal{H}_k\,,$$

$$\overset{\circ}{\mathcal{H}}_i = \mathcal{H}_i + \mathcal{O}(\dot{x}^2/c^2)\,, \tag{2.5.15}$$

$$\overset{\circ}{\mathcal{J}}_i = \mathcal{J}_i + \mathcal{O}(\dot{x}^2/c^2)\,,$$

$$\overset{\circ}{\mathcal{Q}} = \mathcal{Q} - \frac{1}{c^2}\dot{x}_i\mathcal{J}_i + \mathcal{O}(\dot{x}^2/c^2)\,.$$

From these definitions as well as (2.2.24) it follows that $\mathcal{B}_i, \mathcal{E}_i, \mathcal{D}_i$ etc. are the fields B_i, E_i, D_i etc. as measured by an observer travelling with the particle. Furthermore, under rigid rotations they transform like scalars and vectors, and, as is shown in the theory of relativity (see Møller, [158], p. 199), they form within the semi-relativistic approximation the first three components of a set of four-vectors. Hence $\mathcal{B}_i, \mathcal{E}_i, \mathcal{D}_i, \mathcal{H}_i, \mathcal{J}_i$ and $\overset{\circ}{\mathcal{Q}}$ are vectors and scalars, objective under LORENTZ-transformations to within the semi-relativistic approximation.

Mere inspection of (2.5.15) shows, however, that the variables $\mathcal{B}_i$ and $\mathcal{D}_i$ are not objective under GALILEIan- or EUCLIDian transformations, because according to (2.5.12) B_i and D_i are. But, in the non-relativistic limit, where c^{-2}-terms are dropped

$$\mathcal{D}_i \approx D_i \qquad \text{and} \qquad \mathcal{B}_i \approx B_i\,, \tag{2.5.16}$$

and D_i and B_i are under EUCLIDian transformations an objective vector and an objective axial vector, respectively.

Multiplying both sides of equation $(2.5.15)_2$ with μ_0^{-1} shows that

$$\frac{1}{\mu_0}\mathcal{B}_i = \frac{1}{\mu_0}B_i - \varepsilon_0 e_{ijk}\dot{x}_j E_k\,, \tag{2.5.17}$$

so that the second term on the right-hand side is no longer of order c^{-2}. Hence, B_i/μ_0 cannot be objective non-relativistically (although B_i is). The objective quantity must be $\mathcal{B}_i/\mu_0$ and indeed it can be shown (see Appendix A) that $\mathcal{B}_i/\mu_0$ is objective under EUCLIDian transformations. A similar argument holds for D_i, because

$$\frac{1}{\varepsilon_0}\mathcal{D}_i = \frac{1}{\varepsilon_0}D_i + \mu_0 e_{ijk}\dot{x}_j H_k \ . \tag{2.5.18}$$

Here again, $\mathcal{D}_i/\varepsilon_0$ is the quantity that is objective non-relativistically and not D_i/ε_0, although D_i is. Needless to say that EUCLIDian invariance of $\mathcal{D}_i/\varepsilon_0$ also follows from the general relativistic transformation properties (see Appendix A). This is why we have called theories non-relativistic which use the principle of material objectivity under EUCLIDian transformations.

Hence, we may conclude by stating that in the non-relativistic approximation (neglecting c^{-2}-terms) the requirement of invariance of the material response under GALILEI transformations (a special EUCLIDian transformation) is equivalent to the requirement of LORENTZ invariance . The difference is at most a philosophical one.

2.6 Constitutive Equations

The balance laws (2.2.9), (2.2.15) and (2.2.21) express the common physical properties enjoyed by a material exhibiting electromagnetic mechanical interactions. They comprise a set of differential equations for many more variables than there are equations. So they are no field equations from which the field quantities could uniquely be determined. In order to become field equations the balance laws (2.2.9), (2.2.15) and (2.2.21) must be complemented by constitutive equations. Of course, in this regard various degrees of complexity are possible. In PART I, we shall restrict ourselves to *magnetizable* and *polarizable solids*, which deform elastically under the action of electromagnetic and thermal fields and which exhibit electrical and thermal conduction. Mechanical dissipation is left out of consideration as is the exchange interaction and magnetic spin. In PART II, Chap. 8, viscous and plastic effects will also be included.

To obtain field equations it must first be decided which physical variables we suppose to be the independent fields. With regard to thermo-mechanical variables these fields are generally the motion $\mathcal{X}(X,t)$ and the temperature $\Theta(X,t)$. The basic electromagnetic field variables are generally two electromagnetic field vectors and the free charge. Any constitutive relation must be expressed therefore as a functional of the motion, the temperature, the free charge and two electromagnetic field variables and derivatives thereof. Taking for instance $\mathcal{E}_i$ and $\mathcal{H}_i$ as the basic fields the conceivably simplest constitutive class exhibiting the above mentioned properties may have the form.

$$\mathsf{C}(t) = \bar{\mathsf{C}}(x_i(t), \dot{x}_i(t), F_{i\alpha}(t), \mathcal{E}_i(t), \mathcal{H}_i(t), \Theta(t), \Theta_{,i}(t), \mathcal{Q}(t)) \ . \tag{2.6.1}$$

Here, C stands for any scalar, vector or tensor valued quantity for which a constitutive equation is established, and $\bar{C}(\cdot)$ is a function of the indicated variables. If the value of C at instant t depends on the motion x_i, the temperature Θ and the electromagnetic fields $\mathcal{E}_i, \mathcal{H}_i$ and $\mathcal{Q}$ at the same instant only, we call the material response to have no memory. Now the function $\bar{C}(\cdot)$ also depends on $\dot{x}_i(t)$. Hence, with regard to x_i the material appears to remember the past history of a process it has undergone for a very short time. Indeed knowing x_i and $\dot{x}_i$ at a time t allows us to approximate the motion arbitrarily close to times $\tau = t$ as close as we please. We shall show, however, that an explicit dependence on $\dot{x}_i$ is not possible.

As is usual in continuum mechanics, we require the constitutive relations to be independent of the observer. In terms of the above quantities C this principle of material objectivity requires that the quantities C do not only transform as objective scalars, vectors and tensors, respectively, but that $\bar{C}(\cdot)$ is frame independent under the transformation group considered. In view of the fact that we shall restrict ourselves in the following chapters to a non-relativistic approach we require the constitutive equations to be frame independent scalar, vector and tensor valued functions under the group of EUCLIDian transformations. (Actually invariance under GALILEI transformations is sufficient for constitutive relations of the form (2.6.1)). Then it is straightforward to shown that the constitutive quantities C cannot depend on x_i and $\dot{x}_i$ explicitly. Furthermore, we have seen in the preceding section that $\mathcal{E}_i, \mathcal{H}_i$ and $\mathcal{Q}$ are objective, and thus we may set

$$C = \bar{C}(F_{i\alpha}, \mathcal{E}_i, \mathcal{H}_i, \Theta, \Theta_{,i}, \mathcal{Q}) . \tag{2.6.2}$$

If viscous or plastic effects are included then the adequate constitutive postulate is

$$C = \bar{C}(F_{i\alpha}, \dot{F}_{i\alpha}, \mathcal{E}_i, \mathcal{H}_i, \Theta, \Theta_{,i}, \mathcal{Q}) . \tag{2.6.3}$$

Of course, since D_i and B_i are also objective, it is always possible to replace in (2.6.2) $\mathcal{E}_i$ by D_i and/or $\mathcal{H}_i$ by B_i.

Moreover, one can formally introduce polarization P_i and magnetization $\mathcal{M}_i$ by

$$P_i := D_i - \varepsilon_0 \mathcal{E}_i , \quad \text{and} \quad \mu_0 \mathcal{M}_i := B_i - \mu_0 \mathcal{H}_i , \tag{2.6.4}$$

and then, assuming for the moment that P_i and $\mu_0 \mathcal{M}_i$ are an objective vector and an objective axial vector under EUCLIDian transformations, respectively, we could also use P_i and $\mu_0 \mathcal{M}_i$ as basic (i.e. independent) field variables. In this way we can choose from nine possibilities for the constitutive relations of the kind (2.6.2). However, we shall not discuss them all here, but list below only those combinations, which are physically relevant and will appear in the next chapter. These are

$$C = \hat{C}(F_{i\alpha}, \frac{P_i}{\rho}, \frac{\mu_0 \mathcal{M}_i}{\rho}, \Theta, \Theta_{,i}, \mathcal{Q}) \, , \qquad (a)$$

$$C = \tilde{C}(F_{i\alpha}, \mathcal{E}_i, \frac{\mu_0 \mathcal{M}_i}{\rho}, \Theta, \Theta_{,i}, \mathcal{Q}) \, , \qquad (b)$$

$$C = \bar{C}(F_{i\alpha}, \mathcal{E}_i, \mathcal{H}_i, \Theta, \Theta_{,i}, \mathcal{Q}) \, , \qquad (c) \qquad (2.6.5)$$

$$C = \overset{+}{C}(F_{i\alpha}, \frac{P_i}{\rho}, B_i, \Theta, \Theta_{,i}\mathcal{Q}) \, , \qquad (d)$$

$$C = \check{C}(F_{i\alpha}, \mathcal{E}_i, B_i, \Theta, \Theta_{,i}, \mathcal{Q}) \, . \qquad (e)$$

Although all the arguments appearing in the constitutive relations (2.6.5) are objective quantities, we have not satisfied yet the requirement that $\hat{C}(\cdot), \tilde{C}(\cdot), \bar{C}(\cdot), \overset{+}{C}(\cdot)$ and $\check{C}(\cdot)$ must be objective tensorial, vectorial and scalar functions of their variables with respect to the EUCLIDian group.

The explicit form of these expressions depends on whether $\hat{C}(\cdot)$ etc. is a scalar, vector or tensor valued function. The representations in all these cases are well-known and are due to NOLL. The reader may consult TRUESDELL and NOLL's treatise [243] for an account on the history and for a proof.

If C is an objective scalar under the EUCLIDian transformation group it may be shown that its constitutive function is frame independent if

$$C = \hat{C}(C_{\alpha\beta}, \mathsf{P}_\alpha, \mathsf{M}_\alpha, \Theta, \Theta_{,\alpha}, \mathcal{Q}) \, , \qquad (a)$$

$$C = \tilde{C}(C_{\alpha\beta}, \mathsf{E}_\alpha, \mathsf{M}_\alpha, \Theta, \Theta_{,\alpha}, \mathcal{Q}) \, , \qquad (b)$$

$$C = \bar{C}(C_{\alpha\beta}, \mathsf{E}_\alpha, \mathsf{H}_\alpha, \Theta, \Theta_{,\alpha}, \mathcal{Q}) \, , \qquad (c) \qquad (2.6.6)$$

$$C = \overset{+}{C}(C_{\alpha\beta}, \mathsf{P}_\alpha, \mathsf{B}_\alpha, \Theta, \Theta_{,\alpha}, \mathcal{Q}) \, , \qquad (d)$$

$$C = \check{C}(C_{\alpha\beta}, \mathsf{E}_\alpha, \mathsf{B}_\alpha, \Theta, \Theta_{,\alpha}, \mathcal{Q}) \, . \qquad (e)$$

Here, $C_{\alpha\beta}$ is the right CAUCHY-GREEN deformation tensor, defined by $(2.1.10)_1$, and

$$\mathsf{P}_\alpha := \frac{1}{\rho}P_i F_{i\alpha} \, , \qquad \mathsf{M}_\alpha := \frac{\mu_0}{\rho}\mathcal{M}_i F_{i\alpha} \mathrm{sgn} J \, ,$$

$$\mathsf{E}_\alpha := \mathcal{E}_i F_{i\alpha} \, , \quad \mathsf{H}_\alpha := \mathcal{H}_i F_{i\alpha} \mathrm{sgn} J \, , \quad \mathsf{B}_\alpha := \frac{1}{\mu_0}B_i F_{i\alpha} \mathrm{sgn} J \, . \qquad (2.6.7)$$

If C_i is an objective polar vector, then

$$C_i = F_{i\gamma}\hat{C}_\gamma(C_{\alpha\beta}, \mathsf{P}_\alpha, \mathsf{M}_\alpha, \Theta, \Theta_{,\alpha}, \mathcal{Q}), \text{ etc.} \qquad (a) \qquad (2.6.8)$$

If C_i is an objective axial vector, then

$$C_i = F_{i\gamma}\hat{C}_\gamma(C_{\alpha\beta}, \mathsf{P}_\alpha, \mathsf{M}_\alpha, \Theta, \Theta_{,\alpha}, \mathcal{Q})\mathrm{sgn} J, \text{ etc.} \qquad (a) \qquad (2.6.9)$$

Finally, for a second-order tensor C_{ij} one obtains

$$\mathsf{C}_{ij} = F_{i\gamma} F_{j\delta} \hat{\mathsf{C}}_{\gamma\delta}(C_{\alpha\beta}, \mathsf{P}_\alpha, \mathsf{M}_\alpha, \Theta, \Theta_{,\alpha}, \mathcal{Q}), \quad \text{etc.} \qquad (a) \qquad (2.6.10)$$

We now shall outline the procedure that will be used to bring the constitutive equations in their ultimate form. First, we substitute the constitutive relations (2.6.5) for the dependent variables into the MAXWELL equations and into the balance laws of mass, linear and angular momentum and energy. The resulting equations are the *field equations* for $\rho, \chi_i, \Theta, \mathcal{Q}$ and the two basic electromagnetic fields (e.g. P_i and $\mathcal{M}_i$ in case a)). Any solution of these field equations is called a *thermodynamic process*. Following COLEMAN and NOLL, [44] or more generally MÜLLER [160], we request the entropy inequality, or any inequality derived from it, to hold for all smooth thermodynamic processes. This implies that at a particle we may freely choose the independent variables and derivatives thereof as long as the field equations are not violated thereby. In an *open system*, that is for a body with arbitrary external body force ρF_i^{ext} and energy supply ρr^{ext} no contributions are provided by the momentum and energy equations, because to any process there exist appropriate force and energy supply terms such that the momentum and energy equations are satisfied identically. Special care should be observed with regard to the electromagnetic variables, however, since, as can easily be deduced from the MAXWELL equations, not all the gradients of the basic electromagnetic fields are arbitrary. Such gradients may occur in the entropy inequality, and if they do, the relations implied by the MAXWELL equations must be fulfilled along with the exploitation of the entropy inequality. A detailed explanation of this point is given by HUTTER [96]. In any case, once the constitutive relations are substituted into the entropy inequality and all the above mentioned side conditions are properly taken into account, an inequality results with terms that are explicitly linear in variables that, in a thermodynamic process, may have any arbitrarily assigned value. Therefore, since otherwise this inequality would be violated, each of the coefficients of these variables must be identically zero. These conditions then imply the constitutive equations in their ultimate form. This procedure, which we assume to be familiar to the reader, will be applied (and described in greater detail) in the next chapter.

Part I

Equivalence of Different Electromagnetic Formulations in Thermoelastic Solids

3 A Survey of Electromagneto-Mechanical Interaction Models

3.1 Preview

The subject of electrodynamics of moving media has always been a controversial one. This book will not end or resolve all controversies, because we can answer some, but not all the relevant questions in connection with a complete thermodynamic theory of electromagnetism.

The basic difficulties in the description of electromechanical interaction models are manifold. A first difficulty is concerned with the invariance properties of the electromagnetic field equations. As is well-known, MAXWELL's equations are invariant under LORENTZ transformations, while the balance laws of classical mechanics are invariant under EUCLIDian transformations and frame indifferent under GALILEI transformations. Clearly, a proper derivation should also treat the mechanical equations relativistically. This is true, but for most problems of technical relevance, relativistic effects are negligible. It is therefore customary, in general, to treat the mechanical equations classically, while the equations of electrodynamics are handled relativistically. In so doing it might in these theories become uncertain what transformation properties some variables are based upon. However, knowledge of such transformation properties is important, because they give us indications as to what variables are comparable among different theories.

A second and even more serious difficulty can be found in the definitions of electromagnetic body force, body couple and energy supply. The roots of this difficulty lie in the separation of the electromagnetic field quantities in near and far field effects. This separation has been and still is the root of controversies, because almost every author separates the total fields differently. In other words, near and far fields are not unique.

A third difficulty is connected with the MAXWELL equations, which for deformable moving matter were first derived by MINKOWSKI. Apart from the MAXWELL equations in MINKOWSKI's form there exists a variety of other forms of the MAXWELL equations in deformable media, all of which are motivated from particular models. The "action" of the electromagnetic fields upon the material is described hereby by quantities referred to as polarization and magnetization. However, dependent upon the model of derivation, polarization and magnetization of one theory may be and in general are different from polarization and magnetization of another theory. Hence, while all

K. Hutter et al.: *Electromagnetic Field Matter Interaction in Thermoelastic Solids and Viscous Fluids*, Lect. Notes Phys. **710**, 33–88 (2006)
DOI 10.1007/3-540-37240-7_3

formulations of electromagnetism of deformable continua are equally valid – we know of five different descriptions – special care must be observed that variables of one theory are not confused with those of another.

As a final difficulty, we mention that the equations of a theory of electro-mechanical interaction are highly nonlinear. Generally, they defy any exact analysis even for the most simple problems that are of physical relevance. As a result, linearization procedures are needed.

To render the above statements more precise, consider the equations of motion which may be derived by formulating the balance law of momentum to an arbitrary part of the body. The local form of this balance law states that "mass times acceleration equals divergence of stress plus body force". In ordinary classical mechanics the body force is either set equal to zero, or else given by the gravitational force. A body couple hardly occurs in applications, in which case the balance law of moment of momentum implies the symmetry of the (CAUCHY) stress tensor. When the body under consideration is interacting with electromagnetic fields, however, body force and body couple are given by electromagnetic quantities.

The total force and the total moment exerted on a body by electromagnetic fields may be separated into a long range and a short range effect. The long range effect is expressed as a body force and body couple. The short range effect, on the other hand, manifests itself as surface tractions, which can be combined with the mechanical tractions giving rise thereby to the definition of the stress tensor. This decomposition is not unique, thus leading to non-unique body force expressions and non-unique stress tensors. As a consequence, the electromagnetic body couple cannot be unique either.

Although this non-uniqueness might be quite striking to the novel reader it is nonetheless not disturbing at all if looked upon from the right point of view. Indeed, it is not important that the above separation into force and stress is unique, because differences in the body forces can always be absorbed in the stress tensors, provided that they are expressible as a divergence of a stress. A variety of mutually incompatible formulas for the force expressible in terms of stresses are therefore equivalent with respect to the total force. Only this force is physically observable. Thus the incompatibility is not physical, but metaphysical or semantic.

The incompatibility expressed above also occurs in the energy equation (first law of thermodynamics). This equation states that the time rate of change of the internal energy is balanced by the power of working of the stresses, the divergence of the heat flux, and the energy supply due to electromagnetic effects and due to heat. Since stress was already said to be non-unique, it follows that internal energy, heat flux and electromagnetic energy supply cannot be determined uniquely either. Likewise, the electromagnetic energy supply might contain a term that is the divergence of a vector, which could be absorbed in the heat flux vector. As an immediate consequence, it cannot be assured that heat flux is energy flux of thermal nature. We shall therefore prefer the term energy flux instead.

As was the case for the momentum equation, seemingly incompatible expressions for internal energy, energy flux and energy supply of electromagnetic origin do not prevent two theories from being equivalent. However, it is easily understandable that a proof of equivalence must be difficult in general, for stress, internal energy and entropy (and also some electromagnetic field vectors) are interrelated by thermodynamic conditions. More explicitly, thermodynamic requirements make stress and entropy (and other quantities) derivable from a so-called free energy. If two theories are different in the body force, body couple and energy supply, therefore, the condition that the momentum and energy equations remain the same must amount to an interrelation between the free energies of two theories. Hence, equivalence of two theories of electromechanical interactions is a thermodynamic statement in general.

The question of equivalence of two theories lies at the center of the different formulations of electromechanical interaction theories. Although there is a valid point behind the statement that equivalence of different theories need not be proved, because these theories describe different physical situations, we nevertheless take the position that different formulations of electromechanical interaction theories should yield the same results for physically measurable quantities, if the theories are claimed to be applicable to a certain class of material response. For instance, if we call a material a thermoelastic polarizable and magnetizable solid and if there is more than one formulation for such a solid, one should expect that, irrespective of all differences in the details, these theories will in any initial boundary value problem deliver the same results for physically measurable quantities. Measurable or observable quantities are all those which can be measured uniquely by two different observers. All kinematical quantities that are derivable from the motion are measurable in principle and so is the (empirical) temperature. Regarding electromagnetic field quantities, we take the position that they are not measurable except *in vacuo* where they can be observed by measuring the force on a test charge. There exist variables not observable by any means. These are all those which are not defined except by the mathematical properties laid down for them.

To demonstrate the equivalence of the different formulations of electromechanical interaction theories it is necessary to prove that physically measurable quantities in two formulations assume the same values in every point of the body for any initial boundary value problem. This does not only mean that the field equations of one theory must be transformable into those of the other, but this condition also includes the boundary and initial conditions. One of the major goals of this monograph is to give an exposition of the existing theories of polarizable and magnetizable electrically and thermally conducting materials and to show in what sense they can be called equivalent.

The reasons behind this non-uniqueness of electrodynamics in moving media are twofold. For one, the action of the body on the electromagnetic fields is generally described by adding to the field variables occurring *in vacuo* two other electromagnetic field vectors. This addition is not unique and results in different forms of the MAXWELL equations. Second, even when we restrict ourselves to a particular form of the MAXWELL equations, the electromagnetic forces, couples and energy supply terms need not be unique. More precisely, we mention that the two electromagnetic field vectors describing the interaction of a ponderable body can be introduced for instance by postulating that every material point is equipped with a number of non-interacting *electric* and *magnetic dipoles*. These dipole moments then form the two additional electromagnetic field vectors, which are called polarization and magnetization.

When the calculations with these postulated dipoles are carried through consistently, a certain set of MAXWELL equations (now called the Chu formulation) emerges. These equations are different from those which follow from the postulation that *magnetization is modeled as an electric circuit*, which follows the motion of the material particle in question (statistical and LORENTZ formulations).

As far as electromagnetic body forces are concerned these are not even unique when one is restricting oneself to a particular interaction model. Indeed, in the Chu formulation we shall present two versions of body force expressions and we shall prove that the two are not distinguishable by any measurements. This proof will also be given for all other formulations. However, we shall not present the models as such, because they are amply treated in the pertinent literature.

Although the proof of the equivalence of various theories of electrodynamics in deformable continua is a very important achievement, we want to state here clearly that we have performed this proof only on the level of non-relativistic theories. The exact definitions of the term "non-relativistic" will be made precise in the respective chapters. It may suffice to mention that it essentially means that in MKSA-units terms containing a c^{-2}-factor are neglected. Here, c is the speed of light in vacuo. There exists a number of other theories of electromechanical interactions in which it is claimed that terms of order V^2/c^2 are neglected (V = velocity of the particle in the body) while those containing c^{-2}-factor are kept. We term such approximations "semi-relativistic". Quasistatic theories (terms, containing a c^{-1}-factor are neglected) will not be treated here.

It is a well-known fact that fluids are best handled in the spatial description. It is also known that electrodynamics is usually only formulated in the spatial description. However, for a theory of solids it would be advantageous when all equations could be referred to the reference configuration. This is indeed possible and it essentially amounts to the introduction of new electromagnetic field variables as introduced in Sect. 2.2.3 of Chap. 2, see formulas (2.2.39). It turns out that these so-called LAGRANGEan field variables are much more convenient to describe the theory of solids, because many thermodynamic formulas appear in a more condensed form this way (see e.g.

the MAXWELL equations (2.2.40)). Another reason for the introduction of the material description is its advantage in the linearization of the governing equations. This linearization procedure is substantially easier when performed in the material rather than in the spatial description.

This brings us naturally to the linearization procedure of the various theories. In principle, there are two alternatives open to extract some useful information from these complicated equations. One is to find numerical solutions for the nonlinear equations and the other is to linearize the equations on the basis of a sequence of consistent approximations. We shall follow the latter, because it provides a better access to the real physics of the problem. The linearization procedure is analogous to situations referred to as "small fields superimposed upon large fields". The difference between these general treatments and ours is that the deformations are assumed to be small. This assumption is not necessary, and indeed the formal expansion procedures we shall apply also hold true for the general case. When the restriction to small deformations is used, however, it means physically that large external fields primarily induce strong electromagnetic fields within the body, but only small deformations. Therefore, in the first step of evaluating the induced electromagnetism, the deformations may be neglected alltogether. A set of zeroth-order equations, which formally agrees with rigid-body electrodynamics, is thus obtained. In the second step small strains are considered which add small but important corrections to the zeroth-order electromagnetic fields. Thus, the second set consists of linear field equations, the coefficients of which generally depend upon the zeroth-order electromagnetic fields. These field equations may then be applied to solve problems such as magnetoelastic buckling, wave propagation in a material subject to electromagnetic fields, etc.

Clearly, because we shall prove that all electromechanical interaction theories of polarizable and magnetizable solids are equivalent, the linearization procedure mentioned above need only be performed for one particular theory, which can be selected according to our needs. Moreover, it should be clear that this equivalence must amount in the statement how the free energy as a function of its independent variables in one theory is related to the free energy of another theory. The set of independent variables in this second free energy may very well be different from the first one. In other words, the correspondence relations for equivalence of various theories are dependent on which set of independent variables is chosen in the constitutive relations, but it is quite clear that the equivalence as such should not depend on the choice of the independent fields. From a theoretical point of view the problem just raised is not a serious one, because, in principle, equivalence of different constitutive formulations in one single theory can be established quite easily. It then suffices to prove equivalence of two different electromagnetic descriptions of deformable bodies with the aid of just one constitutive formulation in each of them.

To find the free energy of a particular formulation from that of another one is a very difficult problem in practice, however. It amounts to solving a functional differential equation the solutions to which are not known to date. Nevertheless, special cases are straightforward to handle. They serve as explicit examples, which should demonstrate that equivalence is possible. Mathematically this is important, because it serves as an explicit demonstration that the functional differential equations mentioned above do admit exact solutions. That these correspond to a reasonable physical situation is a nice additional property. The mathematical question of existence and non-existence of solutions will not be attacked here. Instead we look at the approximations in the way described below.

It is customary in applications to write for the free energies polynomial expressions, and it is generally assumed that these polynomials can be truncated at a certain level. When polarization and magnetization are amongst the independent constitutive variables the free energy will be a polynomial expression in the deformation tensor, the temperature, polarization and magnetization, and the coefficients in this polynomial expression give rise to effects such as magnetic and electric anisotropy, magnetostriction, electrostriction etc. The coefficients bear the names electric and magnetic susceptibilities etc. The same theory could be derived also with the electric field strength and the magnetic induction as independent fields instead of magnetization and polarization. The free energy of this formulation would again be expressed as a polynomial of its variables and it would again be truncated at a certain level. This polynomial would again give rise to effects such as electric and magnetic anisotropies, magnetostriction and electrostriction etc, but it is evident that the coefficients of this polynomial must be different from those of the other, if the two formulations aim at describing the same phenomena. The literature is full of confusion in this regard, mainly because different coefficients bear the same name. From the above it is, however, quite clear that there must be relations between the above mentioned coefficients. We shall show how these relations look like and in what sense the emerging approximate theories can be regarded to be equivalent. The findings can be summarized as follows: Two formulations, in which the free energies are represented by polynomials of a certain order in their variables, can only be equivalent to within terms that were omitted in the expansion process. Only on the basis that these terms are negligibly small can we claim two theories to be equivalent. An analogous statement also holds for one single formulation in which certain constitutive quantities are interchanged as dependent and independent variables.

3.2 Scope of the Survey

In this chapter we make an attempt to surveying various electromagnetic interaction models known to date. We do not present all the descriptions of magnetizable and polarizable deformable bodies, but list the ones that have received considerable attention in the recent literature only. A comparison of these models with still other ones will be given in Sect. 3.6.

Any continuum theory of deformable bodies subject to electromagnetic fields amounts to the presentation of the basic electromagnetic field variables, their relations to the other fields, as well as to the postulation of electromagnetic body force, body couple and energy supply. Once this is done, the MAXWELL equations and the balance laws of mechanics and thermodynamics can be expressed in terms of the variables of the model in question. Using thermodynamic arguments, it is then a routine matter to reduce a postulated set of constitutive equations to a form compatible with the second law of thermodynamics.

It is the purpose of this chapter to present the various models of electromechanical interactions, to scrutinize their invariance properties, to derive the constitutive theory in each peculiar case and to present each model such that, firstly, a comparison of one model with any other can be achieved fairly straightforwardly and, secondly, an initial boundary value problem can be solved, at least in principle.

3.3 The Two–Dipole Models

As discussed by FANO, CHU and ADLER [73], or PENFIELD and HAUS [177], in the CHU *formulation* the MAXWELL equations for moving matter are expressed in terms of Q, J_i, the velocity field $\dot{x}_i$, and four electromagnetic field quantities E_i^C, H_i^C, P_i^C and M_i^C, which are related to the fields $D_i, \mathcal{E}_i, B_i$ and $\mathcal{H}_i$ by the following transformation rules

$$
\begin{aligned}
D_i &= \varepsilon_0 E_i^C + P_i^C , & \mathcal{E}_i &= E_i^C + \mu_0 e_{ijk} \dot{x}_j H_k^C , \\
B_i &= \mu_0 H_i^C + \mu_0 M_i^C , & \mathcal{H}_i &= H_i^C - \varepsilon_0 e_{ijk} \dot{x}_j E_k^C .
\end{aligned}
\tag{3.3.1}
$$

For reasons that will become apparent shortly we shall occasionally also make use of two auxiliary fields defined by

$$
B_i^{\mathrm{a}} := \mu_0 H_i^C , \qquad D_i^{\mathrm{a}} := \varepsilon_0 E_i^C .
\tag{3.3.2}
$$

As usual, E_i^C and H_i^C are the electric and magnetic field strengths, and P_i^C and M_i^C are the polarization per unit volume and the magnetization per unit volume, respectively. In order to differentiate these fields from those occurring in other formulations, we have used a superscript C to indicate that these fields are those as defined by CHU. Substituting (3.3.1) and (3.3.2) into the MAXWELL equations (2.2.15) and (2.2.21), we obtain

$$B_{i,i}^{a} = -\mu_0 M_{i,i}^{C} \, ,$$

$$e_{ijk}E_{k,j}^{C} + \frac{\partial B_i^{a}}{\partial t} = -\mu_0 \frac{\partial M_i^{C}}{\partial t} + \mu_0 e_{ijk}(e_{klm}\dot{x}_l M_m^{C})_{,j} \, ,$$

$$D_{i,i}^{a} = \mathcal{Q} - P_{i,i}^{C} \, ,$$

$$e_{ijk}H_{k,j}^{C} - \frac{\partial D_i^{a}}{\partial t} = J_i + \frac{\partial P_i^{C}}{\partial t} - e_{ijk}(e_{klm}\dot{x}_l P_m^{C})_{,j} \, .$$

(3.3.3)

In the above equations μ_0 and ε_0 are universal constants with $\varepsilon_0\mu_0 = c^{-2}$, c being the speed of light in vacuo. If we regard E_i^{C} and H_i^{C} as the basic electromagnetic fields, then, apart from J_i and $\mathcal{Q}$, the terms on the right-hand sides of (3.3.3) may be interpreted as charge and current densities due to polarization and magnetization. By assuming that positive and negative electric and magnetic charges exist and may be combined to electric and magnetic dipole moments, the above charge and current densities due to polarization and magnetization can easily be derived, PAO & HUTTER [171], PAO [172]. Equations (3.3.3) are therefore completely symmetric in the electric and magnetic field quantities except that magnetic monopols are assumed not to exist, so that the corresponding charge and current densities are absent in (3.3.3). This interpretation is helpful for the derivation of electromagnetic body force, body couple and energy supply.

The above set of electromagnetic variables is based on the postulations that:

(i) only two vector quantities are necessary to describe the electromagnetic fields in free space, and

(ii) material bodies contribute toward the electromagnetic fields by acting as sources for these fields.

These sources are usually interpreted in terms of electric and magnetic dipoles, each such doublet being composed of a negative and positive electric and magnetic monopole. We would like to de-emphasize this interpretation, mainly because of the well-known objections physicists may rise against it. Nevertheless we accept the CHU formulation as a proper description, but view it as obtained from the definitions (3.3.1) by mere variable transformations.

Before we proceed it is advantageous to investigate how the CHU-variables behave under the transformations discussed in Sect. 1.6. Following an approach analogous to that outlined by TRUESDELL and TOUPIN in [244], Sect. 283, it is not difficult to show that under the EUCLIDIAN transformation group

$$\mathcal{Q} \qquad \text{transforms as an objective scalar ,}$$

$$\mathcal{E}_i, P_i^{C}, \mathcal{J}_i \qquad \text{transform as objective vectors ,} \qquad (3.3.4)$$

$$\mathcal{H}_i, \mu_0 M_i^{C} \qquad \text{transform as objective axial vectors .}$$

It can also be shown that

D_i^{a} transforms as an objective vector, while

B_i^{a} transforms as an objective axial vector ,

$$(3.3.5)$$

under the EUCLIDian transformation group. For the main lines of the proof the reader may consult Appendix A. It is clear that the above transformation rules also apply under the slightly less general GALILEI group. The properties laid down in (3.3.4) and (3.3.5) are exact in the sense that the MAXWELL equations can exactly (that is without the neglect of c^{-2}-terms) be rendered EUCLIDian invariant. However, relations (3.3.2) must also be satisfied, and they are not EUCLIDian invariant, but LORENTZ invariant or GALILEI invariant only to within terms containing a c^{-2}-factor (see also eqs. $(3.3.6)_{7,8}$).

Therefore, the CHU formulation of electromagnetism consisting of the MAXWELL equations (3.3.3) together with the aether relations (3.3.2) is invariant under the GALILEI group only within the non-relativistic approximation. Hence, in order to remain consistent, we must neglect in the following interaction models all terms preceded by a factor c^{-2}.

In order to investigate semi-relativistic objectivity properties, let us consider the extended LORENTZ group. Under these transformations the CHU variables transform (to within terms of order $\mathcal{O}(V^2/c^2)$) according to

$$\overset{\circ}{\mathcal{Q}} = \mathcal{Q} - \frac{1}{c^2}\dot{x}_i\mathcal{J}_i + \mathcal{O}(V^2/c^2) , \qquad \overset{\circ}{\mathcal{J}}_i = \mathcal{J}_i + \mathcal{O}(V^2/c^2) ,$$

$$\overset{\circ}{\mathcal{E}}_i = \mathcal{E}_i + \mathcal{O}(V^2/c^2) , \qquad \overset{\circ}{\mathcal{H}}_i = \mathcal{H}_i + \mathcal{O}(V^2/c^2) ,$$

$$\overset{\circ}{P}_i^C = P_i^C + \mathcal{O}(V^2/c^2) , \qquad \overset{\circ}{M}_i^C = M_i^C + \mathcal{O}(V^2/c^2) , \qquad (3.3.6)$$

$$\overset{\circ}{D}_i^{\mathrm{a}} = D_i^{\mathrm{a}} + \frac{1}{c^2}e_{ijk}\dot{x}_j H_k^C + \mathcal{O}(V^2/c^2) ,$$

$$\overset{\circ}{B}_i^{\mathrm{a}} = B_i^{\mathrm{a}} - \frac{1}{c^2}e_{ijk}\dot{x}_j E_k^C + \mathcal{O}(V^2/c^2) .$$

Thus it is obvious that the transformation rules (3.3.4) also hold under the extended LORENTZ group, if terms of $\mathcal{O}(V^2/c^2)$ are dropped. There is only one exception, namely that $\mathcal{Q}$ has to be replaced by $\overset{\circ}{\mathcal{Q}}$. In the non-relativistic approximation ($c^{-2} = 0$) this difference is negligible, however, and hence in this approximation we may everywhere replace $\mathcal{Q}$ by $\overset{\circ}{\mathcal{Q}}$. Moreover, there are reasons to assume that $\|\mathcal{J}_i\|$ is of the order of $\|\mathcal{Q}V\|$. Then, $\mathcal{Q}$ may even be replaced by $\overset{\circ}{\mathcal{Q}}$ in the semi-relativistic approximation. As a consequence the CHU formulation of electromagnetism has the nice feature to obey the non-relativistic as well as the semi-relativistic invariance requirements. Note that, in contrast to (3.3.5), the auxiliary fields D_i^{a} and B_i^{a} do not behave as objective quantities under the LORENTZ transformation group, not even in a semi-relativistic sense.

A complete theory of deformable, polarizable and magnetizable bodies consists of a set of electromagnetic and thermomechanical equations, the latter including expressions for the body force, body couple and energy supply due to the electromagnetic fields. On the non-relativistic level there are essentially two distinct formulations both of which use the dipole model not only for the polarization but also for the magnetization.

3.3.1 The Two–Dipole Model with a Nonsymmetric Stress Tensor (Model I)

As mentioned above, to complete the description of deformable continua in the electromagnetic fields, expressions for ρF_i^e, ρL_{ij}^e and ρr^e are needed. PENFIELD and HAUS [177] and PAO and HUTTER [171] have motivated and derived the following expressions using the two-dipole model

$$\rho F_i^e = Q E_i^C + \mu_0 e_{ijk} J_j H_k^C$$

$$+ P_j^C E_{i,j}^C + \mu_0 e_{ijk} \dot{x}_j H_{k,l}^C P_l^C + \rho \mu_0 e_{ijk} \frac{d}{dt} \left(\frac{P_j^C}{\rho} \right) H_k^C$$

$$+ \mu_0 M_j^C H_{i,j}^C - \varepsilon_0 e_{ijk} \dot{x}_j E_{k,l}^C \mu_0 M_l^C - \rho \varepsilon_0 e_{ijk} \frac{d}{dt} \left(\frac{\mu_0 M_j^C}{\rho} \right) E_k^C ,$$

$$\rho L_{ij}^e = P_{[i}^C \mathcal{E}_{j]} + \mu_0 M_{[i}^C \mathcal{H}_{j]} ,$$

$$\rho r^e = \mathcal{J}_i \mathcal{E}_i + \rho \mathcal{E}_i \frac{d}{dt} \left(\frac{P_i^C}{\rho} \right) + \rho \mathcal{H}_i \frac{d}{dt} \left(\frac{\mu_0 M_i^C}{\rho} \right) .$$

$$(3.3.7)$$

The complete derivation of (3.3.7) as given by PAO and HUTTER [171], is based on the assumptions that:

(i) each material particle is equipped with a number of mutually noninteracting electric and magnetic dipoles,

(ii) each monopole suffers an electromagnetic body force as described by the LORENTZ force (see below), and

(iii) the monopoles of a particular dipole are only a small distance apart so that TAYLOR series expansions are justified.

The LORENTZ force mentioned in assumption (ii) is hereby taken in the form

$$\rho F_i^{\text{Lorentz}} = Q^e E_i^C + \mu_0 e_{ijk} Q^e \dot{x}_j H_k^C + Q^m H_i^C - \varepsilon_0 e_{ijk} Q^m \dot{x}_j E_k^C , \quad (3.3.8)$$

where Q^e and Q^m are the electric and magnetic charges of the monopoles. Again we do not wish to emphasize the physical aspects of this approach, but we would like to view (3.3.7) as possible interaction postulates. We refer to the paper of PAO and HUTTER [171], for the details of the derivation. Substituting (3.3.7) into the balance laws (2.2.9) yields the local balance laws of mass, linear and angular momentum and energy in the form

$$\dot{\rho} \ + \rho \dot{x}_{i,i} = 0 \ ,$$

$$\rho \ddot{x}_i = t_{ij,j} + \rho F_i^{\text{ext}} + Q E_i^C + \mu_0 e_{ijk} J_j H_k^C$$

$$+ P_j^C E_{i,j}^C + \mu_0 e_{ijk} \dot{x}_j H_{k,l}^C P_l^C + \rho \mu_0 e_{ijk} \frac{d}{dt}\left(\frac{P_j^C}{\rho}\right) H_k^C$$

$$+ \mu_0 M_j^C H_{i,j}^C - \varepsilon_0 e_{ijk} \dot{x}_j E_{k,l}^C \mu_0 M_l^C - \rho \varepsilon_0 e_{ijk} \frac{d}{dt}\left(\frac{\mu_0 M_j^C}{\rho}\right) E_k^C$$

$$t_{[ij]} \ = P_{[i}^C \mathcal{E}_{j]} + \mu_0 M_{[i}^C \mathcal{H}_{j]} \ ,$$

$$\rho \dot{U} \ = t_{ij} \dot{x}_{i,j} - q_{i,i} + \mathcal{J}_i \mathcal{E}_i + \rho \mathcal{E}_i \frac{d}{dt}\left(\frac{P_i^C}{\rho}\right) + \rho \mathcal{H}_i \frac{d}{dt}\left(\frac{\mu_0 M_i^C}{\rho}\right) + \rho r^{\text{ext}} \ .$$

$$(3.3.9)$$

It is a routine though rather elaborate matter to transform (3.3.7) to the form listed in (2.4.7). One obtains

$$^I t_{ij}^M = \varepsilon_0 E_i^C E_j^C + \mu_0 H_i^C H_j^C + \mathcal{E}_i P_j^C + \mu_0 \mathcal{H}_i M_j^C$$

$$- \frac{1}{2}\delta_{ij}\left(\varepsilon_0 E_k^C E_k^C + \mu_0 H_k^C H_k^C\right) \ ,$$

$$^I g_i = -\frac{1}{c^2} e_{ijk} E_j^C H_k^C \ ,$$

$$(3.3.10)$$

$$^I \pi_i = -e_{ijk} E_j^C H_k^C + \left(P_i^C E_j^C + \mu_0 M_i^C H_j^C\right)\dot{x}_j \ ,$$

$$^I \omega = -\frac{1}{2}\left(\varepsilon_0 E_i^C E_i^C + \mu_0 H_i^C H_i^C\right) \ .$$

Here we have used a left upper index to distinguish these quantities from those of other interaction models.

With the expressions (3.3.10) the jump conditions of momentum and energy as listed in (2.4.11) are readily derived. For the sake of easy reference they are given below together with the jump conditions of electromagnetic fields and mass

$$[\![\mu_0 H_i^C + \mu_0 M_i^C]\!] n_i = 0 \ , \qquad [\![\varepsilon_0 E_i^C + P_i^C]\!] n_i = 0 \ ,$$

$$[\![e_{ijk} E_j^C n_k + \mu_0 \dot{x}_i M_k^C n_k + \mu_0 H_i^C \mathrm{w}_n - \mu_0 M_i^C (\dot{x}_k n_k - \mathrm{w}_n)]\!] = 0 \ ,$$

$$[\![e_{ijk} H_j^C n_k - \dot{x}_i P_k^C n_k - \varepsilon_0 E_i^C \mathrm{w}_n + P_i^C (\dot{x}_k n_k - \mathrm{w}_n)]\!] - 0 \ ,$$

$$[\![\rho(\dot{x}_i n_i - \mathrm{w}_n)]\!] = 0 \ ,$$

$$[\![t_{ij} + \varepsilon_0 E_i^C E_j^C + \mu_0 H_i^C H_j^C + \mathcal{E}_i P_j^C + \mu_0 \mathcal{H}_i M_j^C$$

$$- \tfrac{1}{2}\delta_{ij}\left(\varepsilon_0 E_k^C E_k^C + \mu_0 H_k^C H_k^C\right) + \frac{1}{c^2} e_{ikl} E_k^C H_l^C \dot{x}_j]\!] n_j \qquad (3.3.11)$$

$$- [\![(\rho \dot{x}_i + \frac{1}{c^2} e_{ikl} E_k^C H_l^C)(\dot{x}_j n_j - \mathrm{w}_n)]\!] = 0 \ ,$$

$$\left[\!\left[\dot{x}_i t_{ij} - q_j - \mathrm{e}_{ijkl} E^C_k H^C_l + \left(P^C_{ij} E^C_i + \mu_0 M^C_j H^C_i \right) \dot{x}_i \right.\right.$$
$$\left. + \tfrac{1}{2} \left(\varepsilon_0 E^C_k E^C_k + \mu_0 H^C_k H^C_k \right) \dot{x}_j \right]\!\right] n_j$$
$$- \left[\!\left[\left\{ \tfrac{1}{2} \rho \dot{x}_i \dot{x}_i + \rho U + \tfrac{1}{2} \left(\varepsilon_0 E^C_i E^C_i + \mu_0 H^C_i H^C_i \right) \right\} (\dot{x}_j n_j - \mathrm{w}_n) \right]\!\right] = 0 \ .$$

Note that the jump conditions (3.3.11), specialized for material surfaces, have also been obtained by PAO and HUTTER, [171]. Their jump conditions for the stress, however, does not contain the term

$$\left[\!\left[\frac{1}{c^2} \mathrm{e}_{ikl} E^C_k H^C_l \dot{x}_j \right]\!\right] n_j \ ,$$

but, as we shall soon see, this term is negligible in a non-relativistic theory as it contains a prefactor c^{-2}.

Before discussing the constitutive equations, it is interesting to look at the invariance properties of the electromagnetic body force, body couple and energy supply. At the beginning of this section the invariance properties of the electromagnetic field variables were listed. From these it is immediately seen that L^e_{ij} transforms under the EUCLIDian group as an objective skew-symmetric tensor. Under the extended LORENTZ group L^e_{ij} still transforms as an objective skew-symmetric tensor, but only to within terms of order $\mathcal{O}(V^2/c^2)$. Hence, L^e_{ij} is objective non-relativistically as well as semi-relativistically. If we write the energy supply term $(3.3.7)_3$ in the form

$$\rho r^e = (\mathcal{J}_i + \overset{*}{P}_i)\mathcal{E}_i + \mu_0 \overset{*}{\mathcal{M}}_i \, \mathcal{H}_i + (\mathcal{E}_i P_j + \mathcal{H}_i \mu_0 \mathcal{M}_j)\dot{x}_{i,j} \ , \qquad (3.3.12)$$

then we recognize that ρr^e is not an objective scalar under the EUCLIDian transformation group because of the term involving $\dot{x}_{i,j}$. Incidentally, in (3.3.12) we have written P_i and $\mu_0 \mathcal{M}_i$ for P^C_i and $\mu_0 M^C_i$ in order to stress the invariance properties of P^C_i and $\mu_0 M^C_i$. Moreover, it should be noted that the material time derivative of an objective vector is not an objective vector under the EUCLIDian group; but the convective time derivative is objective. This is the reason why we have tried to use convective time derivatives in (3.3.12). The non-objective part of (3.3.12) is now given by

$$(\mathcal{E}_i P_j + \mu_0 \mathcal{H}_i \mathcal{M}_j)\dot{x}_{[i,j]} \ ,$$

which, in view of $(3.3.9)_3$, equals

$$- t_{[ij]}\dot{x}_{[i,j]} \ .$$

With use of this result, (3.3.12) shows that

$$\rho r^e + t_{[ij]}\dot{x}_{[i,j]} = (\mathcal{J}_i + \overset{*}{P}_i)\mathcal{E}_i + \mu_0 \overset{*}{\mathcal{M}}_i \, \mathcal{H}_i + (\mathcal{E}_i P_j + \mu_0 \mathcal{H}_i \mathcal{M}_j)\dot{x}_{(i,j)} \ , \quad (3.3.13)$$

is an objective scalar. Hence, the balance of internal energy can be written as

$$\rho\dot{U} = t_{(ij)}\dot{x}_{(i,j)} - q_{i,i} + (\mathcal{J}_i + \overset{\star}{P}_i)\mathcal{E}_i + \mu_0 \overset{\star}{M}_i \mathcal{H}_i + (\mathcal{E}_i P_j + \mu_0 \mathcal{H}_i M_j)\dot{x}_{(i,j)},$$

$$(3.3.14)$$

a form in which each term is an objective scalar under the EUCLIDian transformation group. In much the same way it can be shown that each term in the above equation is – to within terms of order $\mathcal{O}(V^2/c^2)$ – also an objective scalar under the extended LORENTZ group. Hence, the energy balance law is an invariant equation in the non-relativistic and semi-relativistic sense. Finally, we also investigate the transformation properties of the electromagnetic body force $(3.3.7)_1$. A straightforward calculation shows that ρF_i^e can be written in the following form

$$\rho F_i^\mathrm{e} = (\mathcal{Q} - P_{j,j}^C)\mathcal{E}_i + \mathrm{e}_{ijk}(\mathcal{J}_j + \overset{\star}{P}_j^C)B_k^\mathrm{a}$$

$$(3.3.15)$$

$$-\mu_0 M_{j,j}^C \mathcal{H}_i - \mathrm{e}_{ijk}\mu_0 \overset{\star}{M}_j^C D_k^\mathrm{a} + (\mathcal{E}_i P_j^C + \mathcal{H}_i \mu_0 M_j^C)_{,j}.$$

Using the transformation properties stated in (3.3.4) and (3.3.5), we readily see that each term on the right-hand side of this equation is an objective (polar) vector under the EUCLIDian transformation group. Because of the transformation properties of the auxiliary fields D_i^a and B_i^a, the body force does not enjoy the invariance properties of semi-relativistically correct LORENTZ transformations, however.

The above expression of the electromagnetic body force is written in a form which still contains the auxiliary fields. These fields may be eliminated with the aid of relations (3.3.2), which, however, are only invariant equations in the non-relativistic approximation. A straightforward calculation, in which the MAXWELL equations (3.3.3) are used and in which terms of order $\mathcal{O}(V^2/c^2)$ are neglected, then shows that ρF_i^e can be written in the form

$$\rho F_i^\mathrm{e} = \mathcal{Q}\mathcal{E}_i + \mu_0 \mathrm{e}_{ijk}\mathcal{J}_j\mathcal{H}_k + P_j\mathcal{E}_{i,j} + \mu_0 \mathrm{e}_{ijk}\overset{\star}{P}_j \mathcal{H}_k$$

$$+\mu_0 M_j\mathcal{H}_{i,j} + \frac{1}{c^2}[(\mathcal{J}_j + \overset{\star}{P}_j)(\dot{x}_i\mathcal{E}_j - \dot{x}_j\mathcal{E}_i)$$

$$(3.3.16)$$

$$+\mathrm{e}_{ikl}M_j\mathcal{E}_l\dot{x}_{k,j} + \rho\mathrm{e}_{ijk}\frac{d}{dt}(M_j/\rho)E_k^C].$$

In this form the body force is no longer an objective vector under EUCLIDian transformations; this objectivity is destroyed by use of relations (3.3.2) for the elimination of D_i^a and B_i^a. However, in a non-relativistic approximation the term involving a c^{-2}-term must be dropped. Hence, within a consistent non-relativistic approximation the momentum equation $(3.3.9)_2$ must be written as

$$\rho\ddot{x}_i = t_{ij,j} + \rho F_i^\mathrm{ext} + \mathcal{Q}\mathcal{E}_i + \mu_0 \mathrm{e}_{ijk}\mathcal{J}_j\mathcal{H}_k + P_j\mathcal{E}_{i,j}$$

$$(3.3.17)$$

$$+\mu_0 \mathrm{e}_{ijk}\overset{\star}{P}_j \mathcal{H}_k + \mu_0 M_j\mathcal{H}_{i,j},$$

or since

$$QE_i + \mu_0 e_{ijk} \mathcal{J}_j \mathcal{H}_k + P_j \mathcal{E}_{i,j} + \mu_0 e_{ijk} \overset{\star}{P}_j \mathcal{H}_k + \mu_0 \mathcal{M}_j \mathcal{H}_{i,j}$$
$$= \{\varepsilon_0 \mathcal{E}_i \mathcal{E}_j + \mu_0 \mathcal{H}_i \mathcal{H}_j + \mathcal{E}_i P_j + \mu_0 \mathcal{H}_i \mathcal{M}_j \qquad (3.3.18)$$
$$- \tfrac{1}{2} \delta_{ij} (\varepsilon_0 \mathcal{E}_k \mathcal{E}_k + \mu_0 \mathcal{H}_k \mathcal{H}_k)\}_{,j} + \mathcal{O}(V^2/c^2)$$

as

$$\rho \ddot{x}_i = t_{ij,j} + \rho F_i^{\text{ext}} + \{\varepsilon_0 \mathcal{E}_i \mathcal{E}_j + \mu_0 \mathcal{H}_i \mathcal{H}_j + \mathcal{E}_i P_j + \mu_0 \mathcal{H}_i \mathcal{M}_j$$
$$- \tfrac{1}{2} \delta_{ij} (\varepsilon_0 \mathcal{E}_k \mathcal{E}_k + \mu_0 \mathcal{H}_k \mathcal{H}_k)\}_{,j} . \qquad (3.3.19)$$

Moreover, the c^{-2}-terms in (3.3.10) and (3.3.11) must also be neglected so that $^I g_i$ vanishes, i.e.

$$^I g_i = 0 . \qquad (3.3.20)$$

Thus the jump conditions for momentum and energy reduce to

$$[\![t_{ij} + \varepsilon_0 \mathcal{E}_i \mathcal{E}_j + \mu_0 \mathcal{H}_i \mathcal{H}_j + \mathcal{E}_i P_j + \mu_0 \mathcal{H}_i \mathcal{M}_j$$
$$- \tfrac{1}{2} \delta_{ij} (\varepsilon_0 \mathcal{E}_k \mathcal{E}_k + \mu_0 \mathcal{H}_k \mathcal{H}_k)]\!] n_j - [\![\rho \dot{x}_i (\dot{x}_j n_j - \mathrm{w}_n)]\!] = 0 , \qquad (3.3.21)$$

and

$$[\![\dot{x}_i t_{ij} - q_j - e_{jkl} \mathcal{E}_k \mathcal{H}_l + (\varepsilon_0 \mathcal{E}_j \mathcal{E}_k + \mu_0 \mathcal{H}_j \mathcal{H}_k) \dot{x}_k$$
$$+ (P_j \mathcal{E}_k + \mu_0 \mathcal{M}_j \mathcal{H}_k) \dot{x}_k - \tfrac{1}{2} (\varepsilon_0 \mathcal{E}_k \mathcal{E}_k + \mu_0 \mathcal{H}_k \mathcal{H}_k) \dot{x}_j]\!] n_j \qquad (3.3.22)$$
$$- [\![\{\tfrac{1}{2} \rho \dot{x}_i \dot{x}_i + \rho U + \tfrac{1}{2} (\varepsilon_0 \mathcal{E}_i \mathcal{E}_i + \mu_0 \mathcal{H}_i \mathcal{H}_i)\} (\dot{x}_j n_j - \mathrm{w}_n)]\!] = 0 .$$

To summarize, we may state that the MAXWELL equations together with the balance laws of mass, momentum (in the original not approximated version), angular momentum and energy presented here are EUCLIDian invariant. However, since these equations must be supplemented by the MAXWELL-LORENTZ aether relations (3.3.2), the complete set of electromagnetic and mechanical field equations is EUCLIDian invariant only in the non-relativistic approximation.

With this digression on invariance properties, which in our opinion are crucial, we now continue with the constitutive theory, which in accordance with the preceding conclusions must also be EUCLIDian invariant. The conditions for these requirements to be satisfied are given in Sect. 1.7, and in equation (2.6.5) some possible constitutive relations are presented. In the CHU formulation the B-field does not occur and therefore case d) must be excluded here. As we are interested in a comparison of different formulations of thermoelastic, polarizable and magnetizable bodies, we should present the constitutive theory for each case in (2.6.5). We shall discuss the details for case a) only, but the results for the two other cases will also be listed.

Case a):

$$C = \hat{C}(F_{i\alpha}, P_i/\rho, (\mu_0 \mathcal{M}_i/\rho), \Theta, \Theta_{,i}, \mathcal{Q}) . \tag{3.3.23}$$

Following the usual procedure, we introduce the HELMHOLTZ free energy ψ by

$$\psi = U - \Theta \eta , \tag{3.3.24}$$

and eliminate with the aid of $(3.3.7)_3$ r^{ext} from (2.3.6). This yields

$$-\rho\dot{\psi} - \rho\eta\dot{\Theta} + t_{ij}F_{\alpha j}^{-1}\dot{F}_{i\alpha} + \rho\mathcal{E}_i\frac{d}{dt}(P_i/\rho) + \rho\mathcal{H}_i\frac{d}{dt}(\mu_0\mathcal{M}_i/\rho)$$
$$+ \mathcal{J}_i\mathcal{E}_i - q_{i,i} + \Theta\phi_{i,i} \geq 0 . \tag{3.3.25}$$

Constitutive equations are established for the (dependent) variables

$$t_{ij} , \quad \mathcal{E}_i , \quad \mathcal{H}_i , \quad \eta , \quad \phi_i , \quad q_i \quad \text{and} \quad \mathcal{J}_i ,$$

which we all assume to be functions of the form (3.3.23). Furthermore, we assume the classical form of the entropy flux, namely

$$\phi_i = \frac{q_i}{\Theta} . \tag{3.3.26}$$

Actually this need not be done. Instead one could also postulate a general constitutive equation for ϕ_i and then prove the relation (3.3.26). For this formulation such a proof has not been given. However, in view of the equivalence of this theory with the next one (Model II) in which the result (3.3.26) was rigorously established, we may regard (3.3.26) as a proven statement.

We now proceed in the way as described in Sect. 1.7. Substituting the constitutive equations for the dependent variables into the MAXWELL equations (3.3.3) and into the balance laws of mass, momenta and energy (3.3.9), we arrive at the *field equations* for $x_i, P_i, \mathcal{M}_i, \Theta$ and $\mathcal{Q}$. Any solution to given initial data is called a thermodynamic process. Following COLEMAN and NOLL [44], we request the entropy inequality (or any inequality derived from it) to hold for any smooth thermodynamic process. Since ρF_i^{ext} and ρr^{ext} may have any arbitrarily assigned values, this implies that we may freely choose $F_{i\alpha}, P_i/\rho, \mu_0\mathcal{M}_i/\rho, \Theta, \mathcal{Q}$, their material time derivatives and $(\Theta_{,i})^{\cdot}$ and do not violate the field equations. Using the constitutive relation (3.3.23) for ψ, substituting the expression (3.3.26) for the entropy flux ϕ_i into (3.3.25) and performing all indicated differentiations, we obtain in the inequality

$$\left(t_{ij}F_{\alpha j} - \rho\frac{\partial\hat{\psi}}{\partial F_{i\alpha}}\right)\dot{F}_{i\alpha} + \rho\left(\mathcal{E}_i - \frac{\partial\hat{\psi}}{\partial(P_i/\rho)}\right)\left(\frac{P_i}{\rho}\right)^{\cdot}$$
$$+ \rho\left(\mathcal{H} - \frac{\partial\hat{\psi}}{\partial(\mu_0\mathcal{M}_i/\rho)}\right)\left(\frac{\mu_0\mathcal{M}_i}{\rho}\right)^{\cdot}$$
$$- \rho\left(\eta + \frac{\partial\hat{\psi}}{\partial\Theta}\right)\dot{\Theta} - \rho\frac{\partial\hat{\psi}}{\partial\Theta_{,i}}(\Theta_{,i})^{\cdot} - \rho\frac{\partial\hat{\psi}}{\partial\mathcal{Q}}\dot{\mathcal{Q}} + \mathcal{J}_i\mathcal{E}_i - \frac{q_i\Theta_{,i}}{\Theta} \geq 0 ,$$
$$\tag{3.3.27}$$

which is explicitly linear in

$$\dot{F}_{i,\alpha}, (P_i/\rho)\dot{}, (\mu_0\mathcal{M}_i/\rho)\dot{}, \dot{\Theta}, (\Theta_{,i})\dot{} , \quad \text{and} \quad \dot{\mathcal{Q}} .$$

Since all these variables may have arbitrary values, their coefficients must vanish, which implies that $\hat{\psi}$ cannot depend on $\mathcal{Q}$ and $Q_{,i}$,

$$\psi = \hat{\psi}(F_{i\alpha}, P_i/\rho, \mu_0\mathcal{M}_i/\rho, \Theta) , \tag{3.3.28}$$

and that

$$\mathcal{E}_i = \frac{\partial\hat{\psi}}{\partial(P_i/\rho)} , \quad \mathcal{H}_i = \frac{\partial\hat{\psi}}{\partial(\mu_0\mathcal{M}_i/\rho)} , \tag{3.3.29}$$

$$\eta = -\frac{\partial\hat{\psi}}{\partial\Theta} , \quad t_{ij} = \rho\frac{\partial\hat{\psi}}{\partial F_{i\alpha}}F_{j\alpha} .$$

Hence, the HELMHOLTZ free energy cannot depend on the free charge and the temperature gradient.

Of (3.3.25) there remains the *residual inequality*

$$\mathcal{J}_i\mathcal{E}_i - \frac{\Theta_{,i}q_i}{\Theta} \geq 0 , \tag{3.3.30}$$

where $\mathcal{J}_i$ and q_i are still general functions of the type (3.3.23).

Since $\phi, \eta, \mathcal{E}_i, \mathcal{H}_i$ and t_{ij} must be objective scalar, vector and tensor valued fields under the full EUCLIDian group, ψ should have the form

$$\psi = \hat{\psi}(C_{\alpha\beta}, \mathsf{P}_\alpha, \mathsf{M}_\alpha, \Theta) , \tag{3.3.31}$$

with $C_{\alpha\beta}, \mathsf{P}_\alpha$ and M_α as given in (2.1.10) and (2.6.7). With this choice, the constitutive equations (3.3.29) become

$$\eta = -\frac{\partial\hat{\psi}}{\partial\Theta} ,$$

$$\mathcal{E}_i = \frac{\partial\hat{\psi}}{\partial\mathsf{P}_\alpha}F_{i\alpha} ,$$

$$\mathcal{H}_i = \frac{\partial\hat{\psi}}{\partial\mathsf{M}_\alpha}F_{i\alpha}\mathrm{sgn}J , \tag{3.3.32}$$

$$t_{ij} = 2\rho\frac{\partial\hat{\psi}}{\partial C_{\alpha\beta}}F_{i\alpha}F_{j\beta} + P_i\mathcal{E}_j + \mu_0\mathcal{M}_i\mathcal{H}_j$$

$$= \rho\left[2\frac{\partial\hat{\psi}}{\partial C_{\alpha\beta}} + \frac{\partial\hat{\psi}}{\partial\mathsf{P}_\beta}\mathsf{P}_\gamma C_{\alpha\gamma}^{-1} + \frac{\partial\hat{\psi}}{\partial\mathsf{M}_\beta}\mathsf{M}_\gamma C_{\alpha\gamma}^{-1}\right]F_{i\alpha}F_{j\beta} .$$

Conversely, it is easily shown that with this choice $\eta, \mathcal{E}_i, \mathcal{H}_i$ and t_{ij} indeed behave as objective scalars, vectors and tensors, respectively (note that they are in correspondence with (2.6.8), (2.6.9) and (2.6.10)). Moreover, $(3.3.32)_4$

implies that the balance law of angular momentum $(3.3.9)_3$ is identically satisfied.

The above derivation makes no direct use of the GIBBS relation, as is usually done in irreversible thermodynamics. This must be so, because quite contrary to irreversible thermodynamics, the GIBBS relation is a proven statement here, valid not only in thermostatic equilibrium but also in a general thermodynamic process. Indeed, (3.3.32) together with (3.3.24) imply

$$
\begin{aligned}
\frac{\partial \eta}{\partial \Theta} &= \frac{1}{\Theta} \frac{\partial U}{\partial \Theta} \\
\frac{\partial \eta}{\partial \mathsf{P}_\alpha} &= \frac{1}{\Theta} \left\{ \frac{\partial U}{\partial \mathsf{P}_\alpha} - F_{\alpha i}^{-1} \mathcal{E}_i \right\}, \\
\frac{\partial \eta}{\partial \mathsf{M}_\alpha} &= \frac{1}{\Theta} \left\{ \frac{\partial U}{\partial \mathsf{M}_\alpha} - F_{\alpha i}^{-1} \mathcal{H}_i \mathrm{sgn} J \right\}, \\
\frac{\partial \eta}{\partial C_{\alpha\beta}} &= \frac{1}{\Theta} \left\{ \frac{\partial U}{\partial C_{\alpha\beta}} - \frac{1}{2\rho} (t_{ij} - P_i \mathcal{E}_j - \mu_0 \mathcal{M}_i \mathcal{H}_j) F_{\alpha i}^{-1} F_{\beta j}^{-1} \right\} \\
&= \frac{1}{\Theta} \left\{ \frac{\partial U}{\partial C_{\alpha\beta}} - \frac{1}{2} [\frac{1}{\rho_0} T_{\beta\alpha}^P - (\mathsf{P}_\gamma \mathsf{E}_\delta + \mathsf{M}_\gamma \mathsf{H}_\delta) C_{\alpha\gamma}^{-1} C_{\beta\delta}^{-1}] \right\},
\end{aligned}
\tag{3.3.33}
$$

from which we easily obtain the so-called GIBBS *relation*

$$
\begin{aligned}
d\eta = \frac{1}{\Theta} \Big\{ dU - \frac{1}{2} \left[\frac{1}{\rho_0} T_{\beta\alpha}^P - (\mathsf{P}_\gamma \mathsf{E}_\delta + \mathsf{M}_\gamma \mathsf{H}_\delta) C_{\alpha\gamma}^{-1} C_{\beta\delta}^{-1} \right] dC_{\alpha\beta} \\
- \mathsf{E}_\beta C_{\alpha\beta}^{-1} d\mathsf{P}_\alpha - \mathsf{H}_\beta C_{\alpha\beta}^{-1} d\mathsf{M}_\alpha \Big\},
\end{aligned}
\tag{3.3.34}
$$

where $T_{\alpha\beta}^P$, E_α and H_α are defined by (2.2.33) and (2.6.7), respectively. Mathematically the terms in curly brackets are called a PFAFFian form.

Equation (3.3.34) delivers in a well-known manner integrability conditions for the coefficient functions. These will not be derived here, because picking a particular functional representation for the HELMHOLTZ free energy $\tilde\psi(\cdot)$ and evaluating the fields η, $\mathcal{E}_i$, $\mathcal{H}_i$ and t_{ij} according to (3.3.32) guarantees the GIBBS relation to be satisfied identically. We do not say that these integrability conditions must not be known in general. On the contrary, for experimentalists these relations are easier to be determined than the free energies themselves, which may then be obtained by integration. If the electromagnetic fields vanish, the above PFAFFian form reduces to

$$
d\eta = \frac{1}{\Theta} \left\{ dU - \frac{1}{2\rho_0} T_{\alpha\beta}^P dC_{\alpha\beta} \right\},
\tag{3.3.35}
$$

the GIBBS relation as proved correct by CARATHÉODORY [37] for thermostatic processes of thermoelastic materials. Irreversible thermodynamics postulates (3.3.35) to be the GIBBS relation for all those thermodynamic processes of thermoelastic materials which deviate only slightly from thermostatic equilibrium. Here (3.3.34) holds for all thermodynamic processes in the presence of the electromagnetic fields. Needless to say that, starting with a given electromechanical interaction model, it would probably be rather difficult for an irreversible thermodynamicist to guess a GIBBS relation of the form (3.3.34).

Case b): We now proceed with case b), which only differs from case a) in that the independent variable P_i/ρ is replaced by $\mathcal{E}_i$. In this case the constitutive equations can easily be obtained from the preceding ones by the LEGENDRE transformation

$$\tilde{\psi} = \psi - \frac{1}{\rho}\mathcal{E}_i P_i = U - \Theta\eta - \frac{1}{\rho}\mathcal{E}_i P_i \ , \tag{3.3.36}$$

where

$$\tilde{\psi} = \tilde{\psi}(F_{i\alpha}, \mathcal{E}_i, \mu_0\mathcal{M}_i/\rho, \Theta) = \tilde{\psi}(C_{\alpha\beta}, \mathsf{E}_\alpha, \mathsf{M}_\alpha, \Theta) \ . \tag{3.3.37}$$

With (3.3.36) and (3.3.37) the constitutive equations (3.3.32) must now be replaced by

$$\eta = -\frac{\partial\tilde{\psi}}{\partial\Theta} \ ,$$

$$P_i = -\rho\frac{\partial\tilde{\psi}}{\partial\mathsf{E}_\alpha}F_{i\alpha} \ ,$$

$$\mathcal{H}_i = \frac{\partial\tilde{\psi}}{\partial\mathsf{M}_\alpha}F_{i\alpha}\mathrm{sgn}J \ , \tag{3.3.38}$$

$$t_{ij} = 2\rho\frac{\partial\tilde{\psi}}{\partial C_{\alpha\beta}}F_{i\alpha}F_{j\beta} - \mathcal{E}_i P_j + \mu_0\mathcal{M}_i\mathcal{H}_j \ .$$

With the definition

$$\tilde{\varepsilon} := U - \frac{1}{\rho}\mathcal{E}_i P_i = U - \mathsf{E}_\alpha\mathsf{P}_\beta C_{\alpha\beta}^{-1} \ , \tag{3.3.39}$$

the GIBBS relation becomes

$$d\eta = \frac{1}{\Theta}\left\{d\tilde{\varepsilon} - \frac{1}{2}\left[\frac{1}{\rho_0}T_{\alpha\beta}^P + (\mathsf{E}_\gamma\mathsf{P}_\delta - \mathsf{M}_\gamma\mathsf{H}_\delta)C_{\alpha\gamma}^{-1}C_{\beta\delta}^{-1}\right]dC_{\alpha\beta} \right. $$
$$\left. + \mathsf{P}_\beta C_{\alpha\beta}^{-1}d\mathsf{E}_\alpha - \mathsf{H}_\beta C_{\alpha\beta}^{-1}d\mathsf{M}_\alpha\right\} \ . \tag{3.3.40}$$

In a similar way we can derive constitutive equations for

Case c): Here we define

$$\bar{\psi} := U - \Theta\eta - \frac{\mu_0}{\rho}\mathcal{H}_i\mathcal{M}_i = \bar{\psi}(F_{i\alpha}, \mathcal{E}_i, \mathcal{H}_i, \Theta) = \bar{\psi}(C_{\alpha\beta}, \mathsf{E}_\alpha, \mathsf{H}_\alpha, \Theta) \ . \tag{3.3.41}$$

The constitutive equations pertinent to this case then become

$$\eta \ \ = -\frac{\partial \tilde{\psi}}{\partial \Theta} \ ,$$

$$P_i \ \ = -\rho \frac{\partial \tilde{\psi}}{\partial \mathsf{E}_\alpha} F_{i\alpha} \ ,$$

$$\mathcal{M}_i = -\rho \frac{\partial \tilde{\psi}}{\partial \mathsf{H}_\alpha} F_{i\alpha} \mathrm{sgn} J \ ,$$

$$t_{ij} \ = 2\rho \frac{\partial \tilde{\psi}}{\partial C_{\alpha\beta}} F_{i\alpha} F_{j\beta} - \mathcal{E}_i P_j - \mu_0 \mathcal{H}_i \mathcal{M}_j \ .$$

(3.3.42)

Furthermore, after the introduction of

$$\bar{\varepsilon} := U - \frac{1}{\rho}\mathcal{E}_i P_i - \frac{\mu_0}{\rho}\mathcal{H}_i \mathcal{M}_i = U - (\mathsf{E}_\alpha \mathsf{P}_\beta + \mathsf{H}_\alpha \mathsf{M}_\beta)C_{\alpha\beta}^{-1} \ , \qquad (3.3.43)$$

the GIBBS relation may be transformed into

$$d\eta = \frac{1}{\Theta}\left\{ d\bar{\varepsilon} - \frac{1}{2}\left[\frac{1}{\rho_0}T_{\alpha\beta}^P + (\mathsf{E}_\gamma \mathsf{P}_\delta + \mathsf{H}_\gamma \mathsf{M}_\delta)C_{\alpha\gamma}^{-1}C_{\beta\delta}^{-1}\right] dC_{\alpha\beta}\right.$$
$$\left. +\mathsf{P}_\beta C_{\alpha\beta}^{-1} d\mathsf{E}_\alpha + \mathsf{M}_\beta C_{\alpha\beta}^{-1} d\mathsf{H}_\alpha\right\} \ .$$

(3.3.44)

Before we proceed to other electromechanical interaction models, we would like to mention that all the above constitutive models are equally possible ones. Case a) is particularly fashionable among applied physicists and has for the case that the material under consideration is polarizable-only or magnetizable-only been used by, amongst others, TOUPIN [241] and BROWN [32], respectively. The formulation of case c) is used by PAO and HUTTER [171]. There are reasons for preference of case c). Indeed, since any electromagnetic problem must be solved in the entire space, including the vacuum, any theory must ultimately be expressed in terms of $\mathcal{E}_i$ and $\mathcal{H}_i$. It follows that in the cases a) and b) the constitutive equations for $\mathcal{E}_i$ and/or $\mathcal{H}_i$ must be invertible in the sense that

$$\frac{P_i}{\rho} = \mathfrak{f}_P(F_{i\alpha}, \mathcal{E}_i, \mathcal{H}_i, \Theta) \ , \qquad \frac{\mu_0 \mathcal{M}_i}{\rho} = \mathfrak{f}_M(F_{i\alpha}, \mathcal{E}_i, \mathcal{H}_i, \Theta) \ . \qquad (3.3.45)$$

For linear constitutive relations such inversions are trivial; they become more difficult or impossible analytically when dealing with nonlinear theories.

3.3.2 The Two–Dipole Model with a Symmetric Stress Tensor (Model II)

The above two-dipole model is not the only one that has been proposed. There is another one, leading to a symmetric stress tensor. This model has been investigated by HUTTER [95], but it is already suggested by FANO, CHU and ADLER [73]. It is the two-dipole counterpart of Model V, treated below. The MAXWELL equations are the same as stated for model I. Electromagnetic body force, couple and energy supply, however, are taken as follows

$$\rho F_i^e = \mathcal{Q}^e E_i^C + e_{ijk} J_j^e \mu_0 H_k^C + \mathcal{Q}^m H_i^C - e_{ijk} J_j^m \varepsilon_0 E_k^C ,$$

$$\rho L_{ij}^e = 0 , \qquad\qquad\qquad (3.3.46)$$

$$\rho r^e = \mathcal{J}_i^e \mathcal{E}_i + \mathcal{J}_i^m \mathcal{H}_i .$$

Here,

$$\mathcal{J}_i^e = J_i^e - \mathcal{Q}^e \dot{x}_i \quad \text{and} \quad \mathcal{J}_i^m = J_i^m - \mathcal{Q}^m \dot{x}_i , \qquad (3.3.47)$$

and the superscripts e and m indicate that the corresponding current or charge densities are those due to electric and magnetic charges, respectively. Formally, $(3.3.46)_1$ is simply the electric and magnetic LORENTZ force. The quantities can easily be read off from the MAXWELL equations (3.3.3):

$$\mathcal{Q}^e = \mathcal{Q} - P_{i,i}^C , \qquad J_i^e = J_i + \rho \frac{d}{dt}\left(\frac{P_i^C}{\rho}\right) - (P_j^C \dot{x}_i)_{,j} ,$$

$$\mathcal{Q}^m = -\mu_0 M_{i,i}^C , \qquad J_i^m = \rho \frac{d}{dt}\left(\frac{\mu_0 M_i^C}{\rho}\right) - (\mu_0 M_j^C \dot{x}_i)_{,j} , \qquad (3.3.48)$$

$$\mathcal{J}_i^e = \mathcal{J}_i + \overset{\star}{P}_i{}^C , \qquad \mathcal{J}_i^m = \mu_0 \overset{\star}{M}_i{}^C .$$

Substitution of (3.3.48) into (3.3.46) yields

$$\rho F_i^e = \mathcal{Q} E_i^C + \mu_0 e_{ijk} J_j H_k^C$$

$$+ P_j^C E_{i,j}^C + \mu_0 e_{ijk} \dot{x}_j H_{k,l}^C P_l^C + \rho\mu_0 e_{ijk} \frac{d}{dt}\left(\frac{P_j^C}{\rho}\right) H_k^C$$

$$+ \mu_0 M_j^C H_{i,j}^C - \varepsilon_0 e_{ijk} \dot{x}_j E_{k,l}^C \mu_0 M_l^C - \rho\varepsilon_0 e_{ijk} \frac{d}{dt}\left(\frac{\mu_0 M_j^C}{\rho}\right) E_k^C$$

$$- (P_j \mathcal{E}_i + \mu_0 \mathcal{M}_j \mathcal{H}_i)_{,j} ,$$

$$\rho L_{ij}^e = 0 ,$$

$$\rho r^e = \mathcal{J}_i \mathcal{E}_i + \overset{\star}{P}_i \mathcal{E}_i + \mu_0 \overset{\star}{M}_i \mathcal{H}_i .$$

$$(3.3.49)$$

Before we proceed, let us look at the invariance properties of (3.3.49). To this end, recall that $\mathcal{E}_i, \mathcal{H}_i, P_i = P_i^C, \mu_0 \mathcal{M}_i = \mu_0 M_i^C, \mathcal{Q}$ and $\mathcal{J}_i$ are objective quantities with respect to the EUCLIDIAN transformation group. Hence, the electromagnetic energy supply is an objective scalar under this group. Incidentally, it should be noted that ρr^e is also an objective scalar under the extended LORENTZ group, provided we restrict ourselves to the semi-relativistic approximation.

Regarding the electromagnetic body force $(3.3.49)_1$, we see that the above expression differs from $(3.3.7)_1$ only in a term which is objective under EUCLIDian and under LORENTZ transformations in the semi-relativistic sense. Hence, all conclusions drawn in Model I do apply also here, and we refer the reader to the pertinent discussion in Model I. In conclusion we state the governing equations of this theory consistently to within the order of exactness of the non-relativistic approximation. Hence, all terms preceded by a c^{-2} factor must be dropped; as a result we arrive at the following balance laws for mass, linear and angular momenta and energy:

$$\dot{\rho} \;\; + \rho\dot{x}_{i,i} = 0 \, ,$$

$$\rho\ddot{x}_i = t_{ij,j} + \rho F_i^{\text{ext}} + \mathcal{Q}\mathcal{E}_i + \mu_0 e_{ijk}\mathcal{J}_j\mathcal{H}_k - P_{j,j}\mathcal{E}_i$$

$$+\mu_0 e_{ijk}\overset{\star}{P}_j\,\mathcal{H}_k - \mu_0\mathcal{M}_{j,j}\mathcal{H}_i \, , \tag{3.3.50}$$

$$t_{[ij]} = 0 \, ,$$

$$\rho\dot{U} = t_{ij}\dot{x}_{i,j} - q_{i,i} + \mathcal{J}_i\mathcal{E}_i + \overset{\star}{P}_i\,\mathcal{E}_i + \mu_0\,\overset{\star}{\mathcal{M}}_i\,\mathcal{H}_i + \rho r^{\text{ext}} \, .$$

Obviously the fact that ρr^{e} is by itself an objective scalar is corroboration of the assumption of symmetry of the stress tensor (compare the relevant discussion in Model I).

Furthermore, a long but straightforward calculation shows that, within the non-relativistic approximation (compare these expressions with model I)

$$^{II}t_{ij}^M = \varepsilon_0\mathcal{E}_i\mathcal{E}_j + \mu_0\mathcal{H}_i\mathcal{H}_j - \tfrac{1}{2}\delta_{ij}(\varepsilon_0\mathcal{E}_k\mathcal{E}_k + \mu_0\mathcal{H}_k\mathcal{H}_k) \, ,$$

$$^{II}g_i = 0 \, ,$$

$$^{II}\pi_i = -e_{ijk}E_j^C H_k^C - e_{ijk}\mathcal{E}_j\mathcal{H}_k - (\varepsilon_0\mathcal{E}_k\mathcal{E}_k + \mu_0\mathcal{H}_k\mathcal{H}_k)\dot{x}_i \tag{3.3.51}$$

$$+(\varepsilon_0\mathcal{E}_i\mathcal{E}_j + \mu_0\mathcal{H}_i\mathcal{H}_j)\dot{x}_j \, ,$$

$$^{II}\omega = -\tfrac{1}{2}(\varepsilon_0\mathcal{E}_k\mathcal{E}_k + \mu_0\mathcal{H}_k\mathcal{H}_k) \, .$$

Hence, the jump conditions for momentum and energy become

$$[\![t_{ij} + \varepsilon_0\mathcal{E}_i\mathcal{E}_j + \mu_0\mathcal{H}_i\mathcal{H}_j - \tfrac{1}{2}\delta_{ij}(\varepsilon_0\mathcal{E}_k\mathcal{E}_k + \mu_0\mathcal{H}_k\mathcal{H}_k)]\!]n_j$$

$$-[\![\rho\dot{x}_i(\dot{x}_j n_j - \text{w}_n)]\!] = 0 \, ,$$

$$[\![\dot{x}_i t_{ij} - q_j - e_{jkl}\mathcal{E}_k\mathcal{H}_l + (\varepsilon_0\mathcal{E}_j\mathcal{E}_k + \mu_0\mathcal{H}_j\mathcal{H}_k)\dot{x}_k \tag{3.3.52}$$

$$-\tfrac{1}{2}(\varepsilon_0\mathcal{E}_k\mathcal{E}_k + \mu_0\mathcal{H}_k\mathcal{H}_k)\dot{x}_j]\!]n_j$$

$$-[\![\{\tfrac{1}{2}\rho\dot{x}_i\dot{x}_i + \rho U + \tfrac{1}{2}(\varepsilon_0\mathcal{E}_i\mathcal{E}_i + \mu_0\mathcal{H}_i\mathcal{H}_i)\}\,(\dot{x}_j n_j - \text{w}_n)]\!] = 0 \, .$$

The jump conditions of electromagnetic fields and mass are the same as those for Model I (see (3.3.11)) and will not be repeated here.

We now turn to the constitutive theory. We shall not present all the details here, but list the results only. We start by eliminating ρr^{ext} from the entropy inequality (2.3.6) and the energy balance (3.3.50)$_4$. The result reads

$$-\rho \dot{U} + \rho \Theta \dot{\eta} + \rho \mathcal{E}_i \frac{d}{dt}\left(\frac{P_i}{\rho}\right) + \rho \mathcal{H}_i \frac{d}{dt}\left(\frac{\mu_0 \mathcal{M}_i}{\rho}\right)$$
$$+(t_{ij} - \mathcal{E}_i P_j - \mu_0 \mathcal{H}_i \mathcal{M}_j)\dot{x}_{i,j} + \mathcal{J}_i \mathcal{E}_i - \frac{\Theta_{,i} q_i}{\Theta} \geq 0 \, , \tag{3.3.53}$$

where we have also used the DUHEM relation for the entropy flux,

$$\phi_i = \frac{q_i}{\Theta} \, . \tag{3.3.54}$$

In [95] HUTTER dealt with a theory of this complexity and proved correct the form of the entropy flux vector introduced here as a postulate.

As before, the constitutive theory depends on the selection of the independent variables. We may again choose constitutive relations of the form (2.6.5), but we shall give them for case a) only. The equations for the other cases can be derived in the same way as was done for Model I.

When the constitutive relations are of the form

$$\mathsf{C} = \hat{\mathsf{C}}(F_{i\alpha}, P_i/\rho, \mu_0 \mathcal{M}_i/\rho, \Theta, \Theta_{,i}, \mathcal{Q}) \, , \tag{3.3.55}$$

the HELMHOLTZ free energy

$$\psi := U - \eta \Theta = \hat{\psi}(C_{\alpha\beta}, \mathsf{P}_\alpha, \mathsf{M}_\alpha, \Theta) \, , \tag{3.3.56}$$

does not depend on $\mathcal{Q}$ and $\Theta_{,i}$ and serves as potential for $\eta, \mathcal{E}_i, \mathcal{H}_i$ and t_{ij} as follows

$$\eta = -\frac{\partial \hat{\psi}}{\partial \Theta} \, ,$$
$$\mathcal{E}_i = \frac{\partial \hat{\psi}}{\partial \mathsf{P}_\alpha} F_{i\alpha} \, ,$$
$$\mathcal{H}_i = \frac{\partial \hat{\psi}}{\partial \mathsf{M}_\alpha} F_{i\alpha} \mathrm{sgn} J \, , \tag{3.3.57}$$
$$t_{ij} = 2\left\{ \rho \frac{\partial \hat{\psi}}{\partial C_{\alpha\beta}} F_{i\alpha} F_{j\beta} + \mathcal{E}_{(i} P_{j)} + \mu_0 \mathcal{H}_{(i} \mathcal{M}_{j)} \right\} \, .$$

There remains

$$\mathcal{J}_i \mathcal{E}_i - \frac{\Theta_{,i} q_i}{\Theta} \geq 0 \, , \tag{3.3.58}$$

as the reduced entropy inequality.

Finally, the GIBBS relation takes the form

$$d\eta = \frac{1}{\Theta}\left\{ dU - \left[\frac{1}{2\rho_0}T_{\alpha\beta}^P - C_{\alpha\gamma}^{-1}C_{\beta\delta}^{-1}(\mathsf{E}_\gamma\mathsf{P}_\delta + \mathsf{E}_\delta\mathsf{P}_\gamma + \mathsf{H}_\gamma\mathsf{M}_\delta + \mathsf{H}_\delta\mathsf{M}_\gamma) \right] dC_{\alpha\beta} \right.$$
$$\left. -C_{\alpha\beta}^{-1}\mathsf{E}_\beta d\mathsf{P}_\alpha - C_{\alpha\beta}^{-1}\mathsf{H}_\beta d\mathsf{M}_\alpha \right\}.$$

$$(3.3.59)$$

For the reader's sake of curiosity, we mention that this theory will be shown to be fully equivalent to the theory of model I.

This completes the description of the two-dipole models. Both descriptions outlined above use a dipole model not only for polarization but also for magnetization. There are no others as far as we know and all the remaining models bear to a lesser or larger extent the notion of dipole structure for polarization and the structure of electric circuits for magnetization.

In conclusion we wish to mention a paper by ALBLAS [9], who also used the CHU formulation and who derived a theory which completely agrees with our model I. The derivation is on the basis of a global energy balance law (see also Sects. 2.3 and 2.8) and for a more general problem, including spin- and dissipation effects.

In what follows, we shall discuss theories presented by VAN DE VEN [249], DE GROOT & SUTTORP [53], MÜLLER [160] and HUTTER [92].

3.4 The Maxwell–Minkowski Formulation (Model III)

The foregoing two interaction models were constructed by postulating body force, body couple and energy supply terms due to the electromagnetic fields. The approach in this section is different. Basically, the balance laws of mass and momenta are derived from a global energy balance law by postulating certain invariance properties. In [249] VAN DE VEN follows this method, first formulated by GREEN and RIVLIN [77], and developed for a ferromagnetic solid by ALBLAS [8]. By postulating that the energy balance is invariant under rigid-body motions and by making some a priori assumptions concerning the invariance properties of the quantities involved, the equations of balance of mass and momenta are derived. In this derivation all terms preceded by a factor c^{-2} are neglected.[1]

---

[1] Clearly, the outcome of this approach essentially depends on what a priori assumptions regarding the invariance properties of the various quantities are made. Moreover, special care must be observed when neglecting terms which contain a c^{-2}-factor. Indeed, there is an essential difference between the approximations performed in the above mentioned references and the non-relativistic approximation in Sect. 1.6. While we apply SI-units, GAUSSIAN units are used in [249], and it is a well-known fact that a c^{-2}-term in one system of units is not necessarily a c^{-2}-term in the other system of units as well. For instance

$$\frac{1}{c}\boldsymbol{E}\times\boldsymbol{H}$$

Before proceeding the following remark seems to be in order: It is conceivable that one might object to associate the names of MAXWELL and MINKOWSKI with the formulation presented below, and indeed neither MAXWELL nor MINKOWSKI ever presented a formulation of this kind. However, for the special case of a magnetostatic problem, the magnetoelastic stresses that will be obtained here, are identical with those used by W.F. BROWN Jr. in his monograph, [32]; and BROWN stated that these stresses could be derived by taking over from MAXWELL a formula for the magnetic force. This is justification for us to associate this (dynamic) theory with the name of MAXWELL. Moreover, since the MAXWELL equations will be used in the MINKOWSKI formulation, we shall refer to the formulation presented in this section as the MAXWELL-MINKOWSKI formulation. Clearly, the association of names with certain theories or formulations of theories bears its well-known disadvantages and a reader not willing to accept our proposal may reject it and invent his own name for it. As an example the reader may recall that there is no unique version of the MAXWELL-MINKOWSKI stress tensor either. We shall come back to this subject at the end of this Section.

As said above, the MAXWELL equations will be expressed in terms of the MINKOWSKI field variables E_i^M, D_i^M, H_i^M, and B_i^M, which are related to the field variables introduced in (2.2.12) and (2.2.20) by

$$\mathcal{E}_i = E_i^M + \mathrm{e}_{ijk}\dot{x}_j B_k^M, \qquad D_i = D_i^M,$$
$$\mathcal{H}_i = H_i^M - \mathrm{e}_{ijk}\dot{x}_j D_k^M, \qquad B_i = B_i^M. \tag{3.4.1}$$

Accordingly, the following set of MAXWELL equations is obtained:

$$B_{i,i} = 0, \qquad \mathrm{e}_{ijk}E_{k,j}^M + \frac{\partial B_i}{\partial t} = 0,$$
$$D_{i,i} = \mathcal{Q}, \qquad \mathrm{e}_{ijk}H_{k,j}^M - \frac{\partial D_i}{\partial t} = J_i. \tag{3.4.2}$$

Here and henceforth, since no confusion is possible, the superscript M for D_i and B_i is omitted. In a purely formal way polarization per unit volume and magnetization per unit volume can be introduced via the definitions

---

in GAUSSian units becomes

$$\frac{1}{c^2}\boldsymbol{E} \times \boldsymbol{H}$$

in SI-units. As a consequence of both, the a priori assumptions on invariance that are not in conformity with the results of Sect. 1.6 and the discrepancy in the neglect of terms like that demonstrated above, the equations derived in [249] are not EUCLIDian invariant in the sense as defined in Sect. 1.6. However, it is possible to derive from an analogous energy balance as was postulated in [249] local balance equations of mass, linear and angular momentum and energy that are invariant under EUCLIDian transformations by simply applying in a consistent way non-relativistic invariance requirements. This then fully justifies the a priori assumptions we will impose.

$$P_i^M := D_i - \varepsilon_0 E_i^M , \qquad \mu_0 M_i^M := B_i - \mu_0 H_i^M , \qquad (3.4.3)$$

and it is convenient to introduce also the variable

$$\mathcal{M}_i := M_i^M + \mathrm{e}_{ijk} \dot{x}_j P_k^M . \qquad (3.4.4)$$

It is now a routine matter (see for instance Appendix A) to prove that under the EUCLIDian transformation group

$$\mathcal{E}_i, \ D_i \text{ and } P_i^M \text{ transform as objective vectors, and}$$

$$(3.4.5)$$

$$\mathcal{H}_i, \ B_i \text{ and } \mathcal{M}_i \text{ transform as objective axial vectors },$$

respectively.

In particular, and as an easy calculation also shows directly, under the rigid-body translation

$$x_i' = x_i - b_i(t) , \qquad t' = t \qquad (3.4.6)$$

the following transformation laws must hold

$$
\begin{aligned}
D_i' &= D_i , & E_i^{M'} &= E_i^M + \mathrm{e}_{ijk} \dot{b}_j B_k , \\
B_i' &= B_i , & H_i^{M'} &= H_i^M - \mathrm{e}_{ijk} \dot{b}_j D_k , \\
\mathcal{Q}' &= \mathcal{Q} , & J_i' &= J_i - \mathcal{Q} \dot{b}_i , \\
P_i^{M'} &= P_i^M , & M_i^{M'} &= M_i^M + \mathrm{e}_{ijk} \dot{b}_j P_k^M .
\end{aligned}
\qquad (3.4.7)
$$

On the other hand, the MINKOWSKI fields can also be subjected to the special LORENTZ transformation (2.5.4), which when written in the semi-relativistic approximation, becomes

$$x_i^\star = x_i - V_i t , \qquad t^\star = t - \frac{1}{c^2} x_i V_i . \qquad (3.4.8)$$

Then, the transformation rules are

$$
\begin{aligned}
D_i^\star &= D_i + \frac{1}{c^2} \mathrm{e}_{ijk} V_j H_k^M , & E_i^{M\star} &= E_i^M + \mathrm{e}_{ijk} V_j B_k , \\
B_i^\star &= B_i - \frac{1}{c^2} \mathrm{e}_{ijk} V_j E_k^M , & H_i^{M\star} &= H_i^M - \mathrm{e}_{ijk} V_j D_k , \\
\mathcal{Q}^\star &= \mathcal{Q} - \frac{1}{c^2} J_i V_i , & J_i^\star &= J_i - \mathcal{Q} V_i , \\
(P_i^M)^\star &= P_i^M - \frac{1}{c^2} \mathrm{e}_{ijk} V_j M_k^M , & (M_i^M)^\star &- M_i^M + \mathrm{e}_{ijk} V_j P_k^M .
\end{aligned}
\qquad (3.4.9)
$$

Comparison of (3.4.7) and (3.4.9), which constitute the transformation rules for special EUCLIDian and LORENTZ transformations, shows that the primed and stared quantities equal only in the non-relativistic approximation. This conclusion is also correct if the MINKOWSKI-fields are subject to general EUCLIDian or LORENTZ transformations. In view of the results of Sect. 1.6 this must be so expected.

A nice application of equations (3.4.7) and (3.4.9) is obtained if the invariance properties of the MAXWELL equations are investigated. To this end, they are best expressed in terms of the objective fields $\mathcal{E}_i, D_i, \mathcal{H}_i, B_i, \mathcal{J}_i$ and $\mathcal{Q}$ and read then (see (2.2.15) and (2.2.21))

$$B_{i,i} = 0 \,, \qquad e_{ijk}\mathcal{E}_{k,j} + \overset{\star}{B}_i = 0 \,,$$

$$D_{i,i} = \mathcal{Q} \,, \qquad e_{ijk}\mathcal{H}_{k,j} - \overset{\star}{D}_i = \mathcal{J}_i \,. \tag{3.4.10}$$

If all the fields in these equations were independent it would follow at once from the above listed transformation properties, that the equations (3.4.10) would be invariant under EUCLIDian transformations. However, relations (3.4.3), which replace the MAXWELL-LORENTZ aether relations in the previous discussions, are not invariant under EUCLIDian transformations, for according to (3.4.7) and (3.4.3) we have

$$P_i^{M'} = P_i^M = D_i - \varepsilon_0 E_i^M = D_i' - \varepsilon_0 E_i^{M'} + \varepsilon_0 e_{ijk}\overset{\cdot}{b}_j B_k$$

$$= D_i' - \varepsilon_0 E_i^{M'} + \frac{1}{c^2}e_{ijk}\overset{\cdot}{b}_j(H_k + M_k) \,,$$

$$\mu_0 M_i^{M'} = \mu_0 M_i^M + \mu_0 e_{ijk}\overset{\cdot}{b}_j P_k^M = B_i - \mu_0 H_i^M + \mu_0 e_{ijk}\overset{\cdot}{b}_j P_k^M \tag{3.4.11}$$

$$= B_i' - \mu_0 H_i^{M'} - \mu_0 e_{ijk}\overset{\cdot}{b}_j(D_k - P_k)$$

$$= B_i' - \mu_0 H_i^{M'} - \frac{1}{c^2}e_{ijk}\overset{\cdot}{b}_j E_k \,.$$

Because equation (3.4.6) is a special EUCLIDian transformation, we conclude that relations (3.4.3) are not invariant under EUCLIDian transformations in general except in the non-relativistic sense.

A similar calculation can also be performed with (3.4.9). The result is that (3.4.3) is a LORENTZ invariant equation. We leave this proof to the reader. One further particular point in (3.4.11) must be mentioned. In reaching the non-relativistic invariance property the relation

$$B_i = \mu_0(H_i^M + M_i^M) \,, \tag{3.4.12}$$

must be used and only afterwards terms containing a c^{-2}-factor can be dropped. This procedure is tantamount to assuming B_i to be proportional to μ_0. Apparently this should be done consistently and hence we make the following statements that will be observed as basic rules henceforth:

(i) B_i must be considered to be proportional to μ_0 .

(ii) Under this provision terms containing a c^{-2}-factor should be dropped.

When these rules are observed the relations (3.4.3) and (3.4.4) may be replaced by the very suggestive formulas

$$D_i = \varepsilon_0 \mathcal{E}_i + P_i , \qquad \text{and} \qquad B_i = \mu_0 \mathcal{H}_i + \mu_0 \mathcal{M}_i , \qquad (3.4.13)$$

in which we have set

$$P_i = P_i^M . \qquad (3.4.14)$$

These preliminary remarks may suffice, and so we proceed to derive the balance laws of mass, linear and angular momentum and energy by starting with a global balance law of energy that is subjected to superimposed rigid-body motions (EUCLIDian transformations). Although our approach is different in detail we shall in the following derivation essentially follow VAN DE VEN [249]. Starting equation is a global energy balance law of the form

$$\frac{d}{dt} \int_{\mathcal{V}} \{\rho U + \tfrac{1}{2}(\varepsilon_0 \mathcal{E}_i \mathcal{E}_i + \mu_0 \mathcal{H}_i \mathcal{H}_i) + \tfrac{1}{2}\rho \dot{x}_i \dot{x}_i + \rho T\} d\nu$$

$$= \int_{\mathcal{V}} \{\rho r^{\text{ext}} + \rho F_i^{\text{ext}} \dot{x}_i\} d\nu + \int_{\partial \mathcal{V}} \{t_{ij}\dot{x}_i - q_j - e_{jkl}E_k^M H_l^M \qquad (3.4.15)$$

$$+ \tfrac{1}{2}(\varepsilon_0 \mathcal{E}_k \mathcal{E}_k + \mu_0 \mathcal{H}_k \mathcal{H}_k)\dot{x}_j + R_j\} da_j .$$

This equation, in which ρT and R_i are still to be determined as functions of the electromagnetic field quantities and the motion, is a generalization of the purely mechanical energy balance law. Indeed, when all electromagnetic quantities are set to zero, what results is the classical non-relativistic energy balance law. It is also easy to interpret in (3.4.15) the terms of electromagnetic origin. Firstly,

$$\tfrac{1}{2}(\varepsilon_0 \mathcal{E}_i \mathcal{E}_i + \mu_0 \mathcal{H}_i \mathcal{H}_i)$$

is an electromagnetic energy density, and, secondly

$$\boldsymbol{E}^M \times \boldsymbol{H}^M$$

is the Poynting vector. As is well-known, both are not unique and we shall see that other electromagnetic models correspond to different postulates of the electromagnetic energy density and the Poynting vector. This does not mean, however, that these models will also yield different results for physically measurable quantities.

A clue as to what should be chosen for the as yet undetermined quantities ρT and R_i is obtained, if (3.4.15) is written for a body whose mass density is vanishingly small. In this case we expect (3.4.15) to become the energy equations for the electromagnetic fields in vacuo. This energy equation is a consequence of the MAXWELL equations (3.4.10), and it can be brought into the form (3.4.15) (with vanishing $t_{ij}, q_i, \dot{x}_i, U, r^{\text{ext}}, F_i^{\text{ext}}$). It is not hard to show that in this case and to within terms of order c^{-2}

$$\rho T = 0 \qquad \text{and} \qquad R_i = 0 . \tag{3.4.16}$$

In other words, in a vacuum ρT and $\boldsymbol{R}$ must vanish.

Next we determine ρT and $\boldsymbol{R}$ in a ponderable body, and for that purpose (3.4.15) must be written in local form. Assuming sufficient smoothness of the fields involved this yields

$$(\dot{\rho} + \rho \dot{x}_{i,i})(U + \tfrac{1}{2}\dot{x}_i\dot{x}_i + T) + \rho \dot{T} - R_{i,i} + (e_{ijk}E_j^M H_k^M)_{,i}$$

$$+\{\rho\ddot{x}_i - \rho F_i^{\text{ext}} - t_{ij,j} - \tfrac{1}{2}(\varepsilon_0\mathcal{E}_k\mathcal{E}_k + \mu_0\mathcal{H}_k\mathcal{H}_k)_{,i}\}\dot{x}_i \tag{3.4.17}$$

$$+\rho\dot{U} - t_{ij}\dot{x}_{i,j} + q_{i,i} - \rho r^{\text{ext}} + \tfrac{1}{2}(\varepsilon_0\mathcal{E}_k\mathcal{E}_k + \mu_0\mathcal{H}k\mathcal{H}_k)^{\cdot} = 0 .$$

This equation is now subjected to EUCLIDian transformations and it is postulated that it is invariant under these transformations. The details of these calculations are somewhat tedious and are presented in full detail in Appendix B. They deliver expressions for ρT and $\boldsymbol{R}$ as well as local balance laws of mass, linear and angular momentum and energy. The results for ρT and $\boldsymbol{R}$ in Appendix B are

$$\rho T = 0 , \qquad \text{and} \qquad R_i = (P_j\mathcal{E}_j + \mu_0\mathcal{M}_j\mathcal{H}_j)\dot{x}_i + e_{jkl}P_k B_l\dot{x}_j\dot{x}_i . \tag{3.4.18}$$

We note that only the second term in the expression for R_i follows from invariance requirements, whereas the first term is chosen arbitrarily to simplify certain formulas. This seems to be a disadvantage of this method of derivation, however a change in $\boldsymbol{R}$ only results in a different stress tensor t_{ij} (and, eventually, a different energy flux q_i) and it can be shown that this has no effect in the ultimate form of the balance laws.

If (3.4.18) is substituted into (3.4.15) we obtain as global energy balance

$$\frac{d}{dt}\int\limits_{\mathcal{V}}\left\{\rho U + \tfrac{1}{2}(\varepsilon_0\mathcal{E}_i\mathcal{E}_i + \mu_0\mathcal{H}_i\mathcal{H}_i) + \tfrac{1}{2}\rho\dot{x}_i\dot{x}_i\right\}d\nu$$

$$= \int\limits_{\mathcal{V}}\left\{\rho r^{\text{ext}} + \rho F_i^{\text{ext}}\dot{x}_i\right\}d\nu + \int\limits_{\partial\mathcal{V}}\left\{t_{ij}\dot{x}_i - q_j - e_{jkl}\mathcal{E}_k\mathcal{H}_l\right. \tag{3.4.19}$$

$$\left. +(D_j\mathcal{E}_k + B_j\mathcal{H}_k)\dot{x}_k - \tfrac{1}{2}(\varepsilon_0\mathcal{E}_k\mathcal{E}_k + \mu_0\mathcal{H}_k\mathcal{H}_k)\dot{x}_j\right\}da_j ,$$

from which a local balance law can be derived, which with the use of (3.4.10) and (3.4.13) may be written in the form

$$(\dot{\rho} + \rho\dot{x}_{i,i})(U + \tfrac{1}{2}\dot{x}_i\dot{x}_i) + \rho\dot{U} - \mathcal{J}_i\mathcal{E}_i - \mathcal{E}_i\overset{\star}{P}_i - \mu_0\mathcal{H}_i\overset{\star}{\mathcal{M}}_i$$

$$+q_{i,i} - \rho r^{\text{ext}} - [t_{ij} + \mathcal{E}_iP_j + \mu_0\mathcal{H}_i\mathcal{M}_j]\dot{x}_{i,j}$$

$$+[\rho\ddot{x}_i - t_{ij,j} - \rho F_i^{\text{ext}} - \mathcal{Q}\mathcal{E}_i - e_{ijk}\mathcal{J}_jB_k - P_j\mathcal{E}_{j,i}$$

$$-\mu_0\mathcal{M}_j\mathcal{H}_{j,i} - e_{ijk}(D_j\overset{\star}{B}_k + \overset{\star}{D}_j\ B_k)]\dot{x}_i = 0\ .$$

(3.4.20)

The invariance requirements under which

$$\rho, U, t_{ij}, (\ddot{x}_i - F_i^{\text{ext}}), q_i, \rho r^{\text{ext}}, \mathcal{Q}, \mathcal{J}_i, \mathcal{E}_i, P_i, \mathcal{H}_i, \mathcal{M}_i, D_i, B_i$$

are assumed to transform as objective quantities then yield the local balance laws of

mass

$$\dot{\rho} + \rho\dot{x}_{i,i} = 0\ . \tag{3.4.21}$$

momentum

$$\rho\ddot{x}_i - t_{ij,j} - \rho F_i^{\text{ext}} = \rho F_i^{\text{e}} = \mathcal{Q}\mathcal{E}_i + e_{ijk}\mathcal{J}_jB_k$$

$$+P_j\mathcal{E}_{j,i} + \mu_0\mathcal{M}_j\mathcal{H}_{j,i} + e_{ijk}(D_j\overset{\star}{B}_k + \overset{\star}{D}_j\ B_k)\ , \tag{3.4.22}$$

angular momentum

$$t_{[ij]} = \rho L_{ij}^{\text{e}} = P_{[i}\mathcal{E}_{j]} + \mu_0\mathcal{M}_{[j}\mathcal{H}_{j]}\ , \tag{3.4.23}$$

energy

$$\rho\dot{U} - t_{ij}\dot{x}_{i,j} + q_{i,i} - \rho r^{\text{ext}} = \rho r^{\text{e}}$$

$$\rho r^{\text{e}} = \mathcal{J}_i\mathcal{E}_i + \rho\mathcal{E}_i\frac{d}{dt}(P_i/\rho) + \rho\mu_0\mathcal{H}_i\frac{d}{dt}(\mathcal{M}_i/\rho)\ . \tag{3.4.24}$$

The source terms in (3.4.22) and (3.4.24) can be transformed into the form (2.4.7) whereby, to within terms of order c^{-2},

$$^{III}t_{ij}^M = \mathcal{E}_iD_j + \mathcal{H}_iB_j - \tfrac{1}{2}\delta_{ij}(\varepsilon_0\mathcal{E}_k\mathcal{E}_k + \mu_0\mathcal{H}_k\mathcal{H}_k)\ ,$$

$$^{III}g_i = 0\ ,$$

$$^{III}\pi_i = -e_{ijk}\mathcal{E}_j\mathcal{H}_k + (D_i\mathcal{E}_j + B_i\mathcal{H}_j)\dot{x}_j - (\varepsilon_0\mathcal{E}_j\mathcal{E}_j + \mu_0\mathcal{H}_j\mathcal{H}_j)\dot{x}_i\ ,$$

$$^{III}\omega = -\tfrac{1}{2}(\varepsilon_0\mathcal{E}_i\mathcal{E}_i + \mu_0\mathcal{H}_i\mathcal{H}_i)\ ,$$

(3.4.25)

is obtained.

Substitution of (3.4.1) into (2.4.3) and of (3.4.25) into (2.4.11) yields the following set of jump conditions

$$\llbracket e_{ijk}E_j^M n_k + B_i \mathrm{w}_n \rrbracket = 0 \,, \qquad \llbracket D_i \rrbracket n_i = 0 \,,$$

$$\llbracket e_{ijk}H_j^M n_k - D_i \mathrm{w}_n \rrbracket = 0 \,, \qquad \llbracket B_i \rrbracket n_i = 0 \,,$$

$$\llbracket \rho(\dot{x}_i n_i - \mathrm{w}_n) \rrbracket = 0 \,,$$

$$\llbracket t_{ij} + \mathcal{E}_i D_j + \mathcal{H}_i B_j - \tfrac{1}{2}\delta_{ij}(\varepsilon_0 \mathcal{E}_k \mathcal{E}_k + \mu_0 \mathcal{H}_k \mathcal{H}_k) \rrbracket n_j$$

$$- \llbracket \rho \dot{x}_i (\dot{x}_j n_j - \mathrm{w}_n) \rrbracket = 0 \,, \tag{3.4.26}$$

$$\llbracket \dot{x}_i t_{ij} - q_j - e_{jkl}\mathcal{E}_k \mathcal{H}_l + (D_j \mathcal{E}_k + B_j \mathcal{H}_k)\dot{x}_k$$

$$- \tfrac{1}{2}(\varepsilon_0 \mathcal{E}_k \mathcal{E}_k + \mu_0 \mathcal{H}_k \mathcal{H}_k)\dot{x}_j \rrbracket n_j - \llbracket \{\tfrac{1}{2}\rho \dot{x}_i \dot{x}_i + \rho U$$

$$+ \tfrac{1}{2}(\varepsilon_0 \mathcal{E}_i \mathcal{E}_i + \mu_0 \mathcal{H}_i \mathcal{H}_i)\}(\dot{x}_j n_j - \mathrm{w}_n) \rrbracket = 0 \,.$$

We note that $(3.4.26)_{6,7}$ can still be written in a somewhat different form. As described in [249] it can be shown that *for a material surface* relations $(3.4.26)_{6,7}$ are equivalent to

$$\llbracket t_{ij} \rrbracket n_j = \tfrac{1}{2} \llbracket (\mu_0 \mathcal{M}_j n_j)^2 + (P_j n_j)^2 \rrbracket n_i \,,$$

$$\llbracket q_j \rrbracket n_j = \llbracket \dot{x}_i t_{ij} - \tfrac{1}{2}\{(\mu_0 \mathcal{M}_k n_k)^2 + (P_k n_k)^2\}\dot{x}_j \rrbracket n_j \,. \tag{3.4.27}$$

If the singular surface is a boundary separating a ponderable body from a vacuum, it is easily shown that equation $(3.4.27)_2$ reduces to $\llbracket q_j \rrbracket n_j = 0$. In this special case, therefore, the normal component of the energy flux vector must vanish.

In order to make the theory complete, the balance equations must be supplemented by constitutive equations. Our presentation will be brief as it follows exactly the approach in the preceding sections. As independent variables we choose

$$C_{\alpha\beta}, \mathsf{E}_\alpha, \mathsf{M}_\alpha, \Theta, \Theta_{,\alpha} \quad \text{and} \quad \mathcal{Q} \,, \tag{3.4.28}$$

where $C_{\alpha\beta}$, E_α and M_α have already been introduced in (2.1.10) and (2.6.7). Defining the energy functional $\tilde{\psi}$ by (see (3.3.36))

$$\tilde{\psi} := U - \eta\Theta - \frac{P_i}{\rho}\mathcal{E}_i \,, \tag{3.4.29}$$

and using, as before, for the entropy flux the classical relation

$$\phi_i = \frac{q_i}{\Theta} \,,$$

we deduce from the reduced entropy inequality that $\tilde{\psi}$ can neither depend on $\Theta_{,\alpha}$ nor $\mathcal{Q}$; hence

$$\tilde{\psi} = \tilde{\psi}(C_{\alpha\beta}, \mathsf{E}_\alpha, \mathsf{M}_\alpha, \Theta) \,. \tag{3.4.30}$$

Moreover,

$$\eta = -\frac{\partial\tilde{\psi}}{\partial\Theta}\ ,$$

$$P_i = -\rho\frac{\partial\tilde{\psi}}{\partial\mathsf{E}_\alpha}F_{i\alpha}\ ,$$

$$\mathcal{H}_i = \frac{\partial\tilde{\psi}}{\partial\mathsf{M}_\alpha}F_{i\alpha}\mathrm{sgn}J\ ,$$

$$t_{ij} = 2\rho\frac{\partial\tilde{\psi}}{\partial C_{\alpha\beta}}F_{i\alpha}F_{j\beta} - \mathcal{E}_iP_j + \mu_0\mathcal{M}_i\mathcal{H}_j\ .$$

(3.4.31)

Of the reduced entropy inequality there remains the residual inequality

$$\mathcal{J}_i\mathcal{E}_i - \frac{\Theta_{,i}q_i}{\Theta} \geq 0\ ,$$

(3.4.32)

and the GIBBS relation becomes

$$dn = \frac{1}{\Theta}\left\{d[U - \mathsf{P}_\alpha\mathsf{E}_\beta C_{\alpha\beta}^{-1}]\right.$$
$$\left. -\frac{1}{2}\left[\frac{1}{\rho_0}T_{\alpha\beta}^P - (\mathsf{P}_\gamma\mathsf{E}_\delta + \mathsf{M}_\gamma\mathsf{H}_\delta)C_{\alpha\gamma}^{-1}C_{\beta\delta}^{-1}\right]dC_{\alpha\beta} + \mathsf{P}_\alpha d\mathsf{E}_\alpha - \mathsf{H}_\alpha d\mathsf{M}_\alpha\right\}\ ,$$

(3.4.33)

with definitions of E_α, H_α, P_α and M_α as given in (2.6.7).

We still wish to point out that with the constitutive equation $(3.4.31)_4$ the angular momentum equation (3.4.23) is satisfied identically.

At this point we have set up a complete theory, consisting of balance laws, constitutive equations and jump conditions, for the interactions of electromagnetic and thermoelastic fields in solids based on the MINKOWSKIan formulation of electrodynamics. However, there is a certain arbitrariness in the basic postulate of the theory, as manifest by the energy balance (3.4.19). As a result, the so called MAXWELL stress tensor (here denoted by t_{ij}^M) does not appear in a unique form. Therefore, we shall give here some other formulations for this global energy balance which all lead to seemingly different theories, which, however, in the end all turn out to be equivalent.

First, we wish to consider a MINKOWSKIan formulation as given by PENFIELD and HAUS ([177], Sect. 7.3). Taking as energy balance the expression

$$\frac{d}{dt}\int_V\left\{\rho U + \tfrac{1}{2}(\varepsilon_0\mathcal{E}_i\mathcal{E}_i + \mu_0\mathcal{H}_i\mathcal{H}_i) + \tfrac{1}{2}\rho\dot{x}_i\dot{x}_i\right\}dv$$
$$= \int_V\left\{\rho r^{\mathrm{ext}} + \rho F_i^{\mathrm{ext}}\dot{x}_i\right\}dv + \int_{\partial V}\left\{t_{ij}\dot{x}_i - q_j - \mathrm{e}_{jkl}\mathcal{E}_k\mathcal{H}_l\right.$$
$$\left. +(D_j\mathcal{E}_k\ |\ B_j\mathcal{H}_k)\dot{x}_k - (\mathcal{E}_kD_k + \mathcal{H}_kB_k)\dot{x}_j\right\}da_j\ ,$$

(3.4.34)

which differs from (3.4.15) or (3.4.19) only in the choice of R_i, we derive the local balance equations:

$$\rho \ddot{x}_i - t_{ij,j} - \rho F_i^{\text{ext}} = \rho F_i^{\text{e}}$$

$$\rho F_i^{\text{e}} = Q\mathcal{E}_i + e_{ijk}\mathcal{J}_j B_k - B_{j,i}\mathcal{H}_j - D_{j,i}\mathcal{E}_j + e_{ijk}(D_j \overset{\star}{B}_k + \overset{\star}{D}_j B_k)$$

$$= [\mathcal{E}_i D_j + \mathcal{H}_i B_j - \delta_{ij}(\mathcal{E}_k D_k + \mathcal{H}_k B_k)]_{,j} = t_{ij,j}^M \; , \tag{3.4.35}$$

$$t_{[ij]} = \rho L_{ij}^{\text{e}} = P_{[i}\mathcal{E}_{j]} + \mu_0 \mathcal{M}_{[i}\mathcal{H}_{j]} \; , \tag{3.4.36}$$

$$\frac{\partial W_M}{\partial t} + (W_M \dot{x}_i)_{,i} - t_{ij}\dot{x}_{i,j} + q_{i,i} - \rho r^{\text{ext}} = \rho r^{\text{e}} \tag{3.4.37}$$

$$\rho r^{\text{e}} = \mathcal{J}_i \mathcal{E}_i + \mathcal{E}_i \dot{D}_i + \mathcal{H}_i \dot{B}_i \; ,$$

where

$$W_M := \rho U + \tfrac{1}{2}(\varepsilon_0 \mathcal{E}_k \mathcal{E}_k + \mu_0 \mathcal{H}_k \mathcal{H}_k) \; . \tag{3.4.38}$$

We note that these balance laws equal those of PENFIELD and HAUS, if the latter are taken in the non-relativistic approximation (cf. eqs. (7.39)–(7.43) of [177]). Furthermore, one can easily show that the only difference with the previous formulation lies in the stress tensor. Indeed, when we require the momentum equations in the two formulations to be the same, what results is:

$$t_{ij}(3.4.35) - t_{ij}(3.4.22) = \delta_{ij}[D_k\mathcal{E}_k + B_k\mathcal{H}_k - \tfrac{1}{2}(\varepsilon_0 \mathcal{E}_k \mathcal{E}_k + \mu_0 \mathcal{H}_k \mathcal{H}_k)] \; . \tag{3.4.39}$$

Once this difference is taken into account the two systems based on (3.4.19) and (3.4.34) are equivalent.

Still another possible form of the energy balance is

$$\frac{d}{dt}\int_{\mathcal{V}} \left\{\rho U + \tfrac{1}{2}(\mathcal{E}_i D_i + \mathcal{H}_i B_i) + \tfrac{1}{2}\rho \dot{x}_i \dot{x}_i\right\}d\nu$$

$$= \int_{\mathcal{V}} \left\{\rho r^{\text{ext}} + \rho F_i^{\text{ext}}\dot{x}_i\right\}d\nu + \int_{\partial\mathcal{V}} \left\{t_{ij}\dot{x}_i - q_j - e_{jkl}\mathcal{E}_k\mathcal{H}_l \right. \tag{3.4.40}$$

$$\left. +(D_j\mathcal{E}_k + B_j\mathcal{H}_k)\dot{x}_k - \tfrac{1}{2}(\mathcal{E}_k D_k + \mathcal{H}_k B_k)\dot{x}_j\right\}da_j \; ,$$

and it leads to the following balance equations of momentum and energy

$$\rho \ddot{x}_i - t_{ij,j} - \rho F_i^{\text{ext}} = \rho F_i^{\text{e}}$$

$$\rho F_i^{\text{e}} = [\mathcal{E}_i D_j + \mathcal{H}_i B_j - \tfrac{1}{2}\delta_{ij}(\mathcal{E}_k D_k + \mathcal{H}_k B_k)]_{,j} \; , \tag{3.4.41}$$

and

$$\rho \dot{U} - t_{ij}\dot{x}_{i,j} + q_{j,j} - \rho r^{\text{ext}} = \rho r^{\text{e}}$$

$$\rho r^{\text{e}} = \mathcal{J}_i \mathcal{E}_i - \tfrac{1}{2}(\dot{\mathcal{E}}_i D_i - \mathcal{E}_i \dot{D}_i + \dot{\mathcal{H}}_i B_i - \mathcal{H}_i \dot{B}_i) \; . \tag{3.4.42}$$

These equations resemble in a way a formulation in which the expressions for the electromagnetic body force and energy supply are derived from a four-dimensional formulation of the MAXWELL equations, as outlined by Møller ([158], Ch. 7) and in the report of PAO, ([172], Sect. 6). However, there is one essential difference, as in (3.4.41) and (3.4.42) the rest-frame fields $\mathcal{E}_i$ and $\mathcal{H}_i$ are used, whereas in [158] and [172] the laboratory fields E_i and H_i are employed. Since the balance laws as presented in [158] or [172] are not invariant in the non-relativistic sense, they can never be deduced from an energy balance in the way described above. This makes it very questionable whether the formulation, discussed in [158], describes the interactions between matter and field in any meaningful way (see also [177], p. 202 and [172], pp. 121–122). It is not too difficult to show that the results corresponding to (3.4.40) and those based upon (3.4.19) are in correspondence provided that

$$t_{ij}(3.4.40) - t_{ij}(3.4.19) = \tfrac{1}{2}\delta_{ij}(\mathcal{E}_k P_k + \mathcal{H}_k \mathcal{M}_k) \,,$$

$$\rho U(3.4.40) - \rho U(3.4.19) = -\tfrac{1}{2}(\mathcal{E}_i P_i + \mathcal{H}_i \mathcal{M}_i) \,.$$

(3.4.43)

So far we have given three possible MINKOWSKI formulations of electromagnetoelastic interactions. Whenever in the sequel reference is made to the MAXWELL-MINKOWSKI formulation (model III), the first theory of this section is meant (i.e. system (3.4.21)–(3.4.24)).

We conclude by stating that the two theories outlined in the preceding section could, at least for the consistent non-relativistic part, also be based upon a global energy balance. The underlying energy balance for model I is (see also ALBLAS [9])

$$\frac{d}{dt}\int\limits_{\mathcal{V}} \{\rho U + \tfrac{1}{2}\rho\dot{x}_i\dot{x}_i + \tfrac{1}{2}(\varepsilon_0\mathcal{E}_k\mathcal{E}_k + \mu_0\mathcal{H}_k\mathcal{H}_k)\}d\nu$$

$$= \int\limits_{\mathcal{V}} \{\rho r^{\text{ext}} + \rho F_i^{\text{ext}}\dot{x}_i\}d\nu$$

$$+ \int\limits_{\partial\mathcal{V}} \{t_{ij}\dot{x}_i - q_j - e_{jkl}\mathcal{E}_k\mathcal{H}_l - \tfrac{1}{2}(\varepsilon_0\mathcal{E}_k\mathcal{E}_k + \mu_0\mathcal{H}_k\mathcal{H}_k)\dot{x}_j$$

$$+ (\varepsilon_0\mathcal{E}_j\mathcal{E}_k + \mu_0\mathcal{H}_j\mathcal{H}_k)\dot{x}_k + (P_j^C\mathcal{E}_k + \mu_0 M_j^C\mathcal{H}_k)\dot{x}_k\}da_j \,.$$

(3.4.44)

In the usual way this relation leads to expressions for $\rho F_i^e, \rho L_{ij}^e$ and ρr^e as given in (3.3.16), (3.3.7)$_2$ and (3.3.7)$_3$, respectively, except for the terms preceded by a c^{-2}-factor in (3.3.16).

In an analogous way, model II can be based upon the balance law

$$\frac{d}{dt}\int\limits_{\mathcal{V}} \{\rho U + \tfrac{1}{2}\rho\dot{x}_i\dot{x}_i + \tfrac{1}{2}(\varepsilon_0\mathcal{E}_k\mathcal{E}_k + \mu_0\mathcal{H}_k\mathcal{H}_k)\}d\nu$$

$$= \int\limits_{\mathcal{V}} \{\rho r^{\text{ext}} + \rho F_i^{\text{ext}}\dot{x}_i\}d\nu + \int\limits_{\partial\mathcal{V}} \{t_{ij}\dot{x}_i - q_i - e_{jkl}\mathcal{E}_k\mathcal{H}_l$$

$$- \tfrac{1}{2}(\varepsilon_0\mathcal{E}_k\mathcal{E}_k + \mu_0\mathcal{H}_k\mathcal{H}_k)\dot{x}_j + (\varepsilon_0\mathcal{E}_j\mathcal{E}_k + \mu_0\mathcal{H}_j\mathcal{H}_k)\dot{x}_k\}da_j \,.$$

(3.4.45)

In the following sections we shall show that energy balances of this kind can also serve as bases for the theories outlined there.

3.5 The Statistical Formulation (Model IV)

The proper physical approach toward a formulation of polarizable and magnetizable continua is through methods of statistical mechanics. Early attempts go back to ROSENFELDT [204]. A comprehensive treatment - in the light of relativistically covariant statistical mechanics - is given by DE GROOT and SUTTORP [53]. This book may also serve as guideline for a historical account on the subject. In this theory, matter consists of stable groups of electrically charged particles, such as electrons, ions etc. The field effect of these particles within each stable group is represented by electric and magnetic multipoles, the statistical averages of which give rise to the definition of electric polarization P_i and magnetization M_i.

With regard to the statistical derivation of the MAXWELL equations, DE GROOT and SUTTORP are entirely general and they do not introduce any non-relativistic approximations. As a result, the macroscopic MAXWELL equations are *proved* to be LORENTZ invariant, a property hitherto assumed to hold.

We introduce the statistical formulation again formally by transformation rules, viz.

$$B_i = B_i^S \ ,$$

$$\mathcal{E}_i = E_i^S + e_{ijk}\dot{x}_j B_k^S \ ,$$

$$D_i = \varepsilon_0 E_i^S + P_i^S \ , \tag{3.5.1}$$

$$\mathcal{H}_i = \frac{1}{\mu_0} B_i^S - e_{ijk}\dot{x}_j \varepsilon_0 E_k^S - M_i^S - e_{ijk}\dot{x}_j P_i^S \ .$$

Accordingly, the variables of the statistical formulation bear the superscript S. Substituting (3.5.1) into the MAXWELL equations (2.2.15) and (2.2.21), we obtain

$$B_{i,i}^S = 0 \ ,$$

$$e_{ijk}E_{k,j}^S + \frac{\partial B_i^S}{\partial t} = 0 \ ,$$

$$\varepsilon_0 E_{i,i}^S = \mathcal{Q} - P_{i,i}^S \ , \tag{3.5.2}$$

$$\frac{1}{\mu_0} e_{ijk}B_{k,j}^S - \varepsilon_0 \frac{\partial E_i^S}{\partial t} = J_i + \frac{\partial P_i^S}{\partial t} + e_{ijk}M_{k,j}^S \ .$$

Incidentally, PENFIELD and HAUS [177] call this formulation the BOFFI formulation and, as a comparison with (3.4.2) and (3.4.3) shows, these variables agree with those introduced in the preceding section as the MINKOWSKI variables. The equations (3.5.2) are obtained also when the action of matter upon the electromagnetic fields is derived from very crude physical models of stationary rigid bodies. Indeed they can be found in almost any physics book treating electromagnetism (see e.g. FEYNMAN [74]). They hold, however, for a much broader class of physical processes than anticipated in many of these books and embrace all those for which the action of matter on the electromagnetic fields has been discussed above.

The expressions for the electromagnetic force, couple and energy supply have been derived by DE GROOT and SUTTORP in full generality, including relativistic effects. However, they also present two approximate versions, differing in the degree of approximation, which in our terminology are called the non-relativistic and the semi-relativistic approximations. Here, we shall discuss the non-relativistic case only.

In their second chapter, DE GROOT and SUTTORP arrive at the following non-relativistic expressions for body force, body couple and energy supply (cf. [53], pp. 47, 63)

$$
\begin{aligned}
\rho F_i^e &= Q E_i^S + e_{ijk} J_j B_k^S + P_j^S E_{j,i}^S + M_j^S B_{j,i}^S + \rho e_{ijk} \frac{d}{dt}\left(\frac{1}{\rho} P_j^S B_k^S\right) \\
&= Q \mathcal{E}_i + e_{ijk} \mathcal{J}_j B_k + P_j \mathcal{E}_{j,i} + \mathcal{M}_j B_{j,i} + e_{ijk}(\overset{*}{P}_j B_k + P_j \overset{*}{B}_k) , \\
\rho L_{ij}^e &= P_{[i}\mathcal{E}_{j]} + \mathcal{M}_{[i}B_{j]} , \\
\rho r^e &= \mathcal{J}_i \mathcal{E}_i + \rho \mathcal{E}_i \frac{d}{dt}\left(\frac{P_i}{\rho}\right) - \mathcal{M}_i \dot{B}_i ,
\end{aligned}
\tag{3.5.3}
$$

where

$$
B_i = B_i^S , \qquad P_i = P_i^S \qquad \text{and} \qquad \mathcal{M}_i = M_i^S + e_{ijk}\dot{x}_j P_k^S . \tag{3.5.4}
$$

Thus, the conservation of mass, the balance laws of linear and angular momenta and the balance of energy become

$$
\begin{aligned}
\dot{\rho} + \rho \dot{x}_{i,i} &= 0 , \\
\rho \ddot{x}_i &= t_{ij,j} + \rho F_i^{ext} + Q\mathcal{E}_i + e_{ijk}\mathcal{J}_j B_k + P_j \mathcal{E}_{j,i} \\
&\quad + \mathcal{M}_j B_{j,i} + e_{ijk}(\overset{*}{P}_j B_k + P_j \overset{*}{B}_k) , \\
t_{[ij]} &= P_{[i}\mathcal{E}_{j]} + \mathcal{M}_{[i}B_{j]} , \\
\rho \dot{U} &= t_{ij}\dot{x}_{i,j} - q_{i,i} + \mathcal{J}_i \mathcal{E}_i + \rho \mathcal{E}_i \frac{d}{dt}\left(\frac{P_i}{\rho}\right) - \mathcal{M}_i \dot{B}_i + \rho r^{ext} .
\end{aligned}
\tag{3.5.5}
$$

Since, as already said before, the statistical fields are identical to the MINKOWSKI fields, we note that

$$
Q, \mathcal{J}_i, \mathcal{E}_i, P_i, B_i \quad \text{and} \quad \mathcal{M}_i
$$

transform under EUCLIDian transformations as an objective scalar, as objective vectors and as objective axial vectors, respectively. Consequently, the electromagnetic body force and body couple, as given by the above expressions, are an objective vector and a skew-symmetric tensor, respectively. Of course, this holds true also in the non-relativistic sense as defined in Sect. 1.6. As far as the electromagnetic energy supply is concerned, we should note, as we already have seen in Sect. 2.2.1, that ρr^e cannot be an objective scalar, because the stress tensor is not symmetric in this formulation. We should rather look at the expression

$$\rho r^e + t_{[ij]}\dot{x}_{i,j} \,,$$

which, according to $(3.5.3)_3$, $(3.5.5)_3$ and $(3.5.2)$, equals

$$
\begin{aligned}
\rho r^e + t_{[ij]}\dot{x}_{i,j} &= \mathcal{J}_i \mathcal{E}_i + \mathcal{E}_i(\overset{\star}{P}_i + e_{ijk}\mathcal{M}_{k,j}) + (e_{ijk}\mathcal{E}_j \mathcal{M}_k)_{,i} \\
&\quad + (\mathcal{E}_i P_j - \mathcal{M}_i B_j + \delta_{ij}\mathcal{M}_k B_k)\dot{x}_{(i,j)} \,.
\end{aligned}
\tag{3.5.6}
$$

Each term on the right-hand side of (3.5.6) is obviously an objective scalar. Hence, the balance equations are invariant under the EUCLIDian group. However, these laws are not invariant under LORENTZ transformations, not even in the semi-relativistic sense.

As the momentum equation and energy equation are invariant under EUCLIDian transformations it may be expected that the results given above are also derivable from a global energy balance as derived in Sect. 2.3. Indeed this is true and the underlying energy balance reads

$$
\begin{aligned}
\frac{d}{dt}\int_{\mathcal{V}} &\left\{\rho U + \tfrac{1}{2}\left(\varepsilon_0 \mathcal{E}_i \mathcal{E}_i + \frac{1}{\mu_0}B_i B_i\right) + \tfrac{1}{2}\rho \dot{x}_i \dot{x}_i\right\}d\nu \\
&= \int_{\mathcal{V}}\left\{\rho r^{\text{ext}} + \rho F_i^{\text{ext}}\dot{x}_i\right\}d\nu + \int_{\partial\mathcal{V}}\left\{t_{ij}\dot{x}_i - q_j - e_{jkl}\mathcal{E}_k\left(\frac{1}{\mu_0}B_l - \mathcal{M}_l\right)\right. \\
&\quad -\tfrac{1}{2}\left(\varepsilon_0 \mathcal{E}_k \mathcal{E}_k + \frac{1}{\mu_0}B_k B_k\right)\dot{x}_j + \left(\varepsilon_0 \mathcal{E}_j \mathcal{E}_k + \frac{1}{\mu_0}B_j B_k\right)\dot{x}_k \\
&\quad \left.+(P_j \mathcal{E}_k - B_j \mathcal{M}_k + \delta_{jk}\mathcal{M}_l B_l)\dot{x}_k\right\}da_j \,.
\end{aligned}
\tag{3.5.7}
$$

Using an approach analogous to that outlined in Sect. 2.3 and invoking the invariance requirement that (3.5.7) is EUCLIDian invariant, one can show that the above energy balance law implies the local balance equations (3.5.5).

The electromagnetic body force and energy supply can also be expressed in the form (2.4.7). Indeed, straightforward calculations show that

$$^{IV}t_{ij}^M = \varepsilon_0 E_i^S E_j^S + \frac{1}{\mu_0} B_i B_j + E_i^S P_j - M_i^S B_j$$

$$-\tfrac{1}{2}\delta_{ij}\left(\varepsilon_0 E_k^S E_k^S + \frac{1}{\mu_0} B_k B_k - 2B_k M_k^S\right) + e_{ikl}P_k B_l \dot{x}_j$$

$$= \varepsilon_0 \mathcal{E}_i \mathcal{E}_j + \frac{1}{\mu_0} B_i B_j + \mathcal{E}_i P_j - \mathcal{M}_i B_j$$

$$-\tfrac{1}{2}\delta_{ij}\left(\varepsilon_0 \mathcal{E}_k \mathcal{E}_k + \frac{1}{\mu_0} B_k B_k - 2\mathcal{M}_k B_k\right)\,,$$

$$^{IV}g_i = 0\,,$$

$$^{IV}\pi_i = -e_{ijk}E_j^S\left(\frac{1}{\mu_0}B_k - M_k^S\right) + E_j^S P_j \dot{x}_i$$

$$= -e_{ijk}\mathcal{E}_j\left(\frac{1}{\mu_0}B_k - \mathcal{M}_k\right) - \left(\varepsilon_0 \mathcal{E}_j \mathcal{E}_j + \frac{1}{\mu_0}B_j B_j\right)\dot{x}_i$$

$$+ \left(\varepsilon_0 \mathcal{E}_i \mathcal{E}_j + \frac{1}{\mu_0}B_i B_j\right)\dot{x}_j + \left(P_i \mathcal{E}_j - B_i \mathcal{M}_j + \delta_{ij}\mathcal{M}_k B_k\right)\dot{x}_j\,,$$

$$^{IV}\omega = -\tfrac{1}{2}\left(\varepsilon_0 E_k^S E_k^S + \frac{1}{\mu_0}B_k B_k\right) = -\tfrac{1}{2}\left(\varepsilon_0 \mathcal{E}_k \mathcal{E}_k + \frac{1}{\mu_0}B_k B_k\right)\,.$$

$$(3.5.8)$$

We note that in the above calculations $(\varepsilon_0 \boldsymbol{E} \times \boldsymbol{B})$-terms are neglected, which, according to the statement made in the preceding section between (3.4.12) and (3.4.13), is consistent with our requirements.

The jump conditions for the fields, mass, momentum and energy of matter and fields become now

$$[\![B_i]\!]n_i = 0\,, \qquad [\![\varepsilon_0 E_i^S + P_i]\!]n_i = 0\,,$$

$$[\![e_{ijk}E_j^S]\!]n_k + [\![B_i]\!]\mathrm{w}_n = 0\,,$$

$$\left[\!\left[e_{ijk}\left(\frac{1}{\mu_0}B_j - M_j^S\right)\right]\!\right]n_k - [\![\varepsilon_0 E_i^S + P_i^S]\!]\mathrm{w}_n = 0\,,$$

$$[\![\rho(\dot{x}_i n_i - \mathrm{w}_n)]\!] = 0\,,$$

$$\left[\!\left[t_{ij} + \varepsilon_0 \mathcal{E}_i \mathcal{E}_j + \frac{1}{\mu_0}B_i B_j + \mathcal{E}_i P_j - \mathcal{M}_i B_j\right.\right.$$

$$\left.\left.-\tfrac{1}{2}\delta_{ij}\left(\varepsilon_0 \mathcal{E}_k \mathcal{E}_k + \frac{1}{\mu_0}B_k B_k - 2\mathcal{M}_k B_k\right)\right]\!\right]n_j \qquad (3.5.9)$$

$$-[\![\rho \dot{x}_i(\dot{x}_j n_j - \mathrm{w}_n)]\!] = 0\,,$$

$$\left[\!\left[\dot{x}_i t_{ij} - q_j - e_{jkl}\mathcal{E}_k\left(\frac{1}{\mu_0}B_l - \mathcal{M}_l\right) - \tfrac{1}{2}\left(\varepsilon_0 \mathcal{E}_k \mathcal{E}_k + \frac{1}{\mu_0}B_k B_k\right)\dot{x}_j\right.\right.$$

$$\left.\left.+ \left(\varepsilon_0 \mathcal{E}_i \mathcal{E}_j + \frac{1}{\mu_0}B_i B_j\right)\dot{x}_i + (\mathcal{E}_i P_j - \mathcal{M}_i B_j + \delta_{ij}\mathcal{M}_k B_k)\dot{x}_i\right]\!\right]n_j$$

$$-\left[\!\left[\left\{\tfrac{1}{2}\rho\dot{x}_i\dot{x}_i + \rho U + \tfrac{1}{2}\left(\varepsilon_0 \mathcal{E}_i \mathcal{E}_i + \frac{1}{\mu_0}B_i B_i\right)\right\}(\dot{x}_j n_j - \mathrm{w}_n)\right]\!\right] = 0\,.$$

It remains to develop the constitutive theory. For this purpose, we notice that, in a non-relativistic theory, the invariance group which the principle of material frame indifference relies upon, is the GALILEI, or more generally, the EUCLIDian, group. Hence, we may choose the constitutive relations of the form derived in Sect. 1.6. In this section we restrict ourselves to the case

$$\mathsf{C} = \overset{+}{\mathsf{C}}\,(C_{\alpha\beta}, \mathsf{P}_\alpha, \mathsf{B}_\alpha, \Theta, \Theta_{,\alpha}, \mathcal{Q})\,, \tag{3.5.10}$$

where P_α and B_α are defined in (2.6.7). The application of the entropy principle is again a routine matter and we simply state the results. With the HELMHOLTZ free energy

$$\psi = U - \eta\Theta\,, \tag{3.5.11}$$

and with (3.3.26), one finds that

$$\psi = \overset{+}{\psi}\,(C_{\alpha\beta}, \mathsf{P}_\alpha, \mathsf{B}_\alpha, \Theta)\,, \tag{3.5.12}$$

while the entropy η, the electromotive intensity $\mathcal{E}_i$, the rest-frame magnetization $\mathcal{M}_i$ and the stress t_{ij} are given by

$$\eta = -\frac{\partial \overset{+}{\psi}}{\partial \Theta}\,,$$

$$\mathcal{E}_i = \frac{\partial \overset{+}{\psi}}{\partial \mathsf{P}_\alpha} F_{i\alpha}\,,$$

$$\mu_0 \mathcal{M}_i = -\rho \frac{\partial \overset{+}{\psi}}{\partial \mathsf{B}_\alpha} F_{i\alpha}\mathrm{sgn} J\,, \tag{3.5.13}$$

$$t_{ij} = 2\rho \frac{\partial \overset{+}{\psi}}{\partial C_{\alpha\beta}} F_{i\alpha} F_{j\beta} + P_i \mathcal{E}_j - B_i \mathcal{M}_j\,.$$

These results imply that the balance law of moment of momentum is satisfied identically. Furthermore, the reduced entropy inequality is

$$\mathcal{J}_i \mathcal{E}_i - \frac{\Theta_{,i} q_i}{\Theta} \geq 0\,, \tag{3.5.14}$$

and the GIBBS relation becomes

$$d\eta = \frac{1}{\Theta}\Big\{ dU - \frac{1}{2}\Big[\frac{1}{\rho_0}T^P_{\alpha\beta} + C^{-1}_{\alpha\gamma}C^{-1}_{\beta\delta}(\mathsf{P}_\gamma \mathsf{E}_\delta - \mathsf{B}_\gamma \mathsf{M}_\delta)\Big] dC_{\alpha\beta}$$
$$- C^{-1}_{\alpha\beta}\mathsf{E}_\alpha d\mathsf{P}_\beta - C^{-1}_{\alpha\beta}\mathsf{M}_\alpha d\mathsf{B}_\beta \Big\}\,, \tag{3.5.15}$$

where E_α and M_α are given in (2.6.7).

This completes the non-relativistic theory of magnetizable and polarizable solids in the statistical description.

3.6 The Lorentz Formulation (Model V)

The LORENTZ description of electromagnetism is founded on his theory of electrons, originally formulated for dielectric materials only, see [244]. According to this description, the body is supposed to consist of a set of electrically interfering charged particles. These particles respond to their own fields (as well as to possible external fields) and may move rapidly, thereby producing highly fluctuating (microscopic) electromagnetic fields as well. The pertinent field equations are the MAXWELL equations in a vacuum and when averaged, the MAXWELL equations in the LORENTZ formulation emerge (see for instance the booklet by L.ROSENFELDT [204]). In the light of DE GROOT and SUTTORP's statistical description this average is a crude statistical model. This might lead the reader to the conclusion that the LORENTZ formulation is only approximate. This is not so. On the contrary, this theory has been put on a sound, relativistically correct, axiomatics by TRUESDELL and TOUPIN [244]. (For a non-relativistic presentation, though nevertheless relativistically correct formulation, see the book by MÜLLER [160].) That it may be defined also from any set of MAXWELL equations by simply performing a variable transformation should be corroboration of its correctness.

This is exactly what we shall do here. Indeed, we introduce the LORENTZ formulation by using the definitions

$$D_i = \varepsilon_0 E_i^L + P_i^L \,, \qquad \mathcal{E}_i = E_i^L + e_{ijk}\dot{x}_j B_k^L \,,$$

$$B_i = B_i^L \,, \qquad \mathcal{H}_i = \frac{1}{\mu_0} B_i^L - M_i^L - \varepsilon_0 e_{ijk}\dot{x}_j E_k^L \,. \tag{3.6.1}$$

Upon substitution of (3.6.1) into (2.2.15) and (2.2.21), the MAXWELL equations in the LORENTZ formulation are obtained as follows:

$$B_{i,i}^L = 0 \,,$$

$$e_{ijk}E_{k,j}^L + \frac{\partial B_i^L}{\partial t} = 0 \,,$$

$$\varepsilon_0 E_{i,i}^L = \mathcal{Q} - P_{i,i}^L \,, \tag{3.6.2}$$

$$\frac{1}{\mu_0} e_{ijk}B_{k,j}^L - \varepsilon_0 \frac{\partial E_i^L}{\partial t} = J_i + \frac{\partial P_i^L}{\partial t} + e_{ijk}(e_{klm}P_l^L \dot{x}_m)_{,j} + e_{ijk}M_{k,j}^L$$

These equations, introduced here formally, are often applied in the modern literature, as for instance by TOUPIN [241], LIU and MÜLLER [127], BENACH and MÜLLER [25] and HUTTER [92, 95, 96]. The equations are also mentioned by PAO [172], but he states "that for magnetizable materials their validity is questionable".

Comparing (3.6.2) with the MAXWELL equations of the statistical model (3.5.2), we see that

$$E_i^L = E_i^S \, , \qquad\qquad B_i^L = B_i^S = B_i \, ,$$

$$P_i^L = P_i^S = P_i \, , \qquad M_i^L = M_i^S + \mathrm{e}_{ijk}\dot{x}_j P_k^L = \mathcal{M}_i \, .$$

(3.6.3)

The MAXWELL equations in the LORENTZ description can also be deduced from the assumption that every particle is equipped with a number of electric dipoles and with an electric circuit. If this dipole-circuit model is treated non-relativistically, again the MAXWELL equations in the LORENTZ description are obtained. Of course, such a derivation bears inherently the notion of approximation. What is approximated thereby is not the theory as such, however, but the use of the model. That the theory is indeed correct from a relativistic point of view cannot be seen from such a derivation, and this must be regarded as a major disadvantage of the model. This might also be the reason why PENFIELD and HAUS use a model in which each particle is equipped with a number of dipoles and an electric circuit both of which are treated relativistically. What they obtain is different from the LORENTZ description. They call these MAXWELL equations the AMPÈREan model. The difference between the AMPÈREan and the LORENTZean variables is a small difference in the polarization vector of the amount

$$c^{-2}(\dot{\boldsymbol{x}} \times \boldsymbol{M}) \, .$$

In a non-relativistic theory this term is negligible, and it emerges from the fact that in the LORENTZ description

$$P_i^L = P_i^M \, ,$$

while in the AMPÈREan description

$$P_i^A = P_i^C \, ,$$

Nevertheless, both the AMPÈREan and the LORENTZ formulation are relativistically correct. This should make it clear that the approximations must be sought in the model rather than in the basic theory.

With this digression on approximations we proceed and mention that in the LORENTZ formulation

$$- P_{i,i}^L$$

is the charge density due to polarization,

$$\mathrm{e}_{ijk} M_{k,j}^L$$

the current due to magnetization and

$$\frac{\partial P_i^L}{\partial t} + \mathrm{e}_{ijk}(\mathrm{e}_{klm} P_l^L \dot{x}_m)_{,j}$$

the polarization current, all as shown on the right-hand side of (3.6.2). This suggests to postulate as electromagnetic body force, body couple and energy supply the expressions

$$\rho F_i^{\mathrm{e}} = \mathcal{Q}^{\mathrm{tot}} E_i^L + \mathrm{e}_{ijk} J_j^{\mathrm{tot}} B_k = \mathcal{Q}^{\mathrm{tot}} \mathcal{E}_i + \mathrm{e}_{ijk} \mathcal{J}_j^{\mathrm{tot}} B_k \ ,$$

$$\rho L_{ij}^{\mathrm{e}} = 0 \ , \tag{3.6.4}$$

$$\rho r^{\mathrm{e}} = \mathcal{J}_i^{\mathrm{tot}} \mathcal{E}_i J_i^{\mathrm{tot}} E_i^L - \rho F_i^{\mathrm{e}} \dot{x}_i \ ,$$

where $\mathcal{Q}^{\mathrm{tot}}$ and J_i^{tot} are the total charge and total current densities as given on the right-hand side of $(3.6.2)_{3,4}$. (Compare this postulate with that for model II, $(3.3.46)$). The postulates $(3.6.4)$ have, in this generality, first been introduced by TOUPIN [241]. They were then applied by LIU and MÜLLER [127], BENACH and MÜLLER [25], and HUTTER [92] in theories of polarizable and magnetizable fluids, fluid mixtures and solids, respectively. Substituting $\mathcal{Q}^{\mathrm{tot}}$ and J_i^{tot}, as obtained from $(3.6.2)_{3,4}$, into $(3.6.4)$ yields

$$\rho F_i^{\mathrm{e}} = (\mathcal{Q} - P_{j,j})\mathcal{E}_i + \mathrm{e}_{ijk}(\mathcal{J}_j + \overset{\star}{P}_j + \mathrm{e}_{jmn}\mathcal{M}_{n,m})B_k \ ,$$

$$\rho L_{ij}^{\mathrm{e}} = 0 \ , \tag{3.6.5}$$

$$\rho r^{\mathrm{e}} = \mathcal{J}_i \mathcal{E}_i + \mathcal{E}_i(\overset{\star}{P}_i + \mathrm{e}_{ijk}\mathcal{M}_{k,j}) \ ,$$

where the convective derivative, denoted by a superimposed star, is defined in $(2.2.17)$.

With $(3.6.5)$ the balance laws of mass, momentum and energy assume the form

$$\dot{\rho} + \rho \dot{x}_{i,i} = 0 \ ,$$

$$\rho \ddot{x}_i = t_{ij,j} + \rho F_i^{\mathrm{ext}} + (\mathcal{Q} - P_{j,j})\mathcal{E}_i + \mathrm{e}_{ijk}(\mathcal{J}_j + \overset{\star}{P}_j + \mathrm{e}_{jmn}\mathcal{M}_{n,m})B_k \ ,$$

$$t_{[ij]} = 0 \ , \tag{3.6.6}$$

$$\rho \dot{U} = t_{ij}\dot{x}_{i,j} - q_{i,i} + \mathcal{J}_i \mathcal{E}_i + \mathcal{E}_i(\overset{\star}{P}_i + \mathrm{e}_{ijk}\mathcal{M}_{k,j}) + \rho r^{\mathrm{ext}} \ .$$

The expressions $(3.6.5)$ immediately show that ρr^{e} and ρF_i^{e} are an objective scalar and objective vector under the EUCLIDian transformation group. For this to be true, we have used that $\mathcal{Q}$, $\mathcal{J}_i$, $\mathcal{E}_i$, B_i, P_i and $\mathcal{M}_i$ are objective quantities, as we have already seen in the preceding sections. A proof of this can also be found in TRUESDELL and TOUPIN [244], Chap. 7. Hence, the momentum and energy equations enjoy the classical (non-relativistic) invariance requirements. The energy equation is, however, only invariant under EUCLIDian transformations, because we have also taken the stress tensor to be symmetric. This may serve as a justification for the rather unmotivated choice of $\rho L_{ij}^{\mathrm{e}} = 0$.

Just as in the preceding sections, this formulation can also be derived from a global energy balance which in this case reads

$$\frac{d}{dt} \int_{\mathcal{V}} \left\{ \rho U + \tfrac{1}{2}\rho \dot{x}_i \dot{x}_i + \tfrac{1}{2} \left(\varepsilon_0 \mathcal{E}_k \mathcal{E}_k + \frac{1}{\mu_0} B_k B_k \right) \right\} d\nu$$

$$= \int_{\mathcal{V}} \left\{ \rho r^{\mathrm{ext}} + \rho F_i^{\mathrm{ext}} \dot{x}_i \right\} d\nu \qquad (3.6.7)$$

$$+ \int_{\partial \mathcal{V}} \left\{ t_{ij} \dot{x}_i - q_j - e_{jkl} E_k \frac{B_l}{\mu_0} + \tfrac{1}{2} \left(\varepsilon_0 \mathcal{E}_k \mathcal{E}_k + \frac{1}{\mu_0} B_k B_k \right) \dot{x}_j \right\} da_j \, .$$

Furthermore, it may be shown that the expressions (3.6.5) can be written in the form (2.4.7) (with the non-relativistic approximations), yielding

$$^V t_{ij}^M = \varepsilon_0 \mathcal{E}_i \mathcal{E}_j + \frac{1}{\mu_0} B_i B_j - \tfrac{1}{2}\delta_{ij} \left(\varepsilon_0 \mathcal{E}_k \mathcal{E}_k + \frac{1}{\mu_0} B_k B_k \right) \, ,$$

$$^V g_i = 0 \, ,$$

$$^V \pi_i = -e_{ijk} E_j^L \frac{B_k}{\mu_0} = -e_{ijk} \mathcal{E}_j (\mathcal{H}_k + \mathcal{M}_k) \qquad (3.6.8)$$

$$- \left(\varepsilon_0 \mathcal{E}_k \mathcal{E}_k + \frac{1}{\mu_0} B_k B_k \right) \dot{x}_i + \left(\varepsilon_0 \mathcal{E}_i \mathcal{E}_j + \frac{1}{\mu_0} B_i B_j \right) \dot{x}_j \, ,$$

$$^V \omega = -\frac{1}{2} \left(\varepsilon_0 \mathcal{E}_k \mathcal{E}_k + \frac{1}{\mu_0} B_k B_k \right) \, .$$

With these relations, the jump conditions for momentum and energy of matter and fields can be derived. We list them below, together with the jump conditions for the electromagnetic fields and the density

$$[\![B_i]\!] n_i = 0 \, , \qquad [\![\varepsilon_0 E_i^L + P_i]\!] n_i = 0 \, ,$$

$$[\![e_{ijk} E_j^L]\!] n_k + [\![B_i]\!] \mathrm{w}_n = 0 \, ,$$

$$\left[\!\left[e_{ijk} \left(\frac{1}{\mu_0} B_j - \mathcal{M}_j + e_{jlm} \dot{x}_l P_m \right) \right]\!\right] n_k - [\![\varepsilon_0 E_i^L + P_i]\!] \mathrm{w}_n = 0 \, ,$$

$$[\![\rho(\dot{x}_i n_i - \mathrm{w}_n)]\!] = 0 \, ,$$

$$\left[\!\left[t_{ij} + \varepsilon_0 \mathcal{E}_i \mathcal{E}_j + \frac{1}{\mu_0} B_i B_j - \tfrac{1}{2}\delta_{ij} \left(\varepsilon_0 \mathcal{E}_k \mathcal{E}_k + \frac{1}{\mu_0} B_k B_k \right) \right]\!\right] n_j$$
$$- [\![\rho \dot{x}_i (\dot{x}_j n_j - \mathrm{w}_n)]\!] = 0 \, ,$$

$$\left[\!\left[t_{ij} \dot{x}_i - q_j - e_{jkl} \mathcal{E}_k \frac{B_l}{\mu_0} - \tfrac{1}{2} \left(\varepsilon_0 \mathcal{E}_k \mathcal{E}_k + \frac{1}{\mu_0} B_k B_k \right) \dot{x}_j \right.\right.$$
$$\left.\left. + \left(\varepsilon_0 \mathcal{E}_i \mathcal{E}_j + \frac{1}{\mu_0} B_i B_j \right) \dot{x}_i \right]\!\right] n_j$$
$$- \left[\!\left[\left\{ \tfrac{1}{2}\rho \dot{x}_i \dot{x}_i + \rho U + \tfrac{1}{2} \left(\varepsilon_0 \mathcal{E}_k \mathcal{E}_k + \frac{1}{\mu_0} B_k B_k \right) \right\} (\dot{x}_j n_j - \mathrm{w}_n) \right]\!\right] = 0 \, .$$

$$(3.6.9)$$

We wish to note that the jump conditions $(3.6.9)_{6,7}$ and all corresponding ones of the preceding sections, could have been derived from the respective global energy balance laws. To illustrate this, note that from the global energy balance $(3.6.7)$ a jump condition can be derived as demonstrated in Sect. 1.5. This jump condition (given by $(3.6.9)_7$) is then subjected to a rigid-body translation of the form

$$x_i \Rightarrow x_i - b_i(t) \ ,$$

under which $(\dot{x}_j n_j - \mathrm{w}_n)$ remains invariant. Since under such transformations

$$- \mathrm{e}_{ijk} E_j^L B_k / \mu_0$$

changes into (recall that although B_i^L is invariant, B_i^L/μ_0 is not (see Sect. 1.6))

$$- \mathrm{e}_{ijk} E_j^L \frac{B_k}{\mu_0} - \left\{ \left(\varepsilon_0 \mathcal{E}_i \mathcal{E}_j + \frac{1}{\mu_0} B_i B_j \right) - \delta_{ij} \left(\varepsilon_0 \mathcal{E}_k \mathcal{E}_k + \frac{1}{\mu_0} B_k B_k \right) \right\} b_i \ ,$$

$(3.6.9)_7$ then immediately leads to $(3.6.9)_6$.

We now turn our attention to the constitutive theory and for that purpose it is advantageous to transform the energy balance equation $(3.6.6)_4$ into the form

$$\rho \dot{U} = t_{ij} \dot{x}_{i,j} - q_{i,i} + \mathcal{J}_i \mathcal{E}_i - (\mathrm{e}_{ijk} \mathcal{E}_j \mathcal{M}_k)_{,i} + \rho \mathcal{E}_i \frac{d}{dt} \left(\frac{P_i}{\rho} \right)$$
$$- \mathcal{M}_i \dot{B}_i - (\mathcal{E}_i P_j - \mathcal{M}_i B_j + \mathcal{M}_k B_k \delta_{ij}) \dot{x}_{i,j} + \rho r^{\mathrm{ext}} \ . \tag{3.6.10}$$

Then, elimination of ρr^{ext} from $(2.3.6)$ and introduction of the HELMHOLTZ free energy by

$$\psi = U - \eta \Theta \ ,$$

reveals the inequality

$$- \rho \dot{\psi} - \rho \Theta \dot{\eta} + \rho \mathcal{E}_i \frac{d}{dt} \left(\frac{P_i}{\rho} \right) - \mathcal{M}_i \dot{B}_i + \{ t_{ij} - \mathcal{E}_i P_j + \mathcal{M}_i B_j$$
$$- \mathcal{M}_k B_k \delta_{ij} \} \dot{x}_{i,j} + \mathcal{J}_i \mathcal{E}_i + [\Theta \phi_{,i} - q_{i,i} - (\mathrm{e}_{ijk} \mathcal{E}_j \mathcal{M}_k)_{,i}] \geq 0 \ . \tag{3.6.11}$$

Before we proceed, a comment seems to be in order regarding the last, bracketed, term of this inequality. Evidently, there is no unique definition of energy flux and entropy flux in terms of thermodynamic and electromagnetic variables. To see this, let us introduce a new energy flux vector q_i^S by

$$q_i^S := q_i + \mathrm{e}_{ijk} \mathcal{E}_j \mathcal{M}_k \ . \tag{3.6.12}$$

As will be shown in the next chapter, q_i^S agrees with the energy flux vector of the statistical model IV. With the choice

$$\phi_i = q_i^S/\Theta \;, \tag{3.6.13}$$

the bracketed term in (3.6.11) then becomes $[-\Theta_{,i} q_i^S/\Theta]$, and this is what one would expect classically, if one insisted to call q_i^S *the* heat flux vector. However, this is a purely formal definition, and no argument whatsoever justifies such an identification. Moreover, if we substitute (3.6.12) into the classical relation for the entropy flux (3.3.26), we find

$$\phi_i = \frac{q_i}{\Theta} + \frac{1}{\Theta} e_{ijk} \mathcal{E}_j \mathcal{M}_k \;, \tag{3.6.14}$$

and, if we insist to call q_i the heat flux, this equation teaches us that in this case the entropy flux does not obey the classical relation, namely heat flux divided by absolute temperature. This is the reason why we prefer to call q_i the energy flux vector. Moreover, this shows that it is imperative to refrain from setting the entropy flux equal to the heat flux divided by absolute temperature from the outset, because it is not evident what should be understood under heat flux. Instead, one must prescribe ϕ_i as a general constitutive variable, whose form should be determined in due course with the exploitation of the entropy inequality. For one particular theory of the complexity treated here, the relation

$$\phi_i = \frac{q_i^S}{\Theta} = \frac{q_i + e_{ijk} \mathcal{E}_j \mathcal{M}_k}{\Theta} \;, \tag{3.6.15}$$

was proved by HUTTER [92], and the same relation was also found earlier by LIU and MÜLLER [127] for the case of a simple fluid. For the purpose of this monograph we shall regard (3.6.15) as a postulate.

As was done for the other models, we now establish constitutive relations, which in view of (3.6.11) are assumed in the form

$$\mathsf{C} = \overset{+}{\mathsf{C}} \left(F_{i\alpha}, P_i/\rho, B_i, \Theta, \Theta_{,i}, \mathcal{Q} \right) \;, \tag{3.6.16}$$

where C denotes the set $\{U, \psi, \eta, \mathcal{E}_i, \mu_0 \mathcal{M}_i/\rho, q_i, \mathcal{J}_i, t_{ij}\}$. Moreover, we require the constitutive relations to be invariant under the EUCLIDian group of transformations. In view of the transformation rules listed before this is already ascertained.

In the usual way we can derive the following results:

$$\psi = \overset{+}{\psi} \left(C_{\alpha\beta}, \mathsf{P}_\alpha, \mathsf{B}_\alpha, \Theta \right) \;, \tag{3.6.17}$$

and

$$\eta = -\frac{\partial \overset{+}{\psi}}{\partial \Theta} \;,$$

$$\mathcal{E}_i = \frac{\partial \overset{+}{\psi}}{\partial \mathsf{P}_\alpha} F_{i\alpha} \;, \tag{3.6.18}$$

$$\frac{\mu_0 \mathcal{M}_i}{\rho} = -\frac{\partial \overset{+}{\psi}}{\partial \mathsf{B}_\alpha} F_{i\alpha} \mathrm{sgn} J \;,$$

$$t_{ij} = 2\rho \frac{\partial \overset{+}{\psi}}{\partial C_{\alpha\beta}} F_{i\alpha}F_{j\beta} + (P_i\mathcal{E}_j + P_j\mathcal{E}_i) - (\mathcal{M}_iB_j + \mathcal{M}_jB_i) + \delta_{ij}\mathcal{M}_kB_k \ .$$

The reduced entropy balance then reads

$$\mathcal{J}_i\mathcal{E}_i - \frac{\Theta_{,i}}{\Theta}(q_i + e_{ijk}\mathcal{E}_j\mathcal{M}_k) \geq 0 \ , \tag{3.6.19}$$

and the GIBBS relation becomes

$$
\begin{aligned}
d\eta = \frac{1}{\Theta} \Bigg\{ & dU - \frac{1}{2} \bigg[\frac{1}{\rho_0}T^P_{\alpha\beta} - C^{-1}_{\alpha\gamma}C^{-1}_{\beta\delta}(\mathsf{P}_\gamma\mathsf{E}_\delta + \mathsf{P}_\delta\mathsf{E}_\gamma) \\
& -C^{-1}_{\alpha\gamma}C^{-1}_{\beta\delta}(\mathsf{M}_\gamma\mathsf{B}_\delta + \mathsf{M}_\delta\mathsf{B}_\gamma) - C^{-1}_{\alpha\beta}\mathsf{M}_\gamma\mathsf{M}_\gamma \bigg] dC_{\alpha\beta} \\
& -C^{-1}_{\alpha\beta}\mathsf{E}_\beta d\mathsf{P}_\alpha + C^{-1}_{\alpha\beta}\mathsf{M}_\beta d\mathsf{B}_\alpha \Bigg\} \ .
\end{aligned}
\tag{3.6.20}
$$

This completes the presentation of the LORENTZ formulation.

3.7 Thermostatic Equilibrium – Constitutive Equations for Electric Current and Energy Flux

In the preceding sections we determined constitutive equations for entropy, two electromagnetic field vectors and the stress tensor. All of the above mentioned variables are derivable from a thermodynamic potential (the latter being the HELMHOLTZ free energy or one of its LEGENDRE transformations). Given such a thermodynamic potential as a function of its independent variables the constitutive relations for the remaining fields are derivable. However, it is not possible to construct in the same way constitutive relations for the conductive current density and for the energy flux. Constitutive relations for these quantities are merely restricted by the residual entropy inequality, which in all but the LORENTZ formulation may be written as

$$\gamma := \mathcal{J}_i\mathcal{E}_i - \frac{\Theta_{,i}q_i}{\Theta} \geq 0 \ . \tag{3.7.1}$$

In the LORENTZ formulation q_i must be replaced by

$$q_i^S := q_i + e_{ijk}\mathcal{E}_j\mathcal{M}_k \ . \tag{3.7.2}$$

To find the restrictions imposed by the residual entropy inequality we look at *thermostatic* processes, which will be defined as processes with uniform and time-independent temperature and vanishing conductive current. The results depend on whether we are dealing with an electrical insulator or conductor. Hence, we discuss these cases separately.

(a) In an **electrical insulator** the conductive current $\mathcal{J}_i$ vanishes identically. Hence, (3.7.1) becomes

$$\gamma = -\frac{\Theta_{,i} q_i}{\Theta} \geq 0 \,, \tag{3.7.3}$$

and in equilibrium

$$\gamma = \gamma\Big|_E = 0 \,. \tag{3.7.4}$$

We shall characterize thermostatic equilibrium by the index $\Big|_E$. It follows that γ, which is a function of all the independent variables upon which q_i depends, assumes its minimum when $\Theta_{,i} = 0$. Necessary conditions for this to be the case are

$$\frac{\partial \gamma}{\partial \Theta_{,i}}\Big|_E = 0 \,, \quad \text{and}$$

$$\frac{\partial^2 \gamma}{\partial \Theta_{,i} \partial \Theta_{,j}}\Big|_E \quad \text{is positive-semi definite} \,, \tag{3.7.5}$$

or

$$q_i\Big|_E = 0 \,, \quad \text{and}$$

$$\frac{\partial q_{(i}}{\partial \Theta_{,j)}}\Big|_E \quad \text{is negative-semi definite.} \tag{3.7.6}$$

Hence, for an electrical insulator in thermostatic equilibrium the energy flux vector must necessarily vanish. To see what consequences these restrictions impose, let us consider the case for which the constitutive relations are given in the form

$$\mathsf{C} = \check{\mathsf{C}}(F_{i\alpha}, \mathcal{E}_i, B_i, \Theta, \Theta_{,i}, \mathcal{Q}) \,. \tag{3.7.7}$$

A constitutive relation for the energy flux vector that is objective under EUCLIDian transformations and depends on the above variables is of the form

$$q_i = F_{i\alpha} \check{q}_\alpha(C_{\beta\gamma}, \mathsf{E}_\gamma, \mathsf{B}_\gamma, \Theta, \Theta_{,\beta}, \mathcal{Q}) \,, \tag{3.7.8}$$

where E_γ, B_γ and $C_{\beta\gamma}$ have been defined in (2.6.7) and where

$$\Theta_{,\alpha} := F_{i\alpha}\Theta_{,i} \,. \tag{3.7.9}$$

A necessary and sufficient condition for the energy flux vector to vanish in thermostatic equilibrium is that q_i is of the form

$$q_i = -F_{i\alpha} \check{k}_{\alpha\beta}(C_{\gamma\delta}, \mathsf{E}_\gamma, \mathsf{B}_\gamma, \Theta, \Theta_{,\gamma}, \mathcal{Q})\Theta_{,\beta} \,. \tag{3.7.10}$$

This representation is suitable for solids to which we shall restrict ourselves henceforth. In an electrical insulator, (3.7.10) is also the most general constitutive relation for the energy flux vector which vanishes in thermostatic equilibrium. For this form of the constitutive relation, the condition $(3.7.6)_2$ requires that the matrix

$$A_{ij} := F_{(i\alpha} F_{j)\beta} \check{k}_{\alpha\beta} \Big|_E \quad \text{is positive-semi definite.} \tag{3.7.11}$$

In the theory of solids one often restricts oneself to small deformations and small deviations from thermostatic equilibrium. In such special cases, in the constitutive equations all terms that are quadratic or of higher order in the above mentioned small quantities are neglected. Since the constitutive relation for q_i is already explicitly linear in $\Theta_{,i}$ this requires that $\check{k}_{\alpha\beta}$ is independent of $C_{\alpha\beta}$ and $\Theta_{,\alpha}$ and that $F_{i\alpha}$ may be replaced by $\delta_{i\alpha}$. For such a case (3.7.11) reduced to

$$A_{ij} := \check{k}_{(ij)} \Big|_E \quad \text{is negative-semi definite.} \tag{3.7.12}$$

Nothing can be concluded from the equilibrium condition about the skew-symmetric part of $\check{k}_{ij} \Big|_E$. As a consequence of the ONSAGER relations, one usually requires that

$$\check{k}_{[ij]} \Big|_E = 0 \ . \tag{3.7.13}$$

Finally, in this linear approximation it is not difficult to show that

$$q_i = -\check{k}_{ij}(\mathcal{E}_k, B_k, \Theta, \mathcal{Q})\Theta_{,j} \ . \tag{3.7.14}$$

Here, as a consequence of the ONSAGER relations, we require

$$\check{k}_{[ij]} = 0 \ , \tag{3.7.15}$$

and, furthermore,

$$\check{k}_{ij}\Theta_{,i}\Theta_{,j} \geq 0 \ . \tag{3.7.16}$$

In this drastically simplified version the constitutive relation (3.7.14) for the energy flux vector is known as the FOURIER law of heat conduction.

(b)   In an **electrical conductor** one may write

$$\begin{aligned} q_i &= \check{q}_i(F_{j\alpha}, \mathcal{E}_j, B_j, \Theta, \Theta_{,j}, \mathcal{Q}) \ , \\ \mathcal{J}_i &= \check{\mathcal{J}}_i(F_{j\alpha}, \mathcal{E}_j, B_j, \Theta, \Theta_{,j}, \mathcal{Q}) \ , \end{aligned} \tag{3.7.17}$$

and then γ is a function of

$$F_{i\alpha}, \mathcal{E}_i, B_i, \Theta, \Theta_{,i} \quad \text{and} \quad \mathcal{Q} \ .$$

Since the electrical insulator as a special case of an electrical conductor with zero conductivity was already considered on the previous pages, we exclude it here. For the general case, the second of the constitutive relations (3.7.17) may be inverted in the sense that

$$\mathcal{E}_i = \breve{\mathcal{E}}_i(F_{j\alpha}, \mathcal{J}_j, B_j, \Theta, \Theta_{,j}, \mathcal{Q}) \,. \tag{3.7.18}$$

If this relation is substituted into (3.7.17)$_1$, we may regard γ to be a function of

$$F_{i\alpha}, \mathcal{J}_i, B_i, \Theta, \Theta_{,i} \quad \text{and} \quad \mathcal{Q} \,,$$

so that (3.7.1) becomes

$$\gamma = \mathcal{J}_i \breve{\mathcal{E}}_i - \frac{\Theta_{,i}}{\Theta} \breve{q}_i = \breve{\gamma}(F_{i\alpha}, \mathcal{J}_i, B_i, \Theta, \Theta_{,i}, \mathcal{Q}) \geq 0 \,. \tag{3.7.19}$$

Since then

$$\gamma\big|_E = \breve{\gamma}\left(F_{i\alpha}\big|_E, 0, B_i\big|_E, \Theta\big|_E, 0, \mathcal{Q}\big|_E\right) = 0 \,, \tag{3.7.20}$$

we must necessarily have

$$\frac{\partial \breve{\gamma}}{\partial \mathcal{J}_i}\bigg|_E = 0 \,, \qquad \frac{\partial \breve{\gamma}}{\partial \Theta_{,i}}\bigg|_E = 0 \,, \tag{3.7.21}$$

and

$$\begin{pmatrix} \dfrac{\partial^2 \breve{\gamma}}{\partial \mathcal{J}_i \partial \mathcal{J}_j} & \dfrac{\partial^2 \breve{\gamma}}{\partial \mathcal{J}_i \partial \Theta_{,j}} \\[4mm] \dfrac{\partial^2 \breve{\gamma}}{\partial \mathcal{J}_i \partial \Theta_{,j}} & \dfrac{\partial^2 \breve{\gamma}}{\partial \Theta_{,i} \partial \Theta_{,j}} \end{pmatrix}\Bigg|_E \qquad \text{is positive-semi definite} \,. \tag{3.7.22}$$

Of necessity then

$$\breve{\mathcal{E}}_i\big|_E = 0 \,, \quad \text{and} \quad \breve{q}_i\big|_E = 0 \,. \tag{3.7.23}$$

Therefore, the electromotive intensity and the heat flux vector vanish in thermostatic equilibrium. Hence, we could have defined thermostatic equilibrium also as a time-independent process for which $\mathcal{E}_i$ and $\Theta_{,i}$ vanish. Then, the above relations would read

$$\frac{\partial \breve{\gamma}}{\partial \mathcal{E}_i}\bigg|_E 0 \,, \qquad \frac{\partial \breve{\gamma}}{\partial \Theta_{,i}}\bigg|_E = 0 \,, \tag{3.7.24}$$

and

$$\left(\begin{array}{cc} \dfrac{\partial^2\tilde{\gamma}}{\partial\mathcal{E}_i\partial\mathcal{E}_j} & \dfrac{\partial^2\tilde{\gamma}}{\partial\mathcal{E}_i\partial\Theta_{,j}} \\[4mm] \dfrac{\partial^2\tilde{\gamma}}{\partial\mathcal{E}_i\partial\Theta_{,j}} & \dfrac{\partial^2\tilde{\gamma}}{\partial\Theta_{,i}\partial\Theta_{,j}} \end{array}\right)\Bigg|_E \quad \text{is positive-semi definite .} \qquad (3.7.25)$$

The relations (3.7.24), (3.7.25) now yield

$$\check{\mathcal{J}}_i\Big|_E = 0\,, \qquad \check{q}_i\Big|_E = 0\,, \qquad (3.7.26)$$

as well as

$$\left(\begin{array}{cc} \dfrac{\partial\check{\mathcal{J}}_{(i}}{\partial\mathcal{E}_{j)}} & \left(\dfrac{\partial\check{\mathcal{J}}_i}{\partial\Theta_{,j}} - \dfrac{1}{\Theta}\dfrac{\partial\check{q}_j}{\partial\mathcal{E}_i}\right) \\[4mm] \left(\dfrac{\partial\check{\mathcal{J}}_i}{\partial\Theta_{,j}} - \dfrac{1}{\Theta}\dfrac{\partial\check{q}_j}{\partial\mathcal{E}_i}\right) & -\dfrac{1}{\Theta}\dfrac{\partial\check{q}_{(i}}{\partial\Theta_{,j)}} \end{array}\right)\Bigg|_E \quad \text{is positive-semi definite .}$$

$$(3.7.27)$$

The most general form of the constitutive equations (3.7.17) for a solid satisfying the principle of material frame indifference is

$$\begin{aligned} q_i &= F_{i\alpha}\check{\mathsf{q}}_\alpha(C_{\beta\gamma},\mathsf{E}_\beta,\mathsf{B}_\beta,\Theta,\Theta_{,\beta},\mathcal{Q})\,, \\ \mathcal{J}_i &= F_{i\alpha}\check{\mathsf{J}}_\alpha(C_{\beta\gamma},\mathsf{E}_\beta,\mathsf{B}_\beta,\Theta,\Theta_{,\beta},\mathcal{Q})\,. \end{aligned} \qquad (3.7.28)$$

When they are written in the form

$$\begin{aligned} q_i &= F_{i\alpha}\{-\check{k}_{\alpha\beta}\Theta_{,\beta} + \check{\beta}^{(q)}_{\alpha\beta}\mathsf{E}_\beta\}\,, \\ \mathcal{J}_i &= F_{i\alpha}\{\check{\beta}^{(\mathcal{J})}_{\alpha\beta}\Theta_{,\beta} + \check{\sigma}_{\alpha\beta}\mathsf{E}_\beta\}\,, \end{aligned} \qquad (3.7.29)$$

they automatically satisfy the equilibrium conditions (3.7.26). Moreover, in order to satisfy (3.7.27), the two relations (3.7.29) imply that

$$A := \left(\begin{array}{cc} \check{\sigma}_{\alpha\beta}F_{(i\alpha}F_{j)\beta} & (\check{\beta}^{(\mathcal{J})}_{\alpha\beta} - \dfrac{1}{\Theta}\check{\beta}^{(q)}_{\beta\alpha})F_{i\alpha}F_{j\beta} \\[4mm] (\check{\beta}^{(\mathcal{J})}_{\alpha\beta} - \dfrac{1}{\Theta}\check{\beta}^{(q)}_{\beta\alpha})F_{i\alpha}F_{j\beta} & \dfrac{1}{\Theta}\check{k}_{\alpha\beta}F_{(i\alpha}F_{j)\beta} \end{array}\right) \qquad (3.7.30)$$

must be positive-semi definite. Necessary conditions (that are not sufficient however) for this to be satisfied are

$$\check{\sigma}_{\alpha\beta}F_{(i\alpha}F_{j)\beta} \quad \text{and} \quad \frac{1}{\Theta}\check{k}_{\alpha\beta}F_{(i\alpha}F_{j)\beta} \quad \text{are positive-semi definite .}$$

$$(3.7.31)$$

More interesting than the general case is again the linearized version of the theory. Assuming small deformations and small deviations from thermostatic equilibrium, it may be justified to postulate the constitutive relations for q_i and $\mathcal{J}_i$, (3.7.29), to be linear in $\Theta_{,i}$ and $\mathcal{E}_i$ and independent of the deformation. In this case, the coefficient functions $\check{k}_{\alpha\beta}, \check{\beta}_{\alpha\beta}^{(\mathcal{J})}, \check{\beta}_{\alpha\beta}^{(q)}$, and $\check{\sigma}_{\alpha\beta}$ are independent of $C_{\alpha\beta}, \mathsf{E}_\alpha$ and $\Theta_{,\alpha}$, and $F_{i\alpha}$ may be replaced by $\delta_{i\alpha}$. Under these simplified restrictions

$$A \approx \begin{pmatrix} \check{\sigma}_{(ij)} & \left(\check{\beta}_{ij}^{(\mathcal{J})} - \frac{1}{\Theta}\check{\beta}_{ji}^{(q)}\right) \\ \left(\check{\beta}_{ij}^{(\mathcal{J})} - \frac{1}{\Theta}\check{\beta}_{ji}^{(q)}\right) & \frac{1}{\Theta}\check{k}_{(ij)} \end{pmatrix}, \qquad (3.7.32)$$

must be positive-semi definite. Of necessity then, $\check{\sigma}_{(ij)}$ is positive definite (because $\check{\sigma} \equiv 0$ is excluded) and $\check{k}_{(ij)}$ is positive-semi definite. The ONSAGER relations require here that

$$\check{\sigma}_{[ij]} = 0 , \qquad \check{k}_{[ij]} = 0 ,$$

$$\check{\beta}_{ij}^{(\mathcal{J})} = \frac{1}{\Theta}\beta_{ji}^{(q)} =: \frac{1}{\Theta}\beta_{ji} .$$

$$(3.7.33)$$

Of course, $\check{\sigma}_{ij}, \check{k}_{ij}$ and $\check{\beta}_{ij}$ are still functions of B_k, Θ and $\mathcal{Q}$.

When linearized and with the use of (3.7.33), the constitutive equations (3.7.29) become

$$q_i = -\check{k}_{ij}(B_k, \Theta, \mathcal{Q})\Theta_{,j} + \check{\beta}_{ij}(B_k, \Theta, \mathcal{Q})\mathcal{E}_j ,$$

$$\mathcal{J}_i = \check{\beta}_{ji}(B_k, \Theta, \mathcal{Q})\frac{\Theta_{,j}}{\Theta} + \check{\sigma}_{ij}(B_k, \Theta, \mathcal{Q})\mathcal{E}_j .$$

$$(3.7.34)$$

These equations are known as FOURIER's law of heat conduction and OHM's law of electrical conduction, and in this form they appear in usual treatises on crystallography and solid state physics (e.g. [165, 29]). The coefficients $\check{k}_{\alpha\beta}$ and $\check{\sigma}_{\alpha\beta}$ are termed the thermal and electrical conductivity, respectively. The coefficients $\check{\beta}_{\alpha\beta}$ are responsible for thermoelectric effects (e.g. SEEBECK effect, PELTIER heat, THOMSON heat, cf. [165], Chap. 12 or [263], Chap. 23), and in magnetizable materials they also cause galvanomagnetic or thermomagnetic effects (e.g. ETTINGHAUSEN or NERNST effect). These effects are investigated in more detail than is possible within the scope of this monograph, in a series of papers by PIPKIN and RIVLIN, [180] and [181], and BORGHESANI and MORRO, [27] and [28]. PIPKIN and RIVLIN derive general, nonlinear conduction laws for isotropic materials considering separately the effects of deformation on the electric current and the influences of the electric field, the magnetic induction and the temperature gradient on the constitutive equations for the electric current, the heat flux and the magnetic field strength, whereas BORGHESANI and MORRO derive the fourth-order versions of FOURIER's law and OHM's law.

Finally we remark that for most purposes of this monograph the consequences implied by the ONSAGER relations will be of no relevance. In particular, the results of Chap. 3 are independent of such relations. When using the ONSAGER relations we shall in the following therefore explicitly state it.

In summary, we have given an exposure of five different descriptions of deformable polarizable and magnetizable continua. Each model describes in its own way the interactions between the electromagnetic fields and the thermoelastic body. The field equations consist of the following set of equations:

(i)   The MAXWELL equations, which must be counted as seven independent equations (see the remark in Sect. 1.3.2).
(ii) Five balance laws of mass, momentum and energy.
(iii) Sixteen constitutive relations for entropy, two electromagnetic field vectors and the stress tensor (the latter given in a form such that the balance law of moment of momentum is satisfied identically).
(iv) Six constitutive relations for the electric current and the energy flux vector.

These are 34 equations for the unknowns: four electromagnetic field vectors (12), electric current and electric charge density (4), mass density (1), motion χ_i (3), stress (9), temperature (1), energy flux (3) and entropy (1).

These field equations are supplemented by jump (or boundary) conditions for the electromagnetic field variables, mass, momentum and energy of matter and field.

3.8 Discussion

In the preceding sections five different descriptions of deformable polarizable and magnetizable continua were treated. The interaction of the electromagnetic fields with matter was achieved by introducing two additional electromagnetic field vectors to the two basic fields occurring in vacuo. In the five formulations the choice of these field vectors was not unique, however. We did not emphasize the models which lie behind these descriptions, although each of the formulations can be based upon well-defined models. The CHU variables of electrodynamics for instance can be founded on a two-dipole model, and as long as one restricts oneself to the derivation of the MAXWELL equations, such a model leads to unique answers. The model may bear its disadvantages insofar as it is handled non-relativistically, while the resulting equations are postulated to be relativistically correct equations. This does not change the basic fact that the resulting MAXWELL equations are unique.

When electromagnetic body force and energy supply are derived, however, the dipole model is no longer unique and two theories of magnetizable and polarizable bodies can be developed. We have seen this when presenting models I and II.

One can also develop a theory on the basis that polarization is modelled by a dipole while magnetization is treated as an electric circuit. Dependent on the degree of complexity of derivation, different MAXWELL equations emerge (statistical and LORENTZ formulation), and also the expressions for the electromagnetic body force, body couple and energy supply are different in these derivations. The reader could therefore be misled by these derivations as he might conclude that some of the theories are superior to others, because the derivation of the equations resembles a more profound approach. This is not so, and our point of view is different, as we regard all formulations as equally sound as long as none has been proved to be superior to any other one, e.g. via transformation properties under EUCLIDian or LORENTZ transformations. Of course, this requires that we de-emphasize the models behind the equations. This is why we do not even share the viewpoint of many physicists, who would reject the two-dipole description on the basis that magnetic monopoles have never been observed experimentally and that therefore magnetization must be electric circuits. Corroboration to all the above statements will be found in the next chapter of this monograph. For instance, the statistical model and the LORENTZ model (models IV and V) can be derived by methods of statistical physics, but dependent on which author's book one opens, either one appears to be an approximation of the other. The reason that in the most rigorous treatment the LORENTZ formulation is not obtained is a matter of definition of the macroscopic electric and magnetic dipole moments in terms of statistical averages of microscopic quantities. Nonetheless, PAO concludes in his reviewing article [172] that for magnetizable materials the validity of the LORENTZ model is questionable (page 24 of [172]). Yet, we shall prove that the non-relativistic versions of polarizable or magnetizable continua in the models IV and V are equivalent. A similar situation also exists for the expression of the body force as suggested by FANO, CHU and ADLER that agrees with our model II. Again according to PAO (page 82 of [172]), "PEN-FIELD and HAUS [177] have considered these force expressions incomplete". We shall prove that models I and II are equivalent. Needless to say that PENFIELD and HAUS [177] and PAO and HUTTER [171] have advocated for model I.

The above statements should make it clear then that the derivation of the models is not important and that the formulations should be unified by demonstrating their equivalence, rather than emphasizing their differences. For this reason we have presented the five models per se and have studied their invariance properties quite extensively. This led to the approach of model III, in which electromagnetic body force, body couple and energy supply were derived from a general energy expression that is subjected to specific invariance requirements under the general EUCLIDian transformation group. In order to illustrate the above mentioned unification and because some authors seem to have objections against the approach of Sect. 2.3 (as they do not believe that this approach will lead to unique results) we have also given the energy laws which could serve as basis for the other models. For model I this was already done by ALBLAS [9]. This, together with the results of Chap. 3, in which all models are shown to be equivalent, demonstrates that there is a large amount of freedom in the choice of the specific terms occurring in such an energy law, all leading to equivalent results.

We found further that in formulations in which the stress tensor was symmetric the electromagnetic body force and energy supply are objective quantities with respect to the EUCLIDian transformation group. In formulations with non-symmetric stress, body force and body couple turned out to be objective, while the energy supply term first had to be corrected by the power of working of the skew-symmetric part of the stress. As a result, the energy equation remained invariant.

These transformation properties make it possible that the various formulations of electromagnetic interactions with thermoelastic bodies have a chance to be equivalent. Indeed, full equivalence cannot be achieved when corresponding quantities in different formulations do not transform alike under the EUCLIDian transformation group. Similarly, if approximations in the non-relativistic sense are performed, equivalence can only be attempted to be proved within such approximations.

This discussion would not be complete if we would not point out that other descriptions for field matter interactions often agree with our models I-V. We find it important to discuss these descriptions also. In order to narrow the number of formulations down to a reasonable size, we shall not discuss any quasi-static formulation, such as those by TOUPIN [240], ALBLAS [8], TIERSTEN [233, 236], PAO and YEH [170] and others (see e.g. PENFIELD and HAUS [177]), with the exception of the monograph of BROWN [32], however. The reason for this exception is that BROWN already was aware of the non-uniqueness of the magnetoelastic stresses, and he stated explicitly that these stresses are only then completely determined once the total system of momentum and moment of momentum equations, boundary conditions and constitutive equations are given. For a magnetostatic theory, BROWN introduced four different stress tensors, namely: (see Sect. 5.6 of [32])

(i)   The pole-model $\bar{t}_{ij}$ (eqs. (6.2.9)–(6.2.12)).
(ii)   The AMPÈREan current model $\bar{t}'_{ij}$ (eqs. (6.2.9')(6.2.12')).
(iii) The MAXWELL model I t_{ij} (eqs. (6.2.17)–(6.2.23)).
(iv)   The MAXWELL model II t'_{ij} (eqs. (6.2.17')-(6.2.23')).

Comparing these stress models with the ones introduced in this chapter, we see that when the latter are simplified to the magnetostatic case, the following relations between our stresses and those of BROWN hold:

$$^{I}t_{ij} = {}^{III}t_{ij} = t_{ij} , \qquad {}^{II}t_{ij} = \bar{t}_{ij} ,$$

$$\text{(3.8.1)}$$

$$^{IV}t_{ij} = t'_{ij} , \qquad\qquad {}^{V}t_{ij} = \bar{t}'_{ij} .$$

In comparing other authors' work, any complications due to spin interactions will be left aside. This does not mean that spin interaction theories will not be compared here, but any contribution due to magnetic spin will be set to zero without further mentioning. Furthermore, fully relativistic theories will only be mentioned in connection with non-relativistic approximations.

First when the material is polarizable-only, model V is in full agreement with a description of a dynamical theory of elastic dielectrics as presented by TOUPIN [241], and indeed it must be so as our force and energy supply expressions agree with those of TOUPIN. On the other hand, DIXON and ERINGEN attempted to derive a theory of field-matter interaction using a statistical description (which we believe to be incomplete) [59]. They include in their derivation also macroscopic electric quadrupoles, but when these are neglected their MAXWELL equations and body force and energy supply expressions are in agreement with model V. However, the jump condition of energy derived by DIXON and ERINGEN (their equation (5.16)) is not correct. Indeed, their global energy balance contains a volume source which becomes indefinitely large when a surface of discontinuity is approached. Hence, apart from these shortcomings in the derivation, their results are non-relativistically correct. Moreover, ERINGEN in collaboration with GROT [78], and GROT [79] present relativistic formulations of solids and fluids in the electromagnetic fields, in which they postulate an energy-momentum tensor of field-matter interaction which reduces to the force and energy expressions of model V when terms of $\mathcal{O}(V^2/c^2)$ are neglected.

Of a quite different nature is TIERSTEN's approach to describe the interaction of the electromagnetic fields with deformable continua. He uses a mixture concept and describes the field-matter interactions by coupling a so called lattice-continuum with charge-, electronic-, ionic- and spin-(sub)continua, each bearing the notion of a particular physical effect. All work of TIERSTEN, [236], is based on such a mixture concept. In [237] TIERSTEN and TSAI describe the interaction of the electromagnetic fields with heat conducting deformable electric insulators. In a simplified version of this theory, where only a charge continuum is interacting with the lattice continuum, their body force, body couple and energy supply (see equations (3.48), (8.4) and (8.7) in [237]) agree with model IV. An explicit proof for this equivalence can also be found in PAO's review article, [172]. The same comments hold for the paper of LORENTZ and TIERSTEN, [55].

Still other methods of derivation are those in which the governing equations derive from an over-all energy balance law which is postulated to be invariant under EUCLIDIAN transformations. The technique, first introduced by GREEN and RIVLIN [77] for non-relativistic multipolar theories, was applied to describe deformable polarizable and magnetizable continua by ALBLAS

[9], PARKUS [175] and VAN DE VEN [249]. We have shown in this monograph that all formulations could have been derived from such an over-all energy balance law. When we compare the work of ALBLAS [9] (who uses the CHU variables) with our model I complete agreement is found. As concerns the work of PARKUS [175], he treats magnetizable and polarizable materials separately and when dealing with magnetizable materials restricts himself to quasi-static processes. His treatment of dielectrics on the other hand agrees with that of TOUPIN. PARKUS uses HAMILTON's principle and the same is done by VLASOV and ISHMUKHAMETOV [259], who arrive at results which agree with our model III. We could have added other electromagnetic models, if we had wished to do so, and for reasons of completeness these models should also be mentioned. FANO, CHU and ADLER [73] and PENFIELD and HAUS [177] also present the so called AMPÈREan descriptions, as do HUTTER and PAO [91] in a paper dealing with magnetizable elastic solids with thermal and electrical conduction. The AMPÈREan formulation (in which magnetization is modelled by a relativistic electric circuit) can easily be transformed into the LORENTZ- or the statistical formulation and when this is done, it is found that the body force expression agrees with the one presented in model IV. The expressions for the energy supply and the body couple, when restricted to the non-relativistic approximation, fully agree with those of model IV.

MAUGIN on the other hand, partly in collaboration with ERINGEN and COLLET, [131]–[135] uses the principle of virtual power to derive the basic equations for several different theories of magnetoelastic interactions. The interest of these authors is generally more limited, as they treat specialized subjects, such as spin relaxation and surface effects, all in the quasi-static approximation. In [45], COLLET and MAUGIN, and in [135], MAUGIN treat dynamic processes in a non-relativistic approximation. Although there is a slight difference between our non-relativistic approximation and their owns, which can be traced back to the use of GAUSSIAN units as opposed to MKSA-units, the results of [45] and [135] when brought to our non-relativistic approximation completely agree with model IV. MAUGIN and ERINGEN also presented a fully relativistic treatment in [132]. Finally, starting from a four-dimensional relativistic formulation, BOULANGER and MAYNÉ [30] derive expressions for the electromagnetic body force and energy supply. In cooperation with VAN GEEN, they apply their results in [31] for the investigation of magnetooptical, electrooptical and photoelastic effects in elastic polarizable and magnetizable isotropic media. When comparing the relations for momenta and energy following from Eq. (14) of [50] or the balance laws (4)–(6) of [31], no immediate equivalence with one of our models is found. However, their balance laws can easily be related to those of, for instance, model IV by taking in the equations (4)–(6) of [31]

$$\rho\varepsilon = {}^{IV}(\rho U) + B_i \mathcal{M}_i \, ,$$

and

$$\sigma_{ij} = {}^{IV}t_{ij} + \mathcal{E}_i P_j + B_i \mathcal{M}_j - \delta_{ij} B_k \mathcal{M}_k \ .$$

(In this context it should be noted that in Eq. (16) of [30] two terms are missing, which should follow from (14), and which re-appear in Eq. (4) of [31].)

In conclusion we might justly state that the models I to V embrace within the non-relativistic approximation the description of dynamic theories of polarizable and magnetizable materials known to date. There are formulations simpler than those presented above, but these formulations aim at describing more restrictive situations such as static and quasi-static processes. There are also more complicated descriptions, but those include additional phenomena, as for instance spin interaction, polarization gradient effects and the like. These phenomena are outside the scope of this monograph.

Since the completion of this chapter a series of articles did appear which we would have taken into account if being aware of them. To mention are works by ALBLAS [10], MAUGIN and ERINGEN [136], PRECHTL [185] and ROMANO [200]. ALBLAS presents a general exposition of electro- and magnetoelasticity with special topics such as electro- and magnetostatics including constitutive equations and linearization procedures and magnetoelastic stability. His formulation is that of CHU. MAUGIN and ERINGEN formulate a theory of magnetoelastic interaction including electric quadrupoles. Their approach is similar to that of DIXON and ERINGEN [59] and they seem to correct the erroneous jump condition in this work. When discarding the electric quadrupoles they arrive at body force expressions identical to those of our model IV. PRECHTL, on the other hand presents a relativistic CHU formulation; he reduces it to a three dimensional form, which in the non-relativistic approximation is equivalent to our model II. Unfortunately, PRECHTL does not present constitutive relations, so that a full comparison is not achieved. ROMANO gives a quasi-static magnetoelastic CHU formulation.

4 Equivalence of the Models

4.1 Preliminary Remarks

In Chap. 3 various descriptions of electromechanical interaction models were laid down, but no attempt was made to search for interrelations amongst these models. In this chapter we perform the first steps toward a proof or disproof of the equivalence of these models. Specifically, we are going to present the conditions that need to be satisfied in order to render two formulations equivalent.

To begin with, we should point out that *two models are called equivalent provided they deliver the same results for physically observable quantities in any arbitrary initial-boundary-value problem.* By observable (measurable) quantities we mean all those physical quantities that can be measured uniquely by two different independent observers. All kinematical quantities that are derivable from the motion $\chi_i(\boldsymbol{X}, t)$ are measurable in principle. Regarding electromagnetic field quantities we take the viewpoint that they are not measurable except in vacuo, where they can be observed by measuring the force exerted on a test charge. Finally, (empirical) temperature is also regarded as being measurable, because, as is known from thermodynamics, it is a measure for the hotness of a body. The instruments to measure temperature are thermometers. They relate our sensation of the hotness of a body to a physical quantity, say pressure or volume, and the latter are measurable.

Differences in the various electromagnetic models appear in the body force, body couple and energy supply of electromagnetic origin, but as is well-known, such differences may be compensated in the stress tensor, the internal energy and the energy flux. Except for the latter, these quantities are determined by a thermodynamic potential (the HELMHOLTZ free energy or its LEGENDRE transformations). In other words, equivalence of two electromechanical interaction models depends to a large extent on the fact that the free energies of the two formulations can be determined so as to render stress, entropy and electromagnetic field variables compatible with the above mentioned compensation between stress and body force, etc. *Equivalence of two theories is therefore a thermodynamic requirement.*

Practically, two formulations are equivalent if upon transforming one into the other, not only the field equations but also the initial conditions and jump conditions are alike. Here in this chapter we investigate the equivalence of

K. Hutter et al.: *Electromagnetic Field Matter Interaction in Thermoelastic Solids and Viscous Fluids*, Lect. Notes Phys. **710**, 89–102 (2006)
DOI 10.1007/3-540-37240-7_4 © Springer-Verlag Berlin Heidelberg 2006

the models described in the preceding chapter. In particular, the conditions under which this equivalence is achieved will be formulated.

Needless to say that all comparison will be made to within the order of the non-relativistic approximation.

4.2 Comparison of the Models I and II

Here we discuss the two-dipole models, and we shall for this purpose heavily rely upon the results presented in Sect. 3.2.

To begin with, note that both models are based on the same set of electromagnetic field variables and therefore obey the same set of MAXWELL equations (3.3.3). Differences do occur in the electromagnetic body force, body couple and energy supply, which can most easily be identified by comparing the expressions according to (3.3.7) and (3.3.49). From $(3.3.7)_1$ and $(3.3.49)_1$ we read off that

$$^{II}(\rho F_i^{\mathrm{e}}) = {}^{I}(\rho F_i^{\mathrm{e}}) - (P_j \mathcal{E}_i + \mu_0 \mathcal{M}_j \mathcal{H}_i)_{,j} \; . \tag{4.2.1}$$

We conclude that the balance laws of momentum $(3.3.9)_2$ and $(3.3.50)_2$ are identical provided that

$$^{II}t_{ij} = {}^{I}t_{ij} + \mathcal{E}_i P_j + \mu_0 \mathcal{H}_i \mathcal{M}_j \; . \tag{4.2.2}$$

In order to compare the balance laws of moment of momentum and energy, we must compare

$$\rho L_{ij}^{\mathrm{e}} - t_{[ij]} \; , \quad \text{and} \quad \rho r^{\mathrm{e}} + t_{ij}\dot{x}_{i,j} \; ,$$

respectively.

Using ρL_{ij}^{e} as given in $(3.3.7)_2$ and $(3.3.49)_2$ and ρr^{e} according to (3.3.12) and $(3.3.49)_3$, we immediately see that

$$^{I}(\rho L_{ij}^{\mathrm{e}} - t_{[ij]}) = {}^{II}(\rho L_{ij}^{\mathrm{e}} - t_{[ij]}) \; , \tag{4.2.3}$$

and

$$^{I}(\rho r^{\mathrm{e}} + t_{ij}\dot{x}_{i,j}) = {}^{II}(\rho r^{\mathrm{e}} + t_{ij}\dot{x}_{i,j}) \; , \tag{4.2.4}$$

provided that the stresses are related by (4.2.2). Hence, with the conditions (4.2.2), the balance laws of model I and model II have been proved to be equivalent.

It remains to compare the jump conditions. Since the MAXWELL equations are the same, there cannot be a difference in the jumps of the electromagnetic field variables. On the other hand, as can immediately be seen from (3.3.21), (3.3.22) and (3.3.52), the jump conditions for momentum and energy of matter and fields are identical provided that the stress tensors are related according to (4.2.2).

The above comparison makes no use of thermodynamic arguments. Otherwise stated, equivalence of the two theories is guaranteed only if the relations (4.2.2) hold. That such relations indeed can hold follows from thermodynamic arguments, and they are, although trivial in this case, not obvious in general. In order to invoke these conditions, we must compare theories based on the same constitutive postulates, and, of course, in this case we must choose in both models the same set of independent constitutive variables. Explicitly we have done this for case a):

$$C = \hat{C}(C_{\alpha\beta}, P_\alpha, M_\alpha, \Theta, \Theta_{,i}, \mathcal{Q}) \, ,$$

and the results are listed as (3.3.32) and (3.3.57). It follows from these that

$$^I t_{ij} = 2\rho \frac{\partial \, ^I \hat{\psi}}{\partial C_{\alpha\beta}} F_{i\alpha} F_{j\beta} + P_i \mathcal{E}_j + \mu_0 \mathcal{M}_i \mathcal{H}_j \, ,$$

and that

$$^{II} t_{ij} = 2\rho \frac{\partial \, ^{II} \hat{\psi}}{\partial C_{\alpha\beta}} F_{i\alpha} F_{j\beta} + 2P_{(i} \mathcal{E}_{j)} + 2\mu_0 \mathcal{M}_{(i} \mathcal{H}_{j)} \, ,$$

and it is a straightforward matter to show that (4.2.2) is satisfied if we choose

$$^{II} \hat{\psi} = \, ^I \hat{\psi} \, , \tag{4.2.5}$$

or

$$^{II} U = \, ^I U \, . \tag{4.2.6}$$

With this choice, all the other dependent constitutive quantities that derive from the free energy, e.g. entropy, electromotive intensity, and magnetomotive intensity are guaranteed to be the same. It thus only remains to mention, that in order to obtain full agreement the energy flux vectors must also be chosen to be the same, i.e.

$$^{II} q_i = \, ^I q_i \, . \tag{4.2.7}$$

The reader can readily prove the above statements to be correct also for all other constitutive theories that are obtained from the above one by merely interchanging some of the dependent and independent variables. We therefore have proved the

Proposition. *Within the constitutive class of thermoelastic polarizable and magnetizable materials the two-dipole models I and II are equivalent, provided that the free energies and the energy flux vectors are the same functions of their independent variables.* ∎

4.3 Comparison of the Models I and III

Before comparing the models I and III one should realize that in model I electromagnetic fields according to the CHU model are used, whereas in model

III the MINKOWSKI-fields are used. As stated by PENFIELD and HAUS, [177], Ch.7, and as follows from relations (3.3.1) and (3.4.1), these fields are related in the following way

$$E_i^M = E_i^C - \mu_0 e_{ijk}\dot{x}_j M_k^C \; , \qquad D_i^M = \varepsilon_0 E_i^C + P_i^C \; ,$$

$$H_i^M = H_i^C + e_{ijk}\dot{x}_j P_k^C \; , \qquad B_i^M = \mu_0 H_i^C + \mu_0 M_i^C \; . \tag{4.3.1}$$

According to the definitions (3.4.3) for P_i^M and M_i^M we also have

$$P_i^M = P_i^C + \frac{1}{c^2}e_{ijk}\dot{x}_j M_k^C \; , \qquad M_i^M = M_i^C - e_{ijk}\dot{x}_j P_k^C \; . \tag{4.3.2}$$

Hence, to within the non-relativistic approximation,

$$P_i^C = P_i^M = P_i \; , \quad \text{and} \quad M_i^C = M_i^M + e_{ijk}\dot{x}_j P_k^M = \mathcal{M}_i \; , \tag{4.3.3}$$

where both, P_i and $\mathcal{M}_i$ are known to be objective under EUCLIDian transformations.

From the above relations it follows immediately that the rest-frame fields $\mathcal{E}_i$ and $\mathcal{H}_i$ are equal in both formulations, and, moreover, that polarization and magnetization in the CHU formulation are equal to the corresponding rest-frame fields in the MINKOWSKI formulation. Then, by comparing $(3.3.7)_{2,3}$ with the corresponding quantities as given by (3.4.23) and (3.4.24), respectively, it is obvious that the expressions for the body couple and energy supply of model I and model III are identical.

Since, in the non-relativistic approximation, in both models the electromagnetic momentum vector g_i is zero, it is most convenient to relate the respective electromagnetic body forces by comparing the expressions for the MAXWELL stresses t_{ij}^M according to $(3.3.10)_1$ and $(3.4.25)_1$. With the aid of $(4.3.1)_{2,4}$, we may in (3.4.25) replace the MINKOWSKI fields B_i^M and D_i^M by the CHU fields and can then show that to within terms containing a c^{-2}-factor

$$^{III}t_{ij}^M = \mathcal{E}_i D_j^M + \mathcal{H}_i B_j^M - \tfrac{1}{2}\delta_{ij}(\varepsilon_0 \mathcal{E}_k \mathcal{E}_k + \mu_0 \mathcal{H}_k \mathcal{H}_k)$$

$$= \varepsilon_0 E_i^C E_j^C + \mu_0 H_i^C H_j^C + \mathcal{E}_i P_j^C + \mu_0 \mathcal{H}_i M_j^C \tag{4.3.4}$$

$$-\tfrac{1}{2}\delta_{ij}(\varepsilon_0 E_k^C E_k^C + \mu_0 H_k^C H_k^C) = {}^I t_{ij}^M \; .$$

Hence, we have demonstrated that the balance laws (3.3.17), $(3.3.9)_2$, $(3.3.9)_3$ and (3.4.22), (3.4.23), (3.4.24), are equivalent, provided that

$$^{III}t_{ij} = {}^I t_{ij} \; , \qquad {}^{III}q_i = {}^I q_i \; , \quad \text{and} \quad {}^{III}U = {}^I U \; . \tag{4.3.5}$$

Under these conditions, and to within the non-relativistic approximation, also the jump conditions for momentum and energy of matter and fields of model I and model III are equal (cf. (3.3.21), (3.3.22) and $(3.4.26)_{6,7}$).

All the above is true, of course, again with the provision that constitutive relations show the dependent constitutive variables to be the same in the two theories. This can be seen to be true immediately by comparing the constitutive equations (3.3.38) with those of (3.4.31) (thus $^I\psi = {}^{III}\psi$). The reader may also show this to be correct, if any one of the dependent and independent variables are interchanged.

Thus, we have proved the following

Proposition. *Within the constitutive class of thermoelastic polarizable and magnetizable materials the two-dipole model I and the* MAXWELL-MINKOWSKI *model III are non-relativistically equivalent provided that the constitutive relations for the internal energies (or the free energies) and the energy flux vectors are the same functions of their independent variables.* ∎

4.4 Comparison of the Models III and IV

We begin with the observation that the MAXWELL equations in the statistical formulation IV are written in a form which differs from that in the MINKOWSKI formulation. However, if in the equations (3.4.2) the $\boldsymbol{D}^M$ and $\boldsymbol{H}^M$ fields are eliminated by means of (3.4.3) the resulting equations equal (3.5.2). Hence, $\boldsymbol{E}, \boldsymbol{P}, \boldsymbol{B}$ and $\boldsymbol{M}$ are identical in both formulations, and we do not need superscripts to distinguish them. Of course, the same is then true also for the fields $\mathcal{E}$ and $\mathcal{M}$.

When with the use of (3.4.13) in the expression for the body force, (3.4.22), $\mathcal{H}_i$ and D_i are eliminated and in so doing terms of the form $\varepsilon_0 \boldsymbol{E} \times \boldsymbol{B}$ are neglected (as is justified in view of the corresponding remarks made in Sect. 1.6) one obtains

$$^{III}(\rho F_i^{\mathrm{e}}) = \mathcal{Q}\mathcal{E}_i + \mathrm{e}_{ijk}\mathcal{J}_j B_k + P_j\mathcal{E}_{j,i} + M_j B_{j,i}$$

$$-\mu_0\mathcal{M}_j\mathcal{M}_{j,i} + \mathrm{e}_{ijk}(\overset{\star}{P}_j\, B_k + P_j\, \overset{\star}{B}_k)\,, \qquad (4.4.1)$$

or

$$^{III}(\rho F_i^{\mathrm{e}}) = {}^{IV}(\rho F_i^{\mathrm{e}}) - (\tfrac{1}{2}\mu_0\mathcal{M}_j\mathcal{M}_j)_{,i}\,,$$

as follows from a comparison with $(3.5.3)_1$. Hence, the electromagnetic body forces in model III and IV differ in the term

$$\left(\tfrac{1}{2}\mu_0\mathcal{M}_j\mathcal{M}_j\right)_{,i}\,.$$

However, this term is easy to handle, because substitution of the expression for the body force into the balance laws of momentum (3.4.22) and (3.5.5) reveals that the above term leads to a difference in the stress tensors $^{III}t_{ij}$ and $^{IV}t_{ij}$ given by

$$^{IV}t_{ij} = {}^{III}t_{ij} - \tfrac{1}{2}\delta_{ij}\mu_0\mathcal{M}_k\mathcal{M}_k\,. \qquad (4.4.2)$$

This relation does not change the anti-symmetric part of the stresses, and one can see at once that the expressions for the body couples (3.4.23) and (3.5.3)$_2$ are identical, if the relation (3.4.13) is invoked.

As concerns the jump conditions of momentum, it is not difficult to see that (3.4.26)$_6$ is equal to (3.5.9)$_6$, once D_i and $\mathcal{H}_i$ are eliminated and, furthermore, equation (4.4.2) is substituted.

Because of (4.4.2), it may then be expected that the electromagnetic energy supplies ρr^e in the models III and IV will also differ. This is indeed the case and as can easily be seen from (3.4.24) and (3.5.3)$_3$ they are related by

$$^{IV}(\rho r^e) = {}^{III}(\rho r^e) - \rho \frac{d}{dt}(B_i\mathcal{M}_i/\rho) + \rho\mu_0\mathcal{M}_i\frac{d}{dt}(\mathcal{M}_i/\rho) . \quad (4.4.3)$$

Substituting the expressions (4.4.2) and (4.4.3) into the energy balance law for model IV, (3.5.5)$_4$, and comparing the resulting equation with (3.4.24), we see that the two energy balance laws are equivalent, if the pertinent internal energies are related by

$$^{IV}U = {}^{III}U - B_i\frac{\mathcal{M}_i}{\rho} + \frac{1}{2}\rho\mu_0\frac{\mathcal{M}_i}{\rho}\frac{\mathcal{M}_i}{\rho} , \quad (4.4.4)$$

and provided that the heat fluxes $^{III}q_i$ and $^{IV}q_i$ are identical, i.e.

$$^{IV}q_i = {}^{III}q_i . \quad (4.4.5)$$

Furthermore, the same holds true for the jump conditions of energy of matter and fields (3.4.26)$_7$ and (3.5.9)$_7$ as can be proved in the usual way.

What remains, is to find the conditions for which relations (4.4.2)–(4.4.4) can be made compatible with the constitutive equations. For that purpose we first consider the free energies. From equations (4.4.4) and (3.5.11) we obtain

$$
\begin{aligned}
^{IV}\psi = {}^{IV}U - {}^{IV}\eta\Theta &= {}^{III}U - {}^{III}\eta\Theta - B_i\frac{\mathcal{M}_i}{\rho} + \frac{1}{2}\rho\mu_0\frac{\mathcal{M}_i}{\rho}\frac{\mathcal{M}_i}{\rho} \\
&= {}^{III}\psi + \mathsf{P}_\alpha\mathsf{E}_\beta C_{\alpha\beta}^{-1} - \mathsf{B}_\alpha\mathsf{M}_\beta C_{\alpha\beta}^{-1} + \frac{\rho}{2\mu_0}\mathsf{M}_\alpha\mathsf{M}_\beta C_{\alpha\beta}^{-1} ,
\end{aligned}
\quad (4.4.6)
$$

where according to (3.4.29) and (3.4.30)

$$^{III}\psi = \tilde{\psi}(C_{\alpha\beta}, \mathsf{E}_\alpha, \mathsf{M}_\alpha, \Theta) . \quad (4.4.7)$$

Here, an important point must be mentioned, namely that the free energy function $^{IV}\psi$ on the far left in (4.4.6) is a function of $C_{\alpha\beta}, \mathsf{P}_\alpha, \mathsf{B}_\alpha$ and Θ. That this must be so can immediately be seen from (3.5.5)$_4$ and the definition of $^{IV}\psi$ in (3.5.11). On the other hand, $^{III}\psi$ as given in (4.4.7) is a function of $C_{\alpha\beta}, \mathsf{E}_\alpha, \mathsf{M}_\alpha, \Theta$, and, similarly, so must be the expression on the right-hand side of (4.4.6). If $\tilde{\psi}$ is used as free energy functional in a constitutive theory of model IV, then instead of (3.5.13) we obtain

$$\eta = -\frac{\partial \tilde{\psi}}{\partial \Theta} \, , \quad P_i = -\rho \frac{\partial \tilde{\psi}}{\partial \mathsf{E}_\alpha} F_{i\alpha} \, , \quad \mathcal{H}_i = \frac{\partial \tilde{\psi}}{\partial \mathsf{M}_\alpha} F_{i\alpha} \text{sgn } J \, , \qquad (4.4.8)$$

and

$$^{IV}t_{ij} = 2\rho \frac{\partial \tilde{\psi}}{\partial C_{\alpha\beta}} F_{i\alpha} F_{j\beta} - \mathcal{E}_i P_j + \mu_0 \mathcal{M}_i \mathcal{H}_j - \tfrac{1}{2}\delta_{ij}\mu_0 \mathcal{M}_k \mathcal{M}_k \, ,$$

in the derivation of which use has also been made of $(3.4.13)_2$ and of the relations

$$2\frac{\partial \rho}{\partial C_{\alpha\beta}} F_{i\alpha} F_{j\beta} = \frac{\partial \rho}{\partial F_{i\alpha}} F_{j\alpha} = -\rho \delta_{ij} \, , \qquad (4.4.9)$$

and

$$\frac{\partial C_{\gamma\delta}^{-1}}{\partial C_{\alpha\beta}} F_{i\alpha} F_{j\beta} = -\tfrac{1}{2}(F_{\gamma i}^{-1} F_{\delta j}^{-1} + F_{\delta i}^{-1} F_{\gamma j}^{-1}) \, . \qquad (4.4.10)$$

We note that relations (4.4.8) are not only compatible with (3.4.31) but also with (4.4.2); needless to say once more that in all those comparisons the non-relativistic approximation is employed. Moreover, the above comparison has been made for one set of dependent and independent constitutive variables and can of course also be repeated for all other possibilities. In view of the relation (4.4.4) this can be done easily. We shall not repeat the details here and conclude with the

Proposition. *Within the constitutive class of thermoelastic polarizable and magnetizable materials the* MAXWELL–MINKOWSKI *model III and the statistical model IV can be brought to a one-to-one correspondence provided that the internal energies* ^{III}U *and* ^{IV}U *are related by relation (4.4.4) and, furthermore, that the energy fluxes are the same.* ∎

There still remains one practical question. Given a free energy function $^{III}\psi$ and another one, $^{IV}\psi$, each a function of its own variables, how can it be decided that the theories according to the models III and IV are equivalent? To this end, observe that $^{IV}\psi$ as it occurs in (4.4.6) is a function of the variables $C_{\alpha\beta}, \mathsf{P}_\alpha, \mathsf{B}_\alpha$ and Θ, or

$$^{IV}\psi = \overset{+}{\psi}(C_{\alpha\beta}, \mathsf{P}_\alpha, \mathsf{B}_\alpha, \Theta) \, , \qquad (4.4.11)$$

whereas $^{III}\psi$ as given by (4.4.7) is a function of $C_{\alpha\beta}, \mathsf{E}_\alpha, \mathsf{M}_\alpha$ and Θ instead. We now can express $^{IV}\psi$ in terms of the same set of variables as $^{III}\psi$ by substituting into (4.4.11) for P_α the relation

$$\mathsf{P}_\alpha = -\frac{\partial \tilde{\psi}}{\partial \mathsf{E}_\beta} C_{\alpha\beta} \, , \qquad (4.4.12)$$

which follows from $(4.4.8)_2$ and $(2.6.7)_1$, and for B_α the expression

$$B_\alpha = \frac{\partial \tilde{\psi}}{\partial M_\beta} C_{\alpha\beta} + \frac{\rho}{\mu_0} M_\alpha \,, \tag{4.4.13}$$

following from $(4.4.8)_3$, $(3.4.13)_2$ and $(2.6.7)_{2,5}$. In this way we obtain

$$^{IV}\psi = \mathsf{F}\left(C_{\alpha\beta}, E_\alpha, M_\alpha, \Theta\right) := \overset{+}{\psi}\left(C_{\alpha\beta}, -\frac{\partial \tilde{\psi}}{\partial E_\beta} C_{\alpha\beta}, \frac{\partial \tilde{\psi}}{\partial M_\beta} C_{\alpha\beta} + \frac{\rho}{\mu_0} M_\alpha, \Theta\right).$$
$$\tag{4.4.14}$$

If this expression is substituted into (4.4.6) and in so doing an identity is obtained, in other words if

$$\overset{+}{\psi}\left(C_{\alpha\beta}, -\frac{\partial \tilde{\psi}}{\partial E_\beta} C_{\alpha\beta}, \frac{\partial \tilde{\psi}}{\partial M_\beta} C_{\alpha\beta} + \frac{\rho}{\mu_0} M_\alpha, \Theta\right)$$
$$= \tilde{\psi}(C_{\alpha\beta}, E_\alpha, M_\alpha, \Theta) - \frac{\partial \tilde{\psi}}{\partial E_\alpha} E_\alpha - \frac{\partial \tilde{\psi}}{\partial M_\alpha} M_\alpha - \frac{\rho}{2\mu_0} M_\alpha M_\beta C_{\alpha\beta}^{-1} \tag{4.4.15}$$

holds as an identity, then one condition that model III is equivalent to model IV is satisfied. Of course, the energy flux vectors must also be expressed in the same variables and must also be the same functions of these variables.

The above procedure illustrates how, in practice, two theories can be decided to be equivalent. The procedure may in reality be very elaborate, but it shows explicitly that equivalence of theories amounts to comparison of thermodynamic potentials. We have done this here for the sets (P_α, B_α) in theory IV and (E_α, M_α) in theory III. In the table below we show which variable sets of these two formulations naturally correspond to each other. The above investigation can be made for each of them, of course.

Model *IV*	Model *III*
$P_\alpha,\ B_\alpha$	$P_\alpha,\ M_\alpha$
$P_\alpha,\ M_\alpha$	$E_\alpha,\ H_\alpha$
$E_\alpha,\ M_\alpha$	$P_\alpha,\ H_\alpha$
$E_\alpha,\ B_\alpha$	$P_\alpha,\ M_\alpha$

4.5 Comparison of the Models IV and V

In this section we compare the statistical model IV and the LORENTZ model V. With regard to the electromagnetic variables these models differ only in the definitions of the magnetization vectors, which are related by

$$M_i^S = M_i^L - e_{ijk}\dot{x}_j P_k^L = \mathcal{M}_i - e_{ijk}\dot{x}_j P_k \,. \tag{4.5.1}$$

Here, as before we have set

$$M_i^L = \mathcal{M}_i \,, \qquad \text{and} \qquad P_i^L = P_i \,. \tag{4.5.2}$$

If these relations are used, the MAXWELL equations of one formulation transform into those of the other (see (3.5.2) and (3.6.2)). With the transformation rule (4.5.1) it is also straightforward to relate the body forces. This is most easily achieved by comparing the MAXWELL stresses $(3.5.8)_1$ and $(3.6.8)_1$, since in both formulations the electromagnetic momenta are equal to zero. This gives

$$
\begin{aligned}
{}^V t_{ij}^M &= {}^{IV} t_{ij}^M - E_i^S P_j + M_i^S B_j - \delta_{ij} B_k M_k^S - e_{ikl} P_k B_l \dot{x}_j \\
&= {}^{IV} t_{ij}^M - \mathcal{E}_i P_j + \mathcal{M}_i B_j - \delta_{ij} B_k \mathcal{M}_k \,.
\end{aligned} \tag{4.5.3}
$$

In order for the balance laws of momentum of the two formulations to be equivalent, it is thus necessary that the stress tensors be related by

$$
{}^V t_{ij} = {}^{IV} t_{ij} + \mathcal{E}_i P_j - \mathcal{M}_i B_j + \delta_{ij} B_k \mathcal{M}_k \,. \tag{4.5.4}
$$

This also implies that, if balance of moment of momentum is satisfied in model V, in which the stress tensor is symmetric, so it is in model IV.

For a comparison of the energy equations, we first write $(3.5.5)_4$ in the form

$$
\begin{aligned}
{}^{IV}(\rho \dot{U}) &= {}^{IV} t_{(ij)} \dot{x}_{i,j} - {}^{IV} q_{i,i} + \mathcal{J}_i \mathcal{E}_i + \mathcal{E}_i (\overset{*}{P}_i + e_{ijk} \mathcal{M}_{k,j}) \\
&\quad + (e_{ijk} \mathcal{E}_j \mathcal{M}_k)_{,i} + (\mathcal{E}_i P_j - \mathcal{M}_i B_j + \delta_{ij} \mathcal{M}_k B_k) \dot{x}_{(i,j)} \,,
\end{aligned} \tag{4.5.5}
$$

for the derivation of which use has also been made of (3.5.6).

Comparing (4.5.5) with the balance law of energy of model V, $(3.6.6)_4$, using thereby relations (4.5.3) and (4.5.4), we infer that the two balance laws are equivalent if

$$
{}^V(\rho U) = {}^{IV}(\rho U) \,, \tag{4.5.6}
$$

and

$$
{}^V q_i = {}^{IV} q_i - e_{ijk} \mathcal{E}_j \mathcal{M}_k \,. \tag{4.5.7}
$$

Note that this relation between the energy fluxes was introduced already in Sect. 3.5 (eq. (3.6.12)), where it was used to bring the entropy inequality into its expected classical form. With relation (4.5.7) the reduced entropy inequalities (3.5.14) and (3.6.19) are then also identical. Bearing the transformation rules (4.5.3), (4.5.4) and (4.5.7) in mind, we further recognize that the jump conditions (3.5.9) and (3.6.9) are also the same.

In order to obtain complete equivalence, there remains to consider the constitutive equations. In Chap. 3 the constitutive theories for both models were developed using constitutive assumptions of the form

$$
C = \overset{+}{C} (F_{i\alpha}, P_i/\rho, B_i, \Theta, \Theta_{,i}, \mathcal{Q}) \,.
$$

The results are listed in the two sets of equations (3.5.13) and (3.6.18), and they reveal that relations (4.5.3) and (4.5.4) are satisfied, provided that

$$^{V}\overset{+}{\psi}\left(C_{\alpha\beta}, \mathsf{P}_{\alpha}, \mathsf{B}_{\alpha}, \Theta\right) = \,^{IV}\overset{+}{\psi}\left(C_{\alpha\beta}, \mathsf{P}_{\alpha}, \mathsf{B}_{\alpha}, \Theta\right), \tag{4.5.8}$$

which is in accordance with (4.5.6). Furthermore, this simultaneously guarantees that entropy and electromotive intensity are the same in both formulations.

The reader may show himself that these results remain correct when some of the dependent and independent variables are interchanged. We have thus proved

Proposition. *Within the constitutive class of thermoelastic polarizable and magnetizable materials the non-relativistic statistical model (IV) and the* LORENTZ *model (V) are equivalent provided that the two free energies are the same functions of their variables and the energy flux vectors of the two formulations are related by (4.5.7).*   ∎

4.6 Conclusions

In summary, we have shown that all field interaction models presented in Chap. 3 and describing polarizable and magnetizable thermoelastic materials are equivalent to each other and differ only in terms which in the context of the non-relativistic approximation have been considered to be negligibly small anyhow. We may thus justly call these theories to be equivalent. On the other hand, it is true that results obtained for these various models could be different in exactly these neglected terms. They have been assumed to be unimportant in all of the above theories, because each of them neglects terms that are preceded by a c^{-2}-factor. Hence, if experiments for a material obeying our constitutive assumptions should deviate from what any of these formulations predicts, then either the constitutive class must be extended or the non-relativistic approximation must be replaced by a semi-relativistic formulation.

There exists a semi-relativistic version of the statistical model (cf. [53]), whereas a semi-relativistic counterpart of the LORENTZ model may be derived from a fully relativistic theory of GROT and ERINGEN [78] by merely neglecting terms of the order of V^2/c^2. In the same way, still other semi-relativistic formulations can be found in [177], Ch. 7. For all these semi-relativistic formulations the constitutive treatment differs from what we have presented here. Therefore, the equivalence proof of all these models – although claimed to be established by PENFIELD and HAUS,PENFIELD67 – still remains to be done.

(In [177], neither a constitutive theory nor jump conditions are presented.) This is still one of the important future research topics to be attacked.[1]

We have performed the equivalence proof for a polarizable and magnetizable solid only, which must be regarded as a severe restriction. There are more complex material behaviors than that dealt with here. For all those this proof is not established yet. One immediate generalization is the inclusion of viscosity in the sense that apart from $F_{i\alpha}$ its time rate $\dot{F}_{i\alpha}$ may be included amongst the independent constitutive variables (cf. [9, 134] or [92]). When this is done, it turns out that the stress tensor may be decomposed into two parts:

$$t_{ij} = t_{ij}^{th} + t_{ij}^{e} . \tag{4.6.1}$$

One part, namely t_{ij}^{th}, is then given by a thermodynamic potential and is independent of $\dot{F}_{i\alpha}$. The second, dissipative, part t_{ij}^{e} is called *extra stress* and is given by an independent constitutive relation involving $\dot{F}_{i\alpha}$. Since our equivalence proofs have been performed for $t_{ij}^{e} \equiv 0$, it immediately follows that models I-V are also equivalent for a viscous polarizable and magnetizable thermoelastic body if, in addition to the conditions stated in this chapter, the extra stress remains the same in all formulations.

Another possible extension is obtained when for ferromagnetic materials spin interactions are taken into account. These effects can be reckoned with by including in the constitutive theory the magnetization gradients and by introducing an extra moment of momentum density emanating from the magnetic spin action (see e.g. [9, 249, 233]). This results in two extra terms in the angular momentum law, namely the magnetic spin and the gradient of a magnetic couple stress tensor. The latter is determined by the derivative of the thermodynamic potential (free energy) with respect to magnetization gradients. Moreover, the energy equation must also be supplemented by two

---

[1] On the other hand, if the semi-relativistic formulation is considered to be too complicated, still a small improvement of our non-relativistic approximation can be obtained by changing from SI-units to GAUSSian units. In that case, as already said several times before, some c^{-2}-terms that are neglected in our non-relativistic approximation become c^{-1}-terms and must be retained. The most striking effect caused by this change of units, but not the only one, is the fact that the electromagnetic momentum vector $\mathbf{g}$ is retained, which becomes (in GAUSSian units)

$$\frac{1}{c}(\boldsymbol{E} \times \boldsymbol{H}) ,$$

for the models I, II and III, and

$$\frac{1}{c}(\boldsymbol{E} \times \boldsymbol{B}) ,$$

for the models IV and V. If all transformations are executed in a consistent way, all models remain equivalent also in a non-relativistic approximation based on GAUSSian units.

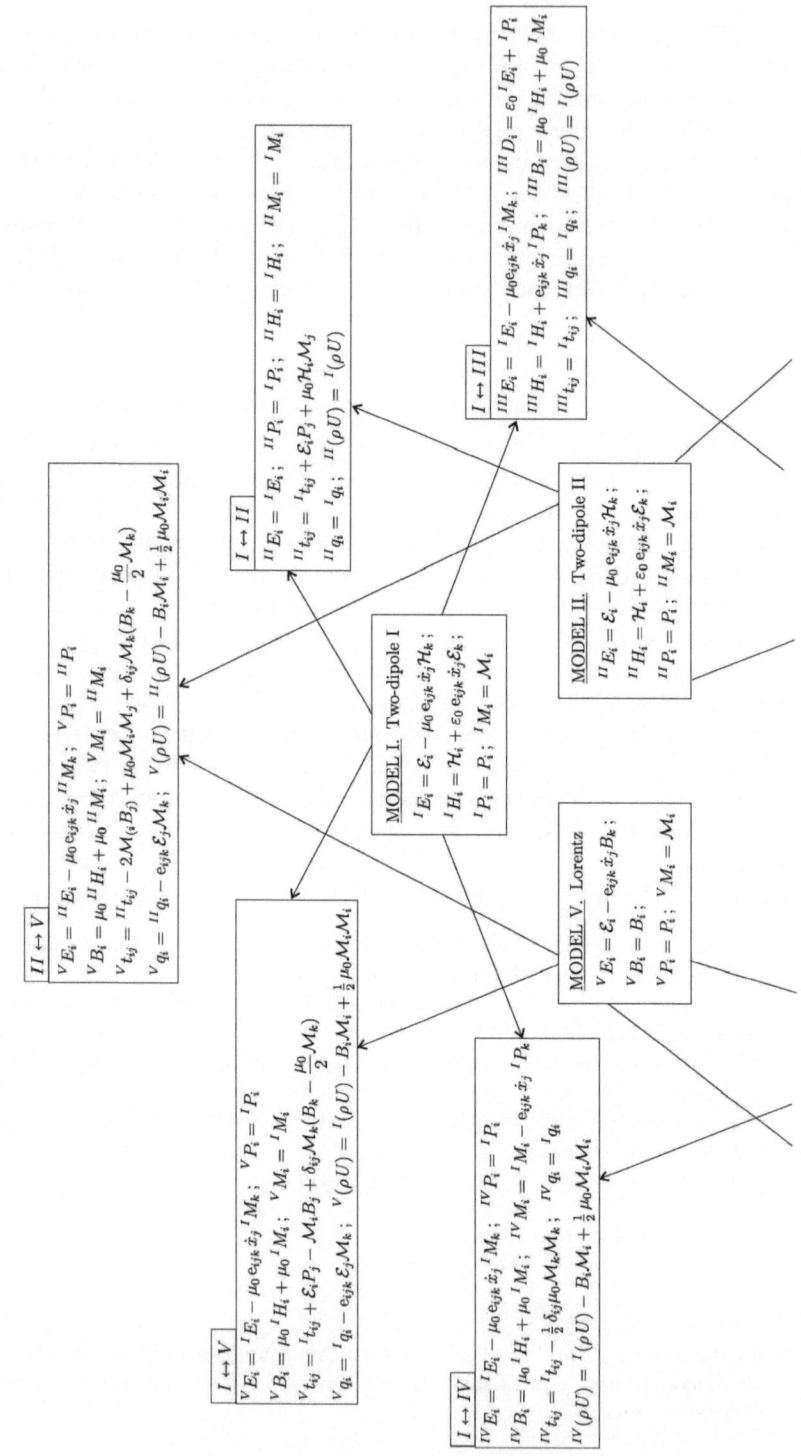

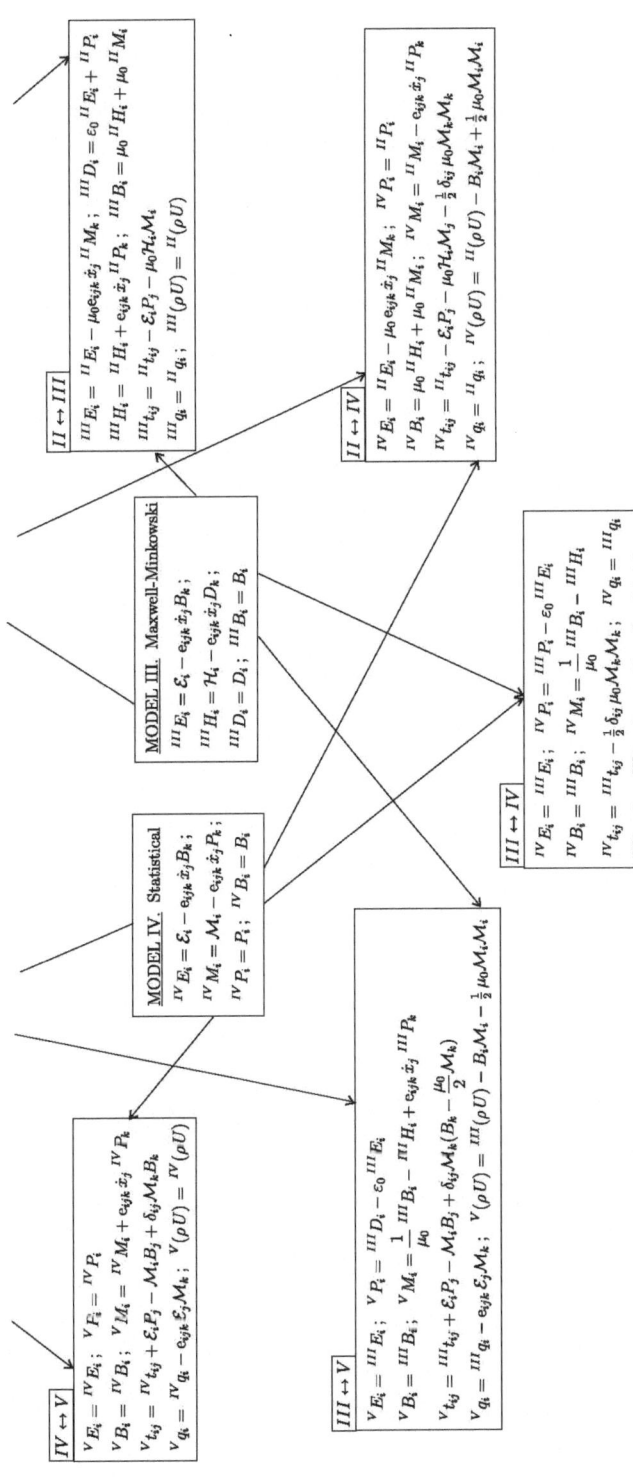

Table 4.1. Definitions and transformations of electromagnetic variables of the five different formulations of electrodynamics and their mutual interrelations

terms, namely the kinetic energy of the magnetic spin and the energy flux due to the magnetic couple stress. Since these terms always seem to be introduced in a unique way, they do not lead to differences in the respective formulations. Analogous remarks hold for constitutive theories in which polarization gradients are taken into account (see e.g. [142, 225] or [42]).

We conclude this chapter with a Table 4.1 that lists all the pertinent interrelations between the various models. In this table, the relations between the four basic electromagnetic fields, between the stresses t_{ij}, the internal energies U and the energy fluxes q_i are listed. The differences between the latter three quantities are expressed in the fields

$$\mathcal{E}_i, P_i, B_i, \mathcal{M}_i \,,$$

which are identical in all formulations. We did not list the remaining variables, which are also the same in all formulations. They are

$$\mathcal{Q}, \mathcal{J}_i, \rho, x_i = \chi_i(\boldsymbol{X}, t), \Theta, \eta \,.$$

5 Material Description

5.1 Motivation

In the last two chapters the governing equations of field matter interaction were chiefly based on the *spatial* or EULER*ian description*. Only very briefly the *material* or LAGRANGE*an formulation* was given. Such a formulation is of advantage in describing the deformation of solids, because the boundary conditions for solids are usually prescribed on the undeformed body, which is generally the body in its reference configuration. As a consequence, any theory describing deformable solids in the electromagnetic fields should from the outset be given in the material description. This is not done in general; on the contrary, in almost all theories the spatial description is applied. The LAGRANGEan formulation is introduced only afterwards, and if so, only by introducing some approximations, e.g. linearizations. These linearization procedures, although being straightforward are nevertheless quite cumbersome. They become an almost trivial matter when the material description of the governing field equations is used from the outset.

It thus should be apparent that the LAGRANGEan formulation is a necessity. The reason for presenting it this late is that it is relatively unknown, so that the equations do at first sight not look familiar. They are not new, however, and have been derived before (see HUTTER [92, 95]). Use of LAGRANGEan variables is also made by ALBLAS, [10], for the linearization of some specialized topics as quasi-electroelastostatics and quasi-magnetoelastostatics. This approach can be seen as an improvement of the method used by TOUPIN, [241]. In his paper on the non-relativistic electrodynamics of deformable media, PRECHTL, [185], devoted one section to a LAGRANGEan formulation of the balance equations and the jump conditions (no constitutive equations are used by him). Both works of ALBLAS and PRECHTL are based on the CHU model. In the following, we shall describe the two-dipole models first and shall then pass on to the description of the statistical and the LORENTZ formulation; we shall conclude with the MAXWELL–MINKOWSKI model.

K. Hutter et al.: *Electromagnetic Field Matter Interaction in Thermoelastic Solids and Viscous Fluids*, Lect. Notes Phys. **710**, 103–146 (2006)
DOI 10.1007/3-540-37240-7_5   © Springer-Verlag Berlin Heidelberg 2006

5.2 Material Description
of the Two–Dipole Models (Models I and II)

In Chap. 2, Sect. 2.3.3, the LAGRANGEan forms of the MAXWELL equations was briefly presented. Those equations appeared in one particular formulation. We could, as was done for the EULERian description, introduce the various models by simply performing variable transformations. However, little insight into the meaning of the new variables would be gained in doing so, whereas a fresh derivation is enlightening.

The CHU formulation is based on the postulations that

(i) only two vector quantities are necessary to describe the electromagnetic fields in free space and,

(ii) that material bodies contribute toward these fields by acting as sources for these fields.

With regard to postulate (i) it is advantageous for the derivation of the LAGRANGEan–MAXWELL equations to work formally with four field vectors and to relate the remaining two to the former ones. Concerning postulate (ii) we assume it to be known that magnetization and polarization act as charge and current distributions. Based on such a conception one arrives at the equations (see HUTTER [95])

$$\int_{\partial V} B_i^a \, d\nu = \int_V \mathcal{Q}^{\mathrm{m}} \, d\nu \,,$$

$$\int_{\partial S} \mathcal{E}_i \, dl_i + \frac{d}{dt} \int_S B_i^a \, da_i = -\int_S \mathcal{J}_i^{\mathrm{m}} \, da_i \,,$$

$$\int_{\partial V} D_i^a \, da_i = \int_V (\mathcal{Q} + \mathcal{Q}^{\mathrm{P}}) \, d\nu \,, \qquad (5.2.1)$$

$$\int_{\partial S} \mathcal{H}_i \, dl_i - \frac{d}{dt} \int_S D_i^a \, da_i = \int_S (\mathcal{J}_i + \mathcal{J}_i^{\mathrm{P}}) \, d\nu \,,$$

$$\frac{d}{dt} \int_V \mathcal{Q} \, d\nu + \int_{\partial V} \mathcal{J}_i \, da_i = 0 \,.$$

Here, $\mathcal{E}_i, \mathcal{H}_i, \mathcal{Q}$ and $\mathcal{J}_i$ are as introduced before; they are the electromotive intensity, the effective magnetic field strength and the charge and conductive current densities due to free charges. Moreover, $\mathcal{Q}^{\mathrm{P}}, \mathcal{J}_i^{\mathrm{P}}, \mathcal{Q}^{\mathrm{m}}$ and $\mathcal{J}_i^{\mathrm{m}}$ are the charge and conductive current densities due to polarization and magnetization, and in the CHU formulation (compare also (3.3.48))

$$\int_{\mathcal{V}} \mathcal{Q}^P d\nu = -\int_{\partial\mathcal{V}} P_i^C da_i \,, \quad \int_{\mathcal{V}} \mathcal{Q}^{\mathrm{m}} d\nu = -\int_{\partial\mathcal{V}} \mu_0 M_i^C da_i \,,$$

$$\int_{S} \mathcal{J}_i^P da_i = \frac{d}{dt}\int_{S} P_i^C da_i \,, \quad \int_{S} \mathcal{J}_i^{\mathrm{m}} da_i = \frac{d}{dt}\int_{S} \mu_0 M_i^C da_i \,. \tag{5.2.2}$$

Finally, we introduced in (5.2.1) the *auxiliary* fields B_i^a and D_i^a, which are related to the electric and magnetic field strength according to

$$B_i^a = \mu_0 H_i^C \,, \qquad D_i^a = \varepsilon_0 E_i^C \,. \tag{5.2.3}$$

These are the MAXWELL–LORENTZ aether relations. To derive a material description from (5.2.1) and (5.2.2) one must by means of relations (2.2.25) simply transform the integrals over spatial volumes and spatial surfaces into integrals over reference volumes and reference surfaces. How this is done is explained in Sect. 2.2.3, and hence we only give results:

$$\int_{\partial V_R} \mathbb{B}_\alpha^a dA_\alpha = \int_{V_R} \mathbb{Q}^{\mathrm{m}} dV \,,$$

$$\int_{\partial S_R} \mathbb{E}_\alpha dL_\alpha + \frac{d}{dt}\int_{S_R} \mathbb{B}_\alpha^a dA_\alpha = -\int_{S_R} \mathbb{J}_\alpha^{\mathrm{m}} dA_\alpha \,,$$

$$\int_{\partial V_R} \mathbb{D}_\alpha^a dA_\alpha = \int_{V_R} (\mathbb{Q} + \mathbb{Q}^{\mathrm{P}}) dV \,, \tag{5.2.4}$$

$$\int_{\partial S_R} \mathbb{H}_\alpha dL_\alpha - \frac{d}{dt}\int_{S_R} \mathbb{D}_\alpha dA_\alpha = \int_{S_R} (\mathbb{J} + \mathbb{J}_\alpha^{\mathrm{P}}) dA_\alpha \,,$$

$$\frac{d}{dt}\int_{V_R} \mathbb{Q} dV + \int_{\partial V_R} \mathbb{J}_\alpha dA_\alpha = 0,$$

with

$$\int_{V_R} \mathbb{Q}^{\mathrm{P}} dV = \int_{\partial V_R} \mathbb{P}_\alpha^C dA_\alpha \,, \quad \int_{V_R} \mathbb{Q}^{\mathrm{m}} dV = -\int_{\partial V_R} \mu_0 \mathbb{M}_\alpha^C dA_\alpha \,,$$

$$\int_{S_R} \mathbb{J}_\alpha^{\mathrm{P}} dA_\alpha = \frac{d}{dt}\int_{S_R} \mathbb{P}_\alpha^C dA_\alpha \,, \quad \int_{S_R} \mathbb{J}_\alpha^{\mathrm{m}} dA_\alpha = \frac{d}{dt}\int_{S_R} \mu_0 \mathbb{M}_\alpha^C dA_\alpha \,. \tag{5.2.5}$$

The following definitions have been used

$$\mathbb{B}_\alpha^a = JF_{\alpha i}^{-1}B_i^a \, , \qquad\qquad B_i^a = J^{-1}F_{i\alpha}\mathbb{B}_\alpha^a \, ,$$

$$\mathbb{J}_\alpha = |J|F_{\alpha i}^{-1}\mathcal{J}_i \, , \qquad\qquad \mathcal{J}_i = |J^{-1}|F_{i\alpha}\mathbb{J}_\alpha \, ,$$

$$\mathbb{P}_\alpha^C = |J|F_{\alpha i}^{-1}P_i^C \, , \qquad\qquad P_i^C = |J^{-1}|F_{i\alpha}\mathbb{P}_\alpha^C \, ,$$

$$\mathbb{Q} = |J|\mathcal{Q} \, , \qquad\qquad\quad \mathcal{Q} = |J^{-1}|\mathbb{Q} \, ,$$

$$\mathbb{E}_\alpha = F_{i\alpha}\mathcal{E}_i \, , \qquad\qquad\quad \mathcal{E}_i = F_{\alpha i}^{-1}\mathbb{E}_\alpha \, , \qquad\qquad (5.2.6)$$

$$\mathbb{D}_\alpha^a = |J|F_{\alpha i}^{-1}D_i^a \, , \qquad\qquad D_i^a = |J^{-1}|F_{i\alpha}\mathbb{D}_\alpha^a \, ,$$

$$\mathbb{H}_\alpha = F_{i\alpha}\mathcal{H}_i \operatorname{sgn}J \, , \qquad\quad \mathcal{H}_i = F_{\alpha i}^{-1}\mathbb{H}_\alpha \operatorname{sgn}J \, ,$$

$$\mu_0\mathbb{M}_\alpha^C = JF_{\alpha i}^{-1}\mu_0 M_i^C \, , \quad \mu_0 M_i^C = J^{-1}F_{i\alpha}\mathbb{M}_\alpha^C \, ,$$

some of which were already defined in (2.2.39) but are repeated here for ease of reference. In the above equations integration is over material parts $V_R, \partial V_R, S_R$ and ∂S_R, and all variables are thought to be functions of X_α and t so that, as usual, d/dt denotes the time derivative at a fixed particle.

The equations (5.2.4) and (5.2.5) may be combined, and for sufficiently smooth fields they yield

$$\mathbb{B}_{\alpha,\alpha}^a + \mu_0\mathbb{M}_{\alpha,\alpha}^C = 0 \, ,$$

$$\dot{\mathbb{B}}_\alpha^a + \mu_0\dot{\mathbb{M}}_\alpha^C + e_{\alpha\beta\gamma}\mathbb{E}_{\gamma,\beta} = 0 \, ,$$

$$\mathbb{D}_{\alpha,\alpha}^a + \mathbb{P}_{\alpha,\alpha} = \mathbb{Q} \, , \qquad\qquad (5.2.7)$$

$$-\dot{\mathbb{D}}_\alpha^a - \dot{\mathbb{P}}_\alpha + e_{\alpha\beta\gamma}\mathbb{H}_{\gamma,\beta} = \mathbb{J}_\alpha \, ,$$

$$\mathbb{J}_{\alpha,\alpha} + \dot{\mathbb{Q}} = 0 \, .$$

Equations (5.2.7) are the MAXWELL equations in the material description. We shall call $\mathbb{E}_\alpha$ and $\mathbb{H}_\alpha$ the LAGRANGEan electric and magnetic field strengths, $\mathbb{P}_\alpha$ and $\mathbb{M}_\alpha^C$ the LAGRANGEan polarization and magnetization of the CHU formulation and $\mathbb{Q}$ and $\mathbb{J}_\alpha$ the LAGRANGEan free charge and current densities. Since, in a non-relativistic approximation, the meaning of $\mathbb{P}_\alpha$ is unique for all formulations the upper index C for $\mathbb{P}_\alpha$ has been dropped. This is not the case for $\mathbb{M}_\alpha$ and, therefore, the index C must be retained there.

Before we proceed, a few comments seem to be in order: Firstly, all quantities listed on the right column of (5.2.6) are objective vectors ($\mathcal{J}_i, P_i^C, \mathcal{E}_i, D_i^a$), objective axial vectors ($B_i^a, M_i^C, \mathcal{H}_i$) and an objective scalar ($\mathcal{Q}$) under the EUCLIDian transformation group. As a consequence, all LAGRANGEan variables (listed on the left column of (5.2.6)) are scalars under this group, as they must. Secondly, we can easily relate the variables occurring in (5.2.7) to those listed in (2.2.39), and indeed, this transformation is achieved by setting

$$\mathbb{B}_\alpha := \mathbb{B}_\alpha^a + \mu_0\mathbb{M}_\alpha^C \, , \qquad \mathbb{D}_\alpha := \mathbb{D}_\alpha^a + \mathbb{P}_\alpha \, , \qquad (5.2.8)$$

where $\mathbb{D}_\alpha$ and $\mathbb{B}_\alpha$ may be defined also in terms of B_i and D_i as was done in (2.2.39). Thirdly, in order to determine the auxiliary fields $\mathbb{B}_\alpha^a$ and $\mathbb{D}_\alpha^a$, we simply must use the MAXWELL–LORENTZ aether relations and must invoke the transformations (5.2.6). When this is done and when terms proportional to c^{-2} are discarded, a straightforward calculation shows that

$$\mathbb{D}_\alpha^a = \varepsilon_0 |J| C_{\alpha\beta}^{-1} \mathbb{E}_\beta \ , \quad \mathbb{B}_\alpha^a = \mu_0 |J| C_{\alpha\beta}^{-1} \mathbb{H}_\beta \ . \tag{5.2.9}$$

These relations thus hold in a non-relativistic theory. When relativistic terms are retained they become much more complicated as can be seen in [95].

Of course, as was the case in the EULERian description, the global equations (5.2.4) and (5.2.5) also imply jump conditions, which follow from (2.4.5) by invoking the definitions (5.2.8). They then read

$$\begin{aligned}
&[\![\mathbb{B}_\alpha^a + \mu_0 \mathbb{M}_\alpha^C]\!] N_\alpha = 0 \ , \\
&[\![\mathbb{D}_\alpha^a + \mathbb{P}_\alpha]\!] N_\alpha = 0 \ , \\
&e_{\alpha\beta\gamma}[\![\mathbb{E}_\beta]\!] N_\gamma + [\![(\mathbb{B}_\alpha^a + \mu_0 \mathbb{M}_\alpha^C) W_N]\!] = 0 \ , \\
&e_{\alpha\beta\gamma}[\![\mathbb{H}_\beta]\!] N_\gamma - [\![(\mathbb{D}_\alpha^a + \mathbb{P}_\alpha) W_N]\!] = 0 \ , \\
&[\![\mathbb{J}_\alpha]\!] N_\alpha + [\![\mathbb{Q} W_N]\!] = 0 \ .
\end{aligned} \tag{5.2.10}$$

To complete the description of the electro-mechanical interaction the balance laws of mass, momentum, moment of momentum and energy must be given. They are listed in (2.2.27)–(2.2.30), and their local forms appear in (2.2.32). It thus suffices to write down the material counterparts of the electromagnetic body force, body couple and energy supply. They are $\rho_0 F_i^e, \rho_0 L_{ij}^e$ and $\rho_0 r^e$, respectively. To calculate them for models I and II we only need to convert the expressions (3.3.15), (3.3.7)$_2$, (3.3.12) and (3.3.49) to LAGRANGEan form omitting thereby c^{-2}-terms. In this calculation it is also advantageous to use relation (4.2.1):

$$\rho^I F_i^e = \rho^{II} F_i^e + (\mathcal{E}_i P_j + \mu_0 \mathcal{H}_i M_j^C)_{,j} \ .$$

The details are tedious even though they are straightforward, and what one obtains reads as follows:

For **model I**

$$\begin{aligned}
{}^I(\rho_0 F_i^e) &= F_{\alpha i}^{-1}\left((\mathbb{Q} - \mathbb{P}_{\beta,\beta})\mathbb{E}_\alpha + e_{\alpha\beta\gamma}(\mathbb{J}_\beta + \dot{\mathbb{P}}_\beta)\mathbb{B}_\gamma^a - \mu_0 \mathbb{M}_{\beta,\beta}^C \mathbb{H}_\alpha\right) \\
&\quad + \left(F_{\beta i}^{-1}(\mathbb{P}_\alpha \mathbb{E}_\beta + \mu_0 \mathbb{M}_\alpha^C \mathbb{H}_\beta)\right)_{,\alpha} \ , \\
{}^I(\rho_0 L_{ij}^e) &= F_{[i\alpha} F_{\beta j]}^{-1}(\mathbb{P}_\alpha \mathbb{E}_\beta + \mu_0 \mathbb{M}_\alpha^C \mathbb{H}_\beta) \ , \\
{}^I(\rho_0 r^e) &= \mathbb{J}_\alpha \mathbb{E}_\alpha + \dot{\mathbb{P}}_\alpha \mathbb{E}_\alpha + \mu_0 \dot{\mathbb{M}}_\alpha^C \mathbb{H}_\alpha + F_{\alpha i}^{-1}(\mathbb{E}_\alpha \mathbb{P}_\beta + \mu_0 \mathbb{H}_\alpha \mathbb{M}_\beta^C)\dot{F}_{i\beta} \ ,
\end{aligned} \tag{5.2.11}$$

and for **model II**

$$^{II}(\rho_0 F_i^e) = F_{\alpha i}^{-1}\{(\mathbb{Q} - \mathbb{P}_{\beta,\beta})\mathbb{E}_\alpha + e_{\alpha\beta\gamma}(\mathbb{J}_\beta + \dot{\mathbb{P}}_\beta)\mathbb{B}_\gamma^a - \mu_0 \mathbb{M}_{\beta,\beta}^C \mathbb{H}_\alpha\},$$
$$^{II}(\rho_0 L_{ij}^e) = 0,$$
$$^{II}(\rho_0 r^e) = \mathbb{J}_\alpha \mathbb{E}_\alpha + \dot{\mathbb{P}}_\alpha \mathbb{E}_\alpha + \mu_0 \dot{\mathbb{M}}_\alpha^C \mathbb{H}_\alpha,$$

(5.2.12)

in the derivation of which also the identities

$$e_{ijk} F_{i\alpha} F_{j\beta} F_{k\gamma} = J e_{\alpha\beta\gamma}, \qquad \left(\frac{1}{J} F_{i\alpha}\right)_{,i} = 0,$$

$$\overset{*}{P_i} = |J^{-1}| F_{i\alpha} \overset{*}{\mathbb{P}}_\alpha, \qquad\qquad P_{i,i} = |J^{-1}| \mathbb{P}_{\alpha,\alpha},$$

(5.2.13)

$$\overset{*}{M_i^C} = J^{-1} F_{i\alpha} \dot{\mathbb{M}}_\alpha^C, \qquad\qquad M_{i,i}^C = J^{-1} \mathbb{M}_{\alpha,\alpha}^C,$$

are used.

Two conclusions are readily drawn from the above expressions. Firstly, mere inspection of (5.2.11) and (5.2.12) shows that the expressions for body force and body couple are (in a non-relativistic sense) an objective vector and an objective skew-symmetric tensor, respectively. It is also seen that the electromagnetic energy supply is an objective scalar only in model II. Needless to state which model seems to be simpler. Secondly, the body force expression $(5.2.12)_1$ is easily interpretable. It is composed of an electric and a magnetic LORENTZ force (compare $(3.3.46)_1$)

$$^{II}(\rho_0 F_i^e) = F_{\alpha i}^{-1} \left((\mathbb{Q}^e \mathbb{E}_\alpha + e_{\alpha\beta\gamma} \mathbb{J}_\beta^e \mathbb{B}_\gamma^a) + (\mathbb{Q}^m \mathbb{H}_\alpha + e_{\alpha\beta\gamma} \mathbb{J}_\beta^m \mathbb{D}_\gamma^a)\right), \quad (5.2.14)$$

where

$$\mathbb{Q}^e = \mathbb{Q} - \mathbb{P}_{\alpha,\alpha}, \qquad \mathbb{J}_\beta^e = \mathbb{J} + \dot{\mathbb{P}}_\beta,$$
$$\mathbb{Q}^m = -\mu_0 \mathbb{M}_{\alpha,\alpha}^C, \qquad \mathbb{J}_\beta^m = \mu_0 \dot{\mathbb{M}}_\beta^C,$$

(5.2.15)

as follows from (5.2.2).

As was done formally in (2.4.10) we can also derive the MAXWELL stress tensor $T_{i\alpha}^M$, the electromagnetic momentum G_i, the energy flux Π_α and the energy density Ω. Formally, this is done by expressing the representations (3.3.10) and (3.3.51) in LAGRANGEan variables and using the transformations (see (2.4.10))

$$T_{i\alpha}^M = |J|(t_{ij}^M - g_i \dot{x}_j) F_{\alpha j}^{-1}, \qquad G_i = |J| g_i,$$
$$\Pi_\alpha = |J|(\pi_i - \omega \dot{x}_i) F_{\alpha i}^{-1}, \qquad \Omega = |J| \omega.$$

(5.2.16)

The calculations are again easy, though rather long, and what one obtains in a non-relativistic formulation can be written as follows:

For **model II**

$$^{II}T^M_{i\alpha} = F^{-1}_{\beta i}(\mathbb{D}^a_\alpha \mathbb{E}_\beta + \mathbb{B}^a_\alpha \mathbb{H}_\beta) - \tfrac{1}{2}F^{-1}_{\alpha i}(\mathbb{D}^a_\beta \mathbb{E}_\beta + \mathbb{B}^a_\beta \mathbb{H}_\beta) \,,$$

$$^{II}G_i = 0 \,,$$

$$^{II}\Pi_\alpha = -e_{\alpha\beta\gamma}\mathbb{E}_\beta \mathbb{H}_\gamma - \tfrac{1}{2}(\mathbb{D}^a_\beta \mathbb{E}_\beta + \mathbb{B}^a_\beta \mathbb{H}_\beta)F^{-1}_{\alpha i}\dot{x}_i \qquad (5.2.17)$$

$$+ (\mathbb{D}^a_\alpha \mathbb{E}_\beta + \mathbb{B}^a_\alpha \mathbb{H}_\beta)F^{-1}_{\beta i}\dot{x}_i = -e_{\alpha\beta\gamma}\mathbb{E}_\beta \mathbb{H}_\gamma + {}^{II}T^M_{i\alpha}\dot{x}_i \,,$$

$$^{II}\Omega = -\tfrac{1}{2}(\mathbb{E}_\alpha \mathbb{D}^a_\alpha + \mathbb{H}_\alpha \mathbb{B}^a_\alpha) \,.$$

For **model I**

$$^{I}T^M_{i\alpha} = {}^{II}T^M_{i\alpha} + F^{-1}_{\beta i}(\mathbb{P}_\alpha \mathbb{E}_\beta + \mu_0 \mathbb{M}^C_\alpha \mathbb{H}_\beta) \,, \qquad {}^{I}G_i = {}^{II}G_i = 0 \,,$$

$$^{I}\Pi_\alpha = {}^{II}\Pi_\alpha + (\mathbb{P}_\alpha \mathbb{E}_\beta + \mu_0 \mathbb{M}^C_\alpha \mathbb{H}_\beta)F^{-1}_{\beta i}\dot{x}_i \,, \qquad {}^{I}\Omega = {}^{II}\Omega \,. \qquad (5.2.18)$$

The jump conditions for the two models can now be obtained by simply substituting (5.2.17) or (5.2.18) into the general jump conditions (2.4.12).

There remains the presentation of the constitutive theory. For that purpose, we first eliminate $\rho_0 r^{\text{ext}}$ from (2.3.7) and (2.2.32)$_3$. When making use of (5.2.14)$_3$, the following entropy inequality for *model II* is obtained:

$$-\rho_0\dot{\psi} - \rho_0\eta\dot{\Theta} + {}^{II}T_{i\alpha}\dot{F}_{i\alpha} + \dot{\mathbb{P}}_\alpha \mathbb{E}_\alpha + \mu_0\dot{\mathbb{M}}^C_\alpha \mathbb{H}_\alpha + \mathbb{J}_\alpha \mathbb{E}_\alpha - \frac{Q_\alpha \Theta_{,\alpha}}{\Theta} \geq 0 \,. \quad (5.2.19)$$

Here we have introduced the HELMHOLTZ free energy

$$\psi = U - \eta\Theta \,, \qquad (5.2.20)$$

and have also set

$$\Phi_\alpha = \frac{Q_\alpha}{\Theta} \,. \qquad (5.2.21)$$

The entropy inequality for model I can be found from (5.2.19) by introducing into the latter the following relation between the PIOLA-KIRCHHOFF stress tensors of the two models (compare (4.2.2))

$$^{I}T_{i\alpha} = {}^{II}T_{i\alpha} - F^{-1}_{\beta i}(\mathbb{P}_\alpha \mathbb{E}_\beta + \mu_0 \mathbb{M}^C_\alpha \mathbb{H}_\beta) \,. \qquad (5.2.22)$$

Inequality (5.2.19) suggests to establish constitutive relations of the form

$$\mathsf{C} = \hat{\mathsf{C}}\left(C_{\alpha\beta}, \frac{\mathbb{P}_\alpha}{\rho_0}, \frac{\mu_0 \mathbb{M}^C_\alpha}{\rho_0}, \Theta, \Theta_{,\alpha}, \mathbb{Q}\right) \,. \qquad (5.2.23)$$

All variables in the above list are objective *per se*, although they are different from the objective variables used in Chaps. 3 and 4 (say $\mathbb{P}_\alpha$ and $\mathbb{M}_\alpha$ etc.). That the LAGRANGEan field variables are the *natural* objective combinations of deformation and electromagnetic fields will become apparent shortly. However, before deducing constitutive equations, we first replace in (5.2.19) the first PIOLA–KIRCHHOFF stress tensor $T_{i\alpha}$ by the second one, which according to (2.2.33) is defined as

$$T_{\alpha\beta}^P := T_{i\alpha} F_{\beta i}^{-1} \ . \tag{5.2.24}$$

Once constitutive relations of the form (5.2.23) are postulated for $\psi, \eta, T_{\alpha\beta}^P, \mathbb{E}_\alpha,$ $\mathbb{H}_\alpha, \mathbb{J}_\alpha$ and Q_α, which all must be objective scalars, we may derive from (5.2.19) in the usual way that i) the free energy cannot depend on $\Theta_{,\alpha}$ and $\mathbb{Q}$,

$$\psi = \hat{\psi} \left(C_{\alpha\beta}, \frac{\mathbb{P}_\alpha}{\rho_0}, \frac{\mu_0 \mathbb{M}_\alpha^C}{\rho_0}, \Theta \right) \ , \tag{5.2.25}$$

and ii) that

$$\eta = -\frac{\partial \hat{\psi}}{\partial \Theta} \ , \qquad\qquad \mathbb{E}_\alpha = \frac{\partial \hat{\psi}}{\partial \mathbb{P}_\alpha / \rho_0} \ ,$$
$$\mathbb{H}_\alpha = \frac{\partial \hat{\psi}}{\partial \mu_0 \mathbb{M}^C / \rho_0} \ , \qquad {}^{II}T_{\alpha\beta}^P = 2\rho_0 \frac{\partial \hat{\psi}}{\partial C_{\alpha\beta}} \ , \tag{5.2.26}$$

so that the reduced entropy inequality becomes

$$\mathbb{J}_\alpha \mathbb{E}_\alpha - \frac{Q_\alpha \Theta_{,\alpha}}{\Theta} \geq 0 \ . \tag{5.2.27}$$

The second PIOLA–KIRCHHOFF stress tensor of model I on the other hand obeys the relation

$${}^{I}T_{\alpha\beta}^P = 2\rho_0 \frac{\partial \hat{\psi}}{\partial C_{\alpha\beta}} - C_{\beta\gamma}^{-1} (\mathbb{P}_\alpha \mathbb{E}_\gamma + \mu_0 \mathbb{M}_\alpha^C \mathbb{H}_\gamma) \ , \tag{5.2.28}$$

which, with the aid of (5.2.24) and (5.2.26)$_4$, can be derived from (5.2.22). Hence, this stress tensor, in contrast to ${}^{II}T_{\alpha\beta}^P$, is not symmetric. However, in view of (5.2.28) and (5.2.11)$_2$ the balance law of momentum of momentum, which reads (see (2.2.32))

$${}^{I}(\rho_0 L_{ij}^e) = {}^{I}T_{[i\alpha} F_{j]\alpha} = -F_{i\alpha} F_{j\beta} \, {}^{I}T_{[\alpha\beta]}^P \tag{5.2.29}$$

is satisfied identically.

When in the energy equation (2.2.32)$_3$ use is made of relations (5.2.20), (5.2.26) and (5.2.12)$_3$, the latter may be written as

$$\rho_0 \Theta \dot{\eta} = \mathbb{J}_\alpha \mathbb{E}_\alpha - Q_{\alpha,\alpha} + \rho_0 r^{\text{ext}} \ . \tag{5.2.30}$$

It should be noted that this relation is derived for model II, but holds for model I as well. It is this form of the energy equation which normally is used in applications.

One advantage of the LAGRANGEan variables introduced in (5.2.6) is the simplicity which the constitutive equation for stress assumes. Corroboration is provided by a comparison of the formulas (3.3.57)$_4$ and (5.2.26)$_4$. Another advantage is the form of the GIBBS relation when written in terms of these variables. Indeed, it follows from the definition of the HELMHOLTZ free energy ψ, (5.2.20), and the results (5.2.26) that

$$\frac{\partial \eta}{\partial \Theta} = \frac{1}{\Theta} \frac{\partial U}{\partial \Theta} \, ,$$

$$\frac{\partial \eta}{\partial \mathbb{P}_\alpha / \rho_0} = \frac{1}{\Theta} \left\{ \frac{\partial U}{\partial \mathbb{P}_\alpha / \rho_0} - \mathbb{E}_\alpha \right\} \, ,$$

$$\frac{\partial \eta}{\partial \mu_0 \mathbb{M}^C_\alpha / \rho_0} = \frac{1}{\Theta} \left\{ \frac{\partial U}{\partial \mu_0 \mathbb{M}^C_\alpha / \rho_0} - \mathbb{H}_\alpha \right\} \, , \qquad (5.2.31)$$

$$\frac{\partial \eta}{\partial C_{\alpha\beta}} = \frac{1}{\Theta} \left\{ \frac{\partial U}{\partial C_{\alpha\beta}} - \frac{1}{2\rho_0} {}^{II}T^P_{\alpha\beta} \right\} \, ,$$

from which one readily deduces that

$$d\eta = \frac{1}{\Theta} \left\{ dU - \frac{1}{2\rho_0} {}^{II}T^P_{\alpha\beta} dC_{\alpha\beta} - \mathbb{E}_\alpha d \left(\frac{\mathbb{P}_\alpha}{\rho_0} \right) - \mathbb{H}_\alpha d \left(\frac{\mu_0 \mathbb{M}^C_\alpha}{\rho_0} \right) \right\} \, . \quad (5.2.32)$$

Next we would like to explore the consequences, which follow from the GIBBS relation, but were not determined in the previous chapter. The reasons for this postponement are formal ones, for the GIBBS relation assumes a particularly simple form when written in the LAGRANGEan variables (compare (5.2.32) with (3.3.59)).

From a practical point of view, that is from a viewpoint of an applied physicist, who must determine actual constitutive relations by performing appropriate experiments, the HELMHOLTZ free energy is not necessarily the most convenient variable to match experiments with theory. These are rather the internal energy, the stress tensor, the polarization and magnetization per unit mass. In what follows, we shall demonstrate, firstly, how a consistent constitutive theory can be developed when starting from this end. Secondly, this approach will show that, ultimately one searches for the free energy also when using this more physical approach. The method will further demonstrate that the constitutive relations for internal energy, entropy, free energy and stress can all be separated into two parts, one of which is of purely thermoelastic origin and can therefore be determined from thermoelastic experiments in the absence of electromagnetic fields. The second parts are then the respective effects due to the electromagnetic fields. We shall outline the procedure for the case that $\mathbb{P}_\alpha$ and $\mathbb{M}_\alpha$ are chosen as independent electromagnetic field variables. At the center of the following derivation lies the GIBBS relation (5.2.32) or the identities (5.2.31), which are the basis for the derivation of the latter.

Note that relations (5.2.31) imply integrability conditions, which can easily be derived by cross differentiations. If these differentiations are performed the following chain of identities is obtained: (here $\mathbb{M}_\alpha \equiv \mathbb{M}^C_\alpha$)

$$\frac{1}{\Theta} = \frac{-\dfrac{\partial \mathbb{E}_\alpha}{\partial \Theta}}{\dfrac{\partial U}{\partial \mathbb{P}_\alpha / \rho_0} - \mathbb{E}_\alpha} = \frac{-\dfrac{\partial \mathbb{H}_\alpha}{\partial \Theta}}{\dfrac{\partial U}{\partial \mu_0 \mathbb{M}_\alpha / \rho_0} - \mathbb{H}_\alpha} = \frac{-\dfrac{\partial \,^{II}T_{\alpha\beta}^{P}}{\partial \Theta}}{2\rho_0 \dfrac{\partial U}{\partial C_{\alpha\beta}} - {}^{II}T_{\alpha\beta}^{P}} ,$$

$$\frac{\partial \mathbb{E}_\alpha}{\partial \mu_0 \mathbb{M}_\beta / \rho_0} = \frac{\partial \mathbb{H}_\beta}{\partial \mathbb{P}_\alpha / \rho_0} , \quad \frac{\partial \mathbb{E}_\alpha}{\partial \mathbb{P}_\beta / \rho_0} = \frac{\partial \mathbb{E}_\beta}{\partial \mathbb{P}_\alpha / \rho_0} , \quad \frac{\partial \mathbb{H}_\alpha}{\partial \mu_0 \mathbb{M}_\beta / \rho_0} = \frac{\partial \mathbb{H}_\beta}{\partial \mu_0 \mathbb{M}_\alpha / \rho_0} ,$$

$$2\rho_0 \frac{\partial \mathbb{E}_\alpha}{\partial C_{\beta\gamma}} = \frac{\partial \,^{II}T_{\beta\gamma}^{P}}{\partial \mathbb{P}_\alpha / \rho_0} , \quad 2\rho_0 \frac{\partial \mathbb{H}_\alpha}{\partial C_{\beta\gamma}} = \frac{\partial \,^{II}T_{\beta\gamma}^{P}}{\partial \mu_0 \mathbb{M}_\alpha / \rho_0} , \quad \frac{\partial \,^{II}T_{\alpha\beta}^{P}}{\partial C_{\gamma\delta}} = \frac{\partial \,^{II}T_{\gamma\delta}^{P}}{\partial C_{\alpha\beta}} .$$

$$(5.2.33)$$

It follows that the relations on the right of $(5.2.33)_{1,2,3}$ are all functions of the temperature alone. All the more, they are exactly $1/\Theta$ and are equal. Such relations, of course, reduce the effective labor of the experimentalist considerably.

The next step consists in the integration of the identities (5.2.33). To this end, notice that $(5.2.33)_{1,2}$ can also be written in the form

$$\frac{\partial U}{\partial \mathbb{P}_\alpha / \rho_0} = -\Theta^2 \frac{\partial}{\partial \Theta} \left(\frac{\mathbb{E}_\alpha}{\Theta} \right) , \qquad \frac{\partial U}{\partial \mu_0 \mathbb{M}_\alpha / \rho_0} = -\Theta^2 \frac{\partial}{\partial \Theta} \left(\frac{\mathbb{H}_\alpha}{\Theta} \right) . \quad (5.2.34)$$

As we shall see in a moment, these equations allow us to decompose U into two parts of which one is due to the electromagnetic fields, whereas the other is the internal energy of the body in the absence of the fields. To see this, we introduce the 6-tuples

$$\mathrm{x}_A = (\mathbb{P}_\alpha / \rho_0, \mu_0 \mathbb{M}_\alpha / \rho_0), \ \ \mathrm{f}_A = (\mathbb{E}_\alpha, \mathbb{H}_\alpha), (A = 1, 2, \dots, 6; \alpha = 1, 2, 3) .$$
$$(5.2.35)$$

The two equations (5.2.34) then combine to give

$$\frac{\partial U}{\partial \mathrm{x}_A} = -\Theta^2 \frac{\partial}{\partial \Theta} \left(\frac{\mathrm{f}_A}{\Theta} \right) = F_A , \qquad A = 1, 2, \dots, 6 . \quad (5.2.36)$$

If the functions f_A (or F_A) are known functions of their arguments $\mathrm{x}_A, C_{\alpha\beta}$ and Θ, equation (5.2.36) is a set of six partial differential equations for U. Of course, in order that such a system is integrable the functions f_A must satisfy integrability conditions, namely

$$\frac{\partial \mathrm{f}_A}{\partial \mathrm{x}_B} = \frac{\partial \mathrm{f}_B}{\partial \mathrm{x}_A} , \qquad \text{or} \qquad \frac{\partial F_A}{\partial \mathrm{x}_B} = \frac{\partial F_B}{\partial \mathrm{x}_A} . \quad (5.2.37)$$

In the above case these are identical with $(5.2.33)_{4,5,6}$ and consequently, (5.2.36) can indeed be integrated. To construct the solution of (5.2.36) we write it in vector form

$$\nabla U = \boldsymbol{F} ,$$

which is more suggestive, because it shows the resemblance with the relation between the potential energy U and a conservative force $\boldsymbol{F}$, the conservativism of which is assured by (5.2.37). The value of U in a point $\mathbf{x}$ can then be found from the line integral

$$U(\mathbf{x}) = \int_0^{\mathbf{x}} \boldsymbol{F}(\mathbf{x}') \cdot d\mathbf{x}' + \text{constant} ,$$

which is independent of the path transversed from $\mathbf{0}$ to $\mathbf{x}$. Choosing the straight line from $\mathbf{0}$ to $\mathbf{x}$, we thus can take

$$\mathbf{x}' = \mathbf{x}s , \quad 0 \leq s \leq 1 ,$$

and then obtain

$$U(\mathbf{x}) = \mathbf{x} \cdot \int_0^1 \boldsymbol{F}(\mathbf{x}s)ds + \text{constant}.$$

Proceeding in an analogous way with (5.2.36), we find the following solution:

$$U = U(\mathbf{x}_A, C_{\alpha\beta}, \Theta) = -\Theta^2 \frac{\partial}{\partial \Theta} \left(\frac{I}{\Theta} \right) + U^0(C_{\alpha\beta}, \Theta) , \qquad (5.2.38)$$

where $U^0(C_{\alpha\beta}, \Theta)$ replaces the integration constant in the preceding analysis and I stands for

$$I = I(\mathbf{x}_A, C_{\alpha\beta}, \Theta) = \mathbf{x}_A \int_0^1 \mathbf{f}_A(\mathbf{x}_B s, C_{\alpha\beta}, \Theta)ds . \qquad (5.2.39)$$

From this relation it is obvious that I is zero for vanishing electromagnetic fields ($\mathbf{f}_A = 0$). Furthermore, by differentiating (5.2.39) with respect to $\mathbf{x}_A$ it immediately follows that

$$\mathbf{f}_A = \frac{\partial I}{\partial \mathbf{x}_A} . \qquad (5.2.40)$$

With the representation (5.2.39) part of our goal is achieved, namely that the internal energy U is separated into two parts. Here, U^0 is the specific internal energy when the electromagnetic fields vanish. The term involving I is the correction due to the presence of the electromagnetic fields.

Not all the identities (5.2.33) have been exploited when constructing the solution (5.2.38), (5.2.39). For instance, we still must explore conditions $(5.2.33)_3$. They can also be written in the form

$$\frac{1}{\Theta^2} \frac{\partial U^0}{\partial C_{\alpha\beta}} = \frac{\partial}{\partial \Theta} \left\{ \frac{1}{\Theta} \frac{\partial I}{\partial C_{\alpha\beta}} - \frac{{}^{II}T^P_{\alpha\beta}}{2\rho_0\Theta} \right\} . \qquad (5.2.41)$$

This is simplified if we introduce the purely thermoelastic part of the second
PIOLA–KIRCHHOFF stress tensor $^{II}\overset{\circ}{T}{}^P_{\alpha\beta}$ by

$$^{II}\overset{\circ}{T}{}^P_{\alpha\beta} = {}^{II}\overset{\circ}{T}{}^P_{\alpha\beta}\,(C_{\gamma\delta},\Theta) = {}^{II}T^P_{\alpha\beta}(\mathrm{x}_A = 0, C_{\gamma\delta},\Theta)\,. \qquad (5.2.42)$$

Then (5.2.41) with $I = 0$ delivers the relation

$$\frac{1}{\Theta^2}\frac{\partial U^0}{\partial C_{\alpha\beta}} = -\frac{\partial}{\partial\Theta}\left(\frac{^{II}\overset{\circ}{T}{}^P_{\alpha\beta}}{2\rho_0\Theta}\right)\,,$$

whence follows

$$\frac{\partial}{\partial\Theta}\left\{\frac{1}{\Theta}\left[\frac{\partial I}{\partial C_{\alpha\beta}} + \frac{1}{2\rho_0}\left(^{II}\overset{\circ}{T}{}^P_{\alpha\beta} - {}^{II}T^P_{\alpha\beta}\right)\right]\right\} = 0\,,$$

or after integration

$$^{II}T^P_{\alpha\beta} = {}^{II}\overset{\circ}{T}{}^P_{\alpha\beta} + 2\rho_0\frac{\partial I}{\partial C_{\alpha\beta}} + \Theta\tau_{\alpha\beta}(\mathrm{x}_A, C_{\gamma\delta})\,. \qquad (5.2.43)$$

With the aid of $(5.2.33)_{7,8}$ and (5.2.40), $\tau_{\alpha\beta}$ can be shown to be independent
of x_A, so that
$$\tau_{\alpha\beta} = \tau_{\alpha\beta}(C_{\gamma\delta})\,.$$
Taking this result into account in equation (5.2.43) and evaluating the latter
at zero magnetic fields leads us, with the aid of (5.2.42), to the conclusion
that $\tau_{\alpha\beta}$ must be zero. Hence,

$$^{II}T^P_{\alpha\beta} = {}^{II}\overset{\circ}{T}{}^P_{\alpha\beta} + 2\rho_0\frac{\partial I}{\partial C_{\alpha\beta}}\,, \qquad (5.2.44)$$

an equation which separates the stress into a thermoelastic and a field part.
Equations (5.2.40) and (5.2.44) may now be used to write the GIBBS relation
(5.2.32) in the form

$$d\left\{\eta + \frac{\partial I}{\partial\Theta}\right\} = \frac{1}{\Theta}\left\{dU^0 - \frac{1}{2\rho_0}\,{}^{II}\overset{\circ}{T}{}^P_{\alpha\beta}\,dC_{\alpha\beta}\right\}\,,$$

whence, since the right-hand side is independent of the electromagnetic fields,
it follows that
$$\eta = \eta^0(C_{\alpha\beta},\Theta) - \frac{\partial I}{\partial\Theta}\,. \qquad (5.2.45)$$

Thus, it has also been possible to separate the field part of the entropy from
the corresponding thermoelastic part.

After the introduction of

$$\psi^0 = \psi^0(C_{\alpha\beta}, \Theta) \, ,$$

as the free energy in case of vanishing electromagnetic fields, a nice interpretation of I follows from $(5.2.26)_1$ and $(5.2.45)$. Accordingly,

$$\frac{\partial}{\partial\Theta}(\hat{\psi} - I) = -\eta^0 = \frac{\partial\psi^0}{\partial\Theta} \, , \tag{5.2.46}$$

and, hence,

$$\hat{\psi}(\mathbf{x}_A, C_{\alpha\beta}, \Theta) = \psi^0(C_{\alpha\beta}, \Theta) + I(\mathbf{x}_A, C_{\alpha\beta}, \Theta) + g(\mathbf{x}_A, C_{\alpha\beta}) \, .$$

Use of $(5.2.40)$, $(5.2.26)_{2,3}$ and the definition of ψ^0 then shows that $g \equiv 0$, so that

$$\hat{\psi}(\mathbf{x}_A, C_{\alpha\beta}, \Theta) = \psi^0(C_{\alpha\beta}, \Theta) + I(\mathbf{x}_A, C_{\alpha\beta}, \Theta) \, . \tag{5.2.47}$$

Here, ψ^0 denotes the free energy, when there are no electromagnetic fields present, and I is therefore the field part of the HELMHOLTZ free energy.

There is another possible separation of the internal energy, entropy, free energy and the stresses, which, from a practical point of view is as important as the above one. We mean the separation into contributions due to "rigid-body processes" and the corrections due to deformations. This approach gives a natural separation of rigid-body electrodynamics from that of deformable bodies. The idea is to commence the integration of $(5.2.33)_{1,2,3}$ with $(5.2.33)_3$, which, with the definitions

$$\mathbf{y}_A = (C_{11}, C_{22}, C_{33}, C_{23}, C_{31}, C_{12}) \, ,$$

$$g_A = (\,{}^{II}T_{11}^P, \, {}^{II}T_{22}^P, \, {}^{II}T_{33}^P, \, {}^{II}T_{23}^P, \, {}^{II}T_{31}^P, \, {}^{II}T_{12}^P), \quad (A = 1, 2, \ldots, 6) \tag{5.2.48}$$

may be written as

$$\frac{\partial U}{\partial \mathbf{y}_A} = -\Theta^2 \frac{\partial}{\partial\Theta}\left(\frac{g_A}{2\rho_0\Theta}\right) \, , \tag{5.2.49}$$

where in view of $(5.2.33)$

$$\frac{\partial g_A}{\partial \mathbf{y}_B} = \frac{\partial g_B}{\partial \mathbf{y}_A} \, ,$$

holds. Using the same approach as before, we derive the following representation:

$$U(\mathbf{x}_A, \mathbf{y}_A, \Theta) = U^R(\mathbf{x}_A, \Theta) - \Theta^2 \frac{\partial}{\partial\Theta}\left(\frac{J}{\Theta}\right) \, , \tag{5.2.50}$$

where

$$J = J(\mathbf{x}_A, \mathbf{y}_A, \Theta) = \frac{\mathbf{y}_A}{2\rho_0}\int_0^1 g_A(\mathbf{x}_B, \mathbf{y}_B, s, \Theta)ds \, , \tag{5.2.51}$$

and, moreover,

$$g_A = 2\rho_0 \frac{\partial J}{\partial y_A} \, ,$$

$$f_A = f_A^R(x_A, \Theta) + \frac{\partial J}{\partial gx_A} \, , \qquad \eta = \eta^R(x_A, \Theta) - \frac{\partial J}{\partial \Theta} \, , \qquad (5.2.52)$$

$$\hat{\psi}(x_A, y_A, \Theta) = \psi^R(x_A, \Theta) + J(x_A, y_A, \Theta) \, ,$$

where we still note that for the derivation of (5.2.52) the GIBBS relation (5.2.32) was written in the form

$$d\left(\eta + \frac{\partial J}{\partial \Theta}\right) = \frac{1}{\Theta}(dU^R - f_A^R dx_A) \, .$$

In these equations the quantities carrying a superscript R are the internal energy U^R, the electromagnetic fields f_A^R, the entropy η^R, and the free energy ψ^R for zero deformation. They constitute the rigid-body contributions and form the constitutive relations of rigid-body electrodynamics. All constitutive properties that can be traced back to deformation are contained in the function J.

This completes the constitutive theory. The integration procedure has shown that all difficulties rest on an appropriate determination of the functions I or J (or more generally of ψ). Of course, the same procedure can also be taken when some of the dependent and independent variables are interchanged. Since the details of the pertinent calculations are the same as above we leave them to the reader.

Before we proceed we would like to mention that there are still further possibilities of separating the various effects. These can easily be derived if either only (5.2.33)$_1$, or else (5.2.33)$_2$, is used in the integration process. If the results contained in all these combinations are put together, the free energy can, for instance, be written as a sum of four parts, one of which is the field-free energy. The second term involves, apart from the deformation, only the electric effects (polarization); in the third term then only magnetization appears. Only in the fourth term does there occur electromagnetic coupling. Hence, we have shown that the separation of the interaction phenomena into the four physically clearly defined parts is indeed possible – and justified.

For later reference, we would like to list here also the constitutive equations for the case that, instead of $\mathbb{P}_\alpha/\rho_0$ and $\mu_0 \mathbb{M}_\alpha/\rho_0, \mathbb{E}_\alpha$ and $\mathbb{H}_\alpha$ are the independent variables. In other words, we shall postulate constitutive equations of the form

$$\mathsf{C} = \bar{\mathsf{C}}(C_{\alpha\beta}, \mathbb{E}_\alpha, \mathbb{H}_\alpha, \Theta, \Theta_{,\alpha}, \mathbb{Q}) \, , \qquad (5.2.53)$$

which can easily be obtained from the preceding ones by the LEGENDRE transformation

$$\bar{\psi} = U - \eta\Theta - \frac{1}{\rho_0}\mathbb{E}_\alpha\mathbb{P}_\alpha - \frac{\mu_0}{\rho_0}\mathbb{H}_\alpha\mathbb{M}_\alpha^C = \bar{\varepsilon} - \eta\Theta = \bar{\psi}(C_{\alpha\beta}, \mathbb{E}_\alpha, \mathbb{H}_\alpha, \Theta) \, , \quad (5.2.54)$$

which leads to the relations

$$\eta = -\frac{\partial \bar{\psi}}{\partial \Theta} \, , \qquad\qquad \mathbb{P}_\alpha = -\rho_0 \frac{\partial \bar{\psi}}{\partial \mathbb{E}_\alpha} \, ,$$

$$\mu_0 \mathbb{M}_\alpha^C = -\rho_0 \frac{\partial \bar{\psi}}{\partial \mathbb{H}_\alpha} \, , \qquad {}^{II}T_{\alpha\beta}^P = 2\rho_0 \frac{\partial \bar{\psi}}{\partial C_{\alpha\beta}} \, , \qquad (5.2.55)$$

and to the GIBBS relation

$$d\eta = \frac{1}{\Theta} \left\{ d\bar{\varepsilon} - \frac{1}{2\rho_0} \, {}^{II}T_{\alpha\beta}^P dC_{\alpha\beta} + \frac{\mathbb{P}_\alpha}{\rho_0} d\mathbb{E}_\alpha + \frac{\mu_0 \mathbb{M}_\alpha}{\rho_0} d\mathbb{H}_\alpha \right\} .$$

We conclude this section with a few complementary remarks. Firstly, in the above constitutive theory we have introduced the classical expression for the entropy flux, and we have done this already when treating this model in the EULERian description. That this is correct was proved by HUTTER [95], and indeed he obtains exactly the same GIBBS relation. Secondly, taking for ψ, or $\bar{\psi}$, in both models the same function of the form (5.2.25), or (5.2.54), and for the energy flux the same function of the form (5.2.23), or (5.2.53), we automatically guarantee that the two models are thermodynamically equivalent. Finally, we mention that it is particularly easy to linearize the equations in the LAGRANGEan formulation. We shall demonstrate this in Chap. 6.

If we were to follow the order of presentation of the previous chapter, the LAGRANGEan description of the MAXWELL–MINKOWSKI formulation should now follow. For didactic reasons, we shall treat this formulation last.

5.3 Material Description of the Statistical and the Lorentz Formulation (Models IV and V)

As we have seen in Chap. 3 already, the statistical formulation and the LORENTZ formulation are very similar, and indeed in Chap. 4 we demonstrated how the two models could be brought into a one-to-one correspondence. We therefore expect the two models to show a close interrelation in the LAGRANGEan formulation as well. This is indeed the case and we shall give corroboration for this below.

To begin with, recall that the MAXWELL equations in Chap. 2 were stated in terms of the variables $\mathcal{E}_i, D_i, \mathcal{H}_i$ and B_i and that these quantities were related to the statistical and LORENTZian variables in (3.5.1) and (3.6.1), respectively. Using also the relations

$$P_i^S = P_i^L = P_i \, , \qquad \mathcal{M}_i = M_i^S + e_{ijk}\dot{x}_j P_k = M_i^L \, , \qquad (5.3.1)$$

it is then easy to show that the MAXWELL equations (2.2.12), (2.2.13), (2.2.19) and (2.2.20) may be written in the form

$$\frac{d}{dt}\int_S B_i da_i + \int_{\partial S} \mathcal{E}_i dl_i = 0 \,,$$

$$\int_{\partial V} B_i da_i = 0 \,,$$

$$-\frac{d}{dt}\int_S D_i^a da_i + \int_{\partial S} \mathcal{H}_i^a dl_i = \int_S \mathcal{J}_i da_i + \frac{d}{dt}\int_S P_i da_i + \int_{\partial S} M_i^L dl_i \,,$$

$$\int_{\partial V} D_i^a da_i = \int_V \mathcal{Q} d\nu - \int_{\partial V} P_i da_i \,,$$

(5.3.2)

where we have introduced the auxiliary quantities

$$D_i^a := \varepsilon_0 E_i \,, \qquad \mathcal{H}_I^a := H_I^a - e_{ijk}\dot{x}_j D_k^a \,, \qquad (5.3.3)$$

with

$$H_i^a := \frac{1}{\mu_0} B_i \,. \qquad (5.3.4)$$

Note that even in a non-relativistic formulation

$$H_i^a \neq \mathcal{H}_I^a \,,$$

but that in this approximation (recall the rules stated just after (3.4.12))

$$D_i^a = \varepsilon \mathcal{E}_i \,, \qquad \mu_0 H_i^a = \mu_0 \mathcal{H}_i^a \,,$$

$$\frac{1}{\mu_0} B_i B_i = H_i^a B_i = \mathcal{H}_i^a B_i \,. \qquad (5.3.5)$$

We remind the reader that E_i and B_i are the electric field strength and magnetic flux density as they occur in the MINKOWSKI, statistical, and LORENTZ formulations. Furthermore, as it was convenient to introduce auxiliary fields for the two dipole models, so it is here, and the equations $(5.3.3)_1$ and $(5.3.4)$ represent nothing but the familiar MAXWELL–LORENTZ aether relations.

Before we pass on to the presentation of the material description of the equations (5.3.2) a few words are in order regarding the interpretation of the various terms in (5.3.2). Firstly, there is a set of homogenous equations and another one that is inhomogeneous through the presence of free charge and free current terms. As is seen from these, both polarization and magnetization manifest themselves as distributions of surface charges and surface currents, respectively. Secondly, we have expressed the integral laws (5.3.2) in terms of LORENTZian variables, but it is not difficult to write them in terms of the variables of the statistical description by simply invoking the transformations (5.3.1). The only term that changes in this formal substitution is

$$\int_{\partial S} M_i^L dl_i = \int_{\partial S} M_i^S dl_i + \int_{\partial S} e_{ijk}\dot{x}_j P_k dl_i \,. \qquad (5.3.6)$$

It is not convenient to absorb the second member on the right-hand side in the term

$$\int\limits_{\partial S} \mathcal{H}_i^a \, da_i \, ,$$

as it occurs on the left-hand side of $(5.3.2)_3$, because in that case $\mathcal{H}_i^a$ would also be expressed in terms which describe the material behavior. Thus, we conclude that in both the statistical and the LORENTZ formulations

$$\int\limits_{\partial S} M_i^L \, dl_i$$

is the proper (magnetization) current, and this is one reason for us to regard the LORENTZian description as more advantageous than the statistical description, for it is the former, which will directly lead to a material counterpart of the EULERian variables. In the statistical formulation this current term must be attributed to magnetization as well as polarization.

To arrive at the LAGRANGEan counterpart of equations (5.3.2) one needs only to transform the integrals back to the reference configuration. If this is done, one obtains

$$\frac{d}{dt} \int\limits_{S_R} \mathbb{B}_\alpha dA_\alpha + \int\limits_{\partial S_R} \mathbb{E}_\alpha dL_\alpha = 0 \, ,$$

$$\int\limits_{\partial V_R} \mathbb{B}_\alpha dA_\alpha = 0 \, ,$$

$$-\frac{d}{dt} \int\limits_{S_R} \mathbb{D}_\alpha^a dA_\alpha + \int\limits_{\partial S_R} \mathbb{H}_\alpha^a dL_\alpha = \int\limits_{S_R} \mathbb{J}_\alpha dA_\alpha + \frac{d}{dt} \int\limits_{S_R} \mathbb{P}_\alpha dA_\alpha + \int\limits_{\partial S_R} \mathbb{M}_\alpha^L dL_\alpha \, ,$$

$$\int\limits_{\partial V_R} \mathbb{D}_\alpha^a dA_\alpha = \int\limits_{V_R} \mathbb{Q} \, dV - \int\limits_{\partial V_R} \mathbb{P}_\alpha dA_\alpha \, ,$$

$$(5.3.7)$$

and from these one may derive in the usual way the conservation law of charges in the form

$$\frac{d}{dt} \int\limits_{V_R} \mathbb{Q} \, dV + \int\limits_{\partial V_R} \mathbb{J}_\alpha dA_\alpha = 0 \, . \qquad (5.3.8)$$

All newly introduced variables are already defined in (5.2.6), except for $\mathbb{B}_\alpha, \mathbb{H}_\alpha^a$ and $\mathbb{M}_\alpha^L$, which are given by

$$\mathbb{B}_\alpha = J F_{\alpha i}^{-1} B_l \, , \qquad B_l - J^{-1} F_{i\alpha} \mathbb{B}_\alpha \, ,$$

$$\mathbb{H}_\alpha^a = F_{i\alpha} \mathcal{H}_i^a \mathrm{sgn} J \, , \qquad \mathcal{H}_i^a = F_{\alpha i}^{-1} \mathbb{H}_\alpha^a \mathrm{sgn} J \, , \qquad (5.3.9)$$

$$\mathbb{M}_\alpha^L = F_{i\alpha} M_i^L \mathrm{sgn} J \, , \qquad M_i^L = F_{\alpha i}^{-1} \mathbb{M}_\alpha^L \mathrm{sgn} J \, .$$

Specifically, we wish to point out the difference between M_α^L and M_α^C; these two fields are related by

$$M_\alpha^L = \frac{1}{|J|}C_{\alpha\beta}M_\beta^C .$$

(5.3.10)

For sufficiently smooth fields equations (5.3.7) become

$$\mathbb{B}_{\alpha,\alpha} = 0 ,$$

$$\dot{\mathbb{B}}_\alpha + e_{\alpha\beta\gamma}\mathbb{E}_{\gamma,\beta} = 0 ,$$

$$\mathbb{D}_{\alpha,\alpha}^a = \mathbb{Q} - \mathbb{P}_{\alpha,\alpha} ,$$

(5.3.11)

$$-\dot{\mathbb{D}}_\alpha^a + e_{\alpha\beta\gamma}\mathbb{H}_{\gamma,\beta}^a = \mathbb{J} + \dot{\mathbb{P}}_\alpha + e_{\alpha\beta\gamma}M_{\gamma,\beta}^L ,$$

$$\dot{\mathbb{Q}} + \mathbb{J}_{\alpha,\alpha} = 0 .$$

These equations are the MAXWELL equations in the material description, as they naturally emerge from the LORENTZ or the statistical formulations. As was the case in the CHU formulation, they contain two auxiliary variables which can be expressed in terms of $\mathbb{B}_\alpha$ and $\mathbb{E}_\alpha$. Indeed, on using relations (5.3.3) and the transformation rules for $\mathcal{E}_i, D_i^a, \mathcal{H}_i^a$ and B_i, a straightforward calculation shows that in a nonrelativistic approximation

$$\mathbb{D}_\alpha^a = \varepsilon_0|J|C_{\alpha\beta}^{-1}\mathbb{E}_\beta ,$$

$$\mathbb{H}_\alpha^a = \frac{1}{|J|}\left\{ \frac{1}{\mu_0}C_{\alpha\beta}\mathbb{B}_\beta - \varepsilon_0 e_{\mu\beta\gamma}C_{\alpha\beta}F_{j\gamma}\dot{x}_j\mathbb{E}_\mu \right\} .$$

(5.3.12)

Recognize, as was already the case in the material description of the two dipole models, that it is through the MAXWELL-LORENTZ aether relations that the formal linearity of the equations (5.3.11) is destroyed. Nevertheless, equations (5.3.11) appear in a form, which is identical to that in the statistical description when spatial coordinates are used (see (3.5.2)). Variables are, however, different and so are the configurations.

One could, if one so desired, write (5.3.11) also as

$$\mathbb{B}_{\alpha,\alpha} = 0 ,$$

$$\dot{\mathbb{B}}_\alpha + e_{\alpha\beta\gamma}\mathbb{E}_{\gamma,\beta} = 0 ,$$

$$\mathbb{D}_{\alpha,\alpha} = \mathbb{Q} ,$$

(5.3.13)

$$-\dot{\mathbb{D}}_\alpha + e_{\alpha\beta\gamma}\mathbb{H}_{\gamma,\beta} = \mathbb{J}_\alpha ,$$

$$\dot{\mathbb{Q}} + \mathbb{J}_{\alpha,\alpha} = 0 .$$

where

$$\mathbb{D}_\alpha := \mathbb{D}_\alpha^a + \mathbb{P}_\alpha \qquad \text{and} \qquad \mathbb{H}_\alpha := \mathbb{H}_\alpha^a - \mathbb{M}_\alpha^L \, , \qquad (5.3.14)$$

and would in this way formally arrive at a material MINKOWSKI formulation. We shall come back to this in the next section. Note that (5.3.13) agrees with (2.2.40) and thus $\mathbb{D}_\alpha$ and $\mathbb{H}_\alpha$ also agree with $\mathbb{D}_\alpha$ and $\mathbb{H}_\alpha$ introduced there.

The balance laws (5.3.2) imply also jump conditions, which can easily be derived from (2.4.5) by simply invoking (5.3.14). This yields

$$[\![\mathbb{B}_\alpha]\!]\mathbb{N}_\alpha = 0 \, ,$$

$$e_{\alpha\beta\gamma}[\![\mathbb{E}_\beta]\!]\mathbb{N}_\gamma + [\![\mathbb{B}_\alpha W_N]\!] = 0 \, ,$$

$$[\![\mathbb{D}_\alpha^a + \mathbb{P}_\alpha]\!]\mathbb{N}_\alpha = 0 \, , \qquad (5.3.15)$$

$$e_{\alpha\beta\gamma}[\![\mathbb{H}_\beta^a - \mathbb{M}_\beta^L]\!]\mathbb{N}_\gamma - [\![(\mathbb{D}_\alpha^a + \mathbb{P}_\alpha)W_N]\!] = 0 \, ,$$

$$[\![\mathbb{J}_\alpha]\!]\mathbb{N}_\alpha + [\![Q W_N]\!] = 0 \, .$$

To complete the description, we must also derive the LAGRANGEan versions of the body force, body couple and energy supply expressions of electromagnetic origin. For this purpose we simply express these quantities, which in (3.5.3) and (3.6.5) are written in terms of the EULERian variables, in the LAGRANGEan fields $\mathbb{E}_\alpha, \mathbb{B}_\alpha, \mathbb{P}_\alpha, \mathbb{M}_\alpha^L$ etc. Starting from (3.6.5), we obtain for the LORENTZ formulation **(model V)**

$$^V(\rho_0 F_i^e) = F_{\alpha i}^{-1}\left((\mathbb{Q} - \mathbb{P}_{\beta,\beta})\mathbb{E}_\alpha + e_{\alpha\beta\gamma}(\mathbb{J}_\beta + \dot{\mathbb{P}}_\beta)\mathbb{B}_\gamma \right.$$
$$\left. + (\mathbb{M}_{\alpha,\beta}^L - \mathbb{M}_{\beta,\alpha}^L)\mathbb{B}_\beta\right) \, ,$$
$$^V(\rho_0 L_{ij}^e) = 0 \, , \qquad (5.3.16)$$
$$^V(\rho_0 r^e) = \mathbb{J}_\alpha \mathbb{E}_\alpha + \dot{\mathbb{P}}_\alpha \mathbb{E}_\alpha + e_{\alpha\beta\gamma}\mathbb{M}_{\gamma,\beta}^L \mathbb{E}_\alpha \, .$$

For the derivation of the corresponding expressions of the statistical model it is convenient to make use of the relation (compare (4.5.4))

$$^{IV}(\rho F_i^e) = {}^V(\rho_0 F_i^e) + (\mathcal{E}_i P_i - \mathcal{M}_i B_j + \delta_{ij}\mathcal{M}_k B_k)_{,j} \, , \qquad (5.3.17)$$

an expression, which can be derived from (3.5.3) and (3.6.5). Alternatively, the expressions for the body couple and energy supply follow from (3.5.3)$_2$ and (3.5.3)$_3$, respectively, the latter written in the form

$$^{IV}(\rho r^e) = \mathcal{J}_i \mathcal{E}_i + (\overset{*}{P}_i + P_j \dot{x}_{i,j})\mathcal{E}_i - (\overset{*}{B}_i - B_i \dot{x}_{j,j} + B_j \dot{x}_{i,j})\mathcal{M}_i \, . \qquad (5.3.18)$$

Transforming the above expressions to referential coordinates, we obtain for *model IV*

$$^{IV}(\rho_0 F_i^e) = F_{\alpha i}^{-1}\Big((\mathbb{Q} - \mathbb{P}_{\beta,\beta})\mathbb{E}_\alpha + e_{\alpha\beta\gamma}(\mathbb{J}_\beta + \dot{\mathbb{P}}_\beta)\mathbb{B}_\gamma$$

$$+(\mathrm{M}_{\alpha,\beta}^L - \mathrm{M}_{\beta,\alpha}^L)\mathbb{B}_\beta\Big)$$

$$+\Big(F_{\alpha i}^{-1}(\mathbb{E}_\alpha\mathbb{P}_\beta - \mathrm{M}_\alpha^L\mathbb{B}_\beta + \delta_{\alpha\beta}\mathrm{M}_\gamma^L\mathbb{B}_\gamma)\Big)_{,\beta}\ ,$$

$$^{IV}(\rho_0 L_{ij}^e) = F_{[i\alpha}F_{\beta j]}^{-1}(\mathbb{P}_\alpha\mathbb{E}_\beta - \mathbb{B}_\alpha\mathrm{M}_\beta^L)\ ,$$

$$^{IV}(\rho_0 r^e) = \mathbb{J}_\alpha\mathbb{E}_\alpha + \dot{\mathbb{P}}_\alpha\mathbb{E}_\alpha - \mathrm{M}_\alpha^L\dot{\mathbb{B}}_\alpha$$

$$+\dot{F}_{k\gamma}\Big((\mathbb{E}_\beta\mathbb{P}_\gamma - \mathrm{M}_\beta^L\mathbb{B}_\gamma)F_{\beta k}^{-1} + \mathrm{M}_\beta^L\mathbb{B}_\beta F_{\gamma k}^{-1}\Big)\ . \tag{5.3.19}$$

We would like to point out that the body force expression $(5.3.16)_1$ is seemingly different from that given in [92]. However, when using the identity

$$F_{\nu i,\mu}^{-1}\mathbb{B}_\mu\mathrm{M}_\nu^L - F_{k\delta}^{-1}F_{\mu i}^{-1}F_{\nu k,\mu}^{-1}\mathbb{B}_\delta\mathrm{M}_\nu^L = F_{k\mu}(F_{\nu i,k}^{-1} - F_{\nu k,i}^{-1})\mathbb{B}_\mu\mathrm{M}_\nu^L = 0\ , \tag{5.3.20}$$

which was not observed in [92], the body force expressions turn out to be identical.

We see that the body force expressions in the two formulations differ by a term that is the divergence of a tensor. It corresponds to the divergence term occurring already in the EULERian body force. Particularly interesting is a comparison of the electromagnetic energy supply terms listed in $(5.3.16)_3$ and $(5.3.19)_3$. Using $(5.3.11)_2$, we can show that $^{IV}(\rho_0 r^e)$ may also be written as

$$^{IV}(\rho_0 r^e) = \mathbb{J}_\alpha\mathbb{E}_\alpha + \dot{\mathbb{P}}_\alpha\mathbb{E}_\alpha + e_{\alpha\beta\gamma}\mathrm{M}_{\gamma,\beta}^L\mathbb{E}_\alpha + (e_{\alpha\beta\gamma}\mathbb{E}_\beta\mathrm{M}_\gamma^L)_{,\alpha}$$

$$+\dot{F}_{k\gamma}\Big((\mathbb{E}_\beta\mathbb{P}_\gamma - \mathrm{M}_\beta^L\mathbb{B}_\gamma)F_{\beta k}^{-1} + \mathrm{M}_\beta^L\mathbb{B}_\beta F_{\gamma k}^{-1}\Big)\ . \tag{5.3.21}$$

Recognizing that, as already said several times before, for a comparison of the energy supplies of two formulations we need not compare the term $\rho_0 r^e$ alone but rather the combination $(\rho_0 r^e + t_{ij}\dot{x}_{i,j})$. Thus when performing this comparison we see that the last term of expression $(5.3.21)$ for $^{IV}(\rho_0 r^e)$ originates from the difference in the stress tensors of the models IV and V. Indeed with the aid of $(4.5.4)$ we may easily show that

$$^{IV}T_{i\alpha} = {}^{V}T_{i\alpha} + F_{\beta i}^{-1}(\mathbb{P}_\alpha\mathbb{E}_\beta - \mathbb{B}_\alpha\mathrm{M}_\beta^L) - \mathrm{M}_\gamma^L\mathrm{M}_\gamma^L F_{\alpha i}^{-1}\ . \tag{5.3.22}$$

However, the term

$$(e_{\alpha\beta\gamma}\mathbb{E}_\beta\mathrm{M}_\gamma^L)_{,\alpha}$$

cannot be explained in this way. Consequently either the energy fluxes Q_α or the entropy fluxes ϕ_α of the two formulations must differ. Indeed, we may choose, as was already done in the EULERian description,

$$Q_\alpha^S = Q_\alpha^L + e_{\alpha\beta\gamma}\mathbb{E}_\beta\mathrm{M}_\gamma^L\ , \tag{5.3.23}$$

and then obtain

$$\phi_\alpha^S = \frac{Q_\alpha^S}{\Theta} = \frac{Q_\alpha^L + e_{\alpha\beta\gamma}\mathbb{E}_\beta\mathbb{M}_\gamma^L}{\Theta} = \phi_\alpha^L \,. \tag{5.3.24}$$

This form of the entropy flux vector was *proved* to be correct by HUTTER [92] in a theory of viscous isotropic thermoelastic, polarizable and magnetizable solids. Here we treat (5.3.23) as a postulate.

When we compare the expressions for the electromagnetic energy supply $(5.3.16)_3$ and $(5.3.19)_3$ with the corresponding EULERian expressions, $(3.5.3)_3$ and $(3.6.5)_3$, we recognize that the LORENTZian expression $^V(\rho_0 r^e)$ is formally the same as its spatial counterpart $^V(\rho r^e)$. This property is not shared by $^{IV}(\rho_0 r^e)$ and $^{IV}(\rho r^e)$.

Finally, we note that the expressions for the electromagnetic body couple and body force are simpler in the LORENTZ formulation than in the statistical formulation. All this, of course, are reasons which make the LORENTZ formulation to be (formally) superior to the statistical one, although, as we have shown, they are equivalent.

We proceed with the jump conditions for momentum and energy of matter and fields as they are listed in their general form in (2.4.12). Therefore, it suffices to evaluate $T_{i\alpha}^M, G_i, \Pi_\alpha$ and Ω. These quantities are obtained if use is made of the transformations (5.2.16) for t_{ij}^M, g_i, π_i and ω, which in (3.5.8) and (3.6.8) are given for the statistical and the LORENTZ formulations, respectively. When the indicated transformations are performed the following relations are obtained:

(i) in the LORENTZ formulation (**model V**)

$$^V T_{i\alpha}^M = F_{\beta i}^{-1}(\mathbb{D}_\alpha^a \mathbb{E}_\beta + \mathbb{B}_\alpha \mathbb{H}_\beta^a) - \tfrac{1}{2}F_{\alpha i}^{-1}(\mathbb{D}_\beta^a \mathbb{E}_\beta + \mathbb{B}_\beta \mathbb{H}_\beta^a) \,,$$

$$^V G_i = 0 \,,$$

$$^V \Pi_\alpha = -e_{\alpha\beta\gamma}\mathbb{E}_\beta\mathbb{H}_\gamma^a + F_{\beta i}^{-1}(\mathbb{D}_\alpha^a \mathbb{E}_\beta + \mathbb{B}_\alpha \mathbb{H}_\beta^a)\dot{x}_i - \tfrac{1}{2}F_{\alpha i}^{-1}(\mathbb{D}_\beta^a \mathbb{E}_\beta + \mathbb{B}_\beta \mathbb{H}_\beta^a)\dot{x}_i$$

$$= -e_{\alpha\beta\gamma}\mathbb{E}_\beta\mathbb{H}_\gamma^a + {}^V T_{i\alpha}^M \dot{x}_i \,,$$

$$^V \Omega = -\tfrac{1}{2}(\mathbb{D}_\alpha^a \mathbb{E}_\alpha + \mathbb{B}_\alpha \mathbb{H}_\alpha^a) \,, \tag{5.3.25}$$

(ii) in the statistical formulation (**model IV**)

$$^{IV} T_{i\alpha}^M = {}^V T_{i\alpha}^M + F_{\beta i}^{-1}(\mathbb{P}_\alpha \mathbb{E}_\beta - \mathbb{B}_\alpha \mathbb{M}_\beta^L + \delta_{\alpha\beta}\mathbb{B}_\gamma \mathbb{M}_\gamma^L) \,,$$

$$^{IV} G_i = 0 \,,$$

$$^{IV} \Pi_\alpha = {}^V \Pi_\alpha + e_{\alpha\beta\gamma}\mathbb{E}_\beta\mathbb{M}_\gamma^L + F_{\beta i}^{-1}(\mathbb{P}_\alpha \mathbb{E}_\beta - \mathbb{B}_\alpha \mathbb{M}_\beta^L + \delta_{\alpha\beta}\mathbb{B}_\gamma \mathbb{M}_\gamma^L)\dot{x}_i$$

$$= -e_{\alpha\beta\gamma}\mathbb{E}_\beta(\mathbb{H}_\gamma^a - \mathbb{M}_\gamma^L) + {}^{IV} T_{i\alpha}^M \dot{x}_i \,,$$

$$^{IV} \Omega = {}^V \Omega. \tag{5.3.26}$$

When we substitute the above expressions into (2.4.12), what emerges are the jump conditions for momentum and energy of matter and fields. They will not be written down explicitly.

There remains to formulate the constitutive theory of the two models. To this end we derive the reduced entropy inequality for model V by eliminating $\rho_0 r^{\text{ext}}$ from the energy equation and the entropy inequality, taking thereby into account that $\rho_0 r^e$ is given by $(5.3.16)_3$ and the entropy flux by (5.3.24), respectively. In this way we obtain

$$- \rho_0\dot{\psi} - \rho_0\eta\dot{\Theta} + {}^VT_{i\alpha}\dot{F}_{i\alpha} + \mathbb{E}_\alpha\dot{\mathbb{P}}_\alpha - \mathbb{M}^L_\alpha\dot{\mathbb{B}}_\alpha + \mathbb{E}_\alpha\mathbb{J}_\alpha - \frac{Q^S_\alpha\Theta_{,\alpha}}{\Theta} \geq 0 \;, \quad (5.3.27)$$

where ${}^VT_{i\alpha}$ is the first PIOLA-KIRCHHOFF stress tensor in the LORENTZ formulation. It should be noticed that in (5.3.27) the heat flux vector Q^S_α is used instead of Q^L_α.

Constitutive relations are now written as

$$\mathsf{C} = \overset{+}{\mathsf{C}}\,(C_{\alpha\beta}, \mathbb{P}_\alpha/\rho_0, \mathbb{B}_\alpha, \Theta, \Theta_{,\alpha}, \mathbb{Q})\;. \qquad (5.3.28)$$

A short calculation then shows, as usual, that the HELMHOLTZ free energy is independent of $\Theta_{,\alpha}$ and $\mathbb{Q}$,

$$\psi = U - \eta\Theta = \overset{+}{\psi}\,(C_{\alpha\beta}, \mathbb{P}_\alpha/\rho_0, \mathbb{B}_\alpha, \Theta)\;, \qquad (5.3.29)$$

and that

$$\eta = -\frac{\partial\overset{+}{\psi}}{\partial\Theta}\;, \qquad\qquad \mathbb{E}_\alpha = \frac{\partial\overset{+}{\psi}}{\partial(\mathbb{P}_\alpha/\rho_0)}\;,$$

$$\mathbb{M}^L_\alpha = -\rho_0\frac{\partial\overset{+}{\psi}}{\partial\mathbb{B}_\alpha}\;, \qquad {}^VT^P_{\alpha\beta} = 2\rho_0\frac{\partial\overset{+}{\psi}}{\partial C_{\alpha\beta}}\;. \qquad (5.3.30)$$

Here, ${}^VT^P_{\alpha\beta}$ is the second PIOLA-KIRCHHOFF stress tensor, which is defined in terms of $T_{i\alpha}$ in (2.2.33).

With these results the energy equation reduces to its ultimate form

$$\rho_0\Theta\dot{\eta} = \mathbb{E}_\alpha\mathbb{J}_\alpha - Q^S_{\alpha,\alpha} + \rho_0 r^{\text{ext}}\;. \qquad (5.3.31)$$

On the other hand, the reduced entropy inequality takes the form

$$\mathbb{E}_\alpha\mathbb{J}_\alpha - \frac{Q^S_\alpha\Theta_{,\alpha}}{\Theta} \geq 0\;, \qquad (5.3.32)$$

and the GIBBS relation reads

$$d\eta = \frac{1}{\Theta}\left\{dU - \frac{1}{2\rho_0}{}^VT^P_{\alpha\beta}dC_{\alpha\beta} - \mathbb{E}_\alpha d\left(\frac{\mathbb{P}_\alpha}{\rho_0}\right) + \frac{\mathbb{M}^L_\alpha}{\rho_0}d\mathbb{B}_\alpha\right\}\;, \qquad (5.3.33)$$

an equation that was also derived in [92].

The results for model IV and V are identical, except for the stress tensors of which the relation is given in (5.3.22). It follows from the latter and $(5.3.30)_4$ that

$$^{IV}T^P_{\alpha\beta} = 2\rho_0 \frac{\partial \overset{+}{\psi}}{\partial C_{\alpha\beta}} + (\mathbb{P}_\alpha \mathbb{E}_\gamma - \mathbb{B}_\alpha \mathbb{M}^L_\gamma)C^{-1}_{\beta\gamma} - \mathbb{M}^L_\gamma \mathbb{M}^L_\gamma C^{-1}_{\alpha\beta} . \quad (5.3.34)$$

With this constitutive relation for the second PIOLA-KIRCHHOFF stress tensor, the balance law of angular momentum for the statistical formulation is satisfied identically.

For later use, we wish to give constitutive relations also of the form

$$\mathsf{C} = \check{\mathsf{C}}(C_{\alpha\beta}, \mathbb{E}_\alpha, \mathbb{B}_\alpha, \Theta, \Theta_{,\alpha}, \mathbb{Q}) . \quad (5.3.35)$$

With the LEGENDRE transformation

$$\check{\psi} = U - \eta\Theta - \frac{1}{\rho_0}\mathbb{E}_\alpha \mathbb{P}_\alpha = \check{\varepsilon} - \eta\Theta = \check{\psi}(C_{\alpha\beta}, \mathbb{E}_\alpha, \mathbb{B}_\alpha, \Theta) , \quad (5.3.36)$$

this immediately leads to

$$\eta = -\frac{\partial\check{\psi}}{\partial\Theta} , \qquad \mathbb{P}_\alpha = -\rho_0 \frac{\partial\check{\psi}}{\partial\mathbb{E}_\alpha} ,$$

$$\mathbb{M}^L_\alpha = -\rho_0 \frac{\partial\check{\psi}}{\partial\mathbb{B}_\alpha} , \qquad {}^V T^P_{\alpha\beta} = 2\rho_0 \frac{\partial\check{\psi}}{\partial C_{\alpha\beta}} , \qquad (5.3.37)$$

$$d\eta = \frac{1}{\Theta}\left\{ d\check{\varepsilon} - \frac{1}{2\rho_0}{}^V T^P_{\alpha\beta}dC_{\alpha\beta} + \frac{\mathbb{P}_\alpha}{\rho_0}d\mathbb{E}_\alpha + \frac{\mathbb{M}^L_\alpha}{\rho_0}d\mathbb{B}_\alpha \right\} .$$

As was done in the CHU formulation we could now, if we so desired, also explore the consequences implied by the GIBBS relation and would then again be able to show that constitutive relations and free energy are separable into a thermoelastic and a field part or else, a rigid-body part and a part due to deformation. The procedure is analogous to that shown before, and therefore we leave the pertinent details to the reader.

5.4 Material Description of the Maxwell–Minkowski Formulation

We saw in Chap. 3, Sects. 3.4 & 3.5 that the spatial electromagnetic field variables in the MAXWELL–MINKOWSKI formulation are formally the same as those in the statistical description, and that the latter are closely related to the variables in the LORENTZ formulation. We therefore adopt the same material electromagnetic variables as in Sect. 5.3 of this Chapter, which are given by (5.3.9). There is only one difference that must be noted. Magnetization and polarization are regarded as auxiliary variables, whilst the dielectric displacement and the magnetic field strength are considered basic. The LA-GRANGEan counterparts of these fields can easily be read off from (5.3.14), and it is not difficult to show that $\mathbb{D}_\alpha$ and $\mathbb{H}_\alpha$ are related to $\mathcal{H}_i$ and D_i according to

$$\mathbb{D}_\alpha = |J|F_{\alpha i}^{-1}D_i \ , \qquad D_i = \frac{1}{|J|}F_{i\alpha}\mathbb{D}_\alpha \ ,$$
$$\mathbb{H}_\alpha = F_{i\alpha}\mathcal{H}_i \mathrm{sgn}J \ , \qquad \mathcal{H}_i = F_{\alpha i}^{-1}\mathbb{H}_\alpha \mathrm{sgn}J \ . \tag{5.4.1}$$

The MAXWELL equations, expressed in $\mathbb{E}_\alpha, \mathbb{B}_\alpha, \mathbb{D}_\alpha$ and $\mathbb{H}_\alpha$ are already given in (5.3.13), and we refrain from repeating them here.

Before passing on to the presentation of electromagnetic body force, body couple and energy supply, one remark concerning the auxiliary fields must be made. In view of the properties of this formulation as just outlined one would expect $\mathbb{P}_\alpha$ and $\mathbb{M}_\alpha^L$ to be the adequate auxiliary variables. However, since the description becomes formally much simpler when $\mathbb{M}_\alpha^C$ is used instead, we shall prefer to use the latter. In that case the auxiliary fields may be obtained from (5.3.14), (5.3.10) and (5.3.12); they are

$$\mathbb{P}_\alpha = \mathbb{D}_\alpha - \varepsilon_0|J|C_{\alpha\beta}^{-1}\mathbb{E}_\beta \ ,$$
$$\mu_0\mathbb{M}_\alpha^C = \mathbb{B}_\alpha - \mu_0|J|C_{\alpha\beta}^{-1}\mathbb{H}_\beta \ . \tag{5.4.2}$$

In Sect. 3.4 the expressions for the electromagnetic body force, body couple and energy supply are derived directly from the global energy balance law (3.4.19). When written in material form the latter reads

$$\frac{d}{dt}\int_{V_R}\left\{\rho_0 U + \tfrac{1}{2}\rho_0\dot{x}_i\dot{x}_i\right\}dV - \int_{V_R}\left\{\rho_0 r^{\mathrm{ext}} + \rho_0 F_i^{\mathrm{ext}}\dot{x}_i\right\}dV$$
$$- \int_{\partial V_R}\{T_{i\alpha}\dot{x}_i - Q_\alpha\}dA_\alpha$$
$$= \frac{d}{dt}\int_{V_R}\left\{-\tfrac{1}{2}|J|C_{\alpha\beta}^{-1}(\varepsilon_0\mathbb{E}_\alpha\mathbb{E}_\beta + \mu_0\mathbb{H}_\alpha\mathbb{H}_\beta)\right\}dV \tag{5.4.3}$$
$$+ \int_{\partial V_R}\left\{-e_{\alpha\beta\gamma}\mathbb{E}_\beta\mathbb{H}_\gamma + F_{\beta i}^{-1}(\mathbb{D}_\alpha\mathbb{E}_\beta + \mathbb{B}_\alpha\mathbb{H}_\beta)\dot{x}_i\right.$$
$$\left.- \tfrac{1}{2}|J|F_{\alpha i}^{-1}C_{\beta\gamma}^{-1}(\varepsilon_0\mathbb{E}_\beta\mathbb{E}_\gamma + \mu_0\mathbb{H}_\beta\mathbb{H}_\gamma)\dot{x}_i\right\}dA_\alpha \ .$$

From this equation the expressions for $^{III}\Omega$ and $^{III}\Pi_\alpha$ can directly be read off. They are equal to the integrands of the first and the second integral on the right-hand side of (5.4.3).

With the aid of the MAXWELL equations (5.3.13) and relations (5.4.2) this balance law can be transformed into the form

$$\int_{V_R} \Big\{ [\rho_0 \ddot{x}_i - \rho_0 F_i^{\text{ext}} - T_{i\alpha,\alpha} - \Big(F_{\beta i}^{-1}(\mathbb{D}_\alpha \mathbb{E}_\beta + \mathbb{B}_\alpha \mathbb{H}_\beta)$$

$$- \tfrac{1}{2}|J|F_{\alpha i}^{-1}C_{\beta\gamma}^{-1}(\varepsilon_0 \mathbb{E}_\beta \mathbb{E}_\gamma + \mu_0 \mathbb{H}_\beta \mathbb{H}_\gamma)\Big)_{,\alpha}]\dot{x}_i + \rho_0 \dot{U}$$

$$-\rho_0 r^{\text{ext}} + Q_{\alpha,\alpha} - \mathbb{J}_\alpha \mathbb{E}_\alpha - \dot{\mathbb{P}}_\alpha \mathbb{E}_\alpha - \mu_0 \dot{\mathbb{M}}_\alpha^C \mathbb{H}_\alpha \tag{5.4.4}$$

$$-[T_{i\alpha} + F_{\beta i}^{-1}(\mathbb{P}_\alpha \mathbb{E}_\beta + \mu_0 \mathbb{M}_\alpha^C \mathbb{H}_\beta)]\dot{F}_{i\alpha} \Big\}\, dV\ .$$

By applying invariance requirements as was done in Sect. 2.3, it is possible to derive from (5.4.4) local balance equations of linear and angular momentum and energy in a material formulation. Comparison of these equations with those given in (2.2.32) then yields the material versions of the electromagnetic body force, body couple and energy supply. In the derivation of the angular momentum equation it must be observed that $\dot{\mathbb{P}}_\alpha$ and $\dot{\mathbb{M}}_\alpha^C$ are objective quantities under the EUCLIDian transformation group (in contrast to $\dot{P}_i$ and $\dot{M}_i$) this because $\mathbb{P}_\alpha$ and $\mathbb{M}_\alpha^C$ are objective scalars. In this way we obtain

$$^{III}(\rho_0 F_i^e) = \Big(F_{\beta i}^{-1}(\mathbb{D}_\alpha \mathbb{E}_\beta + \mathbb{B}_\alpha \mathbb{H}_\beta)$$

$$- \tfrac{1}{2}|J|F_{\alpha i}^{-1}C_{\beta\gamma}^{-1}(\varepsilon_0 \mathbb{E}_\beta \mathbb{E}_\gamma + \mu_0 \mathbb{H}_\beta \mathbb{H}_\gamma)\Big)_{,\alpha}$$

$$= F_{\alpha i}^{-1}\Big(\mathbb{Q}\mathbb{E}_\alpha + e_{\alpha\beta\gamma}\mathbb{J}_\beta \mathbb{B}_\gamma + \mathbb{P}_\beta \mathbb{E}_{\beta,\alpha} + \mu_0 \mathbb{M}_\beta^C \mathbb{H}_{\beta,\alpha} \tag{5.4.5}$$

$$+e_{\alpha\beta\gamma}(\mathbb{D}_\beta \dot{\mathbb{B}}_\gamma + \dot{\mathbb{D}}_\beta \mathbb{B}_\gamma)\Big) + F_{\beta i,j}^{-1}F_{j\alpha}(\mathbb{P}_\alpha \mathbb{E}_\beta + \mu_0 \mathbb{M}_\alpha^C \mathbb{H}_\beta)\ ,$$

$$^{III}(\rho_0 L_{ij}^e) = F_{[i\alpha}F_{\beta j]}^{-1}(\mathbb{P}_\alpha \mathbb{E}_\beta + \mu_0 \mathbb{M}_\alpha^C \mathbb{H}_\beta)\ ,$$

$$^{III}(\rho_0 r^e) = \mathbb{J}_\alpha \mathbb{E}_\alpha + \dot{\mathbb{P}}_\alpha \mathbb{E}_\alpha + \mu_0 \dot{\mathbb{M}}_\alpha^C \mathbb{H}_\alpha + F_{\beta i}^{-1}(\mathbb{P}_\alpha \mathbb{E}_\beta + \mu_0 \mathbb{M}_\alpha^C \mathbb{H}_\beta)\dot{F}_{i\alpha}\ .$$

Of course, these expressions can also be obtained from their spatial versions (3.4.22), (3.4.23) and (3.4.24) by transforming the latter into their LAGRANGEan counterparts.

Furthermore, from (5.4.5)$_1$ and (5.4.3) the electromagnetic momentum G_i, MAXWELL stress $T_{i\alpha}^M$, electromagnetic energy flux Π_α, and energy density Ω are obtained as

$$^{III}T_{i\alpha}^M = F_{\beta i}^{-1}(\mathbb{D}_\alpha \mathbb{E}_\beta + \mathbb{B}_\alpha \mathbb{H}_\beta) - \tfrac{1}{2}|J|F_{\alpha i}^{-1}C_{\beta\gamma}^{-1}(\varepsilon_0 \mathbb{E}_\beta \mathbb{E}_\gamma + \mu_0 \mathbb{H}_\beta \mathbb{H}_\gamma) \, ,$$

$$^{III}G_i = 0 \, ,$$

$$\begin{aligned}
^{III}\Pi_\alpha &= -e_{\alpha\beta\gamma}\mathbb{E}_\beta \mathbb{H}_\gamma + F_{\beta i}^{-1}(\mathbb{D}_\alpha \mathbb{E}_\beta + \mathbb{B}_\alpha \mathbb{H}_\beta)\dot{x}_i \\
&\quad - \tfrac{1}{2}|J|F_{\alpha i}^{-1}C_{\beta\gamma}^{-1}(\varepsilon_0 \mathbb{E}_\beta \mathbb{E}_\gamma + \mu_0 \mathbb{H}_\beta \mathbb{H}_\gamma)\dot{x}_i \\
&= -e_{\alpha\beta\gamma}\mathbb{E}_\beta \mathbb{H}_\gamma + {}^{III}T_{i\alpha}^M \dot{x}_i \, ,
\end{aligned}$$

$$^{III}\Omega = -\tfrac{1}{2}|J|C_{\alpha\beta}^{-1}(\varepsilon_0 \mathbb{E}_\alpha \mathbb{E}_\beta + \mu_0 \mathbb{H}_\alpha \mathbb{H}_\beta) \, .$$

(5.4.6)

Substitution of (5.4.6) into (2.4.12) then yields the jump conditions for momentum and energy of matter and fields.

There remains to formulate the constitutive theory. To this end, the reduced entropy inequality must be derived. Introducing

$$\psi = U - \eta\Theta - \frac{1}{\rho_0}\mathbb{E}_\alpha \mathbb{P}_\alpha \, , \qquad (5.4.7)$$

and

$$\phi_\alpha = \frac{Q_\alpha}{\Theta} \, , \qquad (5.4.8)$$

we obtain, as usual,

$$\begin{aligned}
&-\rho_0 \dot{\psi} - \rho_0 \eta \dot{\Theta} - \mathbb{P}_\alpha \dot{\mathbb{E}}_\alpha + \mu_0 \dot{\mathbb{M}}_\alpha^C \mathbb{H}_\alpha \\
&+[T_{i\alpha} + F_{\beta i}^{-1}(\mathbb{P}_\alpha \mathbb{E}_\beta + \mu_0 \mathbb{M}_\alpha^C \mathbb{H}_\beta)]\dot{F}_{i\alpha} + \mathbb{J}_\alpha \mathbb{E}_\alpha - \frac{Q_\alpha \Theta_{,\alpha}}{\Theta} \geq 0 \, .
\end{aligned}$$

(5.4.9)

Assuming constitutive relations of the form

$$\mathsf{C} = \tilde{\mathsf{C}}\left(C_{\alpha\beta}, \mathbb{E}_\alpha, \frac{\mu_0 \mathbb{M}_\alpha^C}{\rho_0}, \Theta, \Theta_{,\alpha}, \mathbb{Q}\right) \, , \qquad (5.4.10)$$

we can show that

$$\psi = \tilde{\psi}\left(C_{\alpha\beta}, \mathbb{E}_\alpha, \frac{\mu_0 \mathbb{M}_\alpha^C}{\rho_0}, \Theta\right) \, , \qquad (5.4.11)$$

$$\eta = -\frac{\partial\tilde\psi}{\partial\Theta} \, , \quad \mathbb{P}_\alpha = -\frac{\partial\tilde\psi}{\partial\mathbb{E}_\alpha} \, , \quad \mathbb{H}_\alpha = \frac{\partial\tilde\psi}{\partial\mu_0\mathbb{M}_\alpha^C/\rho_0} \, ,$$

$$^{III}T_{i\alpha} = 2\rho_0 \frac{\partial\tilde\psi}{\partial C_{\alpha\beta}}F_{i\beta} - F_{\beta i}^{-1}(\mathbb{P}_\alpha \mathbb{E}_\beta + \mu_0 \mathbb{M}_\alpha^C \mathbb{H}_\beta) \, .$$

(5.4.12)

Moreover, it is easily shown that with the relation $(5.4.12)_4$ the balance law of moment of momentum is satisfied identically. Furthermore,

$$^{III}T_{\alpha\beta}^P = 2\rho_0 \frac{\partial\tilde\psi}{\partial C_{\alpha\beta}} - C_{\beta\gamma}^{-1}(\mathbb{P}_\alpha \mathbb{E}_\gamma + \mu_0 \mathbb{M}_\alpha^C \mathbb{H}_\gamma) \, . \qquad (5.4.13)$$

When use is made of (5.4.12), (5.4.9) reduces to the residual inequality

$$\mathbb{E}_\alpha \mathbb{J}_\alpha - \frac{Q_\alpha \Theta_{,\alpha}}{\Theta} \geq 0 \ .$$

On the other hand, the energy balance reduces in form to relation (5.2.30), and (5.4.7) and (5.4.12) imply

$$\frac{\partial \eta}{\partial \Theta} = \frac{1}{\Theta} \frac{\partial \tilde{\varepsilon}}{\partial \Theta} \ , \qquad \frac{\partial \eta}{\partial \mathbb{E}_\alpha} = \frac{1}{\Theta} \left[\frac{\partial \tilde{\varepsilon}}{\partial \mathbb{E}_\alpha} + \mathbb{P}_\alpha \right] \ ,$$

$$\frac{\partial \eta}{\partial \mu_0 \mathbb{M}_\alpha^C / \rho_0} = \frac{1}{\Theta} \left[\frac{\partial \tilde{\varepsilon}}{\partial \mu_0 \mathbb{M}_\alpha^C / \rho_0} - \mathbb{H}_\alpha \right] \ , \tag{5.4.14}$$

$$\frac{\partial \eta}{\partial C_{\alpha\beta}} = \frac{1}{\Theta} \left\{ \frac{\partial \tilde{\varepsilon}}{\partial C_{\alpha\beta}} - \frac{1}{2\rho_0} \left[{}^{III}T_{\alpha\beta}^P + C_{\beta\gamma}^{-1} (\mathbb{P}_\alpha \mathbb{E}_\gamma + \mu_0 \mathbb{M}_\alpha^C \mathbb{H}_\gamma) \right] \right\} \ ,$$

where

$$\tilde{\varepsilon} := U - \frac{1}{\rho_0} \mathbb{E}_\alpha \mathbb{P}_\alpha = \tilde{\psi} - \eta \Theta \ , \tag{5.4.15}$$

from which the GIBBS relation

$$dη = \frac{1}{\Theta} \left\{ d\tilde{\varepsilon} + \mathbb{P}_\alpha d\mathbb{E}_\alpha - \mathbb{H}_\alpha d \left(\frac{\mu_0 \mathbb{M}_\alpha^C}{\rho_0} \right) \right.$$
$$\left. - \frac{1}{2\rho_0} \left[{}^{III}T_{\alpha\beta}^P + C_{\beta\gamma}^{-1} (\mathbb{P}_\alpha \mathbb{E}_\gamma + \mu_0 \mathbb{M}_\alpha^C \mathbb{H}_\gamma) \right] dC_{\alpha\beta} \right\} \ , \tag{5.4.16}$$

may be derived.

When the functional ψ as defined by (5.4.7) is replaced by the HELMHOLTZ free energy

$$\psi = U - \eta \Theta = \hat{\psi}(C_{\alpha\beta}, \mathbb{P}_\alpha / \rho_0, \mu_0 \mathbb{M}_\alpha^C / \rho_0, \Theta) \ , \tag{5.4.17}$$

the constitutive equations (5.4.12) only change in that $\tilde{\psi}$ becomes $\hat{\psi}$, and that the second equation must be replaced by

$$\mathbb{E}_\alpha = \frac{\partial \hat{\psi}}{\partial \mathbb{P}_\alpha / \rho_0} \ . \tag{5.4.18}$$

Comparison of the results of this section with those of Sect. 4.2 immediately shows that the LAGRANGEan formulations of models I and III are equivalent. As we have already proved the thermodynamic equivalence of models I and II in Sect. 4.2 and of the models IV and V in Sect. 4.3, we need for a comparison of the LAGRANGEan formulations of the various models only consider one model out of each of the following groups:

(i) models I, II and III,
(ii) models IV and V.

This comparison between models II and V will be made in Sect. 5.6.

5.5 Thermostatic Equilibrium – Constitutive Relations for Energy Flux and Electric Current

In the above, we derived constitutive equations for entropy, stress and two electromagnetic field vectors, all of which turned out to be derivable from a free energy. We did not present constitutive relations for the free current $\mathbb{J}_\alpha$ and the energy flux vector Q_α. These must be given separately and they are restricted by the residual inequality

$$\gamma := \mathbb{E}_\alpha \mathbb{J}_\alpha - \frac{Q_\alpha \Theta_{,\alpha}}{\Theta} \geq 0 . \tag{5.5.1}$$

Here, Q_α denotes the heat flux in all but the LORENTZ formulation, where it must be replaced by the right-hand side of (5.3.23).

As was the case in the EULERian description, exploitation of (5.5.1) depends on whether we are dealing with an electrical conductor or insulator. Hence, we shall discuss the two cases separately.

(a) For an **electrical insulator** ($\mathbb{J}_\alpha = 0$) the residual inequality (5.5.1) reduces to

$$\gamma = -\frac{Q_\alpha \Theta_{,\alpha}}{\Theta} \geq 0 . \tag{5.5.2}$$

In thermostatic equilibrium, which will again be characterized by the index $(\cdot)|_E$, that is for time-independent processes with uniform temperature,

$$\gamma|_E = 0 .$$

Since γ is non-negative in general, it thus assumes its minimum for thermostatic equilibrium. Of necessity then

$$\left.\frac{\partial \gamma}{\partial \Theta_{,\alpha}}\right|_E = 0 , \qquad \left.\frac{\partial^2 \gamma}{\partial \Theta_{,\alpha} \partial \Theta_{,\beta}}\right|_E \quad \text{is positive-semi definite} , \tag{5.5.3}$$

or, when expressed in terms of Q_α,

$$Q_\alpha|_E = 0 , \qquad \left.\frac{\partial Q_{(\alpha}}{\partial \Theta_{,\beta)}}\right|_E \quad \text{is negative-semi definite} . \tag{5.5.4}$$

To see what consequences these relations impose on Q_α we consider the case in which constitutive relations are prescribed in the form

$$\mathsf{C} = \check{\mathsf{C}}(C_{\alpha\beta}, \mathbb{E}_\alpha, \mathbb{B}_\alpha, \Theta, \Theta_{,\alpha}, \mathbb{Q}) . \tag{5.5.5}$$

A necessary and sufficient condition for the energy flux vector Q_α to vanish in thermostatic equilibrium is to write

$$Q_\alpha = -\check{\kappa}_{\alpha\beta}(C_{\gamma\delta}, \mathbb{E}_\gamma, \mathbb{B}_\gamma, \Theta, \Theta_{,\gamma}, \mathbb{Q})\Theta_{,\beta} , \tag{5.5.6}$$

and this implies that $\check{\kappa}_{(\alpha\beta)}\big|_E$ must be positive-semi definite. Nothing can be said about the skew-symmetric part $\check{\kappa}_{[\alpha\beta]}\big|_E$, but when one restricts oneself to small deformations and small deviations from thermostatic equilibrium it follows from the ONSAGER *relations* that

$$\check{\kappa}_{[\alpha\beta]}(\delta_{\gamma\delta}, \mathbb{E}_\gamma, \mathbb{B}_\gamma, \Theta, 0, \mathbb{Q}) = 0 . \tag{5.5.7}$$

(b) In an **electrical conductor** we may write

$$Q_\alpha = \check{Q}_\alpha(C_{\beta\gamma}, \mathbb{E}_\beta, \mathbb{B}_\beta, \Theta, \Theta_{,\beta}, \mathbb{Q}) ,$$
$$\mathbb{J}_\alpha = \check{\mathbb{J}}_\alpha(C_{\beta\gamma}, \mathbb{E}_\beta, \mathbb{B}_\beta, \Theta, \Theta_{,\beta}, \mathbb{Q}) . \tag{5.5.8}$$

Thermostatic equilibrium is defined here as a time-independent process with uniform temperature and vanishing electric field strength $\mathbb{E}_\alpha$. Hence,

$$\gamma\big|_E = 0 ,$$

must hold here too, and thus the following conditions emerge

$$\frac{\partial\gamma}{\partial\mathbb{E}_\alpha}\bigg|_E = 0 , \quad \text{and} \quad \frac{\partial\gamma}{\partial\Theta_{,\alpha}}\bigg|_E = 0 ,$$

$$\begin{pmatrix} \dfrac{\partial^2\gamma}{\partial\mathbb{E}_\alpha\partial\mathbb{E}_\beta} & \dfrac{\partial^2\gamma}{\partial\mathbb{E}_\alpha\partial\Theta_{,\beta}} \\ \dfrac{\partial^2\gamma}{\partial\mathbb{E}_\alpha\partial\Theta_{,\beta}} & \dfrac{\partial^2\gamma}{\partial\Theta_{,\alpha}\partial\Theta_{,\beta}} \end{pmatrix}\bigg|_E \quad \text{is positive-semi definite.} \tag{5.5.9}$$

Of necessity then

$$\mathbb{J}_\alpha\big|_E = 0 , \quad \text{and} \quad Q_\alpha\big|_E = 0 , \tag{5.5.10}$$

as well as

$$\begin{pmatrix} \dfrac{\partial\check{\mathbb{J}}_{(\alpha}}{\partial\mathbb{E}_{\beta)}} & \dfrac{\partial\check{\mathbb{J}}_\alpha}{\partial\Theta_{,\beta}} - \dfrac{1}{\Theta}\dfrac{\partial\check{Q}_\beta}{\partial\mathbb{E}_\alpha} \\ \dfrac{\partial\check{\mathbb{J}}_\alpha}{\partial\Theta_{,\beta}} - \dfrac{1}{\Theta}\dfrac{\partial\check{Q}_\beta}{\partial\mathbb{E}_\alpha} & -\dfrac{1}{\Theta}\dfrac{\partial\check{Q}_{(\alpha}}{\partial\Theta_{,\beta)}} \end{pmatrix}\bigg|_E \quad \text{is positive-semi definite .} \tag{5.5.11}$$

These conditions are satisfied provided that

$$Q_\alpha = -\kappa_{\alpha\beta}\Theta_{,\beta} + \beta^{(Q)}_{\alpha\beta}\mathbb{E}_\beta ,$$
$$\mathbb{J}_\alpha = \beta^{(J)}_{\alpha\beta}\Theta_{,\beta} + \sigma_{\alpha\beta}\mathbb{E}_\beta , \tag{5.5.12}$$

and that

$$\left(\begin{array}{cc} \sigma_{(\alpha\beta)} & \beta^{(\mathrm{J})}_{\alpha\beta} - \dfrac{1}{\Theta}\beta^{(Q)}_{\beta\alpha} \\[2ex] \beta^{(\mathrm{J})}_{\alpha\beta} - \dfrac{1}{\Theta}\beta^{(Q)}_{\beta\alpha} & \dfrac{1}{\Theta}\kappa_{(\alpha\beta)} \end{array} \right) \Bigg|_E \quad \text{is positive-semi definite .}$$

$$(5.5.13)$$

Necessary conditions for (5.5.13) to be satisfied are that $\sigma_{(\alpha\beta)}$ and $\kappa_{(\alpha\beta)}$ are positive definite, but these conditions are by no means sufficient. Conditions of sufficiency include for instance also

$$\sigma_{[\alpha\beta]}\big|_E = 0 , \quad \kappa_{[\alpha\beta]}\big|_E = 0 , \quad \text{and} \quad \beta^{(\mathrm{J})}_{\alpha\beta}\big|_E = \frac{1}{\Theta}\beta^{(Q)}_{\beta\alpha}\big|_E . \qquad (5.5.14)$$

These represent the well-known ONSAGER relations.

5.6 Recapitulation and Comparison

In the preceding sections we presented the material versions of the two dipole models, of the statistical and LORENTZ formulations and of the MAXWELL-MINKOWSKI formulation. The basic principle in this derivation consisted in transforming known equations into the LAGRANGEan form by introducing new variables which are particularly convenient in this material description. Since this chapter is heavily loaded with partly complicated formulas, it might be advantageous when the basic ideas are recollected.

Key idea behind the LAGRANGEan description is to derive the equations of electromechanical interactions in a form, which can directly be used in the theory of solid bodies. All equations should therefore be referred to the reference configuration. While such a formulation is well-known in continuum mechanics, it is hardly used in electrodynamics. This is the reason why most theories of solids of electromechanical interactions are treated in the spatial description.

Whereas the advantages of the material description will be described extensively in Chap. 6, we would like to draw the reader's attention here to formal differences and similarities of the various formulations only. To this end, consider the MAXWELL equations first. They are listed in (5.2.8) and (5.3.11) for the two-dipole and the LORENTZ or statistical models, respectively. For clarity of presentation they will be repeated here. In the CHU formulation they read

$$\mathbb{B}^a_{\alpha,\alpha} = -\mu_0 \mathbb{M}^C_{\alpha,\alpha} ,$$

$$\dot{\mathbb{B}}^a_\alpha + e_{\alpha\beta\gamma}\mathbb{E}_{\gamma,\beta} = -\mu_0\dot{\mathbb{M}}^C_\alpha ,$$

$$\mathbb{D}^a_{\alpha,\alpha} = \mathbb{Q} - \mathbb{P}^C_{\alpha,\alpha} , \qquad\qquad (5.6.1)$$

$$-\dot{\mathbb{D}}^a_\alpha + e_{\alpha\beta\gamma}\mathbb{H}_{\gamma,\beta} = \mathbb{J}_\alpha + \dot{\mathbb{P}}^C_\alpha ,$$

$$\dot{\mathbb{Q}} + \mathbb{J}_{\alpha,\alpha} = 0 ,$$

whereas in the statistical and LORENTZ formulations they are

$$\mathbb{B}_{\alpha,\alpha} = 0 \, ,$$

$$\dot{\mathbb{B}}_{\alpha} + e_{\alpha\beta\gamma}\mathbb{E}_{\gamma,\beta} = 0 \, ,$$

$$\mathbb{D}^a_{\alpha,\alpha} = \mathbb{Q} - \mathbb{P}^L_{\alpha,\alpha} \, , \tag{5.6.2}$$

$$-\dot{\mathbb{D}}^a_{\alpha} + e_{\alpha\beta\gamma}\mathbb{H}^a_{\gamma,\beta} = \mathbb{J}_{\alpha} + \dot{\mathbb{P}}^L_{\alpha} - e_{\alpha\beta\gamma}\mathbb{M}^L_{\gamma,\beta} \, ,$$

$$\dot{\mathbb{Q}} + \mathbb{J}_{\alpha,\alpha} = 0 \, .$$

Both sets of equations allow the presentation of the LAGRANGEan version of the MINKOWSKI formulation by introducing either

$$\mathbb{B}_{\alpha} = \mathbb{B}^a_{\alpha} + \mu_0 \mathbb{M}^C_{\alpha}, \qquad \mathbb{D}_{\alpha} = \mathbb{D}^a_{\alpha} + \mathbb{P}^C_{\alpha} \, , \tag{5.6.3}$$

or

$$\mathbb{H}_{\alpha} = \mathbb{H}^a_{\alpha} - \mathbb{M}^L_{\alpha} \, , \qquad \mathbb{D}_{\alpha} = \mathbb{D}^a_{\alpha} + \mathbb{P}^L_{\alpha} \, . \tag{5.6.4}$$

This then yields

$$\mathbb{P}^C_{\alpha} = \mathbb{P}^L_{\alpha} = \mathbb{P}_{\alpha} \, , \quad \text{and} \quad \mathbb{M}^C_{\alpha} = |J| C^{-1}_{\alpha\beta} \mathbb{M}^L_{\beta} \tag{5.6.5}$$

so that the LAGRANGEan form of the MAXWELL equations in the MINKOWSKI formulation becomes

$$\mathbb{B}_{\alpha,\alpha} = 0 \, ,$$

$$\dot{\mathbb{B}}_{\alpha} + e_{\alpha\beta\gamma}\mathbb{E}_{\gamma,\beta} = 0 \, ,$$

$$\mathbb{D}_{\alpha,\alpha} = \mathbb{Q} \, , \tag{5.6.6}$$

$$-\dot{\mathbb{D}}_{\alpha} + e_{\alpha\beta\gamma}\mathbb{H}_{\gamma,\beta} = \mathbb{J}_{\alpha} \, ,$$

$$\dot{\mathbb{Q}} + \mathbb{J}_{\alpha,\alpha} = 0 \, .$$

Equations (5.6.1), (5.6.2) and (5.6.6) are the only ones that emerge in the LAGRANGEan description from all presently known electromagnetic formulations. We have presented in Chap. 3 four different formulations, but there are others (for instance the AMPÈRE formulation of PENFIELD and HAUS [91]) and all these reduce to (5.6.1), (5.6.2) or (5.6.6). The LAGRANGEan description has therefore reduced this number to at most three, and the differences among these models are particularly transparent in the LAGRANGEan description. The CHU formulation differs from the statistical and LORENTZ formulation only in the choice of magnetization. This resulted in the selection of different auxiliary fields, which are related to the basic fields via the MAXWELL-LORENTZ aether relations (5.2.9) and (5.3.12). As a result, $\mathbb{H}_{\alpha}$ and $\mathbb{E}_{\alpha}$ are the two basic electromagnetic field vectors in the CHU formulation, while $\mathbb{B}_{\alpha}$ and $\mathbb{E}_{\alpha}$ are those of the statistical or LORENTZ formulation. In this connection it is interesting to note that the LAGRANGEan electric field $\mathbb{E}_\alpha$ is the same in all formulations. This was not so in the spatial description where E_i^C and $E_i^L = E_i^S$ are different variables. The same is also true for the LAGRANGEan magnetic field strength $\mathbb{H}_\alpha$; it is the same variable in both the CHU and the MINKOWSKI formulation. Another noteworthy point is that the MAXWELL equations are formally the same as those in the spatial version of the statistical or MINKOWSKI formulation. Because the transformations (5.6.3) and (5.6.4) relating the systems (5.6.1) and (5.6.2) to (5.6.6) are so simple, it is now also plausible why the MAXWELL-MINKOWSKI formulation served as connecting piece between the two-dipole and the statistical and the LORENTZ formulations.

In performing actual calculations in electromagnetism it is advantageous to introduce electromagnetic potentials by satisfying those MAXWELL equations identically that do not involve the charge $\mathbb{Q}$ and the current densities $\mathbb{J}_\alpha$. Mathematically, this is achieved by introducing two potentials A_α and Φ, such that

$$\mathbb{B}_\alpha = e_{\alpha\beta\gamma}A_{\gamma,\beta}\,, \quad \text{and} \quad \mathbb{E}_\alpha = \Phi_{,\alpha} - \dot{A}_\alpha\,. \tag{5.6.7}$$

Since $\mathbb{B}_\alpha$ is a basic field in all but the CHU formulation, the replacement of the MAXWELL equations by the corresponding equations for the potentials A_α and Φ and for the charge $\mathbb{Q}$ is easier in the LORENTZ, the statistical and the MAXWELL–MINKOWSKI formulation than it is in the CHU formulation. This may be regarded as a disadvantage of the CHU formulation.

The above discussion is only concerned with the MAXWELL equations and leaves all mechanical balance laws aside. Yet, the basic advantages of the different formulations are drawn from the expressions of electromagnetic body force, body couple and energy supply. These are listed in (5.2.11) and (5.2.12) (two-dipole models), (5.3.16) (LORENTZ formulation), (5.3.19) (statistical formulation) and (5.3.23) (MAXWELL-MINKOWSKI formulation). Of all these formulations the two-dipole model with symmetric stress tensor and the LORENTZ formulation led to electromagnetic body force and energy supply expressions, which are particularly simple and easy to interpret (see the corresponding expressions (5.2.12) and (5.3.16)). This is not so for all other formulations, although it is known that their interaction terms have a clear physical meaning and are based on a clear method of derivation. It is also interesting to note that the two most simple formulations are those with no electromagnetic body couple and with symmetric CAUCHY stress. Hence, since all formulations are non-relativistically equivalent anyhow, future calculations should be performed with either one of these simplest formulations. The argument on electromagnetic potentials given above favors the LORENTZ formulation. We shall come back to this point in Chap. 6.

As was done for the MAXWELL equations, we also wish to recapitulate the mechanical balance equations and to comment on the equivalence properties in the five presented models. Starting with the CHU models, we showed in

Sect. 5.2 already that these two models are equivalent and, therefore, we only list the balance equations for **model II**:

$$\rho_0 \ddot{x}_i = {}^{II}T_{i\alpha,\alpha} + F_{\alpha i}^{-1}\{(\mathbb{Q} - \mathbb{P}_{\beta,\beta})\mathbb{E}_\alpha + e_{\alpha\beta\gamma}(\mathbb{J}_\beta + \dot{\mathbb{P}}_\beta)\mathbb{B}_\gamma^a$$

$$- \mu_0 \mathbb{M}_{\beta,\beta}^C \mathbb{H}_\alpha\} + \rho_0 F_i^{\text{ext}}.$$

$$(5.6.8)$$

$${}^{II}T_{[i\alpha}F_{j]\alpha} = 0\,,$$

$$\rho_0 \Theta \dot{\eta} = \mathbb{J}_\alpha \mathbb{E}_\alpha - Q_{\alpha,\alpha} + \rho_0 r^{\text{ext}}\,.$$

Here, the first and the second equation follow by substituting $(5.2.12)_{1,2}$ into $(2.2.32)_1$ and $(2.2.32)_2$.

The corresponding mechanical balance equations for **model I** are most easily obtained if in (5.6.8) the stress tensor ${}^{II}T_{i\alpha}$ is replaced by ${}^{I}T_{i\alpha}$; this is achieved through use of (5.2.22).

Next we pass on to **model V**, or the LORENTZ model. Its mechanical balance laws emerge if (5.3.16) and (5.3.30) are substituted into (2.2.32). When this is done care must be observed that the heat flux vector Q_α in (5.3.30) is correctly selected; it is Q_α^S of the statistical model, rather than Q_α^L as originally introduced in the LORENTZ model . We thus obtained

$$\rho_0 \ddot{x}_i = {}^{V}T_{i\alpha,\alpha} + F_{\alpha i}^{-1}\{(\mathbb{Q} - \mathbb{P}_{\beta,\beta})\mathbb{E}_\alpha + e_{\alpha\beta\gamma}(\mathbb{J}_\beta + \dot{\mathbb{P}}_\beta)\mathbb{B}_\gamma$$

$$+ (\mathbb{M}_{\alpha,\beta}^L - \mathbb{M}_{\beta,\alpha}^L)\mathbb{B}_\beta\} + \rho_0 F_i^{\text{ext}}\,,$$

$$(5.6.9)$$

$${}^{V}T_{[i\alpha}F_{j]\alpha} = 0\,,$$

$$\rho_0 \Theta \dot{\eta} = \mathbb{J}_\alpha \mathbb{E}_\alpha - Q_{\alpha,\alpha} + \rho_0 r^{\text{ext}}\,.$$

The next model is **model IV**, but again there is no need to list the results for this model explicitly, because the only difference between models IV and V lies in the stress tensors and consequently the balance equations for model IV may be obtained by merely introducing into (5.6.9) ${}^{IV}T_{i\alpha}$ as given by (5.3.26). Needless to state that the two models have been shown to be equivalent already in Sect. 5.3.

There remains the presentation of **model III**. Because of its complexity as compared to other models it is presented last. Its mechanical balance equations emerge when (5.4.5) is substituted into (2.2.32). This yields

$$\rho_0 \ddot{x}_i = {}^{III}T_{i\alpha,\alpha} + F_{\alpha i}^{-1}\{\mathbb{Q}\,\mathbb{E}_\alpha + e_{\alpha\beta\gamma}\mathbb{J}_\beta\mathbb{B}_\gamma + \mathbb{P}_\beta\mathbb{E}_{\beta,\alpha}$$

$$+ \mu_0 \mathbb{M}_\beta^C \mathbb{H}_{\beta,\alpha} + e_{\alpha\beta\gamma}(\mathbb{D}_\beta\dot{\mathbb{B}}_\gamma + \dot{\mathbb{D}}_\beta\mathbb{B}_\gamma)\}$$

$$+ F_{\beta i,j}^{-1}F_{j\alpha}(\mathbb{P}_\alpha\mathbb{E}_\beta + \mu_0\mathbb{M}_\alpha^C\mathbb{H}_\beta) + \rho_0 F_i^{\text{ext}}\,,$$

$$(5.6.10)$$

$${}^{III}T_{[i\alpha}F_{j]\alpha} = F_{[i\alpha}F_{\beta j]}^{-1}(\mathbb{P}_\alpha\mathbb{E}_\beta + \mu_0\mathbb{M}_\alpha^C\mathbb{H}_\beta)\,,$$

$$\rho_0 \Theta \dot{\eta} = \mathbb{J}_\alpha \mathbb{E}_\alpha - Q_{\alpha,\alpha} + \rho_0 r^{\text{ext}}\,.$$

As already stated in Sect. 5.4, model III is identical to model I, and, hence, the only difference in the formulations of models III and II stems from the difference in the definition of the stress tensors, which are related to each other according to (see (5.4.13) or (5.2.22))

$$^{III}T_{i\alpha} = {}^{II}T_{i\alpha} - F_{\beta i}^{-1}(\mathbb{P}_\alpha \mathbb{E}_\beta + \mu_0 \mathbb{M}_\alpha^C \mathbb{H}_\beta) \; . \tag{5.6.11}$$

From the above considerations we conclude that the only necessary step for a completion of the comparison of the five models is a comparison of models II and V. To perform this comparison, notice that in view of (5.6.8) and (5.6.9) the balance laws of moment of momentum and energy are identical in the two formulations. Consequently, at most the balance laws of linear momentum can differ; and the difference can at most be a difference in stress. This difference must, furthermore, be such that the balance of moment of momentum is met; in other words the difference between the second PIOLA-KIRCHHOFF tensors $^{II}T_{\alpha\beta}^P$ and $^{V}T_{\alpha\beta}^P$ must be a symmetric tensor.

In order to find this difference the balance equations $(5.6.8)_1$ and $(5.6.9)_2$ must be used. However, this step becomes much easier if for the electromagnetic body force the respective representations in terms of the MAXWELL stress tensor are used; see $(5.2.17)_1$ and $(5.3.19)_1$. We then obtain

$$\begin{aligned} {}^{V}T_{i\alpha} - {}^{II}T_{i\alpha} &= {}^{II}T_{i\alpha}^M - {}^{V}T_{i\alpha}^M \\ &= F_{\beta i}^{-1}(\mathbb{B}_\alpha^a \mathbb{H}_\beta - \mathbb{B}_\alpha \mathbb{H}_\beta^a) - \tfrac{1}{2}F_{\alpha i}^{-1}(\mathbb{B}_\beta^a \mathbb{H}_\beta - \mathbb{B}_\beta \mathbb{H}_\beta^a) \; . \end{aligned} \tag{5.6.12}$$

To eliminate from this expression the auxiliary fields use must be made of (5.2.9), (5.3.12) and (5.4.2); we may then derive the expressions

$$\mathbb{B}_\alpha^a \mathbb{H}_\beta = \mu_0 |J| C_{\alpha\gamma}^{-1} \mathbb{H}_\gamma \mathbb{H}_\beta \; ,$$

$$\begin{aligned} \mathbb{B}_\alpha \mathbb{H}_\beta^a &= \frac{1}{\mu_0}|J^{-1}|C_{\beta\gamma}\mathbb{B}_\gamma \mathbb{B}_\alpha = \mu_0 |J| C_{\alpha\gamma}^{-1}\mathbb{H}_\gamma \mathbb{H}_\beta + \mu_0 \mathbb{M}_\alpha^C \mathbb{H}_\beta \\ &\quad + \mu_0 C_{\alpha\gamma}^{-1} C_{\beta\delta}\mathbb{H}_\gamma \mathbb{M}_\delta^C + \mu_0 |J^{-1}|C_{\beta\gamma}\mathbb{M}_\alpha^C \mathbb{M}_\gamma^C \; . \end{aligned} \tag{5.6.13}$$

Substitution into (5.6.12) yields

$$\begin{aligned} {}^{V}T_{i\alpha} &= {}^{II}T_{i\alpha} - \mu_0 F_{\beta i}^{-1}\mathbb{M}_\alpha^C \mathbb{H}_\beta - \mu_0 C_{\alpha\gamma}^{-1}F_{i\beta}\mathbb{H}_\gamma \mathbb{M}_\beta^C \\ &\quad - \mu_0 |J^{-1}|F_{i\beta}\mathbb{M}_\alpha^C \mathbb{M}_\beta^C + \mu_0 F_{\alpha i}^{-1}\mathbb{M}_\beta^C \mathbb{H}_\beta \\ &\quad + \tfrac{1}{2}\mu_0 |J^{-1}|F_{\alpha i}^{-1}C_{\beta\gamma}\mathbb{M}_\beta^C \mathbb{M}_\gamma^C \; , \end{aligned} \tag{5.6.14}$$

and after a multiplication with $F_{\beta i}^{-1}$

$$\begin{aligned} {}^{V}T_{\alpha\beta}^P &= {}^{II}T_{\alpha\beta}^P - \mu_0 C_{\beta\gamma}^{-1}\mathbb{M}_\alpha^C \mathbb{H}_\gamma - \mu_0 C_{\alpha\gamma}^{-1}\mathbb{M}_\beta \mathbb{H}_\gamma + \mu_0 C_{\alpha\beta}^{-1}\mathbb{M}_\gamma \mathbb{H}_\gamma \\ &\quad - \mu_0 |J^{-1}|\mathbb{M}_\alpha^C \mathbb{M}_\beta^C + \tfrac{1}{2}\mu_0 |J^{-1}|C_{\alpha\beta}^{-1}C_{\gamma\delta}\mathbb{M}_\gamma^C \mathbb{M}_\delta^C \; , \end{aligned} \tag{5.6.15}$$

from which it is easily seen that the difference between $^{V}T^{P}_{\alpha\beta}$ and $^{II}T^{P}_{\alpha\beta}$ forms indeed a symmetric tensor. Needless to say that this relation corresponds to an equation that was established in Chap. 4 and relates the CAUCHY stresses $^{V}t_{ij}$ and $^{II}t_{ij}$ (see Table 4.1).

In (5.6.15), the LORENTZian stress $^{V}T^{P}_{\alpha\beta}$ is expressed in terms of $^{II}T^{P}_{\alpha\beta}$ and the CHU-variables. The reverse relation expresses $^{II}T^{P}_{\alpha\beta}$ in terms of $^{V}T^{P}_{\alpha\beta}$ and the LORENTZ-variables. This latter expression is obtained from (5.3.10), $(5.3.14)_2$ and $(5.3.12)_2$ and reads

$$
\begin{aligned}
^{II}T^{P}_{\alpha\beta} = {} & ^{V}T^{P}_{\alpha\beta} + C^{-1}_{\alpha\gamma}\mathrm{M}^{L}_{\gamma}\mathbb{B}_{\beta} + C^{-1}_{\beta\gamma}\mathrm{M}^{L}_{\gamma}\mathbb{B}_{\alpha} - C^{-1}_{\alpha\beta}\mathrm{M}^{L}_{\gamma}\mathbb{B}_{\gamma} \\
& - \mu_0|J|C^{-1}_{\alpha\gamma}C^{-1}_{\beta\delta}\mathrm{M}^{L}_{\gamma}\mathrm{M}^{L}_{\delta} + \tfrac{1}{2}\mu_0|J|C^{-1}_{\alpha\beta}C^{-1}_{\gamma\delta}\mathrm{M}^{L}_{\gamma}\mathrm{M}^{L}_{\delta} \;.
\end{aligned}
\tag{5.6.16}
$$

The above equations and similar relations of the previous models are *necessary conditions*, which must be satisfied in order that the models be equivalent. If these relations hold equivalence goes as far as the local balance laws of linear and angular momentum, energy and the corresponding jump conditions are concerned. The conditions are not sufficient, however, because the constitutive relations and therefore also the thermodynamic requirements impose further conditions. In the following these thermodynamic requirements will be discussed.

It follows from the foregoing considerations that in the constitutive theories of models I, II and III the internal energies must be identical, i.e.

$$
^{I}(\rho_0 U) = {}^{II}(\rho_0 U) = {}^{III}(\rho_0 U) \;,
\tag{5.6.17}
$$

if the models are to be equivalent. The same holds for models IV and V:

$$
^{IV}(\rho_0 U) = {}^{V}(\rho_0 U) \;.
\tag{5.6.18}
$$

However, and as could already be expected from the corresponding results of Chap. 4, relation (5.6.15) can only be satisfied simultaneously with the constitutive equations (5.2.26) and (5.3.29) provided that the internal energies for models II and V differ. To derive the corresponding relation note that according to Chap. 4

$$
^{V}(\rho U) = {}^{II}(\rho U) - \mu_0 \mathcal{H}_i \mathcal{M}_i - \tfrac{1}{2}\mu_0 \mathcal{M}_i \mathcal{M}_i \;,
$$

which in LAGRANGEan notation reads

$$
^{V}(\rho_0 U) = {}^{II}(\rho_0 U) - \mu_0 \mathbb{H}_\alpha \mathrm{M}^{C}_\alpha - \tfrac{1}{2}\mu_0|J^{-1}|C_{\alpha\beta}\mathrm{M}^{C}_\alpha \mathrm{M}^{C}_\beta \;.
\tag{5.6.19}
$$

If for models II and V we use constitutive relations of the form (5.2.53) and (5.3.35) and if we introduce the HELMHOLTZ free energies (5.2.54) and (5.3.36), then (5.6.19) is equivalent to

$$
^{V}(\rho_0 \breve{\psi}) = {}^{II}(\rho_0 \bar{\psi}) - \tfrac{1}{2}\mu_0|J^{-1}|C_{\alpha\beta}\mathrm{M}^{C}_\alpha \mathrm{M}^{C}_\beta \;.
\tag{5.6.20}
$$

This equation may be used in (5.2.55) and (5.3.37) to obtain expressions for the entropy, polarization, magnetization and the stress in the respective formulations. These must then satisfy the identities derived above, and if they do, the models are equivalent. The proof on the basis of (5.6.20), (5.2.55) and (5.3.37) is straightforward and thus we leave it to the reader. However, we shall come back to the consequences of this statement when defining material coefficients in the various formulations.

Finally, we would like to point out that a major goal of the derivation of the material description was in the presentation of a correct approach in linearizing field equations. The LAGRANGEan formulation is useful from just this practical point of view. Indeed, once the equations are known in their material description the transformations

$$\frac{\partial}{\partial x_i} \to \frac{\partial}{\partial X_\alpha} \quad \text{and} \quad \frac{\partial}{\partial t} \to \frac{d}{dt}$$

need no longer be performed, because all variables are per se already functions of X_α and t. This means that any perturbation approximation is much easier to be carried out when the equations are written in material rather than spatial coordinates. All linearization procedures performed so far were in the spatial description (see e.g. TOUPIN [241], HUTTER and PAO [91] and VAN DE VEN [249]). The treatments of TOUPIN and HUTTER and PAO are approximate, however, insofar as they contain ad hoc assumptions that cannot be justified on the basis of non-relativistic arguments. This was pointed out by VAN DE VEN, who also presents the correct solution. All the difficulties TOUPIN and HUTTER and PAO were faced with disappear in the material description and no ad hoc assumptions must be introduced here. Corroboration of this will be given in the next chapter.

5.7 Approach to a Unified Constitutive Theory

In the preceding section we demonstrated that all theories of deformable bodies in the electromagnetic fields which have the complexity of thermoelastic polarizable and magnetizable solids are non-relativistically equivalent, provided that the respective constitutive functions for the internal energy or the free energy satisfy certain relationships. Similar equivalence statements were already established in the EULERian description so that a new proof was not a necessity except, perhaps, that it led to equivalence conditions expressed in the LAGRANGEan variables; these are very useful relationships. To recapitulate them briefly, recall that in order to enforce equivalence of formulations I, II and III the free energies must in all these formulations be the same functions of the same variables. A similar statement also holds for models IV and V, but to achieve equivalence between the groups (I, II, III) and (IV,V) the corresponding free energies must be related by (5.6.20), viz.

$$\rho_0 \check{\psi} = \rho_0 \bar{\psi} - \tfrac{1}{2}\mu_0 |J^{-1}| C_{\alpha\beta} \mathrm{M}_\alpha^C \mathrm{M}_\beta^C \ . \tag{5.7.1}$$

Here, $\bar{\psi}$ is the free energy function in any one of the formulations I, II or III, and, correspondingly $\check{\psi}$ is that of the models IV and V; they are defined in (5.2.54) and (5.3.36), respectively. Accordingly, and apart from a dependence on $C_{\alpha\beta}$ and Θ, $\check{\psi}$ is a function of $(\mathbb{E}_\alpha, \mathbb{B}_\alpha)$, whereas $\bar{\psi}$ depends on $(\mathbb{E}_\alpha, \mathbb{H}_\alpha)$ as does M_α^C because (see $(5.2.55)_3$)

$$\mathrm{M}_\alpha^C = -\frac{\rho_0}{\mu_0} \frac{\partial \bar{\psi}}{\partial \mathbb{H}_\alpha}.$$

Equation (5.7.1) can therefore also be written as

$$\check{\psi}(C_{\alpha\beta}, \mathbb{E}_\alpha, \mathbb{B}_\alpha, \Theta) = \bar{\psi}(C_{\alpha\beta}, \mathbb{E}_\alpha, \mathbb{H}_\alpha, \Theta)$$
$$-\frac{1}{2}\frac{\rho_0}{\mu_0} |J^{-1}| C_{\alpha\beta} \left(\frac{\partial \bar{\psi}}{\partial \mathbb{H}_\alpha} \frac{\partial \bar{\psi}}{\partial \mathbb{H}_\beta} \right) (C_{\alpha\beta}, \mathbb{E}_\alpha, \mathbb{B}_\alpha, \Theta) \ . \tag{5.7.2}$$

In this identity, the left-hand side is a function of $\mathbb{B}_\alpha$, in contrast to the expression on the right-hand side, which is a function of $\mathbb{H}_\alpha$. In view of $(5.4.2)_2$ and $(5.2.55)_3$ we may, however, express $\mathbb{B}_\alpha$ as

$$\mathbb{B}_\alpha = -\rho_0 \frac{\partial \bar{\psi}}{\partial \mathbb{H}_\alpha} + \mu_0 |J| C_{\alpha\beta}^{-1} \mathbb{H}_\beta \ , \tag{5.7.3}$$

so that

$$\check{\psi}(C_{\alpha\beta}, \mathbb{E}_\alpha, \mu_0 |J| C_{\alpha\beta}^{-1} \mathbb{H}_\beta - \rho_0 \partial \bar{\psi}/\partial \mathbb{H}_\alpha, \Theta) =$$
$$\bar{\psi}(C_{\alpha\beta}, \mathbb{E}_\alpha, \mathbb{H}_\alpha, \Theta) - \frac{1}{2}\frac{\rho_0}{\mu_0} |J^{-1}| C_{\alpha\beta} \frac{\partial \bar{\psi}}{\partial \mathbb{H}_\alpha} \frac{\partial \bar{\psi}}{\partial \mathbb{H}_\beta} \ , \tag{5.7.4}$$

in which the arguments in $\partial \bar{\psi}/\partial \mathbb{H}_\alpha$ are, of course, the same as in $\bar{\psi}$. For given functions $\check{\psi}$ and $\bar{\psi}$, equation (5.7.4) must be satisfied identically if the formulations (I, II, III) and (IV,V) are to be equivalent. If, on the other hand, only one of the functions $\check{\psi}$ or $\bar{\psi}$ is given, then (5.7.4) is a functional differential equation to determine the other. The solution to this will, in general, be very complex. We shall not try to solve it for a given function $\check{\psi}$, say. We shall rather exploit (5.7.4) as common forms of the free energies $\bar{\psi}$ and $\check{\psi}$. A very popular procedure is to write these functions as polynomial expressions of the independent variables. These polynomials must be regarded as truncated TAYLOR series expansions about a state of zero electromagnetic fields and zero deformation of a general functional relationship for the free energy. Now, the equivalence of the various models was guaranteed in the above for the general theory using a representation for the constitutive equations of the free energy not restricted by any means. Otherwise stated, if one attempts to establish equivalence statements between two given theories for which the free energies are polynomials truncated at the quadratic terms, the two theories might very well be non-equivalent simply because equivalence would in one formulation require the inclusion of cubic, quartic or even higher order terms.

The point just raised is important, because it illustrates that, strictly, fully equivalent theories might become non-equivalent, because in both formulations one insists in too restrictive energy expressions. Nonetheless, we can keep the usual polynomial representations and still claim equivalence, but if we do so this is only in the following *restricted sense*: We formally interpret the polynomials as truncated TAYLOR series expansions. With the use of the latter equivalence of two formulations can be established exactly; it amounts to a comparison of the polynomial coefficients. If in these TAYLOR series expansions we restrict ourselves to terms of a certain order, then equivalence can be established approximately.

To illustrate the above point more clearly, consider the following somewhat academic example: Let

$$f(x,y) = \sum_{\mu,\nu=0}^{\infty} a_{\mu\nu} x^{\mu} y^{\nu} , \qquad g(x,z) = \sum_{\mu,\nu=0}^{\infty} b_{\mu\nu} x^{\mu} z^{\nu} ,$$

and assume that for some reason $z = x + y$ and $f = g$. Then by the binomial theorem we may set

$$z^{\nu} = (x + y)^{\nu} = \sum_{k=0}^{\nu} \binom{\nu}{k} x^{k} y^{\nu-k} ,$$

and therefore

$$\sum_{\mu,\nu=0}^{\infty} a_{\mu\nu} x^{\mu} y^{\nu} = \sum_{\mu,\nu=0}^{\infty} b_{\mu\nu} \sum_{k=0}^{\nu} \binom{\nu}{k} x^{k+\mu} y^{\nu-k} .$$

The latter equation can also be written as

$$\sum_{\mu,\nu=0}^{\infty} \left\{ \sum_{k=0}^{\mu} \binom{\nu+k}{k} b_{(\mu-k)(\nu+k)} - a_{\mu\nu} \right\} x^{\mu} y^{\nu} = 0 ,$$

so that the coefficient functions must satisfy the relations

$$a_{\mu\nu} = \sum_{k=0}^{\mu} \binom{\nu+k}{k} b_{(\mu-k)(\nu+k)} .$$

These identities must hold for all positive integers μ and ν. If we truncate the above polynomial representations at $\nu = N$ and $\mu = M$ they can still be satisfied, but then f_{MN} and g_{MN}, which denote the truncated expressions

$$f_{MN}(x,y) = \sum_{\mu=0}^{M} \sum_{\nu=0}^{N} a_{\mu\nu} x^{\mu} z^{\nu} , \qquad g_{MN}(x,y) = \sum_{\mu=0}^{M} \sum_{\nu=0}^{N} b_{\mu\nu} x^{\mu} z^{\nu} ,$$

are no longer exactly equal, but only in the sense that

$$f_{MN} = g_{MN} + \mathcal{O}(x^{M+1}) + \mathcal{O}(y^{N+1}) \,,$$

holds. For instance, if

$$f_{12}(x,y) = a_{00} + a_{01}y + a_{02}y^2 + a_{10}x + a_{11}xy + a_{12}xy^2 \,,$$

$$g_{12}(x,z) = b_{00} + b_{01}z + b_{02}z^2 + b_{10}x + b_{11}xz + b_{12}xz^2 \,,$$

and g_{12} is expressed in terms of x and y, then g_{12} should contain quadratic and cubic terms in x, whence follows that f_{12} and g_{12} cannot be identical except in the above mentioned approximate sense. In fact, one obtains

$$g_{12} = f_{12} + (b_{11} + b_{02} + 2b_{12}y)x^2 + b_{12}x^3 \,.$$

The above considerations may look somewhat artificial to the novel reader, yet they are important, and we would like to illustrate them using the most simple example that accounts for magnetoelastic interactions. For that purpose we restrict ourselves to conditions of isotropy and to free energies, which are at most of quartic order in the electric and magnetic field quantities and of quadratic order in the temperature difference

$$\theta = \Theta - \Theta_0 \,.$$

Here, Θ_0 denotes a reference temperature. Moreover, we assume small deformations, so that it suffices to write the free energy as a quadratic function of the LAGRANGEan deformation tensor

$$E_{\alpha\beta} = \tfrac{1}{2}(C_{\alpha\beta} - \delta_{\alpha\beta}) \,. \tag{5.7.5}$$

With these limitations we may choose the following representations for the free energy functions: for **model II**

$$\bar{\psi} = \frac{1}{2\rho_0}\, {}_2\bar{\chi}^{(m)}\mathbb{H}_\alpha\mathbb{H}_\alpha + \frac{1}{4\rho_0}\, {}_4\bar{\chi}^{(m)}(\mathbb{H}_\alpha\mathbb{H}_\alpha)^2 + \frac{1}{2\rho_0}\, {}_2\bar{\chi}^{(e)}\mathbb{E}_\alpha\mathbb{E}_\alpha$$

$$+ \frac{1}{4\rho_0}\, {}_4\bar{\chi}^{(e)}(\mathbb{E}_\alpha\mathbb{E}_\alpha)^2 - \tfrac{1}{2}\bar{c}\theta^2 + \frac{1}{2\rho_0}\,\bar{l}^{(m)}\mathbb{H}_\alpha\mathbb{H}_\alpha\theta + \frac{1}{2\rho_0}\,\bar{L}^{(e)}\mathbb{E}_\alpha\mathbb{E}_\alpha\theta$$

$$+ \left[\frac{1}{2\rho_0}\,\bar{b}^{(m)}_{\alpha\beta\gamma\delta}\mathbb{H}_\alpha\mathbb{H}_\beta + \frac{1}{2\rho_0}\,\bar{b}^{(e)}_{\alpha\beta\gamma\delta}\mathbb{E}_\alpha\mathbb{E}_\beta - \bar{\nu}\delta_{\gamma\delta}\theta\right] E_{\gamma\delta}$$

$$+ \frac{1}{2\rho_0}\,\bar{c}_{\alpha\beta\gamma\delta}E_{\alpha\beta}E_{\gamma\delta} \,, \tag{5.7.6}$$

for **model V**

$$\check{\psi} = \frac{1}{2\rho_0}\,{}_2\check{\chi}^{(m)}\mathbb{B}_\alpha\mathbb{B}_\alpha + \frac{1}{4\rho_0}\,{}_4\check{\chi}^{(m)}(\mathbb{B}_\alpha\mathbb{B}_\alpha)^2 + \frac{1}{2\rho_0}\,{}_2\check{\chi}^{(e)}\mathbb{E}_\alpha\mathbb{E}_\alpha$$

$$+ \frac{1}{4\rho_0}\,{}_4\check{\chi}^{(e)}(\mathbb{E}_\alpha\mathbb{E}_\alpha)^2 - \tfrac{1}{2}\check{c}\theta^2 + \frac{1}{2\rho_0}\check{L}^{(m)}\mathbb{B}_\alpha\mathbb{B}_\alpha\theta + \frac{1}{2\rho_0}\check{L}^{(e)}\mathbb{E}_\alpha\mathbb{E}_\alpha\theta$$

$$+ \left[\frac{1}{2\rho_0}\check{b}^{(m)}_{\alpha\beta\gamma\delta}\mathbb{B}_\alpha\mathbb{B}_\beta + \frac{1}{2\rho_0}\check{b}^{(e)}_{\alpha\beta\gamma\delta}\mathbb{E}_\alpha\mathbb{E}_\beta - \check{\nu}\delta_{\gamma\delta}\theta\right]E_{\gamma\delta}$$

$$+ \frac{1}{2\rho_0}\check{c}_{\alpha\beta\gamma\delta}E_{\alpha\beta}E_{\gamma\delta}\,. \tag{5.7.7}$$

Before we proceed it is worthwhile to look at these expressions more closely. The polynomial representations for the free energy start with quadratic terms. A constant term is left out, because it is immaterial, and linear terms are discarded, because at zero deformation, zero temperature difference and zero electromagnetic fields no stress and no polarization and magnetization should be present. In other words we assume the body to possess a natural unstrained state. Cubic terms are also present in (5.7.6) and (5.7.7), but in this regard the polynomials are not complete, since products such as $E_{\alpha\beta}E_{\gamma\delta}\mathbb{H}_\varepsilon$, $E_{\alpha\beta}$, $E_{\gamma\delta}\mathbb{E}_\varepsilon$, and $E_{\alpha\beta}E_{\gamma\delta}\theta$ are missing. We believe that these omissions are justified, because usually the field-free elasticities are much more important than their change by the electromagnetic fields. We have further neglected electromagnetic coupling terms.

In the ensuing analysis our aim is to investigate in what sense the above representations would allow an identical satisfaction of (5.7.1). In view of the above general remarks we expect that a full equivalence cannot be established, so differences will show up in the higher-order terms of (5.7.6) and (5.7.7). To find the equivalence conditions we substitute (5.7.6) into the first term on the right-hand side of (5.7.1) and rewrite the latter by expressing the second term as a function of the independent variables of $\check{\psi}$. To this end, note that

$$|J^{-1}|C_{\alpha\beta} = \delta_{\alpha\beta}(1 - E_{\gamma\gamma}) + 2E_{\alpha\beta} + \mathcal{O}(E^2)\delta_{\alpha\beta} + \mathrm{n}_{\alpha\beta\gamma\delta}E_{\gamma\delta} + \mathcal{O}(E^2)\,, \tag{5.7.8}$$

where

$$\mathrm{n}_{\alpha\beta\gamma\delta} = -\delta_{\alpha\beta}\delta_{\gamma\delta} + \delta_{\alpha\gamma}\delta_{\beta\delta} + \delta_{\alpha\delta}\delta_{\beta\gamma}\,. \tag{5.7.9}$$

In view of $(5.2.55)_3$ and (5.7.6) we may also write

$$\mu_0\mathbb{M}^C_\alpha = -\rho_0\frac{\partial\bar{\psi}}{\partial\mathbb{H}_\alpha}$$

$$= -\,{}_2\bar{\chi}^{(m)}\mathbb{H}_\alpha - {}_4\bar{\chi}^{(m)}\mathbb{H}_\beta\mathbb{H}_\beta\mathbb{H}_\alpha - \bar{L}^{(m)}\mathbb{H}_\alpha\theta - \bar{b}^{(m)}_{\alpha\beta\gamma\delta}\mathbb{H}_\beta E_{\gamma\delta}\,. \tag{5.7.10}$$

Using these relations in (5.7.1) we find that the right-hand side of the latter may be written as

$$
\begin{aligned}
\check{\psi} =\ & \frac{1}{2\rho_0}\left(1 - \frac{2\bar{\chi}^{(\mathrm{m})}}{\mu_0}\right) {}_2\bar{\chi}^{(\mathrm{m})}\mathbb{H}_\alpha\mathbb{H}_\alpha \\
& + \frac{1}{4\rho_0}\,{}_4\bar{\chi}^{(\mathrm{m})}\left(1 - 4\frac{2\bar{\chi}^{(\mathrm{m})}}{\mu_0}\right)(\mathbb{H}_\alpha\mathbb{H}_\alpha)^2 \\
& + \frac{1}{2\rho_0}\,{}_2\bar{\chi}^{(\mathrm{e})}\mathbb{E}_\alpha\mathbb{E}_\alpha + \frac{1}{2\rho_0}\,{}_4\bar{\chi}^{(\mathrm{e})}(\mathbb{E}_\alpha\mathbb{E}\alpha)^2 - \tfrac{1}{2}\bar{c}\theta^2 \\
& + \frac{1}{2\rho_0}\left(1 - 2\frac{2\bar{\chi}^{(\mathrm{m})}}{\mu_0}\right)\bar{L}^{(\mathrm{m})}\mathbb{H}_\alpha\mathbb{H}_\alpha\theta + \frac{1}{2\rho_0}\bar{L}^{(\mathrm{e})}\mathbb{E}_\alpha\mathbb{E}_\alpha\theta \\
& + \frac{1}{2\rho_0}\left[\left(1 - 2\frac{2\bar{\chi}^{(\mathrm{m})}}{\mu_0}\right)\bar{\mathrm{b}}^{(\mathrm{m})}_{\alpha\beta\gamma\delta} - \frac{2\bar{\chi}^{(\mathrm{m})2}}{\mu_0}\mathrm{n}_{\alpha\beta\gamma\delta}\right]\mathbb{H}_\alpha\mathbb{H}_\beta E_{\gamma\delta} \\
& + \frac{1}{2\rho_0}\bar{\mathrm{b}}^{(\mathrm{e})}_{\alpha\beta\gamma\delta}\mathbb{E}_\alpha\mathbb{E}_\beta E_{\gamma\delta} - \bar{\nu}\theta E_{\gamma\gamma} + \frac{1}{2\rho}\bar{c}_{\alpha\beta\gamma\delta}E_{\alpha\beta}E_{\gamma\delta} \\
& - \frac{\bar{L}^{(\mathrm{m})2}}{2\rho_0\mu_0}\mathbb{H}_\alpha\mathbb{H}_\alpha\theta^2 + \frac{\bar{L}^{(\mathrm{m})}}{\rho_0\mu_0}\left(-{}_2\bar{\chi}^{(\mathrm{m})}\mathrm{n}_{\alpha\beta\gamma\delta} - \bar{\mathrm{b}}^{(\mathrm{m})}_{\alpha\beta\gamma\delta}\right)\mathbb{H}_\alpha\mathbb{H}_\beta E_{\gamma\delta}\theta \\
& - \frac{\bar{L}^{(\mathrm{m})2}}{2\rho_0\mu_0}\mathrm{n}_{\alpha\beta\gamma\delta}\mathbb{H}_\alpha\mathbb{H}_\beta E_{\gamma\delta}\theta^2 + \ \text{higher order terms}.
\end{aligned}
\tag{5.7.11}
$$

Here and henceforth all terms of order

$$
\mathbb{H}_\varepsilon E_{\alpha\beta}E_{\gamma\delta}\,, \qquad \mathbb{E}_\varepsilon E_{\alpha\beta}E_{\gamma\delta}\,, \qquad \theta E_{\alpha\beta}E_{\gamma\delta}\,,
$$

and higher are neglected. This is consistent with our basic assumption that third-order terms involving the square of the deformation tensor are discarded. In contrast to (5.7.6) or (5.7.7) the above representation for $\check{\psi}$ contains also mixed fourth-order terms (the last three expressions on the right-hand side). For reasons described above, these should be neglected as well. The occurrence of these higher-order terms is corroboration for our earlier statement that full equivalence might simply be impossible, because the free energies in the respective formulations are too restrictive. However, it is apparent from the above calculation that a full equivalence could be achieved if the entire TAYLOR series expansion would be kept.

The functional relationship for $\check{\psi}$ is still not in the appropriate form for comparison with (5.7.7). For that purpose $\mathbb{B}_\alpha$ in (5.7.7) must be replaced by $\mathbb{H}_\alpha$. This replacement is accomplished by using (5.4.2) and (5.7.10), and it yields

$$
\begin{aligned}
\mathbb{B}_\alpha =\ & \mu_0(1 - {}_2\bar{\chi}^{(\mathrm{m})})\mathbb{H}_\alpha - {}_4\bar{\chi}^{(\mathrm{m})}\mathbb{H}_\beta\mathbb{H}_\beta\mathbb{H}_\alpha - \bar{L}^{(\mathrm{m})}\mathbb{H}_\alpha\theta \\
& - (\mathrm{b}^{(\mathrm{m})}_{\alpha\beta\gamma\delta} + \mu_0\mathrm{n}_{\alpha\beta\gamma\delta})\mathbb{H}_\beta E_{\gamma\delta} \\
=\ & \mu\mathbb{H}_\alpha - {}_4\bar{\chi}^{(\mathrm{m})}\mathbb{H}_\beta\mathbb{H}_\beta\mathbb{H}_\alpha - \bar{L}^{(\mathrm{m})}\mathbb{H}_\alpha\theta \\
& - (\bar{\mathrm{b}}^{(\mathrm{m})}_{\alpha\beta\gamma\delta} + \mu_0\mathrm{n}_{\alpha\beta\gamma\delta})\mathbb{H}_\beta E_{\gamma\delta}\,,
\end{aligned}
\tag{5.7.12}
$$

where we have set

$$\mu_0 - {}_2\bar{\chi}^{(m)} = \mu \, , \qquad \text{or} \qquad {}_2\bar{\chi}^{(m)} = \mu_0 - \mu \, . \tag{5.7.13}$$

If (5.7.12) is substituted into (5.7.7) and the resulting expression is rearranged (thereby neglecting third and fourth order terms as was done above) we obtain

$$
\begin{aligned}
\check{\psi} =\ & \frac{1}{2\rho_0}\mu^2\, {}_2\check{\chi}^{(m)}\mathbb{H}_\alpha\mathbb{H}_\alpha \\
& + \frac{1}{4\rho_0}\left(\mu^4\, {}_4\check{\chi}^{(m)} - 4\mu\, {}_2\check{\chi}_4^{(m)}\check{\chi}^{(m)}\right)(\mathbb{H}_\alpha\mathbb{H}_\alpha)^2 \\
& + \frac{1}{2\rho_0}\, {}_2\check{\chi}^{(e)}\mathbb{E}_\alpha\mathbb{E}_\alpha + \frac{1}{4\rho_0}\, {}_4\check{\chi}^{(e)}(\mathbb{E}_\alpha\mathbb{E}_\alpha)^2 - \tfrac{1}{2}\check{c}\theta^2 \\
& + \frac{1}{2\rho_0}(\mu^2\check{L}^{(m)} - 2\mu\bar{L}^{(m)}\, {}_2\check{\chi}^{(m)})\mathbb{H}_\alpha\mathbb{H}_\alpha\theta + \frac{1}{2\rho_0}\check{L}^{(e)}\mathbb{E}_\alpha\mathbb{E}_\alpha\theta \\
& + \frac{1}{2\rho_0}\left[\mu^2\check{\mathrm{b}}_{\alpha\beta\gamma\delta}^{(m)} - 2\mu\left(\bar{\mathrm{b}}_{\alpha\beta\gamma\delta}^{(m)} + \mu_0\mathrm{n}_{\alpha\beta\gamma\delta}\right){}_2\check{\chi}^{(m)}\right]\mathbb{H}_\alpha\mathbb{H}_\beta E_{\gamma\delta} \\
& + \frac{1}{2\rho_0}\check{\mathrm{b}}_{\alpha\beta\gamma\delta}^{(e)}\mathbb{E}_\alpha\mathbb{E}_\beta E_{\gamma\delta} - \check{\nu}\theta E\gamma\gamma + \frac{1}{2\rho_0}\check{c}_{\alpha\beta\gamma\delta}E_{\alpha\beta}E_{\gamma\delta} \, ,
\end{aligned}
\tag{5.7.14}
$$

which is expressed in terms of the coefficients $(\check{\ })$ as well as $(\bar{\ })$. Identifying (5.7.11) with (5.7.14), we obtain

$$
\begin{aligned}
& \mu_0\mu\, {}_2\check{\chi}^{(m)} = {}_2\bar{\chi}^{(m)} = \mu_0 - \mu =: -\mu_0\, {}_2\chi^{(m)} \, , \\[4pt]
& {}_2\check{\chi}^{(e)} = {}_2\bar{\chi}^{(e)} =: -{}_2\chi^{(e)} \, , \\[4pt]
& {}_4\check{\chi}^{(m)} = \frac{1}{\mu^4}\, {}_4\bar{\chi}^{(m)}, \quad {}_4\check{\chi}^{(e)} = {}_4\bar{\chi}^{(e)} =: -{}_4\chi^{(e)} \, , \\[4pt]
& \check{c} = \bar{c} =: \frac{c_W}{\Theta_0}, \quad \mu^2\check{L}^{(m)} = \bar{L}^{(m)}, \quad \check{L}^{(e)} = \bar{L}^{(e)} =: L^{(e)} \, , \\[4pt]
& \check{\mathrm{b}}_{\alpha\beta\gamma\delta}^{(m)} = \frac{1}{\mu^2}\bar{\mathrm{b}}_{\alpha\beta\gamma\delta}^{(m)} - \frac{\mu_0\, {}_2\chi^{(m)}}{\mu^2}(2 + {}_2\chi^{(m)})\mathrm{n}_{\alpha\beta\gamma\delta} \, , \\[4pt]
& \check{\mathrm{b}}_{\alpha\beta\gamma\delta}^{(e)} = \bar{\mathrm{b}}_{\alpha\beta\gamma\delta}^{(e)} = \mathrm{b}_{\alpha\beta\gamma\delta}^{(e)} \, , \\[4pt]
& \check{\nu} = \bar{\nu} = \nu, \quad \check{c}_{\alpha\beta\gamma\delta} = \bar{c}_{\alpha\beta\gamma\delta} = c_{\alpha\beta\gamma\delta} \, .
\end{aligned}
\tag{5.7.15}
$$

Several of the coefficients occurring in the above equations can, without confusion, be given specific names, as for instance

${}_2\chi^{(m)}$	magnetic susceptibility ,
μ	magnetic permeability ,
${}_2\chi^{(e)}, {}_4\chi^{(e)}$	second- and fourth-order electric susceptibility ,
c_W	specific heat ,
$L^{(e)}$	thermoelectric constant ,

$b^{(e)}_{\alpha\beta\gamma\delta}$ electrostrictive constants ,

ν thermoelastic constant ,

$c_{\alpha\beta\gamma\delta}$ elastic constants .

More difficulties arise, however, for $L^{(m)}$, because it is not clear whether $\bar{L}^{(m)}$ or $\check{L}^{(m)}$ should be called thermomagnetic constant. A similar statement also holds for the anisotropy coefficients $_4\bar{\chi}^{(m)}$ and $_4\check{\chi}^{(m)}$. Since the difference between $\bar{b}^{(m)}_{\alpha\beta\gamma\delta}$ and $\check{b}^{(m)}_{\alpha\beta\gamma\delta}$ is essential, the situation is even more drastic for $b^{(m)}_{\alpha\beta\gamma\delta}$, the coefficients usually attributed with the notion of magnetostriction. In the CHU formulation one is inclined to call $\bar{b}^{(m)}_{\alpha\beta\gamma\delta}$ a magnetostrictive constant; in the LORENTZ formulation this is the case for $\check{b}^{(m)}_{\alpha\beta\gamma\delta}$ instead.

For isotropic materials the relations $(5.7.15)_8$ can still somewhat be simplified, since for this special group of materials

$$b^{(m)}_{\alpha\beta\gamma\delta} = b^{(m)}_1 \delta_{\alpha\beta}\delta_{\gamma\delta} + b^{(m)}_2 (\delta_{\alpha\gamma}\delta_{\beta\delta} + \delta_{\alpha\delta}\delta_{\beta\gamma}) \tag{5.7.16}$$

must hold. Substituting this into $(5.7.15)_8$ we obtain

$$\begin{aligned}
\check{b}^{(m)}_1 &= \frac{1}{\mu^2}\left(\bar{b}^{(m)}_1 + \mu_0\, _2\chi^{(m)}(\,_2\chi^{(m)} + 2)\right) , \\
\check{b}^{(m)}_2 &= \frac{1}{\mu^2}\left(\bar{b}^{(m)}_1 - \mu_0\, _2\chi^{(m)}(\,_2\chi^{(m)} + 2)\right) .
\end{aligned} \tag{5.7.17}$$

From these expressions it is now evident that *a unique definition of magnetostrictive constants is not possible* by merely establishing the respective polynomial representation of the free energy function. What would be needed, is a simple experiment in which magnetostrictive effects could uniquely be defined and in which the magnetostrictive constants could be measured, which then must be independent of the model chosen.

At this point we must warn the reader to take the above conclusions as the ultimate truth. The results are special insofar as they hold for isotropic bodies and no electromagnetic coupling. For instance that all electric coefficients in the energy expressions are identical is a consequence of the omission of electromagnetic coupling terms (these are terms such as $\bar{\chi}^{(em)}\mathbb{E}_\alpha\mathbb{H}_\alpha$ and $\bar{b}^{(em)}_{\alpha\beta\gamma\delta}\mathbb{E}_\alpha\mathbb{H}_\beta E_{\gamma\delta}$). Any generalization to these more complicated polynomial expressions is straightforward, however, and will be left to the reader.

There still remains the evaluation and comparison of the constitutive relations (5.2.55) and (5.3.37). To begin with, let us look more closely at the entropy. According to $(5.2.55)_1$ and $(5.7.6)$, we obtain

$$^{II}\eta = -\frac{\partial\bar{\eta}'}{\partial\theta} = c_W\frac{\theta}{\Theta_0} - \frac{1}{2\rho_0}\bar{L}^{(m)}\mathbb{H}_\alpha\mathbb{H}_\alpha - \frac{L^{(e)}}{2\rho_0}\mathbb{E}_\alpha\mathbb{E}_\alpha + \nu E_{\gamma\gamma} . \tag{5.7.18}$$

On the other hand, straightforward evaluation of $(5.3.37)_1$ on the basis of (5.7.7) gives

$$V_\eta = -\frac{\partial \check{\psi}}{\partial \theta} = c_W \frac{\theta}{\Theta_0} - \frac{1}{2\rho_0} \check{L}^{(m)} \mathbb{B}_\alpha \mathbb{B}_\alpha - \frac{L^{(e)}}{2\rho_0} \mathbb{E}_\alpha \mathbb{E}_\alpha + \nu E_{\gamma\gamma} , \quad (5.7.19)$$

which with the aid of (5.7.12) and (5.7.15) becomes

$$\begin{aligned} V_\eta = c_W \frac{\theta}{\Theta_0} - \frac{1}{2\rho_0} \bar{L}^{(m)} & \left\{ \delta_{\alpha\beta} \left(1 - \frac{2}{\mu} {}_4\bar\chi^{(m)} \mathbb{H}_\gamma \mathbb{H}_\gamma - \frac{2}{\mu} \bar{L}^{(m)} \theta \right) \right. \\ & \left. - \frac{2}{\mu} (\bar{b}^{(m)}_{\alpha\beta\gamma\delta} + \mu_0 n_{\alpha\beta\gamma\delta}) E_{\gamma\delta} \right\} \mathbb{H}_\alpha \mathbb{H}_\beta - \frac{L^{(e)}}{2\rho_0} \mathbb{E}_\alpha \mathbb{E}_\alpha + \nu E_{\gamma\gamma} \\ & + \text{ higher-order terms} . \end{aligned} \quad (5.7.20)$$

Mere comparison of (5.7.18) and (5.7.20) shows that $^{II}\eta$ differs from $^V\eta$. Hence, use of (5.7.15) has not led to identical expressions for the entropy. The difference arises, because of the terms $(\mathbb{H}_\alpha \mathbb{H}_\alpha)^2, \mathbb{H}_\alpha \mathbb{H}_\alpha \theta, \mathbb{H}_\alpha \mathbb{H}_\beta E_{\gamma\delta}$, and still higher-order terms, which in the energy expression can be traced back to the fourth-order terms $(\mathbb{H}_\alpha \mathbb{H}_\alpha)^2 \theta, \mathbb{H}_\alpha \mathbb{H}_\alpha \theta^2$, and $\mathbb{H}_\alpha \mathbb{H}_\beta E_{\alpha\beta} \theta$. In the process of the transformation of the energy expressions these fourth-order terms were omitted (and they must be, if the polynomials are interpreted as truncated TAYLOR series expansions).

Similar discrepancies also occur when the other constitutive quantities derivable from the free energy are determined. These are given by

$$\begin{aligned} \mathbb{P}_\alpha = {}^{II}\mathbb{P}_\alpha = {}^V\mathbb{P}_\alpha = {}_2\chi^{(e)} \mathbb{E}_\alpha + {}_4\chi^{(e)} \mathbb{E}_\beta \mathbb{E}_\beta \mathbb{E}_\alpha - L^{(e)} \mathbb{E}_\alpha \theta \\ - b^{(e)}_{\alpha\beta\gamma\delta} \mathbb{E}_\beta E_{\gamma\delta} , \end{aligned} \quad (5.7.21)$$

$$\begin{aligned} \mu_0 \mathbb{M}^C_\alpha = \mu_0 {}_2\chi^{(m)} \mathbb{H}_\alpha - {}_4\bar\chi^{(m)} \mathbb{H}_\beta \mathbb{H}_\beta \mathbb{H}_\alpha - \bar{L}^{(m)} \mathbb{H}_\alpha \theta \\ - \bar{b}^{(m)}_{\alpha\beta\gamma\delta} \mathbb{H}_\beta E_{\gamma\delta}, \end{aligned} \quad (5.7.22)$$

$$\mathbb{M}^L_\alpha = \frac{2\chi^{(m)}}{\mu} \mathbb{B}_\alpha - {}_4\check\chi^{(m)} \mathbb{B}_\beta \mathbb{B}_\beta \mathbb{B}_\alpha - \check{L}^{(m)} \mathbb{B}_\alpha \theta - \check{b}^{(m)}_{\alpha\beta\gamma\delta} \mathbb{B}_\beta E_{\gamma\delta} , \quad (5.7.23)$$

$$^{II}\mathbb{T}^P_{\alpha\beta} = \bar{b}^{(m)}_{\gamma\delta\alpha\beta} \mathbb{H}_\gamma \mathbb{H}_\delta + b^{(e)}_{\gamma\delta\alpha\beta} \mathbb{E}_\gamma \mathbb{E}_\delta - \nu\delta_{\alpha\beta}\theta + c_{\alpha\beta\gamma\delta} E_{\gamma\delta} , \quad (5.7.24)$$

$$^V\mathbb{T}^P_{\alpha\beta} = \check{b}^{(m)}_{\gamma\delta\alpha\beta} \mathbb{B}_\gamma \mathbb{B}_\delta + b^{(e)}_{\gamma\delta\alpha\beta} \mathbb{E}_\gamma \mathbb{E}_\delta - \nu\delta_{\alpha\beta}\theta + c_{\alpha\beta\gamma\delta} E_{\gamma\delta} . \quad (5.7.25)$$

They can be transformed into each other by neglecting all inconsistent terms as was done above for the entropy.

This completes our transformation of the theories of group (I, II, III) into those of group (IV,V). The calculations show that full equivalence may be destroyed by a too special choice of the energy functions. However, the calculations have simultaneously demonstrated how the material coefficients of one theory can be related to those of another. These questions are of immense practical importance and will be reconsidered in Chap. 6.

6 Linearization

6.1 Statement of the Problem

The governing dynamical equations of field-matter interaction in thermo-elastic materials as outlined in the previous chapters are highly nonlinear. Generally, it is hardly possible to find exact solutions even if the most simple problems that are still of some physical relevance are attacked. As stated in Chap. 5 already, one of the major disadvantages, namely that the equations are given in the spatial description, while boundary conditions for a solid body are usually prescribed in the reference configuration, has been removed by the introduction of a consistent material description for both the field equations and the jump and boundary conditions. Nevertheless, the resulting equations are still highly nonlinear, and this implies that some approximation scheme must be found.

To render the equations amenable to direct analysis, they will be linearized with respect to some *intermediate state*. We suppose that in the intermediate state the position of a material point, initially at $\boldsymbol{X}$, is given by $\boldsymbol{\xi}$ (ξ_α, $\alpha = 1, 2, 3$). The total motion of the particle from its initial position $\boldsymbol{X}$ to its final position $\boldsymbol{x}$ is then decomposed into the motion from the reference state to the intermediate state, characterized by the displacement vector $\bar{\boldsymbol{U}}$,

$$\bar{\boldsymbol{U}} = \boldsymbol{\xi} - \boldsymbol{X} , \qquad (6.1.1)$$

and the motion from the intermediate state to the present state with displacement $\boldsymbol{u}$

$$\boldsymbol{u} = \boldsymbol{x} - \boldsymbol{\xi} , \qquad (6.1.2)$$

(i.e. from now on $\boldsymbol{u}$ is not the total displacement, but the displacement from $\boldsymbol{\xi}$ to $\boldsymbol{x}$ only).

In the above, $\boldsymbol{\xi}$ was not specified. In principle, any continuous map $\boldsymbol{X} \rightarrow \boldsymbol{\xi}$ can be considered suitable for this intermediate configuration of the body. However, it will be assumed that the problem at hand suggests a natural definition of this state. For the time being we assume that it is known or at least determinable. We further assume that the position of a material point in the intermediate state $\boldsymbol{\xi}$ is close to its final position $\boldsymbol{x}$. Based on this closeness the motion and, more generally, all fields may be decomposed into two parts. One of these parts represents the fields when $\boldsymbol{u} = \boldsymbol{0}$, the other one is due to the

K. Hutter et al.: *Electromagnetic Field Matter Interaction in Thermoelastic Solids and Viscous Fluids*, Lect. Notes Phys. **710**, 147–197 (2006)
DOI 10.1007/3-540-37240-7_6　　　　　　　© Springer-Verlag Berlin Heidelberg 2006

perturbations from the $\boldsymbol{\xi}$-state to the $\boldsymbol{x}$-state. Because these perturbations are assumed to be small, the governing equations can be linearized in these perturbations.

This linearization procedure can be performed in a consistent way. For instance, if a body is initially subjected to large biasing electromagnetic fields and wave propagation or vibration properties of this body are investigated, then our linearization procedure will be applicable (e.g. [233, 248] and [93, 94]). Of the same nature are magnetoelastic stability problems [152, 250].

To describe the linearization procedure into more detail, all field variables will be decomposed into two parts. Those in the intermediate state are labelled with an overhead bar, whereas the perturbations on this state are indicated by lower case letters, e.g.

$$\mathbb{B}_\alpha = \bar{\mathbb{B}}_\alpha + b_\alpha , \qquad \mathbb{E}_\alpha = \bar{\mathbb{E}}_\alpha + e_\alpha , \qquad (6.1.3)$$

where

$$\frac{\|b\|}{\|\mathbb{B}\|} = \mathcal{O}(\varepsilon) , \qquad \frac{\|e\|}{\|\mathbb{E}\|} = \mathcal{O}(\varepsilon) , \quad \text{etc.} , \qquad (6.1.4)$$

and where ε denotes a small positive quantity $(0 < \varepsilon \ll 1)$. The norms in (6.1.4) may conveniently be defined as

$$\|a\|^2 := \limsup_{0 \le \tau \le t}(a^2(\tau)) , \qquad (6.1.5)$$

where $\tau = 0$ is the time at which the process started and t is the current time. Moreover, the displacements from $\boldsymbol{\xi}$ to $\boldsymbol{x}$ and their material time derivatives are assumed to be small in the sense that

$$\left\| \frac{\partial \boldsymbol{u}}{\partial \boldsymbol{\xi}} \right\| = \mathcal{O}(\varepsilon) \qquad \text{and} \qquad \frac{\|\dot{\boldsymbol{u}}\|}{v_0} = \mathcal{O}(\varepsilon) \qquad (6.1.6)$$

where v_0 is some characteristic wave speed.

In the LAGRANGEan description any differentiation with respect to $\boldsymbol{X}$ or t falls directly onto the respective variables as decomposed in (6.1.3). For instance

$$\mathbb{B}_{\alpha,\alpha} = \bar{\mathbb{B}}_{\alpha,\alpha} + b_{\alpha,\alpha} .$$

This is particularly easy and convenient, and to see this let us first briefly investigate the EULERian formulation. In this case all equations must be traced back to the intermediate configuration. In particular, all derivatives with respect to the present coordinates must be expressed in terms of the intermediate coordinates by means of the transformation rules, which we shall now briefly outline, (although lateron they will not be used). To this end, let f be any physical quantity. It may be regarded as a function of the variables $(\boldsymbol{X}, t)$, $(\boldsymbol{\xi}, t)$ or $(\boldsymbol{x}, t)$; thus

$$f = \hat{f}(\boldsymbol{X}, t) = \tilde{f}(\boldsymbol{\xi}, t) = \check{f}(\boldsymbol{x}, t) . \qquad (6.1.7)$$

Depending on which representation we choose, we thus have

$$\frac{\partial \check{f}}{\partial x_i} = \frac{\partial \tilde{f}}{\partial \xi_\alpha} \frac{\partial \check{\xi}_\alpha}{\partial x_i} = \left(\delta_{i\alpha} - \frac{\partial \check{u}_\alpha}{\partial x_i} \right) \frac{\partial \tilde{f}}{\partial \xi_\alpha} \ .$$

In particular with $f \equiv u_\alpha$ we obtain

$$\frac{\partial \check{u}_\alpha}{\partial x_i} = \left(\delta_{i\beta} - \frac{\partial \check{u}_\beta}{\partial x_i} \right) \frac{\partial \tilde{u}_\alpha}{\partial \xi_\beta} \cong \delta_{ij} \frac{\partial \tilde{u}_\alpha}{\partial \xi_\beta} \ ,$$

whence follows

$$\frac{\partial \check{f}}{\partial x_i} \cong \delta_{i\beta} \left(\delta_{\alpha\beta} - \frac{\partial \tilde{u}_\alpha}{\partial \xi_\beta} \right) \frac{\partial \tilde{f}}{\partial \xi_\alpha} \ . \tag{6.1.8}$$

Similarly, for the material time derivative we obtain

$$\frac{df}{dt} = \frac{\partial \hat{f}}{\partial t} = \frac{\partial \tilde{f}}{\partial t} + \frac{\partial \tilde{f}}{\partial \xi_\alpha} \frac{\partial \hat{\xi}_\alpha}{\partial t} = \frac{\partial \check{f}}{\partial t} + \frac{\partial \tilde{f}}{\partial x_i} \frac{\partial \hat{x}_i}{\partial t} \ ,$$

from which one easily deduces that

$$\frac{\partial \check{f}}{\partial t} = \frac{\partial \tilde{f}}{\partial t} + \frac{\partial \tilde{f}}{\partial \xi_\alpha} \frac{\partial \hat{\xi}_\alpha}{\partial t} - \frac{\partial \tilde{f}}{\partial \xi_\alpha} \frac{\partial \check{\xi}_\alpha}{\partial x_i} \frac{\partial \hat{x}_i}{\partial t} = \frac{\partial \tilde{f}}{\partial t} - \frac{\partial \tilde{f}}{\partial \xi_\alpha} \left(\frac{\partial \check{\xi}_\alpha}{\partial x_i} \frac{\partial \hat{x}_i}{\partial t} - \frac{\partial \hat{\xi}_\alpha}{\partial t} \right) \ .$$

Here, in each of the occurring functions we have indicated the functional dependencies. With the obvious definitions

$$\frac{\partial \hat{\xi}_\alpha}{\partial t} := \dot{\xi}_\alpha \qquad \text{and} \qquad \frac{\partial \hat{x}_i}{\partial t} := \dot{x}_i = \dot{u}_i + \dot{\xi}_\alpha \delta_{i\alpha}$$

this now becomes

$$\begin{aligned} \frac{\partial \check{f}}{\partial t} &= \frac{\partial \tilde{f}}{\partial t} - \left[\delta_{i\beta} \left(\delta_{\alpha\beta} - \frac{\partial \tilde{u}_\alpha}{\partial \xi_\beta} \right) \left(\dot{u}_i + \delta_{i\gamma} \dot{\xi}_\gamma \right) - \dot{\xi}_\alpha \right] \frac{\partial \tilde{f}}{\partial \xi_\alpha} \\ &\cong \frac{\partial \tilde{f}}{\partial t} - \left[\dot{u}_\alpha - \dot{\xi}_\beta \frac{\partial \tilde{u}_\alpha}{\partial \xi_\beta} \right] \frac{\partial \tilde{f}}{\partial \xi_\alpha} \ . \end{aligned} \tag{6.1.9}$$

The above formulas (6.1.8) and (6.1.9) must be applied in all field equations whenever space and time derivatives of physical quantities occur. That this is very tedious can be seen from the fact that for a dynamical theory it has been tried by several authors in the past as e.g. TOUPIN [241], HUTTER and PAO [91] and VAN DE VEN [249]. However, except for the procedure of VAN DE VEN none of these is completely correct. TOUPIN, TOUPIN63, for instance, seems to replace (6.1.8) and (6.1.9) at certain places by the approximations

$$\frac{\partial}{\partial \boldsymbol{X}} \cong \frac{\partial}{\partial \boldsymbol{x}} \qquad \text{and} \qquad \frac{\partial}{\partial t} \cong \frac{d}{dt} \ .$$

HUTTER and PAO, on the other hand, make full use of (6.1.8), but disregard (6.1.9) all together. This led to inconsistencies, which resulted in other ad hoc assumptions (as, for instance equation (5.7) in [91]).

In the EULERian formulation the linearization of jump conditions is equally tedious, because they are described at the deformed surface in the x-state, which first must be traced back to the surface in the intermediate state. Corroboration for the fact that this procedure is rather cumbersome can be found e.g. in [91] or [249]. As we shall see in Sect. 5.2.4, in a LAGRANGEan description the linearization of jump conditions is straightforward. Several other authors have also published linearization procedures, which are all less general than that we shall present here, because these authors restrict themselves to (quasi-) magneto or electrostatic processes, to static intermediate states, or they simply delete jump conditions, etc. We confine ourselves to mentioning ALBLAS [10], PAO and YEH [171], TIERSTEN [233], BAUMHAUER and TIERSTEN [22] and JORDAN and ERINGEN [104].

The advantage of the LAGRANGEan formulation is that the transformations (6.1.8) and (6.1.9) need not be applied, because in this formulation all operations are already referred to the reference or initial state. Similarly, the jump conditions hold on the undeformed surface. Once this is realized, the linearization procedure turns out to be straightforward.

A second, but less direct advantage of the LAGRANGEan or material formulation is that it allows an immediate introduction of electromagnetic potentials for the perturbed electromagnetic fields, which is, although also possible, more elaborate in the EULERian description. The reader may find corroboration for this by noting (as we shall show lateron) that in the LAGRANGEan description the homogeneous MAXWELL-equations (i.e. (2.2.15)) remain homogeneous in the perturbed state, whereas they become inhomogeneous in the EULERian description (cf. [91], p. 81).

This will then be our procedure. The nonlinear equations will be developed by consistently expanding all variables about the intermediate state and neglecting terms of order $\mathcal{O}(\varepsilon^2)$. Two systems of equations emerge in doing so, one for the equations in the intermediate state and a second one for the perturbed quantities, in which the quantities of the intermediate state appear as coefficients. From a practical point of view the construction of the solution of the perturbed equations is often more important than that of the intermediate state. Often the intermediate state serves only as a (static) biasing state deviations from which can be calculated by solving the perturbed equations. For buckling problems, for instance, the prebuckled state is taken as the intermediate state and its stability follows from the perturbed equations. Since, moreover, the perturbed equations are relatively insensitive to an exact determination of the intermediate fields, it would be advantageous to find an approximation scheme by which the coefficients could be determined to a sufficient degree of accuracy without making use of the exact intermediate equations. This is indeed often possible; we will demonstrate it below by introducing the so-called *rigid-body state*. The purpose of its use is to have a quick access to the perturbed equations without having to solve the equations in the intermediate state exactly. To explain it we consider the special situation where deformations, caused by the electromagnetic fields, are small and where changes in temperature are small as well. If at the same time coupling terms are small, which means that small deformations or small temperature changes result in small changes in the electromagnetic fields, then the intermediate state is close to an undeformed state. This imaginary state will be called *rigid-body state*. In this state the motion of the body, clearly, consists of a pure translation and rotation. The electromagnetic fields and the temperature distributions in this state are determinable from rigid-body electrodynamics, mechanics and thermodynamics. The fields in this state will be denoted by a superscript $(\ )^0$, e.g. $\mathbb{B}_\alpha^0, \mathbb{E}_\alpha^0$, whereas subscripts $(\ )_0$ will indicate initial values prior to any motion. By mere definition we then have

$$\rho^0 = \rho_0 , \qquad (6.1.10)$$

but generally neither $\Theta^0 = \Theta_0$ nor $U^0 = 0$. If $U^0 = 0$, then rigid-body motions are excluded. In most applications this will be the case.

Since deformations are assumed to be small, we may define a measure $E, 0 < E \ll 1$, by

$$E := \|\bar{U}_{\alpha,\beta}\| , \qquad (6.1.11)$$

with the aid of which we may conclude that

$$\frac{|\bar{\Theta} - \Theta^0|}{\Theta^0} = \mathcal{O}(E) , \qquad \frac{\|\bar{B} - B^0\|}{\| B^0 \|} = \mathcal{O}(E) , \qquad \text{etc} . \qquad (6.1.12)$$

Returning to the perturbed equations, we notice that these equations are linear in the perturbed fields u, b, etc., with coefficients, which depend on the values of the fields in the intermediate state. When the intermediate fields in these coefficients are replaced by the rigid-body fields, errors of order $\mathcal{O}(E)$ are introduced. Because the perturbed equations are of order $\mathcal{O}(\varepsilon)$ themselves, the neglect of these terms ultimately means that the resulting equations are correct except for terms of order $\mathcal{O}(E\varepsilon)$. In short: The linearized perturbation equations, in which the coefficients are referred to the rigid-body state, represent an approximation bound to an error not larger than $\mathcal{O}(E\varepsilon)$. It should further be emphasized that it is not justified, in general, to evaluate the intermediate fields themselves by applying the rigid-body approximations. This would result in errors of order $\mathcal{O}(E)$, which is larger than $\mathcal{O}(E\varepsilon)$.

The approach just described to approximate the equations is more consistent than the usual small-strain approximations. In particular there are two immediate advantages:

(i) the consistency of the linearization is a proven property and not an *a priori* assumption (the neglects can be made explicit) and,

(ii) the stresses (and all other fields) in the intermediate state can still be calculated to within any desired degree of exactness.

This latter point is of importance in particular for a consistent derivation of the equations governing magnetoelastic buckling of beams and plates (see [249], Ch. IX).

As said above, we are still free to solve the equations in the intermediate state as accurately as we please. For most practical purposes, however, a small-strain approximation will suffice. We shall return to this point in due course with the developments in this chapter.

6.2 Linearization of the Lorentz Model

6.2.1 Motivation for this Choice – Governing Equations

In this section we shall linearize the field equations, constitutive relations and jump conditions of one particular interaction model, but before we present the details a justification for our choice of the model is in order. We saw that all models are equivalent; as a consequence, only practical considerations and reasons of convenience can guide us to prefer one particular model over any other one.

There are several reasons for the choice of model V. Firstly, the calculations in Chap. 5 have shown that the structure of many formulas is preserved when they are transformed from the spatial description into their material counterpart. Secondly, several thermodynamic relations, such as the relation between stress and free energy or the GIBBS relation, are much more consize for formulations with a symmetric than with an unsymmetric stress tensor. This would leave us with model II and model V, but model V is again computationally advantageous, because the GAUSS–FARADAY law and the GAUSS law are homogeneous equations. This makes the introduction of electromagnetic potentials much easier than it would be otherwise. We shall deviate in one respect from the original LORENTZ-model, however, in that we shall use Q_α^S as energy-flux instead of Q_α^L. This essentially amounts to a different choice of the entropy flux (see Sects. 3.5 and 5.3).[1] Moreover, we shall exclude external sources, so that

$$\rho_0 F_i^{\text{ext}} = \rho_0 r^{\text{ext}} = 0 \,. \tag{6.2.1}$$

In the LAGRANGEan formulation and for the LORENTZ model the governing equations read as follows

---

[1] Compare the formulas (5.3.23) and (5.3.24).

MAXWELL *equations:* (5.3.11)

$$\mathbb{B}_{\alpha,\alpha} = 0 \,,$$

$$\dot{\mathbb{B}}_\alpha + e_{\alpha\beta\gamma}\mathbb{E}_{\gamma,\beta} = 0 \,,$$

$$\mathbb{D}^a_{\alpha,\alpha} = \mathbb{Q} - \mathbb{P}_{\alpha,\alpha} \,, \tag{6.2.2}$$

$$-\dot{\mathbb{D}}^a_\alpha + e_{\alpha\beta\gamma}\mathbb{H}^a_{\gamma,\beta} = \mathbb{J}_\alpha + \dot{\mathbb{P}}_\alpha + e_{\alpha\beta\gamma}\mathbb{M}^L_{\gamma,\beta} \,,$$

$$\dot{\mathbb{Q}} + \mathbb{J}_{\alpha,\alpha} = 0 \,,$$

where

$$\mathbb{D}^a_\alpha = \varepsilon_0 |J| C^{-1}_{\alpha\beta}\mathbb{E}_\beta \,,$$

$$\mathbb{H}^a_\alpha = \frac{1}{|J|}\left\{ \frac{1}{\mu_0}C_{\alpha\beta}\mathbb{B}_\beta - \varepsilon_0 e_{\mu\beta\gamma}C_{\alpha\beta}F_{j\gamma}\dot{x}_j\mathbb{E}_\mu \right\} \,. \tag{6.2.3}$$

Balance of momentum: $((5.6.9)_1$, with $F_i^{\text{ext}} = 0)$

$$\rho_0\ddot{x}_i - T_{i\alpha,\alpha} = \rho_0 F_i^e = F^{-1}_{\alpha i}\left((\mathbb{Q} - \mathbb{P}_{\beta,\beta})\mathbb{E}_\alpha + e_{\alpha\beta\gamma}(\mathbb{J}_\beta + \dot{\mathbb{P}}_\beta)\mathbb{B}_\gamma \right.$$
$$\left. + (\mathbb{M}^L_{\alpha,\beta} - \mathbb{M}^L_{\beta,\alpha})\mathbb{B}_\beta \right) \,. \tag{6.2.4}$$

Balance of energy: $((5.6.9)_3$, with $Q_\alpha = Q^S_\alpha$ and $r^{\text{ext}} = 0)$

$$\rho_0\Theta\dot{\eta} = \mathbb{J}_\alpha\mathbb{E}_\alpha - Q_{\alpha,\alpha} \,. \tag{6.2.5}$$

In the above, all variables are referred to the time-independent reference state characterized by the coordinates X_α. Superimposed dots thus represent time derivatives at fixed particles.

Constitutive Relations:

In accordance with (5.3.35) we choose as independent variables

$$C_{\alpha\beta}, \ \mathbb{E}_\alpha, \ \mathbb{B}_\alpha, \ \Theta, \ \Theta_{,\alpha} \ \text{and} \ \mathbb{Q} \,.$$

Following (5.3.36) and (5.3.37), we then have

$$\check{\psi} = U - \Theta\eta + \frac{1}{\rho_0}\mathbb{E}_\alpha\mathbb{P}_\alpha = \check{\psi}(C_{\alpha\beta}, \mathbb{E}_\alpha, \mathbb{B}_\alpha, \Theta) \,, \tag{6.2.6}$$

and

$$\eta = -\frac{\partial\check{\psi}}{\partial\Theta} \,, \qquad \mathbb{P}_\alpha = -\rho\frac{\partial\check{\psi}}{\partial\mathbb{E}_\alpha} \,,$$
$$\mathbb{M}^L_\alpha = -\rho_0\frac{\partial\check{\psi}}{\partial\mathbb{B}_\alpha} \,, \qquad T^P_{\alpha\beta} = 2\rho_0\frac{\partial\check{\psi}}{\partial C_{\alpha\beta}} \,. \tag{6.2.7}$$

Moreover, for an electrical conductor the constitutive relations for the electric current and the energy flux are of the form (see (5.5.12) with (5.5.14)$_3$)

$$\mathbb{J}_\alpha = \sigma_{\alpha\beta}\mathbb{E}_\beta + \beta_{\beta\alpha}\frac{\Theta_{,\beta}}{\Theta}, \quad Q_\alpha = -\kappa_{\alpha\beta}\Theta_{,\beta} + \beta_{\alpha\beta}\mathbb{E}_\beta, \quad (6.2.8)$$

where, in general, the coefficient matrices are still functions of $C_{\alpha\beta}$, $\mathbb{E}_\alpha$, $\mathbb{B}_\alpha$, Θ, $\Theta_{,\alpha}$ and $\mathbb{Q}$. However, in accordance with the linearization as described in the derivation of (3.7.34) we shall assume these coefficients to be independent of $\mathbb{E}_\alpha$ and $\Theta_{,\alpha}$. Moreover, we shall exclude an explicit occurrence of $\mathbb{Q}$.

Equations (6.2.8) automatically guarantee that the current $\mathbb{J}_\alpha$ and heat flux Q_α vanish in thermostatic equilibrium; in short, if $\mathbb{E}_\alpha = 0$ and $\Theta_{,\alpha} = 0$, then

$$\mathbb{J}_{\alpha|E} = 0 \quad \text{and} \quad Q_{\alpha|E} = 0.$$

Further conditions of thermostatic equilibrium are that $\sigma_{(\alpha\beta)}$ and $-\kappa_{(\alpha\beta)}$ must be positive-semi definite matrices and, if the ONSAGER relations are adopted, that the skew symmetric parts of $\sigma_{\alpha\beta}$ and $\kappa_{\alpha\beta}$ must vanish.

Jump conditions: ((5.3.15), (2.4.12), (5.3.25))

$$[\![\mathbb{B}_\alpha]\!]N_\alpha = 0, \quad e_{\alpha\beta\gamma}[\![\mathbb{E}_\beta]\!]N_\gamma + [\![\mathbb{B}_\alpha W_N]\!] = 0,$$

$$[\![\mathbb{D}_\alpha^a + \mathbb{P}_\alpha]\!]N_\alpha = 0,$$

$$e_{\alpha\beta\gamma}[\![\mathbb{H}_\beta^a - \mathbb{M}_\beta^L]\!]N_\gamma - [\![(\mathbb{D}_\alpha^a + \mathbb{P}_\alpha)W_N]\!] = 0,$$

$$[\![\mathbb{J}_\alpha]\!]N_\alpha - [\![\mathbb{Q}W_N]\!] = 0, \quad (6.2.9)$$

$$[\![\rho_0 W_N]\!] = 0, \quad [\![\rho_0 \dot{x}_i W_N]\!] + [\![\mathbb{T}_{i\alpha} + \mathbb{T}_{i\alpha}^M]\!]N_\alpha = 0,$$

$$[\![(\tfrac{1}{2}\rho_0 \dot{x}_i \dot{x}_i + \rho_0 U - \Omega)W_N]\!]$$

$$+ [\![(\mathbb{T}_{i\alpha} + \mathbb{T}_{i\alpha}^M)\dot{x}_i - Q_\alpha - e_{\alpha\beta\gamma}\mathbb{E}_\beta(\mathbb{H}_\gamma^a - \mathbb{M}_\gamma^L)]\!]N_\alpha = 0.$$

In these relations, N_α is the unit normal vector on the singular surface, W_N is the speed of propagation, and $T_{i\alpha}^M$ and Ω are given in (5.3.25); for ease of reference they will be repeated here:

$$T_{i\alpha}^M = F_{\beta i}^{-1}\left(\mathbb{D}_\alpha^a \mathbb{E}_\beta + \mathbb{B}_\alpha \mathbb{H}_\beta^a - \tfrac{1}{2}\delta_{\alpha\beta}(\mathbb{D}_\gamma^a \mathbb{E}_\gamma + \mathbb{B}_\gamma \mathbb{H}_\gamma^a)\right),$$
$$\Omega = -\tfrac{1}{2}(\mathbb{D}_\alpha^a \mathbb{E}_\alpha + \mathbb{B}_\alpha \mathbb{H}_\alpha^a). \quad (6.2.10)$$

In this chapter we shall from now on restrict ourselves to *singular surfaces of second order* (see e.g. [68], Sect. 2.8). On such a surface, the deformation gradients and the velocity will be continuous, as is the density ρ_0, i.e.

$$[\![F_{i\alpha}]\!] = [\![\dot{x}_i]\!] = [\![\rho_0]\!] = 0.$$

In view of $(6.2.9)_6$ continuity of ρ_0 also implies that the speed of propagation W_N is continuous,

$$[\![W_N]\!] = 0 \ .$$

For the case that the singular surface is a real propagating surface, i.e. $W_N \neq 0$, the associated waves are called *acceleration waves*. Singular surfaces of second order include as a special case *material surfaces* for which $W_N = 0$, and also embrace the boundary of a solid body in a vacuum, if, as is usually the case, the vacuum is considered as a medium with zero density, which admits continuity of the velocities at the boundary. Surfaces on which tangential velocities may jump are, however, excluded.

Under the above restrictions the last two jump conditions can be simplified. In view of the continuity of ρ_0, $\dot{x}_i$ and W_N, $(6.2.9)_7$ becomes

$$[\![T_{i\alpha}]\!] N_\alpha = -[\![T_{i\alpha}^M]\!] N_\alpha \ . \tag{6.2.11}$$

This equation can be written in the form

$$[\![T_{i\alpha}]\!] N_\alpha = F_{\gamma i}^{-1} \left(\langle \mathbb{E}_\gamma \rangle [\![\mathbb{P}_\beta]\!] N_\beta + \langle \mathbb{B}_\beta \rangle ([\![\mathbb{M}_\beta^L]\!] N_\gamma - [\![\mathbb{M}_\gamma^L]\!] N_\beta) \right. \\
\left. + e_{\beta\gamma\delta} \langle \mathbb{B}_\beta \rangle [\![\mathbb{P}_\delta]\!] W_N \right) \ , \tag{6.2.12}$$

in which the symbol $\langle \cdot \rangle$ stands for the arithmetic mean of a quantity over the singular surface, i.e.

$$\langle \mathbb{E}_\alpha \rangle = \tfrac{1}{2}(\mathbb{E}_\alpha^+ + \mathbb{E}_\alpha^-) \ . \tag{6.2.13}$$

To prove (6.2.12), notice that in view of relations (6.2.3) and of the continuity conditions of second-order singular surfaces,

$$\langle \mathbb{D}_\alpha^a \rangle [\![\mathbb{E}_\alpha]\!] = \langle \mathbb{E}_\alpha \rangle [\![\mathbb{D}_\alpha^a]\!] \ , \qquad \langle \mathbb{H}_\alpha^a \rangle [\![\mathbb{B}_\alpha]\!] = \langle \mathbb{B}_\alpha \rangle [\![\mathbb{H}_\alpha^a]\!] \ . \tag{6.2.14}$$

Moreover, the jump of the first term in the outer brackets of $(6.2.10)_1$ may be written as

$$[\![\mathbb{D}_\alpha^a \mathbb{E}_\beta]\!] N_\alpha = (\langle \mathbb{D}_\alpha^a \rangle [\![\mathbb{E}_\beta]\!] + [\![\mathbb{D}_\alpha^a]\!] \langle \mathbb{E}_\beta \rangle) N_\alpha \\
= \langle \mathbb{D}_\alpha^a \rangle N_\alpha [\![\mathbb{E}_\beta]\!] - \langle \mathbb{E}_\beta \rangle [\![\mathbb{P}_\alpha]\!] N_\alpha \ ,$$

in which use has been made of $(6.2.9)_3$. With $(6.2.14)_1$ and $(6.2.9)_4$ it is then straightforward to show that in the non-relativistic approximation

$$[\![\mathbb{D}_\alpha^a \mathbb{E}_\beta]\!] N_\alpha - \tfrac{1}{2} [\![\mathbb{D}_\gamma^a \mathbb{E}_\gamma]\!] \delta_{\alpha\beta} N_\alpha \\
= -\langle \mathbb{E}_\beta \rangle [\![\mathbb{P}_\alpha]\!] N_\alpha + e_{\alpha\beta\gamma} \langle \mathbb{D}_\alpha^a \rangle [\![\mathbb{B}_\gamma]\!] W_N = -\langle \mathbb{E}_\beta \rangle [\![\mathbb{P}_\alpha]\!] N_\alpha \ .$$

In an analogous way the remaining terms on the right-hand side of (6.2.10), (6.2.11) can be handled, so that, at last, (6.2.12) emerges.

On the other hand, the first and fourth term in $(6.2.9)_8$ vanish for acceleration waves; furthermore, with the use of $(6.2.9)_{2,4}$ it can be shown that

$$\llbracket e_{\alpha\beta\gamma} \mathbb{E}_\beta (\mathbb{H}_\gamma^a - \mathbb{M}_\gamma^L) \rrbracket N_\alpha =$$
$$\left(\llbracket \mathbb{B}_\alpha \rrbracket \langle \mathbb{H}_\alpha^a - \mathbb{M}_\alpha^L \rangle + \langle \mathbb{E}_\alpha \rangle \llbracket \mathbb{D}_\alpha^a + \mathbb{P}_\alpha \rrbracket \right) W_N \ . \tag{6.2.15}$$

Finally, from $(6.2.10)_2$ and with the application of relations $(6.2.14)$ we find

$$\llbracket \Omega \rrbracket = -\langle \mathbb{E}_\alpha \rangle \llbracket \mathbb{D}_\alpha^a \rrbracket - \langle \mathbb{H}_\alpha^a \rangle \llbracket \mathbb{B}_\alpha \rrbracket \ . \tag{6.2.16}$$

With these results the jump condition for the energy flux assumes the simple and elegant form

$$\llbracket Q_\alpha \rrbracket N_\alpha = \left(\rho_0 \llbracket U \rrbracket + \langle \mathbb{M}_\alpha^L \rangle \llbracket \mathbb{B}_\alpha \rrbracket - \langle \mathbb{E}_\alpha \rangle \llbracket \mathbb{P}_\alpha \rrbracket \right) W_N \ . \tag{6.2.17}$$

It should be noted here that the right-hand side vanishes when the surface is material.

Before we close, consider a surface separating a body from the vacuum. According to our interpretation of the vacuum as a medium with vanishingly small mass density, and because we are looking at surfaces of second order, the deformation tensor is continuous across such a surface so that on the vacuum side $(+)$ one has

$$\mathbb{D}_\alpha^{a+} = (\varepsilon_0 J C_{\alpha\beta}^{-1})^+ \mathbb{E}_\beta^+ = (\varepsilon_0 J C_{\alpha\beta}^{-1})^- \mathbb{E}_\beta^+ \ .$$

Similar statements also hold for $\mathbb{H}_\alpha^{a+}$. Note also that $\mathbb{E}_\alpha^+ \neq E_i^+ \delta_{i\alpha}$; one rather has

$$\mathbb{E}_\alpha^+ = F_{i\alpha}^- (E_i^+ + e_{ijk} \dot{x}_j^- B_k^+)$$

and

$$\mathbb{B}_\alpha^+ = (J F_{\alpha i}^{-1})^- B_i^+ \ .$$

These facts should be borne in mind, for otherwise incorrect results emerge. This completes the collection of the basic governing equations. Their linearized versions will be derived below.

6.2.2 Decomposition of the Balance Laws

Having presented the governing equations we now proceed with the decomposition of the balance laws. As was said several times before already, this decomposition is particularly easy in the material description. To corroborate this statement, let us consider the first of the MAXWELL equations $(6.2.2)$. Introducing $(6.1.3)$ as decomposition for the magnetic induction $\mathbb{B}_\alpha$, we may write equation $(6.2.2)_1$ as

$$(\bar{\mathbb{B}}_\alpha + b_\alpha)_{,\alpha} = \bar{\mathbb{B}}_{\alpha,\alpha} + b_{\alpha,\alpha} = 0 \ . \tag{6.2.18}$$

In view of the basic separation assumption, according to which the governing equations must be fulfilled for the intermediate state itself, we thus have

$$\bar{\mathbb{B}}_{\alpha,\alpha} = 0 \,, \tag{6.2.19}$$

and, consequently,

$$b_{\alpha,\alpha} = 0 \,. \tag{6.2.20}$$

In these equations, both $\bar{\mathbb{B}}_{\alpha}$ and b_{α} are regarded as functions of X_{α} and t. Proceeding with all MAXWELL equations as was explained above with the GAUSS law, we can easily show that the equations, valid in the intermediate state, are given by

$$\bar{\mathbb{B}}_{\alpha,\alpha} = 0 \,, \qquad\qquad \dot{\bar{\mathbb{B}}}_{\alpha} + e_{\alpha\beta\gamma}\bar{E}_{\gamma,\beta} = 0 \,,$$

$$\bar{\mathbb{D}}^a_{\alpha,\alpha} = \bar{\mathbb{Q}} - \bar{\mathbb{P}}_{\alpha,\alpha} \,, \quad -\dot{\bar{\mathbb{D}}}^a_{\alpha} + e_{\alpha\beta\gamma}\bar{\mathbb{H}}^a_{\gamma,\beta} = \bar{\mathbb{J}}_{\alpha} + \dot{\bar{\mathbb{P}}}_{\alpha} + e_{\alpha\beta\gamma}\bar{\mathbb{M}}^L_{\gamma,\beta} \,, \quad (6.2.21)$$

$$\dot{\bar{\mathbb{Q}}} + \bar{\mathbb{J}}_{\alpha,\alpha} = 0 \,.$$

These equations are obtained from the original MAXWELL equations (6.2.2) by merely replacing in the latter all quantities by those carrying an overhead bar. On the other hand, the perturbed MAXWELL equations are[2]

$$b_{\alpha,\alpha} = 0 \,, \qquad\qquad \dot{b}_{\alpha} + e_{\alpha\beta\gamma}e_{\gamma,\beta} = 0 \,,$$

$$d^a_{\alpha,\alpha} = q - p_{\alpha,\alpha} \,, \quad -\dot{d}^a_{\alpha} + e_{\alpha\beta\gamma}h^a_{\gamma,\beta} = j_{\alpha} + \dot{p}_{\alpha} + e_{\alpha\beta\gamma}m_{\gamma,\beta} \,, \quad (6.2.22)$$

$$\dot{q} + j_{\alpha,\alpha} = 0 \,,$$

in which all lower-case letters denote perturbed LAGRANGEan electromagnetic field quantities, the definitions being analogous to that for b_{α}. For instance,

$$e_{\alpha} = \mathbb{E}_{\alpha} - \bar{\mathbb{E}}_{\alpha} \,, \quad \text{etc.}$$

For convenience and since no confusion is possible here, we have used m_{α} for $(\mathbb{M}^L_{\alpha} - \bar{\mathbb{M}}^L_{\alpha})$ instead of m^L_{α}. Whenever, in the sequel confusion with the CHU-magnetization becomes possible we shall use m^L_{α}, however.

Equations (6.2.22) are formally the same as the original MAXWELL equations. This is no surprise, because equations $(6.2.2)_{1-5}$ are written such that they appear in a linear form. However, they are nevertheless nonlinear; the nonlinearity is only covered by the use of the auxiliary fields $\mathbb{D}^a_{\alpha}$ and $\mathbb{H}^a_{\alpha}$. In the decomposition process of these quantities use must be made of the following relations, of which the proof is straightforward:

---

[2] In this chapter the perturbed fields from the intermediate state will be denoted by Roman letters and not by italics. Alternatively, fields in the intermediate state are identified by an overhead bar. This notation has already been used in (6.2.18)–(6.2.20).

$$x_i = \delta_{i\alpha}(\xi_\alpha + u_\alpha) \,,$$

$$F_{i\alpha} = \frac{\partial x_i}{\partial X_\alpha} = \delta_{i\beta}\left(\frac{\partial \xi_\beta}{\partial X_\alpha} + \frac{\partial u_\beta}{\partial X_\alpha}\right) = \delta_{i\beta}(\bar{F}_{\beta\alpha} + u_{\beta,\alpha}) \,,$$

$$F_{\alpha i}^{-1} = \frac{\partial X_\alpha}{\partial x_i} = \frac{\partial X_\alpha}{\partial \xi_\beta}\frac{\partial \xi_\beta}{\partial x_i} \cong \bar{F}_{\alpha\beta}^{-1}(\delta_{i\beta} - \delta_{i\delta}\bar{F}_{\gamma\delta}^{-1}u_{\beta,\gamma}) \,,$$

$$C_{\alpha\beta} = F_{i\alpha}F_{i\beta} \cong \bar{C}_{\alpha\beta} + \bar{F}_{\gamma\beta}u_{\gamma,\alpha} + \bar{F}_{\gamma\alpha}u_{\gamma,\beta} \,,$$

$$C_{\alpha\beta}^{-1} = F_{\alpha i}^{-1}F_{\beta i}^{-1} \cong \bar{C}_{\alpha\beta}^{-1} - (\bar{C}_{\beta\gamma}^{-1}\bar{F}_{\alpha\delta}^{-1} + \bar{C}_{\alpha\gamma}^{-1}\bar{F}_{\beta\delta}^{-1})u_{\delta,\gamma} \,,$$

$$J = \det(F_{i\alpha}) = \det[\bar{F}_{\beta\alpha}(\delta_{i\beta} + \delta_{i\delta}\bar{F}_{\beta\gamma}^{-1}u_{\delta,\gamma})] \cong \bar{J}(1 + \bar{F}_{\gamma\delta}^{-1}u_{\delta,\gamma}) \,. \tag{6.2.23}$$

Here,

$$\bar{F}_{\alpha\beta} = \frac{\partial \xi_\alpha}{\partial X_\beta} \,, \qquad \bar{F}_{\alpha\beta}^{-1} = \frac{\partial X_\alpha}{\partial \xi_\beta} \,, \qquad \bar{C}_{\alpha\beta} = \bar{F}_{\gamma\alpha}\bar{F}_{\gamma\beta} \,,$$

$$\bar{C}_{\alpha\beta}^{-1} = \bar{F}_{\alpha\gamma}^{-1}\bar{F}_{\beta\gamma}^{-1} \,, \qquad \bar{J} = \det(\bar{F}_{\alpha\beta}) \,. \tag{6.2.24}$$

With these preliminary calculations the decomposition of the auxiliary fields $\mathbb{D}_\alpha^a$ and $\mathbb{H}_\alpha^a$ can now be performed. The results are

$$\bar{\mathbb{D}}_\alpha^a = \varepsilon_0 \bar{J}\bar{C}_{\alpha\beta}^{-1}\bar{\mathbb{E}}_\beta \,,$$

$$\bar{\mathbb{H}}_\alpha^a = \frac{1}{\mu_0 \bar{J}}\bar{C}_{\alpha\beta}\bar{\mathbb{B}}_\beta - \frac{\varepsilon_0}{\bar{J}}e_{\beta\gamma\delta}\bar{C}_{\alpha\beta}\bar{F}_{\varepsilon\gamma}\dot{\xi}_\varepsilon\bar{\mathbb{E}}_\delta \,, \tag{6.2.25}$$

and

$$d_\alpha^a = \varepsilon_0 \bar{J}\left(\bar{C}_{\alpha\beta}^{-1}e_\beta + (\bar{C}_{\alpha\beta}^{-1}\bar{F}_{\gamma\delta}^{-1} - \bar{C}_{\beta\gamma}^{-1}\bar{F}_{\alpha\delta}^{-1} - \bar{C}_{\alpha\gamma}^{-1}\bar{F}_{\beta\delta}^{-1})\bar{\mathbb{E}}_\beta u_{\delta,\gamma}\right) \,,$$

$$h_\alpha^a = \frac{1}{\bar{J}}\left\{\frac{1}{\mu_0}\bar{C}_{\alpha\beta}b_\beta - \varepsilon_0 e_{\beta\gamma\delta}\bar{F}_{\varepsilon\gamma}\dot{\xi}_\varepsilon\bar{C}_{\alpha\beta}e_\delta\right.$$

$$+ \left[\left(\frac{1}{\mu_0}\bar{\mathbb{B}}_\beta - e_{\beta\nu\mu}\bar{F}_{\varepsilon\nu}\dot{\xi}_\varepsilon\bar{\mathbb{E}}_\mu\right)\left(\bar{F}_{\delta\alpha}\delta_{\beta\gamma} + \bar{F}_{\delta\beta}\delta_{\alpha\gamma} - \bar{C}_{\alpha\beta}\bar{F}_{\gamma\delta}\right)\right.$$

$$\left.- \varepsilon_0 e_{\beta\gamma\mu}\bar{\mathbb{E}}_\mu\bar{C}_{\alpha\beta}\dot{\xi}_\delta\right]u_{\delta,\gamma} - \varepsilon_0 e_{\beta\gamma\delta}\bar{\mathbb{E}}_\delta\bar{C}_{\alpha\beta}\bar{F}_{\varepsilon\gamma}\dot{u}_\varepsilon\left.\right\} \,. \tag{6.2.26}$$

Here and henceforth we restrict ourselves to positive values of the Jacobian J and could therefore replace $|J|$ by J whenever it occurred. We further would like to emphasize that the perturbed fields d_α^a and h_α^a are expressed here in terms of the deformation and the electromagnetic fields in the intermediate state, all of which are functions of X_α and t, in general.

The deformation in the intermediate state may, in our approximation, be neglected and the intermediate fields be replaced by the fields in the rigid-body state. If this is the case we may choose

$$\xi_\alpha^0(\boldsymbol{X}, t) = \Xi_\alpha(t) + R_{\alpha\beta}(t)X_\beta \,, \tag{6.2.27}$$

where $\Xi_\alpha(t)$ can be identified with the coordinates of the center of mass of the body in its rigid-body state (the position of the center of mass in the reference state, $t = 0$ is here taken as the origin of our coordinate system), and where $R_{\alpha\beta}(t)$ is a time-dependent proper orthogonal matrix, i.e.

$$R^{-1} = R^T \qquad \text{and} \qquad \det R = +1 \ .$$

Differentiating both sides of (6.2.27) with respect to time, we obtain

$$\dot{\xi}^0_\alpha = \dot{\Xi}_\alpha + \dot{R}_{\alpha\beta} X_\beta = \dot{\Xi}_\alpha + e_{\alpha\beta\gamma} \Omega_\beta R_{\gamma\delta} X_\delta \ , \tag{6.2.28}$$

where $\boldsymbol{\Omega}(t)$ is the angular velocity of the body, which is given by

$$\Omega_\alpha = -\tfrac{1}{2} e_{\alpha\beta\gamma} \dot{R}_{\beta\delta} R_{\gamma\delta} \ . \tag{6.2.29}$$

The approximate versions of equations (6.2.26) can now be obtained by replacing in these equations all variables carrying an overhead bar, $(\bar{\cdot})$, by the variables in the rigid-body state, $(\cdot)^0$. In other words, the following replacements must be made

$$\begin{aligned}
\bar{\mathbb{E}}_\alpha &\to \mathbb{E}^0_\alpha \ , & \bar{\mathbb{B}}_\alpha &\to \mathbb{B}^0_\alpha \ , & \bar{F}_{\alpha\beta} &\to R_{\alpha\beta} \ , & \bar{F}^{-1}_{\alpha\beta} &\to R_{\beta\alpha} \ , \\
\bar{J} &\to 1 \ , & \bar{C}_{\alpha\beta} &\to \delta_{\alpha\beta} \ , & \bar{C}^{-1}_{\alpha\beta} &\to \delta_{\alpha\beta} \ .
\end{aligned} \tag{6.2.30}$$

With these we obtain for (6.2.26)

$$\begin{aligned}
\mathrm{d}^a_\alpha &= \varepsilon_0 \Big(\mathrm{e}_\alpha + (\delta_{\alpha\beta} R_{\delta\gamma} - \delta_{\beta\gamma} R_{\delta\alpha} - \delta_{\alpha\gamma} R_{\delta\beta}) \mathrm{E}^0_\beta \mathrm{u}_{\delta,\gamma} \Big) \ , \\
\mathrm{h}^a_\alpha &= \frac{1}{\mu_0} \mathrm{b}_\alpha - \varepsilon_0 e_{\alpha\beta\gamma} R_{\delta\beta} \dot{\xi}^0_\delta \mathrm{e}_\gamma \\
&\quad + \left[\left(\frac{1}{\mu_0} \mathbb{B}^0_\beta - e_{\beta\nu\mu} R_{\varepsilon\nu} \dot{\xi}^0_\varepsilon \mathbb{E}^0_\mu \right) (R_{\delta\alpha} \delta_{\beta\gamma} + R_{\delta\beta} \delta_{\alpha\gamma} - R_{\gamma\delta} \delta_{\alpha\beta}) \right. \\
&\quad \left. - \varepsilon_0 e_{\alpha\gamma\mu} \mathbb{E}^0_\mu \dot{\xi}^0_\delta \right] \mathrm{u}_{\delta,\gamma} - \varepsilon_0 e_{\alpha\beta\gamma} \mathbb{E}^0_\gamma R_{\delta\beta} \dot{\mathrm{u}}_\delta \ .
\end{aligned} \tag{6.2.31}$$

Still further simplifications can be made, if the intermediate state is not only close to a rigid-body state, but deviates from the initial state by a small amount only (i.e. for a motionless rigid-body state). In that case equations (6.2.31) are further simplified by setting

$$R_{\alpha\beta} = \delta_{\alpha\beta} \qquad \text{and} \qquad \dot{\xi}^0_\alpha = 0 \ .$$

The next step in the simplification of the balance laws is the decomposition of the momentum and energy equations. To this end, (6.2.4) and (6.2.5) will be written as

$$\begin{aligned}
(\rho_0 \ddot{\xi}_\alpha + \rho_0 \ddot{u}_\alpha) \delta_{i\alpha} &= \bar{T}_{i\alpha,\alpha} + \delta_{i\alpha} \mathrm{t}_{\alpha\beta,\beta} + \rho_0 \bar{F}^e_i + \rho_0 \mathrm{f}^e_\alpha \delta_{i\alpha} \ , \\
\rho_0 (\bar{\Theta} + \theta)(\dot{\bar{\eta}} + \dot{s}) &= (\mathbb{J}_\alpha + \mathrm{j}_\alpha)(\mathbb{E}_\alpha + \mathrm{e}_\alpha) - (\bar{Q}_\alpha - \mathrm{q}_\alpha)_{,\alpha} \ .
\end{aligned} \tag{6.2.32}$$

Apart from the perturbation variables already introduced before we have defined here

$$f_\alpha^e = \delta_{i\alpha}(F_i^e - \bar{F}_i^e) , \qquad t_{\alpha\beta} = \delta_{i\alpha}(T_{i\beta} - \bar{T}_{i\beta}) ,$$
$$s = \eta - \bar{\eta} , \qquad \theta = \Theta - \bar{\Theta} , \qquad q_\alpha = Q_\alpha - \bar{Q}_\alpha .$$

(6.2.33)

Because we assume that equations (6.2.32) must hold for the intermediate state, we have

$$\rho_0 \ddot{\bar{\xi}}_\alpha = \delta_{i\alpha}(\bar{T}_{i\beta,\beta} + \rho_0 \bar{F}_i^e) ,$$
$$\rho_0 \bar{\Theta} \dot{\bar{\eta}} = \bar{\mathbb{J}}_\alpha \bar{\mathbb{E}}_\alpha - \bar{Q}_{\alpha,\alpha} ,$$

(6.2.34)

as momentum and energy equations in this state and

$$\rho_0 \ddot{u}_\alpha = t_{\alpha\beta,\beta} + \rho_0 f_\alpha^e ,$$
$$\rho_0(\theta \dot{\bar{\eta}} + \bar{\Theta} \dot{s}) = j_\alpha \bar{\mathbb{E}}_\alpha + \bar{\mathbb{J}}_\alpha e_\alpha - q_{\alpha,\alpha} ,$$

(6.2.35)

as the corresponding equations in the perturbed state.

An explicit representation of the body force expression is obtained from (6.2.4) and (6.2.23)$_3$, namely

$$\rho_0 \bar{F}_i^e = \delta_{i\delta} \bar{F}_{\alpha\delta}^{-1} \{ (\bar{\mathbb{Q}} - \bar{\mathbb{P}}_{\beta,\beta}) \bar{\mathbb{E}}_\alpha + e_{\alpha\beta\gamma}(\bar{\mathbb{J}}_\beta + \dot{\bar{\mathbb{P}}}_\beta) \bar{\mathbb{B}}_\gamma + (\bar{\mathbb{M}}_{\alpha,\beta}^L - \bar{\mathbb{M}}_{\beta,\alpha}^L) \bar{\mathbb{B}}_\beta \} ,$$

(6.2.36)

as the expression for the body force in the intermediate state, and

$$\rho_0 f_\alpha^e = \bar{F}_{\delta\alpha}^{-1} \Big\{ (\bar{\mathbb{Q}} - \bar{\mathbb{P}}_{\beta,\beta}) e_\delta + (q - p_{\beta,\beta}) \bar{\mathbb{E}}_\varepsilon$$
$$+ e_{\beta\gamma\delta} \Big[(\bar{\mathbb{J}}_\beta + \dot{\bar{\mathbb{P}}}_\beta) b_\gamma + (j_\beta + \dot{p}_\beta) \bar{\mathbb{B}}_\gamma \Big]$$
$$+ \bar{\mathbb{B}}_\beta (m_{\delta,\beta} - m_{\beta,\delta}) + b_\beta (\bar{\mathbb{M}}_{\delta,\beta}^L - \bar{\mathbb{M}}_{\beta,\delta}^L) \Big\}$$
$$- \bar{F}_{\beta\gamma}^{-1} \bar{F}_{\delta\alpha}^{-1} \Big\{ (\bar{\mathbb{Q}} - \bar{\mathbb{P}}_{\varepsilon,\varepsilon}) \bar{\mathbb{E}}_\beta + e_{\beta\mu\nu}(\bar{\mathbb{J}}_\mu + \dot{\bar{\mathbb{P}}}_\mu) \bar{\mathbb{B}}_\nu$$
$$+ \bar{\mathbb{B}}_\varepsilon (\bar{\mathbb{M}}_{\beta,\varepsilon}^L - \bar{\mathbb{M}}_{\varepsilon,\beta}^L) \Big\} u_{\gamma,\delta} ,$$

(6.2.37)

as the corresponding expression in the perturbed state. When deformations in the intermediate state are neglected, but rigid motions are allowed, then $\bar{\mathbb{F}}_{\alpha\beta}$ in (6.2.37) must simply be replaced by $R_{\alpha\beta}$ and, furthermore, all variables carrying an overhead bar must be replaced by the variables in the rigid-body state. Moreover, for a motionless rigid-body state $R_{\alpha\beta} = \delta_{\alpha\beta}$ and $\dot{\xi}_\alpha^0 = 0$.

The balance laws of mechanics and electrodynamics are thus decomposed into equations valid in the intermediate state and those in the perturbed state. When in the latter the rigid-body approximation is applied, the corresponding equations in this state are also needed. The governing equations consist of the MAXWELL equations

$$\mathbb{B}^0_{\alpha,\alpha} = 0\,, \qquad\qquad \dot{\mathbb{B}}^0_\alpha + e_{\alpha\beta\gamma}\mathbb{E}^0_{\gamma,\beta} = 0\,,$$

$$\mathbb{D}^{a0}_{\alpha,\alpha} = \mathbb{Q}^0 - \mathbb{P}^0_{\alpha,\alpha}\,, \quad -\dot{\mathbb{D}}^{a0}_\alpha + e_{\alpha\beta\gamma}\mathbb{H}^{a0}_{\gamma,\beta} = \mathbb{J}^0_\alpha + \dot{\mathbb{P}}^0_\alpha + e_{\alpha\beta\gamma}\mathbb{M}^{L0}_{\gamma,\beta}\,, \quad (6.2.38)$$

$$\dot{\mathbb{Q}}^0 + \mathbb{J}^0_{\alpha,\alpha} = 0\,,$$

where

$$\mathbb{D}^{a0}_\alpha = \varepsilon_0 \mathbb{E}^0_\alpha\,, \quad \mathbb{H}^{a0}_\alpha = \frac{1}{\mu_0}\mathbb{B}^0_\alpha - \varepsilon_0 e_{\alpha\beta\gamma}R_{\delta\beta}\dot{\xi}^0_\delta \mathbb{E}^0_\gamma\,, \tag{6.2.39}$$

and of the energy equation

$$\rho_0 \Theta^0 \dot{\eta}^0 = \mathbb{J}^0_\alpha \mathbb{E}^0_\alpha - Q^0_{\alpha,\alpha}\,, \tag{6.2.40}$$

supplemented by the equations for the rigid-body motion (which will be stated below), by the pertinent constitutive relations and by the jump conditions.

Equations (6.2.38)–(6.2.40), together with constitutive relations and jump conditions, suffice to determine the rigid-body fields, provided that the rigid-body state is motionless ($\dot{\xi}^0_\alpha = 0$) or has a prescribed motion. If this is not the case, the motion (i.e. $\Xi_\alpha(t)$ and $R_{\alpha\beta}(t)$ or $\Omega_\alpha(t)$) must be determined from momentum equations, which can best be handled, if the balance laws of linear and angular momentum are recast into global form. We restrict ourselves to a rigid body placed in a vacuum, in which case these relations are

$$\frac{d}{dt}\int_\Omega \rho_0 \dot{\xi}^0_\alpha dV = \int_{\partial\Omega^-} \delta_{i\alpha}(T_{i\beta} + T^M_{i\beta})^0 N_\beta dA + \int_\Omega \rho_0 F^{\mathrm{ext}}_\alpha dV\,,$$

$$\frac{d}{dt}\int_\Omega \rho_0 (\xi^0_{[\alpha} - \Xi_{[\alpha})\dot{\xi}^0_{\beta]} dV = \int_{\partial\Omega^-} (\xi^0_{[\alpha} - \Xi_{[\alpha})(T_{i\gamma} + T^M_{i\gamma})^0 \delta_{i\beta]} N_\gamma dA \qquad (6.2.41)$$

$$+ \int_\Omega \rho_0 \{L^{\mathrm{ext}}_{\alpha\beta} + (\xi^0_{[\alpha} - \Xi_{[\alpha})F^{\mathrm{ext}}_{\beta]}\} dV\,.$$

Here, $T^M_{i\alpha}$ is the MAXWELL stress tensor as defined in $(6.2.10)_1$ and F^{ext}_α and $L^{\mathrm{ext}}_{\alpha\beta}$ are externally applied body forces and body couples with respect to the center of mass, respectively. Ω is the body manifold and $\partial\Omega^-$ the surface of Ω just inside the boundary of the body. According to the jump condition (6.2.11), which is applied for the boundary of the body separating the vacuum, it is possible to simplify the surface terms in (6.2.41), for

$$((T_{i\beta} + T^M_{i\beta})^0 N_\beta)_{\delta\Omega^-} = (t^{(N)}_i + (T^M_{i\beta})^0 N_\beta)_{\partial\Omega^+}\,,$$

where $t^{(N)}_i$ are the surface tractions of other than electromagnetic origin, $(T^M_{i\beta})_{\partial\Omega^+}$ is the MAXWELL stress tensor evaluated just outside the boundary of the body and where $\partial\Omega^+$ represents a surface just outside the boundary of the body.

With the aid of (6.2.28) we may then transform (6.2.41) into

$$M\ddot{\Xi}_\alpha = \mathcal{F}_\alpha^e + \mathcal{F}_\alpha^{ext} \,,$$

$$\frac{d}{dt}(I_{\alpha\beta}\Omega_\beta) = \mathcal{L}_\alpha^e + \mathcal{L}_\alpha^{ext} \,, \tag{6.2.42}$$

where

$$M = \int_\Omega \rho_0 dV \,, \qquad I_{\alpha\beta} = \int_\Omega \rho_0 \Big(\delta_{\alpha\beta}X_\gamma X_\gamma - R_{\alpha\gamma}R_{\beta\delta}X_\gamma X_\delta\Big) dV \,, \tag{6.2.43}$$

are the total mass and the instantaneous central moments of inertia of the body,

$$\mathcal{F}_\alpha^e = \int_{\partial\Omega+} \delta_{i\alpha}(T_{i\beta}^M)^0 N_\beta dA \,,$$

$$\mathcal{L}_\alpha^e = e_{\alpha\beta\gamma}\delta_{i\gamma}\int_{\partial\Omega+} (\xi_\beta^0 - \Xi_\beta)(T_{i\delta}^M)^0 N_\delta dA \,, \tag{6.2.44}$$

are the resultant force and moment of electromagnetic origin relative to the center of mass, and where

$$\mathcal{F}_\alpha^{ext} = \int_{\partial\Omega+} \delta_{i\alpha}t_i^{(N)} dA + \int_\Omega \rho_0 F_\alpha^{ext} dV \,,$$

$$\mathcal{L}_\alpha^e = e_{\alpha\beta\gamma}\int_{\partial\Omega+} (\xi_\beta^0 - \Xi_\beta)t_i^{(N)}\delta_{i\gamma} dA$$

$$+ e_{\alpha\beta\gamma}\int_\Omega \rho_0\{L_{\beta\gamma}^{ext} + (\xi_\beta^0 - \Xi_\beta)F_\gamma^{ext}\} dV \,, \tag{6.2.45}$$

are the resultant force and moment of the externally applied forces relative to the center of mass.

To summarize, the velocity of the center of mass $\dot{\Xi}_\alpha$ and the angular velocity Ω_α of a rigid body moving in an electromagnetic field are governed by equations (6.2.42), in which the electromagnetic surface forces and surface couples acting on the body are determined from the MAXWELL stress tensor in the surrounding vacuum. Clearly, equations (6.2.42) must be solved along with the MAXWELL equations and the energy equation in the rigid-body state, (6.2.38)–(6.2.40). In these equations the angular velocity Ω_α and the matrix $R_{\alpha\beta}$ arise; these themselves are connected by equation (6.2.29).

This completes the decomposition of the electromagnetic and mechanical balance laws. In the following sections the decomposed versions of the constitutive relations and jump conditions will be given. With these the zeroth-order solution can be solved, and once this is done the first-order perturbed problem can be attacked.

6.2.3 Decomposition of the Constitutive Equations

The constitutive equations for entropy, polarization, magnetization and stress are known, once the functional $\check{\psi}$ as defined in (6.2.6) is specified. In order to make the theory complete, constitutive equations for Q_α and $\mathbb{J}_\alpha$ must also be established.

Assuming $\check{\psi}$ to be differentiable, we may expand it in terms of TAY-LOR series about the intermediate state (eventually approximated by the rigid-body state). The series is truncated and substituted into (6.2.7) and what emerges are expressions for entropy, polarization, magnetization and the PIOLA–KIRCHHOFF stress tensor, which all are linear in the perturbed quantities e_α, b_α, u_α, etc. The expansion of the constitutive equations into TAYLOR series is straightforward and could be performed formally without specifying the energy functional $\check{\psi}$. Such expressions are of little use, however, because ultimately one must specify the free energy anyhow. A common procedure, applicable for many practical purposes, is to assume $\check{\psi}$ to be a polynomial in its independent variables, which can be truncated at a certain order.

A reasonable expression for $\check{\psi}$, which contains all interaction effects at least to within first-order terms is

$$
\begin{aligned}
\check{\psi} =\ & \frac{1}{2\rho_0}\, 2\chi^{(m)}_{\alpha\beta}\mathbb{B}_\alpha\mathbb{B}_\beta + \frac{1}{4\rho_0}\, 4\chi^{(m)}_{\alpha\beta\gamma\delta}\mathbb{B}_\alpha\mathbb{B}_\beta\mathbb{B}_\gamma\mathbb{B}_\delta + \frac{1}{2\rho_0}\, 2\chi^{(e)}_{\alpha\beta}\mathbb{E}_\alpha\mathbb{E}_\beta \\
& + \frac{1}{4\rho_0}\, 4\chi^{(e)}_{\alpha\beta\gamma\delta}\mathbb{E}_\alpha\mathbb{E}_\beta\mathbb{E}_\gamma\mathbb{E}_\delta + \frac{1}{\rho_0}\chi^{(em)}_{\alpha\beta}\mathbb{B}_\alpha\mathbb{E}_\beta - \tfrac{1}{2}c(\Theta-\Theta_0)^2 \\
& + \frac{1}{\rho_0}\lambda^{(m)}_\alpha\mathbb{B}_\alpha(\Theta-\Theta_0) + \frac{1}{2\rho_0}L^{(m)}_{\alpha\beta}\mathbb{B}_\alpha\mathbb{B}_\beta(\Theta-\Theta_0) \\
& + \frac{1}{\rho_0}\lambda^{(e)}_\alpha\mathbb{E}_\alpha(\Theta-\Theta_0) + \frac{1}{2\rho_0}L^{(e)}_{\alpha\beta}\mathbb{E}_\alpha\mathbb{E}_\beta(\Theta-\Theta_0) \\
& + \frac{1}{\rho_0}\Big[\varepsilon^{(m)}_{\beta\gamma\delta}\mathbb{B}_\beta + \tfrac{1}{2}b^{(m)}_{\alpha\beta\gamma\delta}\mathbb{B}_\alpha\mathbb{B}_\beta + \varepsilon^{(e)}_{\beta\gamma\delta}\mathbb{E}_\beta + \tfrac{1}{2}b^{(e)}_{\alpha\beta\gamma\delta}\mathbb{E}_\alpha\mathbb{E}_\beta \\
& - \rho_0\nu_{\gamma\delta}(\Theta-\Theta_0)\Big]\mathbb{E}_{\gamma\delta} + \frac{1}{2\rho_0}c_{\alpha\beta\gamma\delta}E_{\alpha\beta}E_{\gamma\delta}\ .
\end{aligned} \tag{6.2.46}
$$

Here, instead of $C_{\alpha\beta}$ we have used

$$
E_{\alpha\beta} := \tfrac{1}{2}(C_{\alpha\beta} - \delta_{\alpha\beta})\ , \tag{6.2.47}
$$

as deformation measure. According to (6.2.23)$_4$ this can be developed as

$$
E_{\alpha\beta} = \bar{E}_{\alpha\beta} + e_{\alpha\beta}\ , \tag{6.2.48}
$$

where

$$
\bar{E}_{\alpha\beta} = \tfrac{1}{2}(\bar{C}_{\alpha\beta} - \delta_{\alpha\beta})\ , \qquad e_{\alpha\beta} = \tfrac{1}{2}(\bar{F}_{\gamma\beta}u_{\gamma,\alpha} + \bar{F}_{\gamma\alpha}u_{\gamma,\beta})\ . \tag{6.2.49}
$$

Neglecting the deformations in the $\boldsymbol{\xi}$-state, we may approximate the latter by

$$e_{\alpha\beta} = \tfrac{1}{2}(R_{\gamma\beta}u_{\gamma,\alpha} + R_{\gamma\alpha}u_{\gamma,\beta}) \ . \tag{6.2.50}$$

The polynomial representation (6.2.46) for the free energy is such that elastic and thermal effects are essentially linear and that interactions with three fields are excluded. Yet, some magnetic and electric nonlinearities (fourth order terms) are nevertheless present. The reasons for this are twofold:

(i) For ferromagnetic materials and for cubic crystals fourth-order anisotropy effects (represented by $_4\chi^{(m)}_{\alpha\beta\gamma\delta}$) are often rather important and may, frequently, even be more important than the second-order terms; for cubic crystals fourth-order terms are the only nontrivial magnetic anisotropy effects, in general.

(ii) For materials with central symmetry (e.g. cubic crystals or isotropic bodies) first- and third-order coefficients in (6.2.46) vanish. In this case second- and fourth-order terms such as $L_{\alpha\beta}$ and $b_{\alpha\beta\gamma\delta}$ must be dominant.

We further note that it is possible to extend the above functional form of the free energy to include still higher-order effects without the result that this would fundamentally influence the subsequent analysis. This will not be done here, and for the time being we also refrain from assigning names to the various coefficients occurring in (6.2.46), simply because the results of Sect. 5.7 have shown that unique interpretations are not possible. The coefficients occurring in (6.2.46) also satisfy certain symmetry requirements, which are readily obtained from (6.2.46). Because they are so obvious, we shall not explicitly state them here.

It is now straightforward to derive explicit expressions for entropy, polarization, magnetization and stress by substituting (6.2.46) into (6.2.7) and performing the respective differentiations. For the *entropy* this leads to

$$
\begin{aligned}
\eta = {} & +c(\Theta - \Theta_0) - \frac{1}{\rho_0}\lambda^{(m)}_\alpha \mathbb{B}_\alpha - \frac{1}{2\rho_0}L^{(m)}_{\alpha\beta}\mathbb{B}_\alpha\mathbb{B}_\beta \\
& -\frac{1}{\rho_0}\lambda^{(e)}_\alpha \mathbb{E}_\alpha - \frac{1}{2\rho_0}L^{(e)}_{\alpha\beta}\mathbb{E}_\alpha\mathbb{E}_\beta + \nu_{\alpha\beta}E_{\alpha\beta} \\
= {} & c(\bar\Theta - \Theta_0) - \frac{1}{\rho_0}\lambda^{(m)}_\alpha \bar{\mathbb{B}}_\alpha - \frac{1}{2\rho_0}L^{(m)}_{\alpha\beta}\bar{\mathbb{B}}_\alpha\bar{\mathbb{B}}_\beta - \frac{1}{\rho_0}\lambda^{(e)}_\alpha \bar{\mathbb{E}}_\alpha \\
& -\frac{1}{2\rho_0}L^{(e)}_{\alpha\beta}\bar{\mathbb{E}}_\alpha\bar{\mathbb{E}}_\beta + \nu_{\alpha\beta}\bar{E}_{\alpha\beta} + c\Theta - \frac{1}{\rho_0}(\lambda^{(m)}_\alpha + L^{(m)}_{\alpha\beta}\bar{\mathbb{B}}_\alpha)b_\alpha \\
& -\frac{1}{\rho_0}(\lambda^{(e)}_\alpha + L^{(e)}_{\alpha\beta}\bar{\mathbb{E}}_\beta)e_\beta + \nu_{\alpha\beta}e_{\alpha\beta} =: \bar\eta + s \ .
\end{aligned}
\tag{6.2.51}
$$

Here, $\mathcal{O}(\varepsilon^2)$-terms are neglected and the entropy of the intermediate state is given by

$$
\begin{aligned}
\bar\eta = {} & c(\bar\Theta - \Theta_0) - \frac{1}{\rho_0}\lambda^{(m)}_\alpha \bar{\mathbb{B}}_\alpha - \frac{1}{2\rho_0}L^{(m)}_{\alpha\beta}\bar{\mathbb{B}}_\alpha\bar{\mathbb{B}}_\beta - \frac{1}{\rho_0}\lambda^{(e)}_\alpha \bar{\mathbb{E}}_\alpha \\
& -\frac{1}{2\rho_0}L^{(e)}_{\alpha\beta}\bar{\mathbb{E}}_\alpha\bar{\mathbb{E}}_\beta + \nu_{\alpha\beta}\bar{E}_{\alpha\beta} \ ,
\end{aligned}
\tag{6.2.52}
$$

whilst the perturbed entropy s may be written as

$$s = s^1_{\alpha\beta} u_{\alpha,\beta} + s^2_\alpha b_\alpha + s^3_\alpha e_\alpha + s^4 \theta \ , \tag{6.2.53}$$

with coefficients, which can easily be derived from (6.2.51) and (6.2.49)$_2$ as

$$
\begin{aligned}
s^1_{\alpha\beta} &= \bar{F}_{\alpha\gamma} \nu_{\beta\gamma} \ , & s^2_\alpha &= -\frac{1}{\rho_0}(\lambda^{(m)}_\alpha + L^{(m)}_{\alpha\beta}\bar{\mathbb{B}}_\beta) \\
s^3_\alpha &= -\frac{1}{\rho_0}(\lambda^{(e)}_\alpha + L^{(e)}_{\alpha\beta}\bar{\mathbb{E}}_\beta) \ , & s^4 &= c \ .
\end{aligned}
\tag{6.2.54}
$$

When deformations of the intermediate state are neglected we simply set

$$\bar{F}_{\alpha\beta} = R_{\alpha\beta} \qquad \text{and} \qquad (\bar{\cdot}) = (\cdot)^0.$$

In a completely analogous way also the constitutive equations for polarization, magnetization, and PIOLA–KIRCHHOFF stress can be deduced. One obtains for the

- *polarization*

$$\mathbb{P}_\alpha = \bar{\mathbb{P}}_\alpha + p_\alpha = \bar{\mathbb{P}}_\alpha + p^1_{\alpha\beta\gamma} u_{\beta,\gamma} + p^2_{\alpha\beta} b_\beta + p^3_{\alpha\beta} e_\beta + p^4_\alpha \theta \ , \tag{6.2.55}$$

where

$$
\begin{aligned}
\bar{\mathbb{P}}_\alpha = &-(\varepsilon^{(e)}_{\alpha\gamma\delta} + b^{(e)}_{\alpha\beta\gamma\delta}\bar{\mathbb{E}}_\beta)\bar{E}_{\gamma\delta} - (\, 2\chi^{(e)}_{\alpha\beta} + {}_4\chi^{(e)}_{\alpha\beta\gamma\delta}\bar{\mathbb{E}}_\gamma\bar{\mathbb{E}}_\delta)\bar{\mathbb{E}}_\beta \ , \\
&-\chi^{(em)}_{\beta\alpha}\bar{\mathbb{B}}_\beta - (\lambda^{(e)}_\alpha + L^{(e)}_{\alpha\beta}\bar{\mathbb{E}}_\beta)(\bar{\Theta} - \Theta_0) \ ,
\end{aligned}
\tag{6.2.56}
$$

and

$$
\begin{aligned}
p^1_{\alpha\beta\gamma} &= -\bar{F}_{\beta\delta}(\varepsilon^{(e)}_{\alpha\gamma\delta} + b^{(e)}_{\alpha\varepsilon\gamma\delta}\bar{\mathbb{E}}_\varepsilon) \ , \\
p^2_{\alpha\beta} &= -\chi^{(em)}_{\beta\alpha} \ , \\
p^3_{\alpha\beta} &= - {}_2\chi^{(e)}_{\alpha\beta} - 3\,{}_4\chi^{(e)}_{\alpha\beta\gamma\delta}\bar{\mathbb{E}}_\gamma\bar{\mathbb{E}}_\delta - b^{(e)}_{\alpha\beta\gamma\delta}\bar{E}_{\gamma\delta} - L^{(e)}_{\alpha\beta}(\bar{\Theta} - \Theta_0) \ , \\
p^4_{\alpha\beta} &= -\lambda^{(e)}_\alpha - L^{(e)}_{\alpha\beta}\bar{\mathbb{E}}_\beta;
\end{aligned}
\tag{6.2.57}
$$

- *magnetization*

$$\mathbb{M}_\alpha = \bar{\mathbb{M}}_\alpha + m_\alpha = \bar{\mathbb{M}}_\alpha + m^1_{\alpha\beta\gamma} u_{\beta,\gamma} + m^2_{\alpha\beta} b_\beta + m^3_{\alpha\beta} e_\beta + m^4_\alpha \theta \ , \tag{6.2.58}$$

where

$$
\begin{aligned}
\bar{\mathbb{M}}_\alpha = &-(\varepsilon^{(m)}_{\alpha\gamma\delta} + b^{(m)}_{\alpha\beta\gamma\delta}\bar{\mathbb{B}}_\beta)\bar{E}_{\gamma\delta} - (\, 2\chi^{(m)}_{\alpha\beta} + {}_4\chi^{(m)}_{\alpha\beta\gamma\delta}\bar{\mathbb{B}}_\gamma\bar{\mathbb{B}}_\delta)\bar{\mathbb{B}}_\beta \ , \\
&-\chi^{(em)}_{\alpha\beta}\bar{\mathbb{E}}_\beta - (\lambda^{(m)}_\alpha + L^{(m)}_{\alpha\beta}\bar{\mathbb{B}}_\beta)(\bar{\Theta} - \Theta_0) \ ,
\end{aligned}
\tag{6.2.59}
$$

and

$$m^1_{\alpha\beta\gamma} = -\bar{F}_{\beta\delta}(\varepsilon^{(m)}_{\alpha\gamma\delta} + b^{(m)}_{\alpha\varepsilon\gamma\delta}\bar{\mathbb{B}}_\varepsilon) \,,$$

$$m^2_{\alpha\beta} = -2\chi^{(m)}_{\alpha\beta} - 3{}_4\chi^{(m)}_{\alpha\beta\gamma\delta}\bar{\mathbb{B}}_\gamma\bar{\mathbb{B}}_\delta - b^{(m)}_{\alpha\beta\gamma\delta}\bar{E}_{\gamma\delta} - L^{(m)}_{\alpha\beta}(\bar{\Theta} - \Theta_0) \,,$$

$$m^3_{\alpha\beta} = -\chi^{(em)}_{\alpha\beta} \,,$$

$$m^4_\alpha = -\lambda^{(m)}_\alpha - L^{(m)}_{\alpha\beta}\bar{\mathbb{B}}_\beta \,.$$

$$\tag{6.2.60}$$

- Piola–Kirchhoff *stress tensor*, which according to $(6.2.7)_4$ and $(5.2.24)$ is given by

$$T_{i\beta} = \rho_0 \frac{\partial\check{\psi}}{\partial E_{\beta\gamma}} F_{i\gamma} \,, \tag{6.2.61}$$

so that

$$T_{i\beta} = \bar{T}_{i\beta} + \delta_{i\alpha}t_{\alpha\beta} = \bar{T}_{i\beta} + \delta_{i\alpha}(t^1_{\alpha\beta\gamma\delta}u_{\gamma,\delta} + t^2_{\alpha\beta\gamma}b_\gamma + t^3_{\alpha\beta\gamma}e_\gamma + t^4_{\alpha\beta}\theta) \,, \tag{6.2.62}$$

where

$$\bar{T}_{i\beta} = \delta_{i\alpha}\bar{F}_{\alpha\gamma}[c_{\beta\gamma\delta\varepsilon}\bar{E}_{\delta\varepsilon} + (\varepsilon^{(m)}_{\delta\beta\gamma} + \tfrac{1}{2}b^{(m)}_{\delta\varepsilon\beta\gamma}\bar{\mathbb{B}}_\varepsilon)\bar{\mathbb{B}}_\delta$$
$$+(\varepsilon^{(e)}_{\delta\beta\gamma} + \tfrac{1}{2}b^{(e)}_{\delta\varepsilon\beta\gamma}\bar{\mathbb{E}}_\varepsilon)\bar{\mathbb{E}}_\delta - \rho_0\nu_{\beta\gamma}(\bar{\Theta} - \Theta_0)] \,, \tag{6.2.63}$$

and

$$t^1_{\alpha\beta\gamma\delta} = \bar{F}_{\alpha\mu}\bar{F}_{\gamma\varepsilon}c_{\beta\mu\varepsilon\delta} + \delta_{\alpha\gamma}[c_{\beta\delta\mu\nu}\bar{E}_{\mu\nu} + (\varepsilon^{(m)}_{\varepsilon\beta\delta} + \tfrac{1}{2}b^{(m)}_{\varepsilon\mu\beta\delta}\bar{\mathbb{B}}_\mu)\bar{\mathbb{B}}_\varepsilon$$
$$+(\varepsilon^{(e)}_{\varepsilon\beta\delta} + \tfrac{1}{2}b^{(e)}_{\varepsilon\mu\beta\delta}\bar{\mathbb{E}}_\mu)\bar{\mathbb{E}}_\varepsilon - \rho_0\nu_{\beta\delta}(\bar{\Theta} - \Theta_0)] \,,$$

$$t^2_{\alpha\beta\gamma} = \bar{F}_{\alpha\delta}(\varepsilon^{(m)}_{\gamma\beta\delta} + b^{(m)}_{\gamma\varepsilon\beta\delta}\bar{\mathbb{B}}_\varepsilon) \,, \tag{6.2.64}$$

$$t^3_{\alpha\beta\gamma} = \bar{F}_{\alpha\delta}(\varepsilon^{(e)}_{\gamma\beta\delta} + b^{(e)}_{\gamma\varepsilon\beta\delta}\bar{\mathbb{E}}_\varepsilon) \,,$$

$$t^4_{\alpha\beta} = -\rho_0\bar{F}_{\alpha\gamma}\nu_{\beta\gamma} \,.$$

The above derivation is perfectly general and applies whether deformations in the intermediate state are small or large. If they are small, all coefficients characterized by a lower case letter (say $p^1_{\alpha\beta\gamma}$, $t^4_{\alpha\beta}$ etc.) may further be simplified by replacing all quantities with an overhead bar by the corresponding rigid-body quantities. The error introduced into the balance laws of the perturbed quantities is then of order $\mathcal{O}(E\varepsilon)$.

Note that the term containing $c_{\beta\mu\varepsilon\delta}$ in the expression for $t^1_{\alpha\beta\gamma\delta}$ is, in practice, always much larger than the remaining terms. This property is the justification for the assumption that the deformations due to the electromagnetic and due to the thermal fields are small. It seems to be reasonable, therefore, to approximate $t^1_{\alpha\beta\gamma\delta}$ by

$$t^1_{\alpha\beta\gamma\delta} \cong R_{\alpha\mu}R_{\gamma\varepsilon}c_{\beta\mu\varepsilon\delta} \,. \tag{6.2.65}$$

Here, we have written $R_{\alpha\mu}$ for $\bar{F}_{\alpha\mu}$ because the error introduced by dropping the remaining terms in $(6.2.64)_1$ is of the same order as that obtained by the replacement $\bar{F}_{\alpha\beta} \rightarrow R_{\alpha\beta}$.

We conclude this section with the presentation of the decomposition of the generalized versions of OHM's and FOURIER's law, as given in (6.2.8). The coefficients in these equations are functions of the variables $E_{\alpha\beta}$, $\mathbb{E}_\alpha$, $\mathbb{B}_\alpha$, Θ and $\Theta_{,\alpha}$.

We shall exclude a possible dependence of these constitutive relations on the free charge $\mathbb{Q}$, and shall, furthermore, restrict ourselves to a linear dependence of current and heat flux on electric field and temperature gradient. We then have

$$\Lambda_{\alpha\beta} = \Lambda_{\alpha\beta}(E_{\gamma\delta}, \mathbb{B}_\gamma, \Theta) , \qquad (6.2.66)$$

where $\Lambda_{\alpha\beta}$ stands for $\sigma_{\alpha\beta}$, $\kappa_{\alpha\beta}$ and $\beta_{\alpha\beta}$, respectively. To be more specific we shall choose the following polynomial expansions

$$\Lambda_{\alpha\beta} = \Lambda_{\alpha\beta}^{(r)} + \Lambda_{\alpha\beta\gamma\delta}^{(d)}E_{\gamma\delta} + {}_3\Lambda_{\alpha\beta\gamma}^{(m)}\mathbb{B}_\gamma + {}_4\Lambda_{\alpha\beta\gamma\delta}^{(m)}\mathbb{B}_\gamma\mathbb{B}_\delta + \Lambda_{\alpha\beta}^{(t)}(\bar{\Theta} - \Theta_0) . \quad (6.2.67)$$

Here we have included second-order terms in $\mathbb{B}_\alpha$, but have deleted third-order terms, because they vanish in a material with point symmetry anyhow. Decomposing the representations (6.2.8) for electric current and energy flux, we find

$$\bar{\mathbb{J}}_\alpha = \bar{\sigma}_{\alpha\beta}\bar{\mathbb{E}}_\beta + \bar{\beta}_{\beta\alpha}\frac{\bar{\Theta}_{,\beta}}{\bar{\Theta}} ,$$

$$\bar{Q}_\alpha = -\bar{\kappa}_{\alpha\beta}\bar{\Theta}_{,\beta} + \bar{\beta}_{\alpha\beta}\bar{\mathbb{E}}_\beta \qquad (6.2.68)$$

as governing equations in the intermediate state and

$$\mathbb{j}_\alpha = \mathbb{j}_{\alpha\beta\gamma}^1 \mathbb{u}_{\beta,\gamma} + \mathbb{j}_{\alpha\beta}^2 \mathbb{b}_\beta + \mathbb{j}_{\alpha\beta}^3 \mathbb{e}_\beta + \mathbb{j}_\alpha^4 \theta + \mathbb{j}_{\alpha\beta}^5 \theta_{,\beta} ,$$

$$\mathbb{q}_\alpha = \mathbb{q}_{\alpha\beta\gamma}^1 \mathbb{u}_{\beta,\gamma} + \mathbb{q}_{\alpha\beta}^2 \mathbb{b}_\beta + \mathbb{q}_{\alpha\beta}^3 \mathbb{e}_\beta + \mathbb{q}_\alpha^4 \theta + \mathbb{q}_{\alpha\beta}^5 \theta_{,\beta} \qquad (6.2.69)$$

as the approximate equations in the perturbed state, in which

$$\mathbb{j}_{\alpha\beta\gamma}^1 = \bar{F}_{\beta\delta}(\sigma_{\alpha\varepsilon\gamma\delta}^{(d)}\bar{\mathbb{E}}_\varepsilon + \beta_{\varepsilon\alpha\gamma\delta}^{(d)}\frac{\bar{\Theta}_{,\varepsilon}}{\bar{\Theta}}) ,$$

$$\mathbb{j}_{\alpha\beta}^2 = \left({}_3\sigma_{\alpha\gamma\beta}^{(m)} + 2\,{}_4\sigma_{\alpha\gamma\beta\delta}^{(m)}\bar{\mathbb{B}}_\delta\right)\bar{\mathbb{E}}_\gamma + \left({}_3\beta_{\gamma\alpha\beta}^{(m)} + 2\,{}_4\beta_{\gamma\alpha\beta\delta}^{(m)}\bar{\mathbb{B}}_\delta\right)\frac{\bar{\Theta}_{,\gamma}}{\bar{\Theta}} ,$$

$$\mathbb{j}_{\alpha\beta}^3 = \sigma_{\alpha\beta}^{(r)} + \sigma_{\alpha\beta\gamma\delta}^{(d)}\bar{E}_{\gamma\delta} + {}_3\sigma_{\alpha\beta\gamma}^{(m)}\bar{\mathbb{B}}_\gamma + {}_4\sigma_{\alpha\beta\gamma\delta}^{(m)}\bar{\mathbb{B}}_\gamma\bar{\mathbb{B}}_\delta + \sigma_{\alpha\beta}^{(t)}(\bar{\Theta} - \Theta_0) ,$$

$$\mathbb{j}_\alpha^4 = -\left(\beta_{\beta\alpha}^{(r)} + \beta_{\beta\alpha\gamma\delta}^{(d)}\bar{E}_{\gamma\delta} + {}_3\beta_{\beta\alpha\gamma}^{(m)}\bar{\mathbb{B}}_\gamma + {}_4\beta_{\beta\alpha\gamma\delta}^{(m)}\bar{\mathbb{B}}_\gamma\bar{\mathbb{B}}_\delta\right.$$

$$\left. + \beta_{\beta\mu}^{(t)}(\bar{\Theta} - \Theta_0)\right)\frac{\bar{\Theta}_{,\beta}}{\bar{\Theta}^2} + \sigma_{\alpha\beta}^{(t)}\bar{\mathbb{E}}_\beta + \beta_{\beta\mu}^{(t)}\frac{\bar{\Theta}_{,\beta}}{\bar{\Theta}} ,$$

$$\mathbb{j}_{\alpha\beta}^5 = \frac{1}{\bar{\Theta}}\left(\beta_{\beta\alpha}^{(r)} + \beta_{\beta\alpha\gamma\delta}^{(d)}\bar{E}_{\gamma\delta} + {}_3\beta_{\beta\alpha\gamma}^{(m)}\bar{\mathbb{B}}_\gamma + {}_4\beta_{\beta\alpha\gamma\delta}^{(m)}\bar{\mathbb{B}}_\gamma\bar{\mathbb{B}}_\delta + \beta_{\beta\alpha}^{(t)}(\bar{\Theta} - \Theta_0)\right) ,$$

$$(6.2.70)$$

and

$$q^1_{\alpha\beta\gamma} = \bar{F}_{\beta\delta}(-\kappa^{(d)}_{\alpha\varepsilon\gamma\delta}\bar{\Theta}_{,\varepsilon} + \beta^{(d)}_{\alpha\varepsilon\gamma\delta}\bar{\mathbb{E}}_\varepsilon) \,,$$

$$q^2_{\alpha\beta} = -\Big(3\kappa^{(m)}_{\alpha\gamma\beta} + 2\,_4\kappa^{(m)}_{\alpha\gamma\beta\delta}\bar{\mathbb{B}}_\delta\Big)\bar{\Theta}_{,\gamma} + \Big(3\beta^{(m)}_{\alpha\gamma\beta} + 2\,_4\beta^{(m)}_{\alpha\gamma\beta\delta}\bar{\mathbb{B}}_\delta\Big)\bar{\mathbb{E}}_\gamma \,,$$

$$q^3_{\alpha\beta} = \beta^{(r)}_{\alpha\beta} + \beta^{(d)}_{\alpha\beta\gamma\delta}\bar{E}_{\gamma\delta} + 3\beta^{(m)}_{\alpha\beta\gamma}\bar{\mathbb{B}}_\gamma + 4\beta^{(m)}_{\alpha\beta\gamma\delta}\bar{\mathbb{B}}_\gamma\bar{\mathbb{B}}_\delta + \beta^{(t)}_{\alpha\beta}(\bar{\Theta} - \Theta_0) \,,$$

$$q^4_\alpha = -\kappa^{(t)}_{\alpha\beta}\bar{\Theta}_{,\beta} + \beta^{(t)}_{\alpha\beta}\bar{\mathbb{E}}_\beta \,,$$

$$q^5_{\alpha\beta} = -\Big(\kappa^{(r)}_{\alpha\beta} + \kappa^{(d)}_{\alpha\beta\gamma\delta}\bar{E}_{\gamma\delta} + 3\kappa^{(m)}_{\alpha\beta\gamma}\bar{\mathbb{B}}_\gamma + 4\kappa^{(m)}_{\alpha\beta\gamma\delta}\bar{\mathbb{B}}_\gamma\bar{\mathbb{B}}_\delta + \kappa^{(t)}_{\alpha\beta}(\bar{\Theta} - \Theta_0)\Big) \,.$$
$$(6.2.71)$$

Specific forms of the coefficients, however only for isotropic materials, have been derived by PIPKIN and RIVLIN, [180, 181], and by BORGHESANI and MORRO, [27, 28]. These authors also discuss the physical significance of the various terms in (6.2.70) and (6.2.71).

At this point we have thus completed the presentation of the linearized field equations. If (6.2.53), (6.2.55), (6.2.58), (6.2.62) and (6.2.69) are substituted into the balance laws (6.2.22) and (6.2.35), what we obtain is a system of 11 independent field equations for the 11 unknowns b_α, e_α, q, u_α and θ. This system could still further by simplified by introducing electromagnetic potentials φ and a_α by (see also (5.6.7))

$$b_\alpha = e_{\alpha\beta\gamma}a_{\gamma,\beta} \,, \qquad e_\alpha = \varphi_{,\alpha} - \dot{a}_\alpha \,, \qquad (6.2.72)$$

where a_α should satisfy a gauge condition; for the LORENTZ gauge

$$a_{\alpha,\alpha} - \mu_0\varepsilon_0\dot{\varphi} = 0 \,. \qquad (6.2.73)$$

For small deformations all quantities referred to the intermediate state that appear in these perturbation equations can be replaced by the corresponding variables in the rigid-body state. To complete the description the form of the constitutive equations in this state will also be given. They are

$$\eta^0 = c(\Theta^0 - \Theta_0) - \frac{1}{\rho_0}\lambda^{(m)}_\alpha\mathbb{B}^0_\alpha - \frac{1}{2\rho_0}L^{(m)}_{\alpha\beta}\mathbb{B}^0_\alpha\mathbb{B}^0_\beta - \frac{1}{\rho_0}\lambda^{(e)}_\alpha\mathbb{E}^0_\alpha$$
$$\qquad - \frac{1}{2\rho_0}L^{(e)}_{\alpha\beta}\mathbb{E}^0_\alpha\mathbb{E}^0_\beta \,,$$

$$\mathbb{P}^0_\alpha = -(2\chi^{(e)}_{\alpha\beta} + 4\chi^{(e)}_{\alpha\beta\gamma\delta}\mathbb{E}^0_\gamma\mathbb{E}^0_\delta)\mathbb{E}^0_\beta - \chi^{(em)}_{\beta\alpha}\mathbb{B}^0_\beta$$
$$\qquad - (\lambda^{(e)}_\alpha + L^{(e)}_{\alpha\beta}\mathbb{E}^0_\beta)(\Theta^0 - \Theta_0) \,,$$

$$(6.2.74)$$

$$\mathbb{M}^{L_0}_\alpha = -(2\chi^{(m)}_{\alpha\beta} + 4\chi^{(m)}_{\alpha\beta\gamma\delta}\mathbb{B}^0_\gamma\mathbb{B}^0_\delta)\mathbb{B}^0_\beta - \chi^{(em)}_{\alpha\beta}\mathbb{E}^0_\beta$$
$$\qquad - (\lambda^{(m)}_\alpha + L^{(m)}_{\alpha\beta}\mathbb{B}^0_\beta)(\Theta^0 - \Theta_0) \,,$$

$$\mathbb{J}^0_\alpha = \sigma^0_{\alpha\beta}\mathbb{E}^0_\beta + \beta^0_{\beta\alpha}\frac{\Theta^0_{,\beta}}{\Theta^0} \,,$$

$$Q^0_\alpha = -\kappa^0_{\alpha\beta}\Theta^0_{,\beta} + \beta^0_{\alpha\beta}\mathbb{E}^0_\beta \,,$$

where the coefficients $\sigma^0_{\alpha\beta}$, $\beta^0_{\alpha\beta}$ and $\kappa^0_{\alpha\beta}$ take the forms

$$\Lambda^0_{\alpha\beta} = \Lambda_{\alpha\beta}(0, \mathbb{B}^0_\gamma, \Theta^0)$$

$$= \Lambda^{(\mathrm{r})}_{\alpha\beta} + 3\Lambda^{(\mathrm{m})}_{\alpha\beta\gamma}\mathbb{B}^0_\gamma + 4\Lambda^{(\mathrm{m})}_{\alpha\beta\gamma\delta}\mathbb{B}^0_\gamma\mathbb{B}^0_\delta + \Lambda^{(\mathrm{t})}_{\alpha\beta}(\Theta^0 - \Theta_0) . \quad (6.2.75)$$

When these constitutive relations are substituted into the balance equations (6.2.38)–(6.2.40) and (6.2.42), the rigid-body problem is reduced to a problem for the 14 unknowns $\mathbb{E}^0_\alpha$, $\mathbb{B}^0_\alpha$, $\mathbb{Q}^0$, Θ^0, Ξ_α and Ω_α. Again this problem could even further be reduced by introducing electromagnetic potentials according to

$$\mathbb{B}^0_\alpha e_{\alpha\beta\gamma} A^0_{\gamma,\beta} , \qquad \mathbb{E}^0_\alpha = \Phi^0_{,\alpha} - \dot{A}^0_\alpha , \qquad (6.2.76)$$

with

$$A^0_{\alpha,\alpha} - \mu_0\varepsilon_0\dot{\Phi}^0 = 0 . \qquad (6.2.77)$$

What remains, therefore, are now the boundary and jump conditions.

6.2.4 Decomposition of the Jump and Boundary Conditions

The decomposition of the jump conditions is a very complex problem, in general, and this is the reason why we restricted ourselves already in Sect. 6.2.1 to singular surfaces of the second order. This restriction will be maintained here too. We shall be even more restrictive and shall assume simultaneously that there are no propagating singular surfaces in the intermediate state. Hence, in this state $\bar{W}_N = 0$. An acceleration wave, if present, can then only exist in the perturbed state; the speed of propagation in this perturbed state then also forms the total wave speed and may without confusion be called W_N. The above special situation does not necessarily require, however, that W_N be small. In fact, W_N is a wave speed the numerical value of which follows from the material properties. In what follows it is advantageous to distinguish between the two different kinds of singular surfaces, namely

i) a propagating singular surface $\Sigma^{(i)}$ (acceleration wave) with velocity W_N;
ii) a material surface $\Sigma^{(ii)}$ for which $W_N = 0$.

Since $\Sigma^{(i)}$ does not exist in the intermediate state, all quantities in this state must be continuous at $\Sigma^{(i)}$, i.e.

$$[\![\bar{\mathbb{E}}_\alpha]\!] = 0 , \quad \text{etc, at } \Sigma^{(i)}.$$

The linearization of the jump conditions is rather simple in a LAGRANGEan formulation, and from (6.2.9), (6.2.11) and (6.2.17) we immediately obtain

$$[\![b_\alpha]\!]N_\alpha = 0 \,, \quad e_{\alpha\beta\gamma}[\![e_\beta]\!]N_\gamma + [\![b_\alpha]\!]W_N = 0 \,,$$

$$[\![d_\alpha^a + p_\alpha]\!]N_\alpha = 0 \,, \quad e_{\alpha\beta\gamma}[\![h_\beta^a - m_\beta]\!]N_\gamma - [\![(d_\alpha^a + p_\alpha)]\!]W_N = 0 \,,$$

$$[\![j_\alpha]\!]N_\alpha - [\![q]\!]W_N = 0 \,,$$

$$[\![t_{\alpha\beta}]\!]N_\beta = \bar{F}_{\gamma\alpha}^{-1}\Big(\bar{\mathbb{E}}_\gamma[\![p_\beta]\!]N_\beta + \bar{\mathbb{B}}_\beta([\![m_\beta]\!]N_\gamma - [\![m_\gamma]\!]N_\beta) \qquad\qquad (6.2.78)$$

$$\qquad + e_{\beta\gamma\delta}\bar{\mathbb{B}}_\beta[\![p_\delta]\!]W_N\Big) \,,$$

$$[\![q_\alpha]\!]N_\alpha = \Big(\rho_0[\![u]\!] + \bar{\mathbb{M}}_\alpha[\![b_\alpha]\!] - \bar{\mathbb{E}}_\alpha[\![p_\alpha]\!]\Big)W_N \,,$$

all valid on $\Sigma^{(i)}$. In the above, u denotes the perturbed internal energy, i.e.

$$u = \{\breve{\psi} - \overline{(\breve{\psi})}\} + \bar{\eta}\theta + \bar{\Theta}s - \frac{1}{\rho_0}\bar{\mathbb{E}}_\alpha p_\alpha - \frac{1}{\rho_0}\bar{\mathbb{P}}_\alpha e_\alpha \,. \qquad (6.2.79)$$

On the other hand, on $\Sigma^{(ii)}$ the jumps in $\bar{\mathbb{E}}_\alpha$, $\bar{\mathbb{B}}_\alpha$ etc. need not be zero, but W_N is. In this case the linearized jump conditions read

$$[\![b_\alpha]\!]N_\alpha = 0 \,, \qquad e_{\alpha\beta\gamma}[\![e_\beta]\!]N_\gamma = 0 \,,$$

$$[\![d_\alpha^a + p_\alpha]\!]N_\alpha = 0 \,, \qquad e_{\alpha\beta\gamma}[\![h_\beta^a - m_\beta]\!]N_\gamma = 0 \,, \qquad [\![j_\alpha]\!]N_\alpha = 0 \,,$$

$$[\![t_{\alpha\beta}]\!]N_\beta = -\bar{F}_{\gamma\delta}^{-1}\bar{F}_{\varepsilon\alpha}^{-1}\Big\{\langle\bar{\mathbb{E}}_\gamma\rangle[\![\bar{\mathbb{P}}_\beta]\!]N_\beta$$

$$\qquad + \langle\bar{\mathbb{B}}_\beta\rangle([\![\bar{\mathbb{M}}_\beta]\!]N_\gamma - [\![\bar{\mathbb{M}}_\gamma]\!]N_\beta)\Big\}u_{\delta,\varepsilon}$$

$$\qquad + \bar{F}_{\gamma\alpha}^{-1}\Big\{\langle\bar{\mathbb{E}}_\gamma\rangle[\![p_\beta]\!]N_\beta + [\![\bar{\mathbb{P}}_\beta]\!]N_\beta\langle e_\gamma\rangle$$

$$\qquad + \langle\bar{\mathbb{B}}_\beta\rangle([\![m_\beta]\!]N_\gamma - [\![m_\gamma]\!]N_\beta) + ([\![\bar{\mathbb{M}}_\beta]\!]N_\gamma - [\![\bar{\mathbb{M}}_\gamma]\!]N_\beta)\langle b_\beta\rangle\Big\} \,,$$

$$[\![q_\alpha]\!]N_\alpha = 0 \,,$$

$$\qquad\qquad\qquad\qquad\qquad\qquad\qquad\qquad\qquad\qquad\qquad\qquad (6.2.80)$$

on $\Sigma^{(ii)}$, whereas the jump conditions in the intermediate state follow from (6.2.9), (6.2.11) and (6.2.17) by invoking $W_N = 0$ and writing all quantities with an overhead bar.

This completes the linearization of the jump conditions. It should be noted that they can still be simplified (formally just a little bit, but for practical calculations enormously) when $\mathcal{O}(E\varepsilon)$-terms are neglected, i.e. when the $(\bar{\cdot})$-fields are replaced by the rigid-body fields $(\cdot)^0$. This step is a trivial one and will therefore be deleted here.

6.3 Linearisation of the Other Models and Comparison

In the preceding section we presented the decomposition and linearization procedure for the LORENTZ model. This would suffice, in general, because the proofs in Chaps. 4 and 5 showed that all formulations are equivalent anyhow. This is true, but since various formulations are used in the literature and because one should be able to compare these with our presentation, we shall in this section collect the balance laws and jump conditions for all five models. The comparison of the constitutive equations will be postponed until Sects. 6.4 and 6.5, however.

To perform the comparison of the various models, recall that the results of Chap. 5 showed us that, in the LAGRANGEan formulation there are essentially only two different formulations. The first is the CHU-formulation (with its two different stress models) the other one is the LORENTZ model, which with a different stress tensor also embraces the statistical model. The MAXWELL–MINKOWSKI model can be connected to either one of these two basic models. In the previous chapter the LAGRANGEan formulation of it was developed with the aid of the CHU-magnetization, rather than the LORENTZ-magnetization, because most formulas took a much simpler form this way. In what follows we shall discuss therefore models (I,II,III) and (IV,V) as the two basic different groups, although the separation of the various models could equally well be made according to (I,II) and (III,IV,V).

We begin our comparison by stating the decoupled *electromagnetic equations*. They are obtained from equations (5.6.1)–(5.6.6) (compare also systems (6.2.21), (6.2.22) of the previous section).

In the CHU *formulation* they read

$$\bar{\mathbb{B}}^{\mathrm{a}}_{\alpha,\alpha} = -\mu_0 \bar{\mathbb{M}}^C_{\alpha,\alpha} \,, \quad \dot{\bar{\mathbb{B}}}^{\mathrm{a}}_\alpha + e_{\alpha\beta\gamma} \bar{\mathbb{E}}_{\gamma,\beta} = -\mu_0 \dot{\bar{\mathbb{M}}}^C_\alpha \,,$$

$$\bar{\mathbb{D}}^{\mathrm{a}}_{\alpha,\alpha} = \bar{\mathbb{Q}} - \bar{\mathbb{P}}_{\alpha,\alpha} \,, \quad -\dot{\bar{\mathbb{D}}}^{\mathrm{a}}_\alpha + e_{\alpha\beta\gamma} \bar{\mathbb{H}}_{\gamma,\beta} = \bar{\mathbb{J}}_\alpha + \dot{\bar{\mathbb{P}}}_\alpha \,, \tag{6.3.1}$$

$$\dot{\bar{\mathbb{Q}}} + \bar{\mathbb{J}}_{\alpha,\alpha} = 0 \,,$$

and

$$\mathrm{b}^{\mathrm{a}}_{\alpha,\alpha} = -\mu_0 \mathrm{m}^C_{\alpha,\alpha} \,, \quad \dot{\mathrm{b}}^{\mathrm{a}}_\alpha + e_{\alpha\beta\gamma} \mathrm{e}_{\gamma,\beta} = -\mu_0 \dot{\mathrm{m}}^C_\alpha \,,$$

$$\mathrm{d}^{\mathrm{a}}_{\alpha,\alpha} = \mathrm{q} - \mathrm{p}_{\alpha,\alpha} \,, \quad -\dot{\mathrm{d}}^{\mathrm{a}}_\alpha + e_{\alpha\beta\gamma} \mathrm{h}_{\gamma,\beta} = \mathrm{j}_\alpha + \dot{\mathrm{p}}_\alpha \,, \tag{6.3.2}$$

$$\dot{\mathrm{q}} + \mathrm{j}_{\alpha,\alpha} = 0 \,,$$

in which $\bar{\mathbb{B}}^{\mathrm{a}}_\alpha$, $\bar{\mathbb{D}}^{\mathrm{a}}_\alpha$, $\mathrm{b}^{\mathrm{a}}_\alpha$ and $\mathrm{d}^{\mathrm{a}}_\alpha$ are auxiliary fields, which, with the aid of (5.2.9) and (6.2.23), become

$$\bar{\mathbb{D}}^{\mathrm{a}}_\alpha = \varepsilon_0 \bar{J} \bar{C}^{-1}_{\alpha\beta} \bar{\mathbb{E}}_\beta \,, \quad \bar{\mathbb{B}}^{\mathrm{a}}_\alpha = \mu_0 \bar{J} \bar{C}^{-1}_{\alpha\beta} \bar{\mathbb{H}}_\beta \,, \tag{6.3.3}$$

and

$$\mathrm{d}^{\mathrm{a}}_\alpha = \varepsilon_0 J \Big(C^{-1}_{\alpha\beta} \mathrm{e}_\beta + (\bar{C}^{-1}_{\alpha\beta} \bar{F}^{-1}_{\gamma\delta} - \bar{C}^{-1}_{\beta\gamma} \bar{F}^{-1}_{\alpha\delta} - \bar{C}^{-1}_{\alpha\gamma} \bar{F}^{-1}_{\beta\delta}) \bar{\mathbb{E}}_\beta \mathrm{u}_{\delta,\gamma} \Big) \,,$$

$$\mathrm{b}^{\mathrm{a}}_\alpha = \mu_0 \bar{J} \Big(\bar{C}^{-1}_{\alpha\beta} \mathrm{h}_\beta + (\bar{C}^{-1}_{\alpha\beta} \bar{F}^{-1}_{\gamma\delta} - \bar{C}^{-1}_{\beta\gamma} \bar{F}^{-1}_{\alpha\delta} - \bar{C}^{-1}_{\alpha\gamma} \bar{F}^{-1}_{\beta\delta}) \bar{\mathbb{H}}_\beta \mathrm{u}_{\delta,\gamma} \Big) \,. \tag{6.3.4}$$

Note the symmetry of the formulas (6.3.4). Its origin is the complete symmetry of the MAXWELL equations in this formulation.

In the *statistical and the* LORENTZ *model* the decomposition of the MAXWELL equations was already made in Sect. 6.2.2. For completeness and for the purpose of comparison they will be repeated here. In the intermediate state they are

$$\bar{\mathbb{B}}_{\alpha,\alpha} = 0 \,,$$

$$\dot{\bar{\mathbb{B}}}_{\alpha} + e_{\alpha\beta\gamma}\bar{\mathbb{E}}_{\gamma,\beta} = 0 \,,$$

$$\bar{\mathbb{D}}^{\mathrm{a}}_{\alpha,\alpha} = \bar{\mathbb{Q}} - \bar{\mathbb{P}}_{\alpha,\alpha} \,, \tag{6.3.5}$$

$$-\dot{\bar{\mathbb{D}}}^{\mathrm{a}}_{\alpha} + e_{\alpha\beta\gamma}\bar{\mathbb{H}}^{\mathrm{a}}_{\gamma,\beta} = \bar{\mathbb{J}}_{\alpha} + \dot{\bar{\mathbb{P}}}_{\alpha} + e_{\alpha\beta\gamma}\bar{\mathbb{M}}^{L}_{\gamma,\beta} \,,$$

$$\dot{\bar{\mathbb{Q}}} + \bar{\mathbb{J}}_{\alpha,\alpha} = 0 \,,$$

whereas in the perturbed state they become

$$\mathrm{b}_{\alpha,\alpha} = 0 \,,$$

$$\dot{\mathrm{b}}_{\alpha} + e_{\alpha\beta\gamma}\mathrm{e}_{\gamma,\beta} = 0 \,,$$

$$\mathrm{d}^{\mathrm{a}}_{\alpha,\alpha} = \mathrm{q} - \mathrm{p}_{\alpha,\alpha} \,, \tag{6.3.6}$$

$$-\dot{\mathrm{d}}^{\mathrm{a}}_{\alpha} + e_{\alpha\beta\gamma}\mathrm{h}_{\gamma,\beta} = \mathrm{j}_{\alpha} + \dot{\mathrm{p}}_{\alpha} + e_{\alpha\beta\gamma}\mathrm{m}^{L}_{\gamma,\beta} \,,$$

$$\dot{\mathrm{q}} + \mathrm{j}_{\alpha,\alpha} = 0 \,,$$

with $\bar{\mathbb{D}}^{\mathrm{a}}_{\alpha}$, $\bar{\mathbb{H}}^{\mathrm{a}}_{\alpha}$, $\mathrm{d}^{\mathrm{a}}_{\alpha}$ and $\mathrm{h}^{\mathrm{a}}_{\alpha}$ as given in (6.2.25) and (6.2.26):

$$\bar{\mathbb{D}}^{\mathrm{a}}_{\alpha} = \varepsilon_0 \bar{J}\bar{C}^{-1}_{\alpha\beta}\bar{\mathbb{E}}_{\beta} \,,$$

$$\bar{\mathbb{H}}^{\mathrm{a}}_{\alpha} = \frac{1}{\mu_0\bar{J}}\bar{C}_{\alpha\beta}\bar{\mathbb{B}}_{\beta} - \frac{\varepsilon_0}{\bar{J}}e_{\beta\gamma\delta}\bar{C}_{\alpha\beta}\bar{F}_{\varepsilon\gamma}\dot{\xi}_{\varepsilon}\bar{\mathbb{E}}_{\delta} \tag{6.3.7}$$

and

$$\mathrm{d}^{\mathrm{a}}_{\alpha} = \varepsilon_0 \bar{J}\{\bar{C}^{-1}_{\alpha\beta}\mathrm{e}_{\beta} + (\bar{C}^{-1}_{\alpha\beta}\bar{F}^{-1}_{\gamma\delta} - \bar{C}^{-1}_{\beta\gamma}\bar{F}^{-1}_{\alpha\delta} - \bar{C}^{-1}_{\alpha\gamma}\bar{F}^{-1}_{\beta\delta})\bar{\mathbb{E}}_{\beta}\mathrm{u}_{\delta,\gamma}\} \,,$$

$$\mathrm{h}^{\mathrm{a}}_{\alpha} = \frac{1}{\bar{J}}\left\{ \frac{1}{\mu_0}\bar{C}_{\alpha\beta}\mathrm{b}_{\beta} - \varepsilon_0 e_{\beta\gamma\delta}\bar{F}_{\varepsilon\gamma}\dot{\xi}_{\varepsilon}\bar{C}_{\alpha\beta}\mathrm{e}_{\delta} \right.$$
$$+ \left[\left(\frac{1}{\mu_0}\bar{\mathbb{B}}_{\beta} - e_{\beta\nu\mu}\bar{F}_{\varepsilon\nu}\dot{\xi}_{\varepsilon}\bar{\mathbb{E}}_{\mu} \right)(\bar{F}_{\delta\alpha}\delta_{\beta\gamma} + \bar{F}_{\delta\beta}\delta_{\alpha\gamma} - \bar{C}_{\alpha\beta}\bar{F}_{\gamma\delta}) \right. \tag{6.3.8}$$
$$\left. - \varepsilon_0 e_{\beta\gamma\mu}\bar{\mathbb{E}}_{\mu}\bar{C}_{\alpha\beta}\dot{\xi}_{\delta} \right] \mathrm{u}_{\delta,\gamma} - \varepsilon_0 e_{\beta\gamma\delta}\bar{\mathbb{E}}_{\delta}\bar{C}_{\alpha\beta}\bar{F}_{\varepsilon\gamma}\dot{\mathrm{u}}_{\varepsilon} \right\} \,.$$

Note that in the above sets of MAXWELL equations we have retained the upper indices C and L for the magnetizations.

Clearly, because polarization and magnetization are based on different conceptions, we cannot expect the formulas (6.3.7) and (6.3.8) to be symmetric. This difference becomes formally particularly apparent if the equations in the intermediate state are approximated by those in the rigid-body state. Then the rigid-body motion does, formally, not enter the CHU-formulation, but it does show up in the LORENTZ- formulation.

We conclude this listing of the various forms of the MAXWELL equations with the MAXWELL–MINKOWSKI formulation. The equations are

$$\bar{\mathbb{B}}_{\alpha,\alpha} = 0 \,, \qquad \dot{\bar{\mathbb{B}}}_\alpha + e_{\alpha\beta\gamma}\bar{\mathbb{E}}_{\gamma,\beta} = 0 \,,$$

$$\bar{\mathbb{D}}_{\alpha,\alpha} = \bar{\mathbb{Q}} \,, \qquad -\dot{\bar{\mathbb{D}}}_\alpha + e_{\alpha\beta\gamma}\bar{\mathbb{H}}_{\gamma,\beta} = \bar{\mathbb{J}}_\alpha \,, \qquad (6.3.9)$$

$$\dot{\bar{\mathbb{Q}}} + \bar{\mathbb{J}}_{\alpha,\alpha} = 0 \,,$$

$$b_{\alpha,\alpha} = 0 \,, \qquad \dot{b}_\alpha + e_{\alpha\beta\gamma}e_{\gamma,\beta} = 0 \,,$$

$$d_{\alpha,\alpha} = q \,, \qquad -\dot{d}_\alpha + e_{\alpha\beta\gamma}h_{\gamma,\beta} = j_\alpha \,, \qquad (6.3.10)$$

$$\dot{q} + j_{\alpha,\alpha} = 0 \,,$$

and must be supplemented by the relations

$$\bar{\mathbb{D}}_\alpha = \varepsilon_0 \bar{J}\bar{C}^{-1}_{\alpha\beta}\bar{\mathbb{E}}_\beta + \bar{\mathbb{P}}_\alpha \,, \quad \bar{\mathbb{B}}_\alpha = \mu_0 \bar{J}\bar{C}^{-1}_{\alpha\beta}\bar{\mathbb{H}}_\beta + \mu_0 \bar{\mathbb{M}}^C_\alpha \,, \qquad (6.3.11)$$

and

$$d_\alpha = \varepsilon_0 \bar{J}\Big(\bar{C}^{-1}_{\alpha\beta}e_\beta + (\bar{C}^{-1}_{\alpha\beta}\bar{F}^{-1}_{\gamma\delta} - \bar{C}^{-1}_{\beta\gamma}\bar{F}^{-1}_{\alpha\delta} - \bar{C}^{-1}_{\alpha\gamma}\bar{F}^{-1}_{\beta\delta})\bar{\mathbb{E}}_\beta u_{\delta,\gamma}\Big) + p_\alpha \,,$$

$$b_\alpha = \mu_0 \bar{J}\Big(\bar{C}^{-1}_{\alpha\beta}e_\beta + (\bar{C}^{-1}_{\alpha\beta}\bar{F}^{-1}_{\gamma\delta} - \bar{C}^{-1}_{\beta\gamma}\bar{F}^{-1}_{\alpha\delta} - \bar{C}^{-1}_{\alpha\gamma}\bar{F}^{-1}_{\beta\delta})\bar{\mathbb{H}}_\beta u_{\delta,\gamma}\Big) + \mu_0 m^C_\alpha \,.$$

$$(6.3.12)$$

Apart from the use of different auxiliary fields the most essential difference in these formulations lies in the choice of the magnetization: $\mathbb{M}^L_\alpha$ against $\mathbb{M}^C_\alpha$, which are related to each other according to (see (5.3.10))

$$\bar{\mathbb{M}}^C_\alpha = \bar{J}\bar{C}^{-1}_{\alpha\beta}\bar{\mathbb{M}}^L_\beta \,,$$
$$(6.3.13)$$
$$m^C_\alpha = \bar{J}\bar{C}^{-1}_{\alpha\beta}m^L_\beta + \bar{J}(\bar{C}^{-1}_{\alpha\beta}\bar{F}^{-1}_{\gamma\delta} - \bar{C}^{-1}_{\beta\gamma}\bar{F}^{-1}_{\alpha\delta} - \bar{C}^{-1}_{\alpha\gamma}\bar{F}^{-1}_{\beta\delta})\bar{\mathbb{M}}^L_\beta u_{\delta,\gamma} \,.$$

Before we pass on to the mechanical balance laws, we would like to draw the reader's attention to the basic differences in the GAUSS and FARADAY laws of the three formulations above. Because these laws are homogeneous in the LORENTZ and in the MAXWELL–MINKOWSKI formulations electromagnetic potentials could easily and straightforwardly be introduced as was shown at the end of Sect. 6.2.4 (cf. (6.2.72), (6.2.73) or (5.6.7)). In the CHU formulation, on the other hand, these equations are inhomogeneous. Therefore, in this formulation the potentials A_α and Φ cannot be introduced in the same way as in Sect. 6.2.4. They must rather be defined as

$$\bar{\mathbb{B}}_\alpha^a = e_{\alpha\beta\gamma}\bar{A}_{\gamma,\beta} - \mu_0\bar{\mathbb{M}}_\alpha^C , \qquad \bar{\mathbb{E}}_\alpha = \bar{\Phi}_{,\alpha} - \dot{\bar{A}}_\alpha ,$$

$$\mathbb{b}_\alpha^a = e_{\alpha\beta\gamma}a_{\gamma,\beta} - \mu_0\mathbb{m}_\alpha^C , \qquad e_\alpha\varphi_{,\alpha} - \dot{a}_\alpha , \tag{6.3.14}$$

with the gauge conditions

$$\bar{A}_{\alpha,\alpha} - \varepsilon_0\mu_0\dot{\bar{\Phi}} = 0 , \qquad a_{\alpha,\alpha} - \varepsilon_0\mu_0\dot{\varphi} = 0 . \tag{6.3.15}$$

The above listed electromagnetic equations must still be supplemented by two sets of constitutive relations for e.g. polarization and magnetization. They are all given in Chap. 5, and since they are directly derivable from a free energy functional the linearization is trivial and will, therefore, be omitted here.

As was done for the MAXWELL equations, we also wish to recapitulate the *mechanical balance laws*. If body forces of electromagnetic origin are not specified the equations can easily be taken over from (6.2.34) and (6.2.35). The momentum equations are

$$\rho_0\ddot{\xi}_\alpha = \delta_{i\alpha}(\bar{T}_{i\beta,\beta} + \rho_0\bar{F}_i^e) , \qquad \rho_0\ddot{u}_\alpha = t_{\alpha\beta,\beta} + \rho_0 f_\alpha^{fe} , \tag{6.3.16}$$

whereas the energy equations in the intermediate and in the perturbed state can be written as

$$\rho_0\bar{\Theta}\dot{\bar{\eta}} = \bar{\mathbb{J}}_\alpha\bar{\mathbb{E}}_\alpha - \bar{Q}_{\alpha,\alpha}$$

$$\rho_0\dot{\bar{\eta}}\theta + \rho_0\bar{\Theta}\dot{s} = \mathbb{j}_\alpha\bar{\mathbb{E}}_\alpha + \bar{\mathbb{J}}_\alpha e_\alpha - q_{\alpha,\alpha} . \tag{6.3.17}$$

All that is needed to complete the above equations for the various formulations is to prescribe the body force; for, all remaining quantities are either independent fields or else given by constitutive relations. The correspondence conditions of the various formulations of the latter were treated in Sect. 5.7 and will again be taken up in the following sections. Incidentally, that angular momentum is satisfied identically by satisfying objectivity requirements in the constitutive relations is reason for us not to list the balance law of moment of momentum. Moreover, it is noted that the reduced energy equation is the same in all formulations.

Let us now list the body force expressions of the various formulations.

In the CHU *formulation* we have, from (5.2.11)$_1$,

for **model I**:

$$^I(\rho_0\bar{F}_i^e) = \delta_{i\delta}\Big\{\bar{F}_{\alpha\delta}^{-1}\Big[(\bar{\mathbb{Q}} - \bar{\mathbb{P}}_{\beta,\beta})\bar{\mathbb{E}}_\alpha + e_{\alpha\beta\gamma}(\bar{\mathbb{J}}_\beta + \dot{\bar{\mathbb{P}}}_\beta)\bar{\mathbb{B}}_\gamma^a$$

$$- \mu_0\bar{\mathbb{M}}_{\beta,\beta}^C\bar{\mathbb{H}}_\alpha + \bar{F}_{\beta\delta}^{-1}(\bar{\mathbb{P}}_\alpha\bar{\mathbb{E}}_\beta + \mu_0\bar{\mathbb{M}}_\alpha^C\bar{\mathbb{H}}_\beta)\Big]_{,\alpha}\Big\} ,$$

$$^I(\rho_0 f_\alpha^{fe}) = {}^{II}(\rho_0 f_\alpha^{fe}) + \Big[\bar{F}_{\beta\alpha}^{-1}(\bar{\mathbb{P}}_\varepsilon e_\beta + p_\varepsilon\bar{\mathbb{E}}_\beta + \mu_0\bar{\mathbb{M}}_\varepsilon^C h_\beta + \mu_0 m_\varepsilon^C\bar{\mathbb{H}}_\beta)$$

$$- \bar{F}_{\beta\gamma}^{-1}\bar{F}_{\delta\alpha}^{-1}(\bar{\mathbb{P}}_\varepsilon\bar{\mathbb{E}}_\beta + \mu_0\bar{\mathbb{M}}_\varepsilon^C\bar{\mathbb{H}}_\beta)u_{\gamma,\delta}\Big]_{,\varepsilon} , \tag{6.3.18}$$

where $^{II}(\rho_0 f_\alpha^{fe})$ is listed in (6.3.19)$_2$. On the other hand, for **model II**: (from $(5.2.12)_1$)

$$^{II}(\rho_0 \bar{F}_i^e) = \delta_{i\delta} \bar{F}_{\alpha\delta}^{-1} \Big[(\bar{\mathbb{Q}} - \bar{\mathbb{P}}_{\beta,\beta}) \bar{\mathbb{E}}_\alpha + e_{\alpha\beta\gamma} (\bar{\mathbb{J}}_\beta + \dot{\bar{\mathbb{P}}}_\beta) \bar{\mathbb{B}}_\gamma^a$$

$$- \mu_0 \bar{\mathbb{M}}_{\beta,\beta}^C \bar{\mathbb{H}}_\alpha \Big] ,$$

$$^{II}(\rho_0 f_\alpha^e) = \bar{F}_{\beta\alpha}^{-1} \Big[(q - p_{\varepsilon,\varepsilon}) \bar{\mathbb{E}}_\beta + (\bar{\mathbb{Q}} - \bar{\mathbb{P}}_{\varepsilon,\varepsilon}) e_\beta + e_{\beta\gamma\delta} (\bar{\mathbb{J}}_\gamma + \dot{\bar{\mathbb{P}}}_\gamma) b_\delta^a$$

$$+ e_{\beta\gamma\delta} (j_\gamma + \dot{p}_\gamma) \bar{\mathbb{B}}_\delta^a - \mu_0 \bar{\mathbb{M}}_{\varepsilon,\varepsilon}^C h_\beta - \mu_0 m_{\varepsilon,\varepsilon}^C \bar{\mathbb{H}}_\beta \Big]$$

$$- \bar{F}_{\beta\gamma}^{-1} \bar{F}_{\delta\alpha}^{-1} \Big[(\bar{\mathbb{Q}} - \bar{\mathbb{P}}_{\varepsilon,\varepsilon}) \bar{\mathbb{E}}_\beta + e_{\beta\mu\nu} (\bar{\mathbb{J}}_\mu + \dot{\bar{\mathbb{P}}}_\mu) \bar{\mathbb{B}}_\nu^a$$

$$- \mu_0 \bar{\mathbb{M}}_{\varepsilon,\varepsilon}^C \bar{\mathbb{H}}_\beta \Big] u_{\gamma,\delta} . \tag{6.3.19}$$

In actual calculations one only needs the force expression for one single model, because the stress tensors of the two models are related by (5.2.22), or

$$^I \bar{T}_{i\alpha} = {}^{II} \bar{T}_{i\alpha} - \delta_{i\alpha} \bar{F}_{\beta\delta}^{-1} (\bar{\mathbb{P}}_\alpha \bar{\mathbb{E}}_\beta + \mu_0 \bar{\mathbb{M}}_\alpha^C \bar{\mathbb{H}}_\beta) ,$$

$$^I \bar{t}_{\alpha\beta} = {}^{II} \bar{t}_{\alpha\beta} - \bar{F}_{\gamma\alpha}^C (p_\beta \bar{\mathbb{E}}_\gamma + \bar{\mathbb{P}}_\beta e_\gamma + \mu_0 m_\beta^C \bar{\mathbb{H}}_\gamma + \mu_0 \bar{\mathbb{M}}_\beta^C h_\gamma) \tag{6.3.20}$$

$$+ \bar{F}_{\varepsilon\gamma}^{-1} \bar{F}_{\delta\alpha}^{-1} (\bar{\mathbb{P}}_\beta \bar{\mathbb{E}}_\varepsilon + \mu_0 \bar{\mathbb{M}}_\beta^C \bar{\mathbb{H}}_\varepsilon) u_{\gamma,\delta} .$$

When the stress tensor and the body force of model I are substituted into the momentum equation what results is the body force of model II and a term which agrees with $^{II} T_{i\alpha}$.

In the MAXWELL–MINKOWSKI **model III** the body force is given by $(5.4.5)_1$. When decomposed this becomes

$$^{III}(\rho_0 \bar{F}_i^e) = \delta_{i\delta} \Big\{ \bar{F}_{\alpha\delta}^{-1} \Big[\bar{\mathbb{Q}} \mathbb{E}_\alpha + e_{\alpha\beta\gamma} \bar{\mathbb{J}}_\beta \bar{\mathbb{B}}_\gamma + \bar{\mathbb{P}}_\beta \bar{\mathbb{E}}_{\beta,\alpha} + \mu_0 \bar{\mathbb{M}}_\beta^C \bar{\mathbb{H}}_{\beta,\alpha}$$

$$+ e_{\alpha\beta\gamma} (\bar{\mathbb{D}}_\beta \dot{\bar{\mathbb{B}}}_\gamma + \dot{\bar{\mathbb{D}}}_\beta \bar{\mathbb{B}}_\gamma) \Big] + \bar{F}_{\beta\delta,\alpha}^{-1} (\bar{\mathbb{P}}_\alpha \bar{\mathbb{E}}_\beta + \mu_0 \bar{\mathbb{M}}_\alpha^C \bar{\mathbb{H}}_\beta) \Big\} ,$$

$$^{III}(\rho_0 f_\alpha^e) = \bar{F}_{\beta\alpha}^{-1} \Big\{ q \bar{\mathbb{E}}_\beta + \bar{\mathbb{Q}} e_\beta + e_{\beta\gamma\delta} (\bar{\mathbb{J}}_\gamma b_\delta + j_\gamma \bar{\mathbb{B}}_\delta) + \bar{\mathbb{P}}_\gamma e_{\gamma,\beta}$$

$$+ p_\gamma \bar{\mathbb{E}}_{\gamma,\beta} + \mu_0 \bar{\mathbb{M}}_\gamma^C h_{\gamma,\beta} + \mu_0 m_\gamma^C \bar{\mathbb{H}}_{\gamma,\beta} + e_{\beta\gamma\delta} (\bar{\mathbb{D}}_\gamma \dot{b}_\delta + \dot{\bar{\mathbb{D}}}_\gamma b_\delta$$

$$+ d_\gamma \dot{\bar{\mathbb{B}}}_\delta + \dot{d}_\gamma \bar{\mathbb{B}}_\delta) \Big\} + \bar{F}_{\beta\alpha,\gamma}^{-1} \Big\{ \bar{\mathbb{P}}_\gamma e_\beta + p_\gamma \bar{\mathbb{E}}_\beta$$

$$+ \mu_0 \bar{\mathbb{M}}_\gamma^C h_\beta + \mu_0 m_\gamma^C \bar{\mathbb{H}}_\beta \Big\} - \bar{F}_{\beta\gamma}^{-1} \bar{F}_{\delta\alpha}^{-1} \Big\{ \bar{\mathbb{Q}} \bar{\mathbb{E}}_\beta + e_{\beta\mu\nu} \bar{\mathbb{J}}_\mu \bar{\mathbb{B}}_\nu$$

$$+ \bar{\mathbb{P}}_\varepsilon \bar{\mathbb{E}}_{\varepsilon,\beta} + \mu_0 \bar{\mathbb{M}}_\varepsilon^C \bar{\mathbb{H}}_{\varepsilon,\beta} + e_{\beta\mu\nu} (\bar{\mathbb{D}}_\mu \dot{\bar{\mathbb{B}}}_\nu + \dot{\bar{\mathbb{D}}}_\mu \bar{\mathbb{B}}_\nu) \Big\} u_{\gamma,\delta}$$

$$- (\bar{F}_{\beta\gamma}^{-1} \bar{F}_{\delta\alpha}^{-1} u_{\gamma,\delta})_{,\varepsilon} (\bar{\mathbb{P}}_\varepsilon \bar{\mathbb{E}}_\beta + \mu_0 \bar{\mathbb{M}}_\varepsilon^C \bar{\mathbb{H}}_\beta) . \tag{6.3.21}$$

We recall that the stress tensors in models I and III are identical, hence

$$^{I}\bar{T}_{i\alpha} = {}^{III}\bar{T}_{i\alpha} \qquad \text{and} \qquad {}^{I}t_{\alpha\beta} = {}^{III}t_{\alpha\beta} . \tag{6.3.22}$$

The relationships between ($^{II}\bar{T}_{i\alpha}$, $^{II}t_{\alpha\beta}$) and ($^{III}\bar{T}_{i\alpha}$, $^{III}t_{\alpha\beta}$) can then easily be read off from (6.3.20).

Next, we list the body forces for the statistical and the LORENTZ-formulations. In the LORENTZ-*formulation* (**model V**) they are already given in (6.2.36) and (6.2.37). For reasons of comparison they will be repeated here:

$$^{V}(\rho_0\bar{F}_i^e) = \delta_{i\delta}\bar{F}_{\alpha\delta}^{-1}\left\{(\bar{\mathbb{Q}} - \bar{\mathbb{P}}_{\beta,\beta})\mathbb{E}_\alpha + e_{\alpha\beta\gamma}(\bar{\mathbb{J}}_\beta + \dot{\bar{\mathbb{P}}}_\beta)\bar{\mathbb{B}}_\gamma\right.$$

$$\left. + (\bar{\mathbb{M}}_{\alpha,\beta}^L - \bar{\mathbb{M}}_{\beta,\alpha}^L)\bar{\mathbb{B}}_\beta\right\},$$

$$^{V}(\rho_0 f_\alpha^e) = \bar{F}_{\beta\alpha}^{-1}\left\{(\bar{\mathbb{Q}} - \bar{\mathbb{P}}_{\varepsilon,\varepsilon})e_\beta + (q - p_{\varepsilon,\varepsilon})\bar{\mathbb{E}}_\beta + e_{\beta\gamma\delta}(\bar{\mathbb{J}}_\gamma + \dot{\bar{\mathbb{P}}}_\gamma)b_\delta\right.$$

$$+e_{\beta\gamma\delta}(j_\gamma + \dot{p}_\gamma)\bar{\mathbb{B}}_\delta + \bar{\mathbb{B}}_\gamma(m_{\beta,\gamma}^L - m_{\gamma,\beta}^L) \tag{6.3.23}$$

$$\left. + b_\gamma(\bar{\mathbb{M}}_{\beta,\gamma}^L - \bar{\mathbb{M}}_{\gamma,\beta}^L)\right\} - \bar{F}_{\beta\gamma}^{-1}\bar{F}_{\delta\alpha}^{-1}\left\{(\bar{\mathbb{Q}} - \bar{\mathbb{P}}_{\varepsilon,\varepsilon})\bar{\mathbb{E}}_\beta\right.$$

$$\left. + e_{\beta\mu\nu}(\bar{\mathbb{J}}_\mu + \dot{\bar{\mathbb{P}}}_\mu)\bar{\mathbb{B}}_\nu + \bar{\mathbb{B}}_\varepsilon(\bar{\mathbb{M}}_{\beta,\varepsilon}^L - \bar{\mathbb{M}}_{\varepsilon,\beta}^L)\right\}u_{\gamma,\delta} .$$

On the other hand, in **model IV** we have (cf. $(5.3.19)_1$)

$$^{IV}(\rho_0\bar{F}_i^e) = {}^{V}(\rho_0\bar{F}_i^e) + \delta_{i\delta}\left\{\bar{F}_{\alpha\delta}^{-1}(\bar{\mathbb{E}}_\alpha\bar{\mathbb{P}}_\beta - \bar{\mathbb{M}}_\alpha^L\bar{\mathbb{B}}_\beta + \delta_{\alpha\beta}\bar{\mathbb{M}}_\gamma^L\bar{\mathbb{B}}_\gamma)\right\}_{,\beta} ,$$

$$^{IV}(\rho_0 f_\alpha^e) = {}^{V}(\rho_0 f_\alpha^e) + \left\{\bar{F}_{\beta\alpha}^{-1}\left[\bar{\mathbb{E}}_\beta p_\varepsilon + e_\beta\bar{\mathbb{P}}_\varepsilon - \bar{\mathbb{M}}_\beta^L b_\varepsilon\right.\right.$$

$$\left. -m_\beta^L\bar{\mathbb{B}}_\varepsilon + \delta_{\beta\varepsilon}(\bar{\mathbb{M}}_\gamma^L b_\gamma + m_\gamma^L\bar{\mathbb{B}}_\gamma)\right] \tag{6.3.24}$$

$$\left. - \bar{F}_{\beta\gamma}^{-1}\bar{F}_{\delta\alpha}^{-1}\left[\bar{\mathbb{E}}_\beta\bar{\mathbb{P}}_\varepsilon - \bar{\mathbb{M}}_\beta^L\bar{\mathbb{B}}_\varepsilon + \delta_{\beta\varepsilon}\bar{\mathbb{M}}_\mu^L\bar{\mathbb{B}}_\mu\right]u_{\gamma,\delta}\right\}_{,\varepsilon} .$$

These forces need not be calculated, however, because the divergence term on the right-hand side of (6.3.24) will, with opposite sign, also arise in the stress tensor of the statistical formulation; hence, the momentum equation remains unchanged. Indeed, (cf. (5.3.22))

$$^{IV}\bar{T}_{i\alpha} = {}^{V}\bar{T}_{i\alpha} - \delta_{i\delta}\bar{F}_{\beta\delta}^{-1}(\bar{\mathbb{P}}_\alpha\bar{\mathbb{E}}_\beta - \bar{\mathbb{B}}_\alpha\bar{\mathbb{M}}_\beta^L + \delta_{\alpha\beta}\bar{\mathbb{B}}_\gamma\bar{\mathbb{M}}_\gamma^L) ,$$

$$^{IV}t_{\alpha\beta} = {}^{V}t_{\alpha\beta} - \bar{F}_{\gamma\alpha}^{-1}\left[p_\beta\bar{\mathbb{E}}_\gamma + \bar{\mathbb{P}}_\beta e_\gamma - b_\beta\bar{\mathbb{M}}_\gamma^L - \bar{\mathbb{B}}_\beta m_\gamma^L\right.$$

$$\left. + \delta_{\beta\gamma}(b_\delta\bar{\mathbb{M}}_\delta^L + \bar{\mathbb{B}}_\delta m_\delta^L)\right] \tag{6.3.25}$$

$$+ \bar{F}_{\varepsilon\gamma}^{-1}\bar{F}_{\delta\alpha}^{-1}\left[\bar{\mathbb{P}}_\beta\bar{\mathbb{E}}_\varepsilon - \bar{\mathbb{B}}_\beta\bar{\mathbb{M}}_\varepsilon^L + \delta_{\beta\varepsilon}\bar{\mathbb{B}}_\mu\bar{\mathbb{M}}_\mu^L\right]u_{\gamma,\delta} .$$

To find the link between the groups (I, II, III) and (IV, V) we also need the decomposed versions of relation (5.6.12) between $^{V}T_{i\alpha}$ and $^{II}T_{i\alpha}$. This relation gives

$$^{V}\bar{T}_{i\alpha} = {}^{II}\bar{T}_{i\alpha} + \delta_{i\delta}\bar{F}_{\beta\delta}^{-1}\left[\bar{\mathbb{B}}_{\alpha}^{a}\bar{\mathbb{H}}_{\beta} - \bar{\mathbb{B}}_{\alpha}\bar{\mathbb{H}}_{\beta}^{a} - \tfrac{1}{2}\delta_{\alpha\beta}(\bar{\mathbb{B}}_{\gamma}^{a}\bar{\mathbb{H}}_{\gamma} - \bar{\mathbb{B}}_{\gamma}\bar{\mathbb{H}}_{\gamma}^{a})\right] ,$$

$$^{V}t_{\alpha\beta} = {}^{II}t_{\alpha\beta} + \bar{F}_{\gamma\alpha}^{-1}\left[b_{\beta}^{a}\bar{\mathbb{H}}_{\gamma} + \bar{\mathbb{B}}_{\beta}^{a}h_{\gamma} - b_{\beta}\bar{\mathbb{H}}_{\gamma}^{a} - \bar{\mathbb{B}}_{\beta}h_{\gamma}^{a}\right.$$

$$\left. - \tfrac{1}{2}\delta_{\beta\gamma}(b_{\delta}^{a}\bar{\mathbb{H}}_{\delta} + \bar{\mathbb{B}}_{\delta}^{a}h_{\delta} - b_{\delta}\bar{\mathbb{H}}_{\delta}^{a} - \bar{\mathbb{B}}_{\delta}h_{\delta}^{a})\right]$$

$$- \bar{F}_{\varepsilon\gamma}^{-1}\bar{F}_{\delta\alpha}^{-1}\left[\bar{\mathbb{B}}_{\beta}^{a}\bar{\mathbb{H}}_{\varepsilon} - \bar{\mathbb{B}}_{\beta}\bar{\mathbb{H}}_{\varepsilon}^{a} - \tfrac{1}{2}\delta_{\beta\varepsilon}(\bar{\mathbb{B}}_{\mu}^{a}\bar{\mathbb{H}}_{\mu} - \bar{\mathbb{B}}_{\mu}\bar{\mathbb{H}}_{\mu}^{a})\right]u_{\gamma,\delta} .$$

$$\text{(6.3.26)}$$

With the aid of the formulas $(6.2.25)_2$, $(6.2.26)_2$, $(6.3.3)_2$ and $(6.3.4)_2$ the auxiliary fields $\bar{\mathbb{B}}_{\alpha}^{a}$, $\bar{\mathbb{H}}_{\alpha}^{a}$, b_{α}^{a} and h_{α}^{a} can be eliminated.

If we wished to do so, we could also give the decoupled constitutive equations for the stresses $^{I}T_{i\alpha}, \ldots, {}^{V}T_{i\alpha}$. We shall not do it here and restrict ourselves to recalling that only $^{II}T_{i\alpha}$ and $^{V}T_{i\alpha}$ are directly derivable from a free energy; hence, their linearization is trivial. However, the free energies occurring in these relations are not identical; they are related according to (5.6.20), which when decoupled yields

$$\overline{^{V}(\rho_0\check{\psi})} = \overline{^{II}(\rho_0\check{\psi})} - \frac{\mu_0}{2\bar{J}}\bar{C}_{\alpha\beta}\bar{\mathbb{M}}_{\alpha}^{C}\bar{\mathbb{M}}_{\beta}^{C} ,$$

$$\left(\overline{^{V}(\rho_0\check{\psi})} - \overline{^{V}(\rho_0\check{\psi})}\right) = \left(\overline{^{II}(\rho_0\check{\psi})} - \overline{^{II}(\rho_0\check{\psi})}\right) - \frac{\mu_0}{\bar{J}}\bar{C}_{\alpha\beta}\bar{\mathbb{M}}_{\alpha}^{C}m_{\beta}^{C} \qquad \text{(6.3.27)}$$

$$+ \frac{\mu_0}{\bar{J}}\bar{\mathbb{M}}_{\alpha}^{C}\bar{\mathbb{M}}_{\beta}^{C}\left[\bar{F}_{\gamma\delta}^{-1}\bar{C}_{\alpha\beta} - \delta_{\alpha\delta}\bar{F}_{\gamma\beta} - \delta_{\beta\delta}\bar{F}_{\gamma\alpha}\right]u_{\gamma,\delta} ,$$

We conclude this section with a survey of the decomposed jump and boundary conditions. As before, the singular surfaces will be assumed to be of *order 2*. Since the deduction of the respective conditions is straightforward and analogous to the methods illustrated in Sect. 6.2.4, we shall only present the final results. In accord with the assumptions laid down in Sect. 6.2.4, $\Sigma^{(i)}$ will denote the propagating surface of acceleration waves across which the intermediate fields $(\bar{\cdot})$ do not suffer a jump ($[\![(\bar{\cdot})]\!] = 0$, on $\Sigma^{(i)}$). On a material surface $\Sigma^{(ii)}$, on the other hand, $W_N = 0$, but the intermediate fields may jump there.

We start with the jump conditions for the electromagnetic fields. One obtains:

for model 1 and model II

$$\left.\begin{aligned}
[\![\bar{\mathbb{B}}_{\alpha}^{a} + \mu_0\bar{\mathbb{M}}_{\alpha}^{C}]\!]N_{\alpha} &= 0 , \\
[\![\bar{\mathbb{D}}_{\alpha}^{a} + \bar{P}_{\alpha}]\!]N_{\alpha} &= 0 , \\
e_{\alpha\beta\gamma}[\![\bar{\mathbb{E}}_{\beta}]\!]N_{\gamma} &= 0 , \\
e_{\alpha\beta\gamma}[\![\bar{\mathbb{H}}_{\beta}]\!]N_{\gamma} &= 0 , \\
[\![\bar{\mathbb{J}}_{\alpha}]\!]N_{\alpha} &= 0 ,
\end{aligned}\right\} \qquad \text{on } \Sigma^{(ii)}, \qquad \text{(6.3.28)}$$

and

$$\left.\begin{aligned}
&[\![b_\alpha^a + \mu_0 m_\alpha^C]\!] N_\alpha = 0\,, \\
&[\![d_\alpha^a + p_\alpha]\!] N_\alpha = 0\,, \\
&e_{\alpha\beta\gamma}[\![e_\beta]\!] N_\gamma + [\![(b_\alpha^a + \mu_0 m_\alpha^C)]\!] W_N = 0\,, \\
&e_{\alpha\beta\gamma}[\![h_\beta]\!] N_\gamma - [\![(d_\alpha^a + p_\alpha)]\!] W_N = 0\,, \\
&[\![j_\alpha]\!] N_\alpha - [\![q]\!] W_N = 0\,,
\end{aligned}\right\} \quad \text{on } \Sigma^{(i)} \text{ and } \Sigma^{(ii)}, \quad (6.3.29)$$

for **model III**

$$\left.\begin{aligned}
&[\![\bar{\mathbb{B}}_\alpha]\!] N_\alpha = 0\,, \qquad [\![\bar{\mathbb{D}}_\alpha]\!] N_\alpha = 0\,, \\
&e_{\alpha\beta\gamma}[\![\bar{\mathbb{E}}_\beta]\!] N_\gamma = 0,\ e_{\alpha\beta\gamma}[\![\bar{\mathbb{H}}_\beta]\!] N_\gamma = 0,\ [\![\bar{\mathbb{J}}_\alpha]\!] N_\alpha = 0\,,
\end{aligned}\right\} \text{on } \Sigma^{(ii)}, \ (6.3.30)$$

and

$$\left.\begin{aligned}
&[\![b_\alpha]\!] N_\alpha = 0\,, \qquad [\![d_\alpha]\!] N_\alpha = 0\,, \\
&e_{\alpha\beta\gamma}[\![e_\beta]\!] N_\gamma + [\![b_\alpha]\!] W_N = 0\,, \\
&e_{\alpha\beta\gamma}[\![h_\beta]\!] N_\gamma - [\![d_\alpha]\!] W_N = 0\,, \\
&[\![j_\alpha]\!] N_\alpha - [\![q]\!] W_N = 0\,;
\end{aligned}\right\} \text{on } \Sigma^{(i)} \text{ and } \Sigma^{(ii)}, \qquad (6.3.31)$$

for **model IV** and **model V**

$$\left.\begin{aligned}
&[\![\bar{\mathbb{B}}_\alpha]\!] N_\alpha = 0\,, \\
&[\![\bar{\mathbb{D}}_\alpha + \bar{\mathbb{P}}_\alpha]\!] N_\alpha = 0\,, \\
&e_{\alpha\beta\gamma}[\![\bar{\mathbb{E}}_\beta]\!] N_\gamma = 0\,, \\
&e_{\alpha\beta\gamma}[\![\bar{\mathbb{H}}_\beta - \bar{\mathbb{M}}_\beta^L]\!] N_\gamma = 0\,, \\
&[\![\bar{\mathbb{J}}_\alpha]\!] N_\alpha = 0\,,
\end{aligned}\right\} \quad \text{on } \Sigma^{(ii)}, \qquad (6.3.32)$$

and

$$\left.\begin{aligned}
&[\![b_\alpha]\!] N_\alpha = 0\,, \qquad [\![d_\alpha^a + p_\alpha]\!] N_\alpha = 0\,, \\
&e_{\alpha\beta\gamma}[\![e_\beta]\!] N_\gamma + [\![b_\alpha]\!] W_N = 0\,, \\
&e_{\alpha\beta\gamma}[\![h_\beta^a - m_\beta^L]\!] N_\gamma - [\![d_\alpha^a + p_\alpha]\!] N_\alpha = 0\,, \\
&[\![j_\alpha]\!] N_\alpha - [\![q]\!] W_N = 0\,,
\end{aligned}\right\} \quad \text{on } \Sigma^{(i)} \text{ and } \Sigma^{(ii)}. \quad (6.3.33)$$

These equations can easily be specified for $\Sigma^{(i)}$ and $\Sigma^{(ii)}$.

For the jump conditions of momentum and energy of matter and fields we must distinguish between $\Sigma^{(i)}$ and $\Sigma^{(ii)}$. We first list those valid at the wave surface $\Sigma^{(i)}$; they only need be given for the perturbed relations. One obtains:

for **model I** and **model III**

$$
\left.
\begin{aligned}
[\![{}^{I}t_{\alpha\beta}]\!]N_\beta &= [\![{}^{III}t_{\alpha\beta}]\!]N_\beta \\
&= \bar{F}_{\gamma\alpha}^{-1}\Big\{-\bar{\mathbb{P}}_\beta[\![e_\gamma]\!]N_\beta - \mu_0\bar{\mathbb{M}}_\beta^C[\![h_\gamma]\!]N_\beta \\
&\quad +e_{\beta\gamma\delta}\bar{\mathbb{B}}_\beta^a[\![p_\delta]\!]W_N\Big\} , \\
[\![q_\alpha]\!]N_\alpha &= \Big\{\rho_0[\![{}^{II}u]\!] - \bar{\mathbb{E}}_\alpha[\![p_\alpha]\!] - \bar{\mathbb{H}}_\alpha[\![\mu_0 m_\alpha^C]\!]\Big\}W_N .
\end{aligned}
\right\} \text{ on } \Sigma^{(i)}. \quad (6.3.34)
$$

We note that in the first of the above conditions the symmetry between the electric and the magnetic fields, which is characteristic for the Chu-formulation, is destroyed, because in the derivation the term

$$
e_{\beta\gamma\delta}\langle\mathbb{D}_\beta^a\rangle[\![\mu_0\bar{\mathbb{M}}_\delta^C]\!]W_N
$$

was dropped as it is proportional to c^{-2}. Furthermore, for model III we have replaced in $(6.3.34)_1$ $\bar{\mathbb{B}}_\beta^a$ by

$$
\bar{\mathbb{B}}_\beta - \mu_0\bar{\mathbb{M}}_\beta^C .
$$

For **model II** we have

$$
\left.
\begin{aligned}
[\![{}^{II}t_{\alpha\beta}]\!]N_\beta &= \bar{F}_{\gamma\alpha}^{-1}\Big\{\bar{\mathbb{E}}_\gamma[\![p_\beta]\!]N_\beta + \bar{\mathbb{H}}_\gamma[\![\mu_0 m_\beta^C]\!]N_\beta \\
&\quad +e_{\beta\gamma\delta}\bar{\mathbb{B}}_\beta^a[\![p_\delta]\!]W_N\Big\} , \\
[\![q_\alpha]\!]N_\alpha &= \Big\{\rho_0[\![{}^{II}u]\!] - \bar{\mathbb{E}}_\alpha[\![p_\alpha]\!] - \bar{\mathbb{H}}_\alpha[\![\mu_0 m_\alpha^C]\!]\Big\}W_N ,
\end{aligned}
\right\} \text{ on } \Sigma^{(i)} . \quad (6.3.35)
$$

Moreover, for **model IV**

$$
\left.
\begin{aligned}
[\![{}^{IV}t_{\alpha\beta}]\!]N_\beta &= \bar{F}_{\gamma\alpha}^{-1}\Big\{-\bar{\mathbb{P}}_\beta[\![e_\gamma]\!]N_\beta - \bar{\mathbb{M}}_\beta^L[\![b_\beta]\!]N_\gamma \\
&\quad +e_{\beta\gamma\delta}\bar{\mathbb{B}}_\beta[\![p_\delta]\!]W_N\Big\} , \\
[\![q_\alpha]\!]N_\alpha &= \Big\{\rho_0[\![{}^{V}u]\!] + \bar{\mathbb{M}}_\alpha[\![b_\alpha]\!] - \bar{\mathbb{E}}_\alpha[\![p_\alpha]\!]\Big\}W_N ,
\end{aligned}
\right\} \text{ on } \Sigma^{(i)} , \quad (6.3.36)
$$

and for **model V**

$$
\left.
\begin{aligned}
[\![{}^{V}t_{\alpha\beta}]\!]N_\beta &= \bar{F}_{\gamma\alpha}^{-1}\Big\{\bar{\mathbb{E}}_\gamma[\![p_\beta]\!]N_\beta + \bar{\mathbb{B}}_\beta([\![m_\beta^L]\!]N_\gamma \\
&\quad -[\![m_\gamma^L]\!]N_\beta) + e_{\beta\gamma\delta}\bar{\mathbb{B}}_\beta[\![p_\delta]\!]W_N\Big\} , \\
[\![q_\alpha]\!]N_\alpha &= \Big\{\rho_0[\![{}^{V}u]\!] + \bar{\mathbb{M}}_\alpha^L[\![b_\alpha]\!] - \bar{\mathbb{E}}_\alpha[\![p_\alpha]\!]\Big\}W_N ,
\end{aligned}
\right\} \text{ on } \Sigma^{(i)} . \quad (6.3.37)
$$

Since in all of the above jump conditions for the energy, q_α is the same function (needless to say once more that in model V q_α stands for q_α^S rather than q_α^L) the right-hand sides of these conditions must be identical also. That this is indeed the case can most easily be seen from (5.6.19), which in perturbed form reads

$$
\rho_0 \, {}^V u = \rho_0 \, {}^{II} u - \mu_0 h_\alpha \bar{M}_\alpha^C - \mu_0 \bar{H}_\alpha m_\alpha^C - \frac{\mu_0}{\bar{J}} \bar{C}_{\alpha\beta} \bar{M}_\alpha^C m_\beta^C
$$
$$
+ \frac{\mu_0}{2\bar{J}} \bar{M}_\alpha^C \bar{M}_\beta^C \left[\bar{F}_{\gamma\delta}^{-1} \bar{C}_{\alpha\beta} - \delta_{\alpha\delta} \bar{F}_{\gamma\beta} - \delta_{\beta\delta} \bar{F}_{\gamma\alpha} \right] u_{\gamma,\delta} \, .
\tag{6.3.38}
$$

Finally, we list the boundary conditions for momentum and energy on material surfaces $\Sigma^{(ii)}$. On these, W_N vanishes but the intermediate fields may jump instead. Consequently ,

for model I and model III

$$
\left[\!\left[{}^I \bar{T}_{i\alpha} \right]\!\right] N_\alpha = \left[\!\left[{}^{III} \bar{T}_{i\alpha} \right]\!\right] N_\alpha
$$
$$
\left.
\begin{aligned}
&= \delta_{i\delta} \bar{F}_{\alpha\delta}^{-1} \left\{ \langle \bar{\mathbb{P}}_\beta \rangle \left[\!\left[\bar{\mathbb{E}}_\alpha \right]\!\right] N_\beta - \langle \mu_0 \bar{\mathbb{M}}_\alpha^C \rangle \left[\!\left[\bar{\mathbb{H}}_\alpha \right]\!\right] N_\beta \right\} \, , \\
\left[\!\left[{}^I t_{\alpha\beta} \right]\!\right] N_\beta &= \left[\!\left[{}^{III} t_{\alpha\beta} \right]\!\right] N_\beta \\
&= \bar{F}_{\gamma\alpha}^{-1} \Big\{ -\langle \bar{\mathbb{P}}_\beta \rangle \left[\!\left[e_\gamma \right]\!\right] N_\beta - \langle p_\beta \rangle \left[\!\left[\bar{\mathbb{E}}_\gamma \right]\!\right] N_\beta \\
&\quad -\langle \mu_0 \bar{\mathbb{M}}_\beta^C \rangle \left[\!\left[h_\gamma \right]\!\right] N_\beta - \langle \mu_0 m_\beta^C \rangle \left[\!\left[\bar{\mathbb{H}}_\gamma \right]\!\right] N_\beta \Big\} \\
&\quad +\bar{F}_{\gamma\delta}^{-1} \bar{F}_{\varepsilon\alpha}^{-1} \left\{ \langle \bar{\mathbb{P}}_\beta \rangle \left[\!\left[\bar{\mathbb{E}}_\gamma \right]\!\right] N_\beta + \langle \mu_0 \bar{\mathbb{M}}_\beta^C \rangle \left[\!\left[\bar{\mathbb{H}}_\gamma \right]\!\right] N_\beta \right\} u_{\delta,\varepsilon} \, ,
\end{aligned}
\right\}
\tag{6.3.39}
$$

on $\Sigma^{(ii)}$, in which for model III, $\bar{\mathbb{B}}_\alpha^a$ and b_α^a must, respectively, be replaced by

$$
(\bar{\mathbb{B}}_\alpha - \mu_0 \bar{\mathbb{M}}_\alpha^C) \qquad \text{and} \qquad (b_\alpha - \mu_0 m_\alpha^C) \, .
$$

For model II

$$
\left.
\begin{aligned}
\left[\!\left[{}^{II} \bar{T}_{i\alpha} \right]\!\right] N_\alpha &= \delta_{i\delta} \bar{F}_{\alpha\delta}^{-1} \left\{ \langle \bar{\mathbb{E}}_\alpha \rangle \left[\!\left[\bar{\mathbb{P}}_\beta \right]\!\right] N_\beta + \langle \bar{\mathbb{H}}_\alpha \rangle \left[\!\left[\mu_0 \bar{\mathbb{M}}_\beta^C \right]\!\right] N_\beta \right\} \, , \\
\left[\!\left[{}^{II} t_{\alpha\beta} \right]\!\right] N_\beta &= \bar{F}_{\gamma\alpha}^{-1} \Big\{ \langle \bar{\mathbb{E}}_\gamma \rangle \left[\!\left[p_\beta \right]\!\right] N_\beta + \langle e_\gamma \rangle \left[\!\left[\bar{\mathbb{P}}_\beta \right]\!\right] N_\beta \\
&\quad +\langle \bar{\mathbb{H}}_\gamma \rangle \left[\!\left[\mu_0 m_\beta^C \right]\!\right] N_\beta + \langle h_\gamma \rangle \left[\!\left[\mu_0 \bar{\mathbb{M}}_\beta^C \right]\!\right] N_\beta \Big\} \\
&\quad -\bar{F}_{\gamma\delta}^{-1} \bar{F}_{\varepsilon\alpha}^{-1} \left\{ \langle \bar{\mathbb{E}}_\gamma \rangle \left[\!\left[\bar{\mathbb{P}}_\beta \right]\!\right] N_\beta + \langle \bar{\mathbb{H}}_\gamma \rangle \left[\!\left[\mu_0 \bar{\mathbb{M}}_\beta^C \right]\!\right] N_\beta \right\} u_{\delta,\varepsilon} \, ,
\end{aligned}
\right\}
\tag{6.3.40}
$$

on $\Sigma^{(ii)}$,

for **model IV**

$$
\left.\begin{aligned}
\left[\!\left[\,{}^{IV}\bar{T}_{i\alpha}\right]\!\right]N_\alpha &= \delta_{i\delta}\bar{F}^{-1}_{\alpha\delta}\Big\{-\langle\bar{\mathbb{P}}_\beta\rangle\left[\!\left[\bar{\mathbb{E}}_\alpha\right]\!\right]N_\beta - \langle\bar{\mathbb{M}}^L_\beta\rangle\left[\!\left[\bar{\mathbb{B}}_\beta\right]\!\right]N_\alpha\Big\}\,, \\[2mm]
\left[\!\left[\,{}^{IV}\mathrm{t}_{\alpha\beta}\right]\!\right]N_\beta &= \bar{F}^{-1}_{\gamma\alpha}\Big\{-\langle\bar{\mathbb{P}}_\beta\rangle\left[\!\left[\mathrm{e}_\gamma\right]\!\right]N_\beta - \langle\mathrm{p}_\beta\rangle\left[\!\left[\bar{\mathbb{E}}_\gamma\right]\!\right]N_\beta \\
&\quad -\langle\bar{\mathbb{M}}^L_\beta\rangle\left[\!\left[\mathrm{b}_\beta\right]\!\right]N_\gamma - \langle\mathrm{m}^L_\beta\rangle\left[\!\left[\bar{\mathbb{B}}_\beta\right]\!\right]N_\gamma\Big\} \\
&\quad +\bar{F}^{-1}_{\gamma\delta}\bar{F}^{-1}_{\varepsilon\alpha}\Big\{\langle\bar{\mathbb{P}}_\beta\rangle\left[\!\left[\bar{\mathbb{E}}_\gamma\right]\!\right]N_\beta + \langle\bar{\mathbb{M}}^L_\beta\rangle\left[\!\left[\bar{\mathbb{B}}_\beta\right]\!\right]N_\gamma\Big\}\mathrm{u}_{\delta,\varepsilon}\,,
\end{aligned}\right\} \tag{6.3.41}
$$

on $\Sigma^{(ii)}$,
and for **model V**

$$
\left.\begin{aligned}
\left[\!\left[\,{}^{V}\bar{T}_{i\alpha}\right]\!\right]N_\alpha &= \delta_{i\delta}\bar{F}^{-1}_{\alpha\delta}\Big\{\langle\bar{\mathbb{E}}_\alpha\rangle\left[\!\left[\bar{\mathbb{P}}_\beta\right]\!\right]N_\beta \\
&\quad +\langle\bar{\mathbb{B}}_\beta\rangle(\left[\!\left[\bar{\mathbb{M}}^L_\beta\right]\!\right]N_\alpha - \left[\!\left[\bar{\mathbb{M}}^L_\alpha\right]\!\right]N_\beta)\Big\}\,, \\[2mm]
\left[\!\left[\,{}^{V}\mathrm{t}_{\alpha\beta}\right]\!\right]N_\beta &= \bar{F}^{-1}_{\gamma\alpha}\Big\{\langle\bar{\mathbb{E}}_\gamma\rangle\left[\!\left[\mathrm{p}_\beta\right]\!\right]N_\beta + \langle\mathrm{e}_\gamma\rangle\left[\!\left[\bar{\mathbb{P}}_\beta\right]\!\right]N_\beta \\
&\quad +\langle\bar{\mathbb{B}}_\beta\rangle(\left[\!\left[\mathrm{m}^L_\beta\right]\!\right]N_\gamma - \left[\!\left[\mathrm{m}^L_\gamma\right]\!\right]N_\beta) \\
&\quad +\langle\mathrm{b}_\beta\rangle(\left[\!\left[\bar{\mathbb{M}}^L_\beta\right]\!\right]N_\gamma - \left[\!\left[\bar{\mathbb{M}}^L_\gamma\right]\!\right]N_\beta)\Big\} \\
&\quad -\bar{F}^{-1}_{\gamma\delta}\bar{F}^{-1}_{\varepsilon\alpha}\Big\{\langle\bar{\mathbb{E}}_\gamma\rangle\left[\!\left[\bar{\mathbb{P}}_\beta\right]\!\right]N_\beta \\
&\quad +\langle\bar{\mathbb{B}}_\beta\rangle(\left[\!\left[\bar{\mathbb{M}}^L_\beta\right]\!\right]N_\gamma - \left[\!\left[\bar{\mathbb{M}}^L_\gamma\right]\!\right]N_\beta)\Big\}\mathrm{u}_{\delta,\varepsilon}\,,
\end{aligned}\right\} \tag{6.3.42}
$$

on $\Sigma^{(ii)}$. The boundary conditions of energy on $\Sigma^{(ii)}$ are equal in all formulations and simply read

$$
\left[\!\left[\bar{Q}_\alpha\right]\!\right]N_\alpha = \left[\!\left[\mathrm{q}_\alpha\right]\!\right]N_\alpha = 0\,, \qquad \text{on } \Sigma^{(ii)}\,. \tag{6.3.43}
$$

Needless to say here that, with the use of relations (6.3.20), (6.3.25), (6.3.26) and (6.3.38), all these jump conditions and boundary conditions can be transformed into each other.

6.4 The Meaning of Interchanging Dependent and Independent Constitutive Variables in one Formulation

In Chaps. 4 and 5 it was demonstrated that all theories of magnetizable and polarizable thermoelastic materials are fully equivalent. The equivalence requirements could be stated as interrelationships between the free energies and heat flux vectors of the formulations being compared. It was further demonstrated in Sect. 5.7 of Chap. 5 that full equivalence of different formulations can be destroyed simply by comparing energy expressions, which are too restrictive to allow complete matching. In particular, using in each theory a polynomial expansion of the free energy in terms of its variables and truncating these expansions at a certain prescribed level, destroyed complete agreement. One can, of course, insist in polynomial representations, but must then accept the fact that equivalence of two formulations is only possible to within terms not being omitted in the expansion process.

A similar situation prevails if one tries to compare theories, which are based on one single model (say the LORENTZ model), in which different dependent and independent constitutive variables are used. The situation is similar as in the nonlinear theory of elasticity, in which there also exist two different constitutive formulations, one in which the free energy is a function of the strains and a second one, whose energy function is obtained from the former through a LEGENDRE transformation. This latter energy, which is called enthalpy (or complementary energy) is a function of the stresses rather than the strains. It gives the strains as functions of the stresses. Inverting the stress-strain relationship, that is expressing stress as a function of strain then effectively amounts to finding the free energy from the enthalpy. If these inversions can be performed, the theories based on enthalpy and free energy, respectively, are fully equivalent. It is known that this step is not a trivial one, except in the linear elasticity theory, in which it amounts to writing the elasticity coefficients in terms of the compliances.

In an electromechanical interaction theory changes of dependent and independent variables are possible also among the electromagnetic constitutive variables. Indeed, it is not the change of strain and stress as dependent and independent constitutive variables, that ordinarily gives rise to different constitutive approaches, but rather that of electromagnetic variables.

We showed in Chap. 3 already that there are four different constitutive theories in just one formulation, not counting that stress and strain and temperature and entropy could also be interchanged as dependent and independent variables. In what follows we shall only be dealing with two possibilities, namely the LORENTZ formulation, in which $(\mathbb{E}_\alpha, \mathbb{B}_\alpha)$ and $(\mathbb{P}_\alpha/\rho_0, \mathbb{M}_\alpha/\rho_0)$ are taken as the respective independent constitutive variables. Results for the CHU-formulation and other variable combinations will be stated, however.

In the LORENTZ model and if $\mathbb{E}_\alpha$ and $\mathbb{B}_\alpha$ are selected we have

$$\check{\psi} = U - \eta\Theta - \frac{1}{\rho_0}\mathbb{E}_\alpha\mathbb{P}_\alpha = \check{\psi}(E_{\alpha\beta}, \mathbb{E}_\alpha, \mathbb{B}_\alpha, \Theta) , \qquad (6.4.1)$$

and

$$\eta = -\frac{\partial\check{\psi}}{\partial\Theta} , \quad \mathbb{P}_\alpha = -\rho_0\frac{\partial\check{\psi}}{\partial\mathbb{E}_\alpha} ,$$

$$\mathbb{M}_\alpha^L = -\rho_0\frac{\partial\check{\psi}}{\partial\mathbb{B}_\alpha} , \quad {}^V T_{\alpha\beta}^P = \rho_0\frac{\partial\check{\psi}}{\partial E_{\alpha\beta}} . \qquad (6.4.2)$$

On the other hand, when $\mathbb{P}_\alpha/\rho_0$ and $\mathrm{M}_\alpha^L/\rho_0$ are the independent fields, we use as energy functional

$$\hat{\psi} = U - \eta\Theta + \frac{1}{\rho_0}\mathrm{M}_\alpha^L\mathbb{B}_\alpha = \hat{\psi}\left(E_{\alpha\beta}, \frac{\mathbb{P}_\alpha}{\rho_0}, \frac{\mathrm{M}_\alpha^L}{\rho_0}, \Theta\right) , \qquad (6.4.3)$$

and from it we obtain

$$\eta = -\frac{\partial\hat{\psi}}{\partial\Theta} , \qquad \mathbb{E}_\alpha = \frac{\partial\hat{\psi}}{\partial\mathbb{P}_\alpha/\rho_0} ,$$

$$\mathbb{B}_\alpha = \frac{\partial\hat{\psi}}{\partial\mathrm{M}_\alpha^L/\rho_0} , \qquad {}^VT_{\alpha\beta}^P = \rho_0\frac{\partial\hat{\psi}}{\partial E_{\alpha\beta}} . \qquad (6.4.4)$$

In view of the definitions of $\check{\psi}$ and $\hat{\psi}$, (6.4.1) and (6.4.3), the two constitutive theories lead to identical results if

$$\hat{\psi} = \check{\psi} + \frac{1}{\rho_0}\mathbb{E}_\alpha\mathbb{P}_\alpha + \frac{1}{\rho_0}\mathbb{B}_\alpha\mathrm{M}_\alpha^L , \qquad (6.4.5)$$

or with the use of (6.4.2), if

$$\hat{\psi}\left(E_{\alpha\beta}, \frac{\mathbb{P}_\alpha}{\rho_0}, \frac{\mathrm{M}_\alpha^L}{\rho_0}, \Theta\right)$$
$$= \check{\psi}\left(E_{\alpha\beta}, \frac{\partial\hat{\psi}}{\partial\mathbb{P}_\alpha/\rho_0}, \frac{\partial\hat{\psi}}{\partial\mathrm{M}_\alpha^L/\rho_0}, \Theta\right) + \frac{\partial\hat{\psi}}{\partial\mathbb{P}_\alpha/\rho_0}\frac{\mathbb{P}_\alpha}{\rho_0} + \frac{\partial\hat{\psi}}{\partial\mathrm{M}_\alpha^L/\rho_0}\frac{\mathrm{M}_\alpha^L}{\rho_0} . \qquad (6.4.6)$$

Of course, there is also a dual relation to (6.4.6), namely

$$\check{\psi}(E_{\alpha\beta}, \mathbb{E}_\alpha, \mathbb{B}_\alpha, \Theta) = \hat{\psi}\left(E_{\alpha\beta}, -\frac{\partial\check{\psi}}{\partial\mathbb{E}_\alpha}, -\frac{\partial\check{\psi}}{\partial\mathbb{B}_\alpha}, \Theta\right) + \mathbb{E}_\alpha\frac{\partial\check{\psi}}{\partial\mathbb{E}_\alpha} + \mathbb{B}_\alpha\frac{\partial\check{\psi}}{\partial\mathbb{B}_\alpha} .$$
$$(6.4.7)$$

For given functionals $\hat{\psi}$ and $\check{\psi}$, equations (6.4.6) and (6.4.7) must be satisfied identically if the two constitutive theories are to be equivalent. If only one of the energy functionals, $\hat{\psi}$ or $\check{\psi}$, is given, then (6.4.6) or (6.4.7) are functional differential equations to determine the other. The solution to these remains an open problem, as we shall not attack it here. Nevertheless, in order to demonstrate that this problem is extremely complex in general, consider, as a more or less arbitrary example, the following energy functional

$$\hat{\psi} = \hat{\psi}(\mathbb{E}_{\alpha\beta}, \mathrm{M}_\alpha) = \mathrm{f}(\mathrm{M}^2) + \frac{1}{2\rho_0}\left(\lambda(E_{\alpha\alpha})^2 + 2GE_{\alpha\beta}E_{\alpha\beta}\right) , \qquad (6.4.8)$$

which may be regarded as the most simple functional form for an isotropic nonlinearly magnetizable body. In (6.4.8)

$$\mathrm{M}^2 := \frac{1}{\rho_0^2}\mathrm{M}_\alpha^L\mathrm{M}_\alpha^L , \qquad (6.4.9)$$

and $f(M^2)$ is a continuous, differentiable function. With the aid of (6.4.4) we may derive the following expression for $\mathbb{B}_\alpha$

$$\mathbb{B}_\alpha = 2\frac{df}{dM^2}\frac{M_\alpha^L}{\rho_0} . \tag{6.4.10}$$

Substitution of this into the identity (6.4.5) allows determination of $\check{\psi}$,

$$\check{\psi}(E_{\alpha\beta}, \mathbb{B}_\alpha) = f(M^2) - 2M^2\frac{df}{dM^2} + \frac{1}{2\rho_0}\{\lambda(E_{\alpha\alpha})^2 + 2GE_{\alpha\beta}E_{\alpha\beta}\} . \tag{6.4.11}$$

The right-hand side of (6.4.11) can be determined as an explicit function of $\mathbb{B}_\alpha$, or more precisely of $\mathbb{B}^2 = \mathbb{B}_\alpha\mathbb{B}_\alpha$, only if relation (6.4.10) is invertible, i.e., only if M^2 is expressible as a function of $\mathbb{B}_\alpha$. This is rarely the case in general. However, the special choice

$$f(M^2) = \tfrac{1}{2}\,{}_2\chi\int_0^{M^2}\left[1 + 2\frac{{}_4\chi}{{}_2\chi}\xi\right]^{1/2}d\xi \tag{6.4.12}$$
$$(\cong \tfrac{1}{2}\,{}_2\chi M^2 + \tfrac{1}{4}\,{}_4\chi M^4 + \cdots) ,$$

which is a nonlinear representation that may be regarded as an extension of the usual functional dependencies of the free energy on magnetization, gives (for ${}_2\chi \neq 0$, ${}_4\chi \neq 0$)

$$M^2 = \frac{{}_2\chi}{4\,{}_4\chi}\left\{-1 + \left[1 + \frac{8\,{}_4\chi}{{}_2\chi^3}\mathbb{B}^2\right]^{1/2}\right\} , \tag{6.4.13}$$

so that $\check{\psi}$ may be written as

$$\check{\psi} = \tfrac{1}{2}\,{}_2\chi\int_0^{M^2}\left[1 + 2\frac{{}_4\chi}{{}_2\chi}\xi\right]^{1/2}d\xi - {}_2\chi\left[1 + 2\frac{{}_4\chi}{{}_2\chi}M^2\right]^{1/2}M^2$$
$$+ \frac{1}{2\rho_0}\left(\lambda(E_{\alpha\alpha})^2 + 2GE_{\alpha\beta}E_{\alpha\beta}\right) , \tag{6.4.14}$$

in which M^2 is given by (6.4.13).

Often the construction of a free energy of one formulation from that of another is a matter of shear patience or simply becomes impossible analytically. This does not mean that equivalence is not possible in these cases; it simply means that a free energy function of the second formulation is given only implicitly. Such is already the case in the above representation if we leave $f(M^2)$ unspecified.

The above construction of a free energy may appear to be rather artificial, because the representation (6.4.8) is of very limited practical applicability. Nevertheless, it is important from a mathematical point of view, because it explicitly demonstrates that the functional differential equations mentioned above do indeed admit exact solutions, at least for the demonstrated case. Conditions imposed on the free energy functions that guarantee the existence of such solutions would be of value, and in particular, it would be valuable if, for instance, free energies could be constructed, which would still be of physical relevance, but would not admit a solution of the functional differential equations. In that case non-equivalence of two theories would be demonstrated. We shall not go any deeper into this subject, but will mention one simple and physically important case, in which existence of solutions of equations (6.4.6) or (6.4.7) can easily be established and for which equivalence of the theories compared is guaranteed. What we have in mind is the case for which the free energy functions can be expressed as TAYLOR series expansions about zero deformation, constant temperature and zero electromagnetic fields. The proof to this case will not be outlined here, but from the procedure explained below the reader should be able to construct his own proof.

With these few comments we shall now leave the subject of an exact determination of the functional $\check{\psi}$ from $\hat{\psi}$ and pass on to an approximate satisfaction of relations (6.4.6) and (6.4.7). To this end, we shall choose truncated polynomials as expressions for the free energies. If the degree of the polynomial representation of $\hat{\psi}$ is known, then with (6.4.6) that of $\check{\psi}$ can be determined. In general the order of truncation of $\check{\psi}$ needed to obtain full equivalence is not the same as that for $\hat{\psi}$. We shall not be so general and choose for both energy functionals the same polynomial expressions. In complexity, we shall be as general as we were in (6.2.46) and thus write

$$
\begin{aligned}
\check{\psi} = {}& \frac{1}{2\rho_0} \, {}_2\check{\chi}^{(m)}_{\alpha\beta} \mathbb{B}_\alpha \mathbb{B}_\beta + \frac{1}{4\rho_0} \, {}_4\check{\chi}^{(m)}_{\alpha\beta\gamma\delta} \mathbb{B}_\alpha \mathbb{B}_\beta \mathbb{B}_\gamma \mathbb{B}_\delta + \frac{1}{\rho_0} \check{\chi}^{(em)}_{\alpha\beta} \mathbb{B}_\alpha \mathbb{E}_\beta \\[2mm]
&+ \frac{1}{2\rho_0} \, {}_2\check{\chi}^{(e)}_{\alpha\beta} \mathbb{E}_\alpha \mathbb{E}_\beta + \frac{1}{4\rho_0} \, {}_4\check{\chi}^{(e)}_{\alpha\beta\gamma\delta} \mathbb{E}_\alpha \mathbb{E}_\beta \mathbb{E}_\gamma \mathbb{E}_\delta - \frac{1}{2}\check{c}(\Theta - \Theta_0)^2 \\[2mm]
&+ \frac{1}{\rho_0}\check{\lambda}^{(m)}_\alpha \mathbb{B}_\alpha (\Theta - \Theta_0) + \frac{1}{2\rho_0}\check{L}^{(m)}_{\alpha\beta} \mathbb{B}_\alpha \mathbb{B}_\beta (\Theta - \Theta_0) \\[2mm]
&+ \frac{1}{\rho_0}\check{\lambda}^{(e)}_\alpha \mathbb{E}_\alpha (\Theta - \Theta_0) + \frac{1}{2\rho_0}\check{L}^{(e)}_{\alpha\beta} \mathbb{E}_\alpha \mathbb{E}_\beta (\Theta - \Theta_0) \\[2mm]
&+ \Big\{ \frac{1}{\rho_0}\check{\mathcal{E}}^{(m)}_{\alpha\beta\gamma} \mathbb{B}_\beta + \frac{1}{2\rho_0}\check{b}^{(m)}_{\alpha\beta\gamma\delta} \mathbb{B}_\alpha \mathbb{B}_\beta + \frac{1}{\rho_0}\check{\mathcal{E}}^{(e)}_{\beta\gamma\delta} \mathbb{E}_\beta \\[2mm]
&+ \frac{1}{2\rho_0}\check{b}^{(e)}_{\alpha\beta\gamma\delta} \mathbb{E}_\alpha \mathbb{E}_\beta - \check{\nu}_{\gamma\delta}(\Theta - \Theta_0) \Big\} E_{\gamma\delta} + \frac{1}{2\rho_0}\check{c}_{\alpha\beta\gamma\delta} E_{\alpha\beta} E_{\gamma\delta} ,
\end{aligned}
\tag{6.4.15}
$$

and

$$\hat{\psi} = \frac{\rho_0}{2} {}_2\hat{\chi}_{\alpha\beta}^{(m)} \frac{\mathbb{M}_{\alpha}^L}{\rho_0} \frac{\mathbb{M}_{\beta}^L}{\rho_0} + \frac{\rho_0^3}{4} {}_4\hat{\chi}_{\alpha\beta\gamma\delta}^{(m)} \frac{\mathbb{M}_{\alpha}^L}{\rho_0} \frac{\mathbb{M}_{\beta}^L}{\rho_0} \frac{\mathbb{M}_{\gamma}^L}{\rho_0} \frac{\mathbb{M}_{\delta}^L}{\rho_0}$$

$$+\rho_0\hat{\chi}_{\alpha\beta}^{(em)} \frac{\mathbb{M}_{\alpha}^L}{\rho_0} \frac{\mathbb{P}_{\beta}}{\rho_0}$$

$$+\frac{\rho_0}{2} {}_2\hat{\chi}_{\alpha\beta}^{(e)} \frac{\mathbb{P}_{\alpha}}{\rho_0} \frac{\mathbb{P}_{\beta}}{\rho_0} + \frac{\rho_0^3}{4} {}_4\hat{\chi}_{\alpha\beta\gamma\delta}^{(e)} \frac{\mathbb{P}_{\alpha}}{\rho_0} \frac{\mathbb{P}_{\beta}}{\rho_0} \frac{\mathbb{P}_{\gamma}}{\rho_0} \frac{\mathbb{P}_{\delta}}{\rho_0}$$

$$-\tfrac{1}{2}\hat{c}(\Theta - \Theta_0)^2 + \hat{\lambda}_{\alpha}^{(m)} \frac{\mathbb{M}_{\alpha}^L}{\rho_0} (\Theta - \Theta_0)$$

$$+\frac{\rho_0}{2} \hat{L}_{\alpha\beta}^{(m)} \frac{\mathbb{M}_{\alpha}^L}{\rho_0} \frac{\mathbb{M}_{\beta}^L}{\rho_0} (\Theta - \Theta_0) + \hat{\lambda}_{\alpha}^{(e)} \frac{\mathbb{P}_{\alpha}}{\rho_0} (\Theta - \Theta_0)$$

$$+\frac{\rho_0}{2} \hat{L}_{\alpha\beta}^{(e)} \frac{\mathbb{P}_{\alpha}}{\rho_0} \frac{\mathbb{P}_{\beta}}{\rho_0} (\Theta - \Theta_0) + \left\{ \hat{\varepsilon}_{\beta\gamma\delta}^{(m)} \frac{\mathbb{M}_{\beta}^L}{\rho_0} + \frac{\rho_0}{2} \hat{b}_{\alpha\beta\gamma\delta}^{(m)} \frac{\mathbb{M}_{\alpha}^L}{\rho_0} \frac{\mathbb{M}_{\beta}^L}{\rho_0} \right.$$

$$\left. +\hat{\varepsilon}_{\beta\gamma\delta}^{(e)} \frac{\mathbb{P}_{\beta}}{\rho_0} + \frac{\rho_0}{2} \hat{b}_{\alpha\beta\gamma\delta}^{(e)} \frac{\mathbb{P}_{\alpha}}{\rho_0} \frac{\mathbb{P}_{\beta}}{\rho_0} - \hat{\nu}_{\gamma\delta}(\Theta - \Theta_0) \right\} E_{\gamma\delta}$$

$$+\frac{1}{2\rho_0} \hat{c}_{\alpha\beta\gamma\delta} E_{\alpha\beta} E_{\gamma\delta} \ .$$

(6.4.16)

With these representations full equivalence is not possible, because the transformations indicated by (6.4.6) and (6.4.7) lead to terms, which have been omitted in the formulation of the respective energy functions. Nevertheless, except for these terms equivalence can be established. If (6.4.15) and (6.4.16) are used to exploit (6.4.6) and (6.4.7) in this approximate sense a series of identities can be derived for the phenomenological coefficients of the two formulations. We have done this; the calculations for the derivation of the corresponding relations are very long. Unfortunately, the emerging identities are much too long, and conclusions that can be drawn from them in this full generality are very meagre in order to justify to list them here. Nevertheless, one result derivable from these identities may be quoted. It reads: If the coefficients $_2\hat{\chi}_{\alpha\beta}^{(m)}$, $_2\hat{\chi}_{\alpha\beta}^{(e)}$, $\hat{\chi}_{\alpha\beta}^{(em)}$ vanish, all remaining coefficients accounting for electromagnetic effects must also vanish if the theories are to be equivalent in the above mentioned sense. Otherwise stated, if $_2\hat{\chi}^{(m)}$, $_2\hat{\chi}^{(e)}$ and $\hat{\chi}^{(em)}$ vanish in one theory and the two theories are to be equivalent, the free energies reduce to

$$\psi = \frac{1}{2\rho_0} c_{\alpha\beta\gamma\delta} E_{\alpha\beta} E_{\gamma\delta} - \nu_{\alpha\beta}(\Theta - \Theta_0) E_{\alpha\beta} - \tfrac{1}{2}c(\Theta - \Theta_0)^2 \ ,$$

and no other terms. In a truly polarizable and magnetizable material at least one of the coefficients $_2\chi_{\alpha\beta}^{(m)}$, $_2\chi_{\alpha\beta}^{(e)}$ or $\chi_{\alpha\beta}^{(em)}$ must therefore be non-zero.

In the following we shall exploit the equations (6.4.6) and (6.4.7) for an isotropic body, in which

$$\lambda_\alpha^{(m)} = \lambda_\alpha^{(e)} = 0 \,, \qquad \varepsilon_{\alpha\beta\gamma}^{(m)}\varepsilon_{\alpha\beta\gamma}^{(e)} = 0 \,,$$

$$2\chi_{\alpha\beta}^{(m)} = {}_2\chi^{(m)}\delta_{\alpha\beta} \,, \qquad 2\chi_{\alpha\beta}^{(e)} = {}_2\chi^{(e)}\delta_{\alpha\beta} \,, \qquad \chi_{\alpha\beta}^{(em)} = 0 \,,$$

$$4\chi_{\alpha\beta\gamma\delta}^{(m)} = {}_4\chi^{(m)}\tfrac{1}{3}(\delta_{\alpha\beta}\delta_{\gamma\delta} + \delta_{\alpha\gamma}\delta_{\beta\delta} + \delta_{\alpha\delta}\delta_{\beta\gamma}) \,,$$

$$4\chi_{\alpha\beta\gamma\delta}^{(e)} = {}_4\chi^{(e)}\tfrac{1}{3}(\delta_{\alpha\beta}\delta_{\gamma\delta} + \delta_{\alpha\gamma}\delta_{\beta\delta} + \delta_{\alpha\delta}\delta_{\beta\gamma}) \,,$$

$$L_{\alpha\beta}^{(m)} = L^{(m)}\delta_{\alpha\beta}, \qquad L_{\alpha\beta}^{(e)} = L^{(e)}\delta_{\alpha\beta} \,, \qquad \nu_{\alpha\beta} = \nu\delta_{\alpha\beta} \,,$$

$$b_{\alpha\beta\gamma\delta}^{(m)} = b_1^{(m)}\delta_{\alpha\beta}\delta_{\gamma\delta} + \tfrac{1}{2}b_2^{(m)}(\delta_{\alpha\gamma}\delta_{\beta\delta} + \delta_{\alpha\delta}\delta_{\beta\gamma}) \,,$$

$$b_{\alpha\beta\gamma\delta}^{(e)} = b_1^{(e)}\delta_{\alpha\beta}\delta_{\gamma\delta} + \tfrac{1}{2}b_2^{(e)}(\delta_{\alpha\gamma}\delta_{\beta\delta} + \delta_{\alpha\delta}\delta_{\beta\gamma}) \,,$$

$$c_{\alpha\beta\gamma\delta} = \lambda\delta_{\alpha\beta}\delta_{\gamma\delta} + G(\delta_{\alpha\gamma}\delta_{\beta\delta} + \delta_{\alpha\delta}\delta_{\beta\gamma}) \,. \tag{6.4.17}$$

Substituting these expressions into (6.4.15) and (6.4.16), we obtain

$$\begin{aligned}
\check{\psi} =& \frac{1}{2\rho_0}\,{}_2\check{\chi}^{(m)}\mathbb{B}_\alpha\mathbb{B}_\alpha + \frac{1}{4\rho_0}\,{}_4\check{\chi}^{(m)}(\mathbb{B}_\alpha\mathbb{B}_\alpha)^2 + \frac{1}{2\rho_0}\,{}_2\check{\chi}^{(e)}\mathbb{E}_\alpha\mathbb{E}_\alpha \\
&+ \frac{1}{4\rho_0}\,{}_4\check{\chi}^{(e)}(\mathbb{E}_\alpha\mathbb{E}_\alpha)^2 - \tfrac{1}{2}\check{c}(\Theta - \Theta_0)^2 \\
&+ \frac{1}{2\rho_0}\check{L}^{(m)}\mathbb{B}_\alpha\mathbb{B}_\alpha(\Theta - \Theta_0) + \frac{1}{2\rho_0}\check{L}^{(e)}\mathbb{E}_\alpha\mathbb{E}_\alpha(\Theta - \Theta_0) \\
&+ \frac{1}{2\rho_0}(\check{b}_1^{(m)}\mathbb{B}_\alpha\mathbb{B}_\alpha + \check{b}_1^{(e)}\mathbb{E}_\alpha\mathbb{E}_\alpha)E_{\beta\beta} \\
&+ \frac{1}{2\rho_0}(\check{b}_2^{(m)}\mathbb{B}_\alpha\mathbb{B}_\beta + \check{b}_2^{(e)}\mathbb{E}_\alpha\mathbb{E}_\beta)E_{\alpha\beta} - \check{\nu}(\Theta - \Theta_0)E_{\alpha\alpha} \\
&+ \frac{1}{2\rho_0}\Big(\check{\lambda}(E_{\alpha\alpha})^2 + 2\check{G}E_{\alpha\beta}E_{\alpha\beta}\Big) \,.
\end{aligned} \tag{6.4.18}$$

and

$$\begin{aligned}
\hat{\psi} =& \frac{\rho_0}{2}\,{}_2\hat{\chi}^{(m)}\frac{\mathbb{M}_\alpha^L}{\rho_0}\frac{\mathbb{M}_\alpha^L}{\rho_0} + \frac{\rho_0^3}{4}\,{}_4\hat{\chi}^{(m)}\left(\frac{\mathbb{M}_\alpha^L}{\rho_0}\frac{\mathbb{M}_\alpha^L}{\rho_0}\right)^2 + \frac{\rho_0}{2}\,{}_2\hat{\chi}^{(e)}\frac{\mathbb{P}_\alpha}{\rho_0}\frac{\mathbb{P}_\alpha}{\rho_0} \\
&+ \frac{\rho_0^3}{4}\,{}_4\hat{\chi}^{(e)}\left(\frac{\mathbb{P}_\alpha}{\rho_0}\frac{\mathbb{P}_\alpha}{\rho_0}\right)^2 - \tfrac{1}{2}\hat{c}(\Theta - \Theta_0)^2 \\
&+ \frac{\rho_0}{2}\hat{L}^{(m)}\frac{\mathbb{M}_\alpha^L}{\rho_0}\frac{\mathbb{M}_\alpha^L}{\rho_0}(\Theta - \Theta_0) + \frac{\rho_0}{2}\hat{L}^{(e)}\frac{\mathbb{P}_\alpha}{\rho_0}\frac{\mathbb{P}_\alpha}{\rho_0}(\Theta - \Theta_0) \\
&+ \frac{\rho_0}{2}\left(\hat{b}_1^{(m)}\frac{\mathbb{M}_\alpha^L}{\rho_0}\frac{\mathbb{M}_\alpha^L}{\rho_0} + \hat{b}_1^{(e)}\frac{\mathbb{P}_\alpha}{\rho_0}\frac{\mathbb{P}_\alpha}{\rho_0}\right)E_{\beta\beta} \\
&+ \frac{\rho_0}{2}\left(\hat{b}_2^{(m)}\frac{\mathbb{M}_\alpha^L}{\rho_0}\frac{\mathbb{M}_\beta^L}{\rho_0} + \hat{b}_2^{(e)}\frac{\mathbb{P}_\alpha}{\rho_0}\frac{\mathbb{P}_\beta}{\rho_0}\right)E_{\alpha\beta} \\
&- \hat{\nu}(\Theta - \Theta_0)E_{\alpha\alpha} + \frac{1}{2\rho_0}\Big(\hat{\lambda}(E_{\alpha\alpha})^2 + 2\hat{G}E_{\alpha\beta}E_{\alpha\beta}\Big) \,.
\end{aligned} \tag{6.4.19}$$

With these exploitation of (6.4.6) stays within reasonable effort. The approach is analogous to that demonstrated in Sect. 5.7 and at the beginning of this section. One simply evaluates $\mathbb{E}_\alpha$ and $\mathbb{B}_\alpha$ according to (6.4.4),

$$\mathbb{B}_\alpha = {}_2\hat{\chi}^{(\mathrm{m})}\mathrm{M}_\alpha^L + {}_4\hat{\chi}^{(\mathrm{m})}\mathrm{M}_\beta^L\mathrm{M}_\beta^L\mathrm{M}_\alpha^L + \hat{L}^{(\mathrm{m})}\mathrm{M}_\alpha^L(\Theta - \Theta_0)$$

$$+\hat{\mathrm{b}}_1^{(\mathrm{m})}\mathrm{M}_\alpha^L E_{\beta\beta} + \hat{\mathrm{b}}_2^{(\mathrm{m})}\mathrm{M}_\beta^L E_{\alpha\beta} ,$$

$$\mathbb{E}_\alpha = {}_2\hat{\chi}^{(\mathrm{e})}\mathbb{P}_\alpha + {}_4\hat{\chi}^{(\mathrm{e})}\mathbb{P}_\beta\mathbb{P}_\beta\mathbb{P}_\alpha + \hat{L}^{(\mathrm{e})}\mathbb{P}_\alpha(\Theta - \Theta_0)$$

$$+\hat{\mathrm{b}}_1^{(\mathrm{e})}\mathbb{P}_\alpha E_{\beta\beta} + \hat{\mathrm{b}}_2^{(\mathrm{e})}\mathbb{P}_\beta E_{\alpha\beta} ,$$

(6.4.20)

expresses with their use $\hat{\psi}$ as a function $\mathbb{P}_\alpha/\rho_0$ and $\mathrm{M}_\alpha^L/\rho_0$ rather than $\mathbb{E}_\alpha$ and $\mathbb{B}_\alpha$ and substitutes the emerging relation together with (6.4.16) into (6.4.6). When this is done, the following identities are obtained:

$$2\hat{\chi}^{(\mathrm{m})} = \frac{1}{2\check{\chi}^{(\mathrm{m})}} = \frac{\mu_0\mu}{\mu - \mu_0} = \frac{\mu}{\chi^{(\mathrm{m})}} , \qquad 2\hat{\chi}^{(\mathrm{e})} = -\frac{1}{2\check{\chi}^{(\mathrm{e})}} = \frac{1}{\chi^{(\mathrm{e})}} ,$$

$$4\hat{\chi}^{(\mathrm{m})} = \left(\frac{\mu}{\chi^{(\mathrm{m})}}\right)^4 {}_4\check{\chi}^{(\mathrm{m})} , \qquad 4\hat{\chi}^{(\mathrm{e})} = \left(\frac{1}{\chi^{(\mathrm{e})}}\right)^4 {}_4\check{\chi}^{(\mathrm{e})} ,$$

$$\hat{L}^{(\mathrm{m})} = \left(\frac{\mu}{\chi^{(\mathrm{m})}}\right)^2 \check{L}^{(\mathrm{m})} , \qquad \hat{L}^{(\mathrm{e})} = \left(\frac{1}{\chi^{(\mathrm{e})}}\right)^2 \check{L}^{(\mathrm{m})} ,$$

$$\hat{\mathrm{b}}_{1,2}^{(\mathrm{m})} = \left(\frac{\mu}{\chi^{(\mathrm{m})}}\right)^2 \check{\mathrm{b}}_{1,2}^{(\mathrm{m})} , \qquad \hat{\mathrm{b}}_{1,2}^{(\mathrm{e})} = \left(\frac{1}{\chi^{(\mathrm{e})}}\right)^2 \check{\mathrm{b}}_{1,2}^{(\mathrm{e})} ,$$

$$\hat{\mathrm{c}} = \check{\mathrm{c}} = \frac{\mathrm{c_W}}{\Theta_0} , \qquad \hat{\nu} = \check{\nu} = \nu , \qquad \hat{\lambda} = \check{\lambda} = \lambda , \qquad \hat{G} = \check{G} = G ,$$

(6.4.21)

with obvious inversions, which we shall not write down. In the above relations, μ is called *magnetic permeability*, $\chi^{(\mathrm{m})}$ *magnetic susceptibility* and $\chi^{(\mathrm{e})}$ *electric susceptibility*; for an ideal medium (i.e. a rigid, isotropic body) they are defined by

$$\boldsymbol{B} = \mu\boldsymbol{H} , \quad \boldsymbol{M} = \chi^{(\mathrm{m})}\boldsymbol{H} , \quad \boldsymbol{P} = \chi^{(\mathrm{e})}\boldsymbol{H} .$$

Further, c_W is the specific heat, ν the thermoelastic constant, and λ and G are the LAMÉ constants. In an isotropic body, therefore, the classical phenomenological coefficients are equal and thus clearly defined. This is not so for all other coefficients, as can be seen from (6.4.21). In fact, the transformations all involve the magnetic and electric susceptibility. The definitions of fourth-order electromagnetic, magnetostrictive, electrostrictive, thermoelectric and thermomagnetic coefficients are not unique in this restricted isotropic theory. Care should therefore be observed with the use of specific names for these effects. Note also that the above described transformation becomes simply impossible whenever $2\chi^{(\mathrm{m})}$ and/or $2\chi^{(\mathrm{e})}$ are zero. In this case, the outlined procedure does not lead to a polynomial expression for $\hat{\psi}$ of the form (6.4.19).

In the above only two possible constitutive formulations were investigated, namely those in which $(\mathbb{E}_\alpha, \mathbb{B}_\alpha)$ and $(\mathbb{P}_\alpha/\rho_0, \mathbb{M}_\alpha/\rho_0)$, respectively, were the independent constitutive variables. Of course, similar calculations can also be performed, if other sets of independent variables are chosen. They also lead to results similar to (6.4.21) relating the phenomenological coefficients of the respective constitutive formulations. Below we shall present the transformations of all possibilities of the statistical and the LORENTZ models (IV,V).

On the other hand, the CHU-formulations (models I,II) and the MAXWELL-MINKOWSKI model (III) have been shown to be equivalent if in each of them the same functional dependencies for the free energy are taken. Hence, similar changes can also be performed for these formulations.

In the remainder of this section we shall list the transformation rules for all these changes of dependent and independent constitutive variables; but we shall restrict ourselves to free energies of the complexity (6.4.18). Thus we write

$$
\begin{aligned}
\psi = {} & \frac{1}{2\rho_0}\, 2\chi^{(m)} V_\alpha V_\alpha + \frac{1}{4\rho_0}\, 4\chi^{(m)}(V_\alpha V_\alpha)^2 \\
& + \frac{1}{2\rho_0}\, 2\chi^{(e)} W_\alpha W_\alpha + \frac{1}{4\rho_0}\, 4\chi^{(m)}(W_\alpha W_\alpha)^2 - \tfrac{1}{2}c(\Theta - \Theta_0)^2 \\
& + \frac{1}{2\rho_0} L^{(m)} V_\alpha V_\alpha(\Theta - \Theta_0) + \frac{1}{2\rho_0} L^{(e)} W_\alpha W_\alpha(\Theta - \Theta_0) \\
& - \nu(\Theta - \Theta_0)E_{\alpha\alpha} + \frac{1}{2\rho_0}\left(\mathrm{b}_1^{(m)} V_\alpha V_\alpha + \mathrm{b}_1^{(e)} W_\alpha W_\alpha\right)E_{\beta\beta} \\
& + \frac{1}{2\rho_0}\left(\mathrm{b}_2^{(m)} V_\alpha V_\beta + \mathrm{b}_2^{(e)} V_\alpha V_\beta\right)E_{\alpha\beta} \\
& + \frac{1}{2\rho_0}\left(\lambda(E_{\alpha\alpha})^2 + 2G E_{\alpha\beta}E_{\alpha\beta}\right).
\end{aligned}
\tag{6.4.22}
$$

Here, (V_α, W_α) stands for the respective pairs of electromagnetic variables which will be chosen as independent variables. In particular, the following choices will be made:

in the models I, II, III		*in the models IV, V*	
$(V_\alpha, W_\alpha) = (\mathbb{E}_\alpha, \mathbb{H}_\alpha):$	$(\overset{\smile}{\cdot})$	$(V_\alpha, W_\alpha) = (\mathbb{E}_\alpha, \mathbb{B}_\alpha):$	$(\overset{\cdot\cdot}{\cdot})$
$(\mathbb{P}_\alpha, \mathbb{M}_\alpha^C):$	$(\overset{\wedge}{\cdot})$	$(\mathbb{P}_\alpha, \mathbb{M}_\alpha^L):$	$(\overset{\cdot}{\cdot})$
$(\mathbb{E}_\alpha, \mathbb{M}_\alpha^C):$	$(\overset{\approx}{\cdot})$	$(\mathbb{E}_\alpha, \mathbb{M}_\alpha^L):$	$(\overset{\sim}{\cdot})$
$(\mathbb{P}_\alpha, \mathbb{H}_\alpha):$	$(\overset{++}{\cdot})$	$(\mathbb{P}_\alpha, \mathbb{B}_\alpha):$	$(\overset{+}{\cdot})$

Correspondingly, ψ in (6.4.22) is the energy functional of the formulation at hand and must be characterized for each of these. The symbols used are also indicated in the above table; clearly, they must also be applied in all coefficients on the right-hand side of (6.4.22). Recall, further, that $\check{\psi}$ and $\hat{\psi}$, for instance, are not the same energy functionals, but they are related to each other by LEGENDRE transformations.

We shall now list all the equivalence relations for the various formulations:

(A): The transformations from the $(\check{\cdot})$- to the $(\hat{\cdot})$-formulation have been listed in (6.4.21) and are labeled with the symbol (A).

(B): The transformations between the constitutive theories $(\hat{\cdot})$ and $(\tilde{\cdot})$ are based on the energy functionals

$$
\begin{aligned}
\hat{\psi} &= U - \eta\Theta + \frac{\mathbb{M}_\alpha^L}{\rho_0}\mathbb{B}_\alpha \ , \\
\tilde{\psi} &= U - \eta\Theta + \frac{\mathbb{M}_\alpha^L}{\rho_0}\mathbb{B}_\alpha - \frac{\mathbb{P}_\alpha}{\rho_0}\mathbb{E}_\alpha \ ,
\end{aligned}
\tag{6.4.23}
$$

and equivalence follows, if the following identities hold

$$
\left.
\begin{aligned}
{}_2\tilde{\chi}^{(m)} &= {}_2\hat{\chi}^{(m)} = \frac{\mu}{\chi^{(m)}} \ , & {}_2\tilde{\chi}^{(e)} &= -\frac{1}{2\hat{\chi}^{(e)}} = -\chi^{(e)} \ , \\
{}_4\tilde{\chi}^{(m)} &= {}_4\hat{\chi}^{(m)} \ , & {}_4\tilde{\chi}^{(e)} &= (\chi^{(e)})^4\,{}_4\hat{\chi}^{(e)} \ , \\
\tilde{L}^{(m)} &= \hat{L}^{(m)} \ , & \tilde{L}^{(e)} &= (\chi^{(e)})^2\hat{L}^{(e)} \ , \\
\tilde{b}_{1,2}^{(m)} &= \hat{b}_{1,2}^{(m)} \ , & \tilde{b}_{1,2}^{(e)} &= (\chi^{(e)})^2\hat{b}_{1,2}^{(e)} \ , \\
\tilde{c} &= \hat{c} = \frac{c_w}{\Theta_0} \ , & \tilde{\lambda} &= \hat{\lambda} = \lambda \ , \\
\tilde{\nu} &= \hat{\nu} = \nu \ , & \tilde{G} &= \hat{G} = G \ ,
\end{aligned}
\right\}
\quad \text{(B) (6.4.24)}
$$

(C): To relate the $(\tilde{\cdot})$- and $(\overset{+}{\cdot})$-formulations one must start with the energy functionals

$$
\tilde{\psi} = U - \eta\Theta + \frac{\mathbb{M}_\alpha^L}{\rho_0}\mathbb{B}_\alpha - \frac{\mathbb{P}_\alpha}{\rho_0}\mathbb{E}_\alpha \ , \quad \overset{+}{\psi} = U - \eta\Theta \ ,
\tag{6.4.25}
$$

and then obtains the equivalence relations

$$
\left.
\begin{aligned}
&2\overset{+}{\chi}{}^{(m)} = -\frac{1}{2\tilde{\chi}^{(m)}} = -\frac{\chi^{(m)}}{\mu}\,,
&&2\overset{+}{\chi}{}^{(e)} = -\frac{1}{2\tilde{\chi}^{(e)}} = \frac{1}{\chi^{(e)}}\,, \\[4pt]
&4\overset{+}{\chi}{}^{(m)} = \left(\frac{\chi^{(m)}}{\mu}\right)^4 4\tilde{\chi}^{(m)}\,,
&&4\overset{+}{\chi}{}^{(e)} = \left(\frac{1}{\chi^{(e)}}\right)^4 4\tilde{\chi}^{(e)} \\[4pt]
&\overset{+}{L}{}^{(m)} = \left(\frac{\chi^{(m)}}{\mu}\right)^2 \tilde{L}^{(m)}\,,
&&\overset{+}{L}{}^{(e)} = \left(\frac{1}{\chi^{(e)}}\right)^2 \tilde{L}^{(e)}, \\[4pt]
&\overset{+}{b}{}_{1,2}^{(m)} = \left(\frac{\chi^{(m)}}{\mu}\right)^2 \tilde{b}_{1,2}^{(m)}\,,
&&\overset{+}{b}{}_{1,2}^{(e)} = \left(\frac{1}{\chi^{(e)}}\right)^2 \tilde{b}_{1,2}^{(e)}\,, \\[4pt]
&\overset{+}{c} = \tilde{c} = \frac{c_w}{\Theta_0}\,,
&&\overset{+}{\lambda} = \tilde{\lambda} = \lambda\,, \\[4pt]
&\overset{+}{\nu} = \tilde{\nu} = \nu\,,
&&\overset{+}{G} = \tilde{G} = G\,,
\end{aligned}
\right\} \quad \text{(C) (6.4.26)}
$$

(D): The energy functionals in the $(\overset{+}{\cdot})$- and $(\overset{\vee}{\cdot})$-formulations have already been stated before, namely in (6.4.1) and (6.4.25)$_2$. Exploiting the identity

$$
\overset{\vee}{\psi} = \overset{+}{\psi} - \frac{\mathbb{P}_\alpha}{\rho_0}\mathbb{E}_\alpha\,,
$$

we obtain

$$
\left.
\begin{aligned}
&2\check{\chi}^{(m)} = 2\overset{+}{\chi}{}^{(m)} = -\frac{\mu}{\chi^{(m)}}\,,
&&2\check{\chi}^{(e)} = -\frac{1}{2\overset{+}{\chi}{}^{(e)}} = -\chi^{(e)}\,, \\[4pt]
&4\check{\chi}^{(m)} = 4\overset{+}{\chi}{}^{(m)}\,,
&&4\check{\chi}^{(e)} = (\chi^{(e)})^4\, 4\overset{+}{\chi}{}^{(e)}\,, \\[4pt]
&\check{L}^{(m)} = \overset{+}{L}{}^{(m)}\,,
&&\check{L}^{(e)} = (\chi^{(e)})^2\, \overset{+}{L}{}^{(e)}\,, \\[4pt]
&\check{b}_{1,2}^{(m)} = \overset{+}{b}{}_{1,2}^{(m)}\,,
&&\check{b}_{1,2}^{(e)} = (\chi^{(e)})^2\, \overset{+}{b}{}_{1,2}^{(e)}\,, \\[4pt]
&\check{c} = \overset{+}{c} = \frac{c_w}{\Theta_0}\,,
&&\check{\lambda} = \overset{+}{\lambda} = \lambda\,, \\[4pt]
&\check{\nu} = \overset{+}{\nu} = \nu\,,
&&\check{G} = \overset{+}{G} = G\,,
\end{aligned}
\right\} \quad \text{(D) (6.4.27)}
$$

With these relations the constitutive theories of models IV and V are all compared. Performing the transformations (A), (B), (C) and (D) must lead to identical free energies. There remains to compare the constitutive theories for models I, II and III.

(AA): We begin with the transformations for the $(\check{})$- and $(\hat{})$-formulations. In these (see (5.2.25) and (6.2.45)) we have

$$\check{\psi} = U - \eta\Theta - \frac{\mathbb{P}_\alpha}{\rho_0}\mathbb{E}_\alpha - \frac{\mu_0 \mathbb{M}_\alpha^C}{\rho_0}\mathbb{H}_\alpha \,,$$

$$\hat{\psi} = U - \eta\Theta. \tag{6.4.28}$$

Note, however, that here $U = {}^{II}U$, whereas in all the foregoing relations one must read ${}^{V}U$ for U. Comparing for both formulations the energy functionals of the complexity (6.4.22) we obtain the following relations:

$$\left.\begin{aligned}
{}_2\hat{\chi}^{(m)} &= -\frac{\mu_0^2}{{}_2\check{\chi}^{(m)}} = \frac{\mu_0}{\chi^{(m)}} \,, & {}_2\hat{\chi}^{(e)} &= -\frac{1}{{}_2\check{\chi}^{(e)}} = \frac{1}{\chi^{(e)}} \,, \\
{}_4\hat{\chi}^{(m)} &= \left(\frac{1}{\chi^{(m)}}\right)^4 {}_4\check{\chi}^{(m)} \,, & {}_4\hat{\chi}^{(e)} &= \left(\frac{1}{\chi^{(e)}}\right)^4 {}_4\check{\chi}^{(e)} \,, \\
\hat{L}^{(m)} &= \left(\frac{1}{\chi^{(m)}}\right)^2 \check{L}^{(m)} \,, & \hat{L}^{(e)} &= \left(\frac{1}{\chi^{(e)}}\right)^2 \check{L}^{(e)} \,, \\
\hat{b}_{1,2}^{(m)} &= \left(\frac{1}{\chi^{(m)}}\right)^2 \check{b}_{1,2}^{(m)} \,, & \hat{b}_{1,2}^{(e)} &= \left(\frac{1}{\chi^{(e)}}\right)^2 \check{b}_{1,2}^{(e)} \,, \\
\hat{c} &= \check{c} = \frac{c_w}{\Theta_0} \,, & \hat{\lambda} &= \check{\lambda} = \lambda \,, \\
\hat{\nu} &= \check{\nu} = \nu \,, & \hat{G} &= \check{G} = G \,,
\end{aligned}\right\} \text{(AA)} \quad (6.4.29)$$

(BB), (CC), (DD): It is now obvious how the transformations must look like for the remaining formulations. For the electrical coefficients the transformations (AA), (BB), (CC), (DD) are identical to the corresponding ones indicated with a single letter, whereas for the magnetic coefficients we must only replace the factor $(\mu/\chi^{(m)})$ in the one-letter transformations by $(1/\chi^{(m)})$ to obtain the double-letter transformations.

What remains is then the transformation of one formulation out of the group (IV, V) to one of (I, II, III). This problem has already been solved in Sect. 5.7 in which the transformation from the $(\cdot)$- to the $(\check{})$-formulation is discussed. (There, $\check{\psi}$ was denoted by $\bar{\psi}$.) For completeness we shall repeat the results here. We shall characterize this transformation, which makes the link between the two distinct groups (IV, V) and (I, II, III) and, therefore, completes the comparison of all possible constitutive theories, with the symbol A. It reads

(A):

$$2\overset{\smallsmile}{\chi}{}^{(m)} = \mu_0\mu\,_2\overset{\smallsmile}{\chi}{}^{(m)} = -\mu_0\chi^{(m)} \qquad\qquad 2\overset{\smallsmile}{\chi}{}^{(e)} = \,_2\overset{\smallsmile}{\chi}{}^{(e)} = -\chi^{(e)} ,$$

$$2\overset{\smallsmile}{\chi}{}^{(m)} = \mu^4\,_4\overset{\smallsmile}{\chi}{}^{(m)} , \qquad\qquad\qquad 4\overset{\smallsmile}{\chi}{}^{(e)} = \,_4\overset{\smallsmile}{\chi}{}^{(e)} ,$$

$$\overset{\smallsmile}{L}{}^{(m)} = \mu^2\overset{\smallsmile}{L}{}^{(m)} , \qquad\qquad\qquad\quad \overset{\smallsmile}{L}{}^{(e)} = \overset{\smallsmile}{L}{}^{(e)} ,$$

$$\overset{\smallsmile}{\mathrm{b}}{}^{(m)}_{\alpha\beta\gamma\delta} = \mu^2\overset{\smallsmile}{\mathrm{b}}{}^{(m)}_{\alpha\beta\gamma\delta} \qquad\qquad\qquad \overset{\smallsmile}{\mathrm{b}}{}^{(e)}_{\alpha\beta\gamma\delta} = \overset{\smallsmile}{\mathrm{b}}{}^{(e)}_{\alpha\beta\gamma\delta} ,$$

$$+\mu_0\chi^{(m)}(2+\chi^{(m)})\mathrm{n}_{\alpha\beta\gamma\delta} ,$$

$$\overset{\smallsmile}{\mathrm{c}} = \overset{\smallsmile}{\mathrm{c}} = \frac{c_W}{\Theta_0} , \qquad\qquad\qquad\quad \overset{\smallsmile}{\lambda} = \overset{\smallsmile}{\lambda} = \lambda ,$$

$$\overset{\smallsmile}{\nu} = \overset{\smallsmile}{\nu} = \nu , \qquad\qquad\qquad\qquad \overset{\smallsmile}{G} = \overset{\smallsmile}{G} = G ,$$

$$\left.\phantom{\begin{matrix}1\\2\\3\\4\\5\\6\\7\end{matrix}}\right\} \text{(A)} \quad (6.4.30)$$

This completes the explicit comparison of all possible constitutive theories in the formulations (IV, V) and (I, II, III) as far as the free energy goes. A complete comparison must, however, also include the constitutive relations for the electric current and the heat flux vector. They must in all formulations be the same functions. For relationships of the complexity of Sect. 5.2.3 this means that relations must hold such that (see (6.2.66))

$$\Lambda_{\alpha\beta} = \overset{\smallsmile}{\Lambda}_{\alpha\beta}\left(E_{\gamma\delta}, \mathbb{B}_\gamma, \Theta\right) = \hat{\Lambda}\left(E_{\gamma\delta}, \frac{\mathrm{M}^L_\gamma}{\rho_0}, \Theta\right) = \hat{\Lambda}(E_{\gamma\delta}, -\frac{\partial\overset{\smallsmile}{\psi}}{\partial\mathbb{B}_\gamma}, \Theta) . \quad (6.4.31)$$

These relations can be elaborated in an analogous way as above and thus lead to relations between coefficients as, for instance, $\overset{\smallsmile}{\Lambda}^{(d)}_{\alpha\beta\gamma\delta}$ and $\hat{\Lambda}^{(d)}_{\alpha\beta\gamma\delta}$ etc. These are, however, at most second order effects, and for most practical problems linear relationships of the form

$$Q_\alpha(= Q^S_\alpha) = -\kappa_{\alpha\beta}\Theta_{,\beta} + \beta_{\alpha\beta}\mathbb{E}_\beta ,$$

$$J_\alpha = \frac{1}{\Theta}\beta_{\alpha\beta}\Theta_{,\beta} + \sigma_{\alpha\beta}\mathbb{E}_\beta , \qquad\qquad (6.4.32)$$

suffice. In (6.4.32), now $\kappa_{\alpha\beta}$, $\beta_{\alpha\beta}$ and $\sigma_{\alpha\beta}$ are constants which correspond to the tensor of heat conduction, the tensor of electrical conductivity and the tensor of the thermoelectric effect. Since the temperature gradient and the LAGRANGEan electromagnetic field are the same variables in all formulations, it suffices for a full equivalence of all theories to choose for $\kappa_{\alpha\beta}$, $\beta_{\alpha\beta}$, $\sigma_{\alpha\beta}$ the same numerical values.

6.5 Discussion

In this chapter, the governing equations of thermoelastic polarizable and magnetizable solids were subjected to a decomposition procedure, which allowed a separation of the general problem, described mathematically by a set of nonlinear partial differential equations, into two simpler problems. One of these involves as unknowns only small quantities, so that all products of these could justly be omitted. The emerging set of equations was linearized this way. We saw that linearization by itself turned out to be a fairly complex problem if not attacked with the proper formulation. The LAGRANGEan description treated in Chap. 5 provided the appropriate vehicle to avoid all complexities which arose in the EULERian description.

The decomposition of the governing field equations into equations valid for the intermediate state and the perturbation state was performed for the LORENTZ model only, because all theories were proved to be equivalent. Hence, and if one so desires, all problems can be solved just with this formulation. However, because all formulations are used in the current literature, the transformations of the phenomenological parameters from one theory to another one should be known, if the link between all these theories should explicitly be possible. With the results derived in Sects. 5.7 and 6.4 this can easily be achieved for the restricted constitutive class investigated there. All necessary informations to perform such transformations are contained in Fig. 6.1, in which the two circles stand for models (I, II, III) and (IV, V), respectively. At the periphery, we show those electromagnetic field variables, which are considered as independent electromagnetic fields of the constitutive theory. The symbols used in these constitutive theories are also indicated in the figure. In Sect. 6.4 we derived the transformation rules necessary for the respective constitutive formulations to lead to equivalent electromagnetic theories. The transformations as listed in Sect. 6.4 are indicated in Fig. 6.1 by an arrow on the periphery connecting the two formulations to which they apply. The arrow is also characterized by a symbol (A), (B), ..., (DD), indicating the transformation rules derived in Sect. 6.4 and valid for the transfer

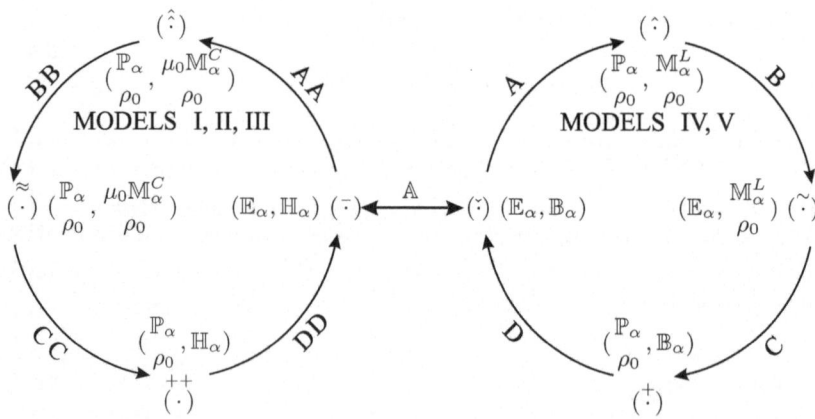

Fig. 6.1. Scheme for the transformation of a constitutive formulation of any model into any other one

of a formulation into the neighboring one on the periphery of the circle. It is evident that to every constitutive theory within a circle any other one corresponds. The missing piece to obtain full correspondence of all presented formulations is the double arrow connecting both circles and characterized by the symbol $(\mathbb{A})$. The transformation from the $(\mathbb{E}_\alpha, \mathbb{H}_\alpha)$-formulation in models (I, II, III) to the $(\mathbb{E}_\alpha, \mathbb{B}_\alpha)$-formulation in models (IV, V) was derived in Sect. 5.7 and recapitulated in Sect. 6.4. It allows a connection between any-one of the constitutive theories in the models (I, II, III) with any other one in the models (IV, V). With this result the main objective is achieved.

We must emphasize once more, however, that conditions of equivalence were derived above for free energy functionals, which were regarded to be truncated polynomial expansions and that equivalence was established to within terms omitted in the respective expansion procedures. Moreover, for reasons of transparency in the presentation and in order to avoid long manipulations, we restricted ourselves to *isotropic* solids for which equivalence conditions turned out to be simple formulas relating the phenomenological coefficients. Of course, the calculations can be performed also for an anisotropic solid, but they are very long and, furthermore, equivalence formulas are complex.

Apart from its practical significance the study of the equivalence conditions has led us to a deeper theoretical understanding at least in the following regard: It is known that the phenomenological constants of the free energy as given in (6.4.15) and (6.4.16) bear standard names. These are as follows:

magnetic constants	$2\chi^{(m)}_{\alpha\beta}$
electric constants	$2\chi^{(e)}_{\alpha\beta}$
electromagnetic coupling constants	$2\chi^{(em)}_{\alpha\beta}$
electric, magnetic anisotropy constants	$4\chi^{(e)}_{\alpha\beta\gamma\delta},\ 4\chi^{(m)}_{\alpha\beta\gamma\delta}$
pyroelectric, pyromagnetic constants	$\lambda^{(e)}_{\alpha},\ \lambda^{(m)}_{\alpha}$
thermoelectric, thermomagnetic constants	$L^{(e)}_{\alpha\beta},\ L^{(m)}_{\alpha\beta}$
piezoelectric, piezomagnetic constants	$\varepsilon^{(e)}_{\alpha\beta\gamma},\ \varepsilon^{(m)}_{\alpha\beta\gamma}$
thermoelastic constants	$\nu_{\alpha\beta}$
thermal constant	c
elastic constants	$c_{\alpha\beta\gamma\delta}$

The association of names with a certain phenomenological constant is not unique, however. In particular, we know that dependent on the constitutive approach the magnetic constants are called magnetic permeability or magnetic susceptibility. Moreover, we saw that for an isotropic body, in which there are no electromagnetic coupling terms and no pyro- and piezoelectric and -magnetic effects, only the last three constants in the above list turned out to remain the same in all formulations. The transformation rules of all other constants involved the magnetic or the electric constants so that the definition of a particular effect depended on the choice of the constitutive variables and on the formulation. For anisotropic bodies the situation becomes even worse, and indeed, one can show that in a body, in which the third-order constants $e_{\alpha\beta\gamma}$ do not all vanish, not even the elastic coefficients are unique. Similarly, in a material with nonvanishing λ_α's thermoelastic constants and the specific heats are not uniquely determinable either. Theoretically, these results are very important ones, because they say, for instance, that in an anisotropic piezomagnetic body the elasticity coefficients depend on the formulation used. Strictly this means that in such a material the elasticity coefficients cannot be determined from an experiment in the absence of electromagnetic fields. Practically, on the other hand, the corrections of the field free elasticity coefficients are so small that the effect of the piezomagnetic constants on them can always be neglected.

The constitutive theories developed in this monograph apply to all bodies which are called magnetizable and polarizable thermoelastic solids. By setting the appropriate phenomenological constants to zero, all special cases of it can also be derived. If, for instance, all coefficients bearing the superscript (m) vanish, the body is called polarizable-only or dielectric. If on the other hand, all coefficients with a superscript (e) vanish, the body is magnetizable-only. This definition is independent of the fact whether electrical conduction is present or not. Often, dielectric substances are electric insulators, however.

Before we close, we would like to draw the readers attention on a few practical problems to which the theories presented in this book and in particular the decomposition procedure of this tractate may be applied. An extensive monograph on such problems is F.C. MOON's "*Problems in Magneto-Solid Mechanics*", [155], so that it suffices if we point out some of the more interesting problems. In principle the problems can be classified into two groups: first, dynamical problems such as magnetoelastic wave propagation and vibration problems, and second, magnetically and electrically induced bifurcation problems. Both classes of problems have been attacked by several authors already, but so far this has only been done in the EULERian description. Early applications to vibration and wave propagation are by DUNKIN and ERINGEN [62], KALISKI and PETYKIEWICZ [105], PARIA [174], ALERS and FLEARY [11] and M.F. McCARTHY [140]. The linearization procedure of these authors is, however, only sketchy and, as far as dynamical equations go the theories are quasi-static. A first attempt to investigate the wave propagation problem with a linearized theory which proceeds along the lines of this article is due to HUTTER, [93, 94]. But his equations are based on those of HUTTER and PAO [91], which have been found not to be completely correct for the reasons explained in Sect. 6.1. Hence, the influence of (polarization and) magnetization on the propagation of waves in solids should still further be attacked, and experiments should be performed, which would corroborate or disprove the theoretical predictions. This has already be done to a certain extent by MOON and CHATTOPADHYAY [154], but an extensive comparison with the results in [93, 94] was not attempted.

Of similar nature are the vibration problems treated by TIERSTEN [234] and VAN DE VEN [248]. Both authors use a linearization procedure equivalent to that presented above, but using the EULERian description and only for a quasi-static theory (in [249], however, it was shown that the results for a dynamic, non-relativistic theory correspond to those of the quasi-static case). TIERSTEN determines the eigenmodes of magnetically saturated plates of cubic Yttrium-Iron-Garnet in a large static transverse magnetic field, perpendicular to which a small time-periodic field is superimposed. VAN DE VEN, on the other hand, discussed vibrations of circular cylinders.

First applications in the bifurcation of electro- and magnetoelasticity trace back to 1967 when MOON and PAO [152] presented experimental and first theoretical results on the buckling of a soft ferromagnetic plate in a homogeneous magnetic field. The discrepancies of the theory and of the experimental evidence initiated further work, notably by WALLERSTEIN and PEACH [260], POPELAR [184], DALRYMPLE, PEACH and VLIEGELAHN [51], PAO and YEH [170] and ALBLAS [10], but with the exception of [170, 10], these articles do not make use of a proper linearization scheme. ALBLAS [10], includes in his analysis the buckling of circular rods based on a LIAPUNOV approach. Since in all of the aforementioned articles the plate boundary conditions were not derived consistently, the entire matter was reinvestigated by VAN DE VEN [250], who also compared the existing formulations of static magnetoelasticity, in particular the MAXWELL–MINKOWSKI and the AMPÈRE-current model, which in the static theory agrees with the LORENTZ model. It turns out that, ultimately all models lead to the same buckling values, as they must; however, the reasons for this buckling can in each case be interpreted differently:

(i) in the MAXWELL–MINKOWSKI formulation buckling is due to distributed magnetic surface forces at the upper and lower surface of the plate and a shear force per unit length of magnetic origin at the boundary of the plate;

(ii) in the AMPÈREan current model, on the other hand, buckling originates from distributed surface moments of magnetic origin at the upper and lower surface of the plate; the boundary is now free of forces.

The difference in the loading could be traced back to differences in the constitutive equations for the shear forces and bending moments in the plate.

Other interesting bifurcation problems are the buckling of superconducting rings and coils which carry a large electric current. A first step towards a solution was done by MOON and CHATTOPADHYAY [153], but further investigations are still needed to improve their solution. A contribution to this field of research is due to ALBLAS. These problems are of extreme practical interest in future reactor technology.

Applications Magnetoelastic (In)stability and Vibrations Electrorheological Fluids

7 Magnetoelastic (In)stability and Vibrations

7.1 Introduction

In modern technological equipments, such as fusion reactors, magnetic energy storage devices, MRI-scanners, superconducting generators, and magnetically levitated trains, huge magnetic fields occur. These fields are generated by high currents through superconducting coils. In these devices, the superconducting currents are so high that the coils are subjected to strong magnetic forces. These forces can cause unwanted vibrations (resulting in loud unpleasant noise as in MRI-scanners) or even collapse (buckling) of structures (coils) in e.g. fusion reactors. Therefore, in the design of these devices the analysis of the vibrations and the stability of magnetic and superconducting structural elements due to electromagnetic forces plays an important role.

In this chapter, we consider the (in)stability of (systems of) ferromagnetic bodies placed in an external magnetic field and of superconducting structures loaded by Lorentz forces due to the electric currents in these conductors. Since a stability problem is always an essentially nonlinear problem, the theory for it must be built upon a nonlinear set of equations for a magnetoelastic model. As seen in the first part of this book, several such models exist and thus one specific model must be chosen first. After that the general approach to the problem could run as follows: the final deformed state is considered as a perturbation of an intermediate state, for which in general the rigid-body state may be taken, and the fields in the deformed state are linearized with respect to the intermediate state. When the resulting homogeneous linear system has a non-trivial solution, we say that the intermediate state is unstable.

We start this chapter with a historical review of magnetoelastic buckling problems, in which we recapitulate some earlier results from the literature. Section 7.3 deals with ferromagnetic systems. We introduce both the so-called *classical method* and a *variational method*. Both methods are illustrated by examples dealing with a cantilevered beam of (narrow) rectangular or elliptic cross-section and a set of two parallel rods. In Sect. 7.4 the buckling of superconducting structures is treated. A variational method is introduced and illustrated by examples such as sets of two parallel rods or two concentric or parallel rings. The results are compared with the results of the so-called *direct* BIOT–SAVARD *method*. It turns out that the latter method delivers a lower bound for the critical buckling current, which however can deviate

K. Hutter et al.: *Electromagnetic Field Matter Interaction in Thermoelastic Solids and Viscous Fluids*, Lect. Notes Phys. **710**, 201–278 (2006)
DOI 10.1007/3-540-37240-7_7 © Springer-Verlag Berlin Heidelberg 2006

substantially from the value obtained by the more exact variational method, when the slender rod-like superconductors are too close to each other. More results for superconducting structures such as helical and spiral coils are presented in Sect. 7.5. In the latter section a somewhat modified method is proposed in which the law of BIOT and SAVARD is used to construct an admissible magnetic field for the variational method; we call this approach the *combined method*. Finally, Sect. 7.6 deals with magnetoelastic vibrations of magnetoelastic or superconducting systems. Eigenfrequencies are determined both by a direct method (comparable with the classical method for buckling) and by a variational method (a generalisation of RAYLEIGH's principle to include magnetoelastic interactions).

7.2 Historical Review of Magnetoelastic Buckling Problems

Magnetoelastic buckling is a phenomenon in which an elastic structure becomes unstable (buckles) under electromagnetic loading. Such a structure can be, for instance, ferromagnetic or (super)conducting. The first investigations of technical relevance in this field are those of MOON [156]. He considered both ferromagnetic and conducting systems, the latter in cooperation with CHATTOPADHYAY; see [152, 40]. A more fundamental theory of magnetoelastic stability was presented by ALBLAS in [9]. In this respect, also the works of ERINGEN [69] and of GOUDJO and MAUGIN [76], who investigated the instability of ferromagnetic plates, should be mentioned. This subject was also studied by VAN DE VEN in [250].

Of more recent date is the paper of ZHOU, ZHENG and MIYA [282], who looked at magnetoelastic buckling of ferromagnetic plates with regard to the safety of the first walls, or blankets, of a fusion reactor. These plates are of ferritic stainless steel, and thus ferromagnetic, and the interactions with the magnetic fields in the reactor are so strong that magnetoelastic buckling comes into sight. In 1997, ZHOU and ZHENG [283] examined the magnetoelastic instability and the increase of natural frequency of a ferromagnetic plate in a magnetic field. They developed a variational formalism by use of which they derived one general expression for the magnetic force that covers both the stability and the frequency problem at one time. YANG [273] considered a special subject, namely the buckling of a piezoelectric plate; he found a buckling load that was greater than the purely mechanical buckling load, implying that neglect of piezoelectric coupling would yield a conservative estimate of the buckling load.

MOON and PAO [156] studied the buckling problem of a cantilevered ferromagnetic beam of narrow rectangular cross-section placed in a transverse magnetic field B_0. They found the magnitude of the buckling field B_{0cr} to be proportional to the (3/2)-power of the thickness-to-length ratio. This result, however, was in disagreement with their own experimental results. This discrepancy started a discussion in the literature between several authors notably among WALLERSTEIN, O'PEACH, BAST, POPELAR, DALRYMPLE, and MIYA et al.; for the latter, see [143]–[145]; for a further list of references, see [252]. Many different approaches were tried, but they did not really result in a much better agreement. The main reason for this disagreement seems to lie in the assumption of an infinite width of the beam-plate. Although the beam-plate was much wider than high, the infinite assumption did yield unacceptable results. Ultimately, a good explanation was found by VAN DE VEN [252, 253], who developed an analysis accounting for the finite width of the rectangular cross-section. The main lines and results of this analysis are presented in the Example in Sect. 7.3.1. YABUNO [271] investigated the bifurcation in buckling of a beam subjected to an electromagnetic force and he proposed a control method to stabilize the magnetoelastic buckling.

TANI and OTOMO, at an IUTAM-conference in Paris 1983, presented a paper on the magnetoelastic buckling of two nearby ferromagnetic panels [231]. As a follow-up, TANI and VAN DE VEN et al. cooperated on the magnetoelastic buckling of two parallel rods. They found a good correspondence between their theoretical and experimental results (obtained in the laboratory of TANI in Sendai, Tohoku University); see [255] or Sect. 7.3.3. This set of two parallel rods will serve as a standard example throughout this chapter.

AMBARTSUMIAN and his coworkers reported on the magnetoelastic interactions in magnetic and current-carrying plates and shells placed in external magnetic fields; for a review of their earlier work, see [12]. Especially the magnetoelastic stability of current-carrying plates and shells was intensively studied in the school of AMBARTSUMIAN,AMB2, [14]; in this respect, we also refer to the work of K.B. and R.A. KAZARIAN [109] and OVAKIMIAN [169] and, of more recent date, MOL'CHENKO [148, 149]. Vibrations and stability of two-layered magnetostrictive and of superconducting plates were studied by BAGDASARIAN et al. [18, 19].

In 1975, CHATTOPADHYAY and MOON [40] were the first to give a closed form solution for the buckling problem of a current-carrying elastic rod in its own field. The stability of conducting strings in a parallel magnetic field were studied by WOLFE [266, 267], and by NOWACKI [164]. CHATTOPADHYAY [41] showed, by numerical means, that a superconducting circular coil in its own field is always stable, a result confirmed by VAN DE VEN and COUWENBERG in [254] via an analytical solution. In MOON's book [156, Ch.5 and 6], a number of other related problems is presented, e.g. circular coils in transverse or toroidal external fields.

For more realistic problems, however, structural problems of greater complexity must be studied. The first example is [84] in which HARA and MOON studied the internal buckling of superconducting solenoid magnets. A special nomination here deserves the work of the group around KENZO MIYA at the University of Tokyo, a.o. [143, 226, 162, 280, 279], who investigated, over a period of more than two decades, theoretically and numerically (finite element analysis) as well as experimentally the stability of ferromagnetic plates and superconducting coils. Their research was especially directed to the design of fusion reactors, specifically with regard to the stability aspects; see [147] and [56] for two survey papers. MIYA presented already in 1982 together with TAKAGI et al. [146] a finite element analysis for the buckling of an eight-coil superconducting full torus. Moreover, together with ZHOU and ZHENG, he analysed the stability of a superconducting three-coil torus in [280, 279] and of a superconducting helix [284]. Impulsive buckling of cylindrical shells was experimentally investigated in cooperation with NEMOTO et al. in [162]. In [282], ZHOU, ZHENG and MIYA investigated the stability of the so-called *first walls* in a fusion reactor. By means of a variational principle the magnetoelastic buckling value for these ferromagnetic plates was found.

In our view, several of the approaches to magnetoelastic buckling mentioned above are rather ad hoc. Moreover, often a number of more or less ad hoc assumptions are needed to keep the analysis in hand. Therefore, we felt a need for a unified approach. To this end, VAN DE VEN with his coworkers LIESHOUT, RONGEN and SMITS constructed a variational principle specifically suited for the solution of magnetoelastic buckling problems for ferromagnetic and superconducting systems [121, 122, 123]. They opted for a variational method in the hope that this would serve as

1. a unified method in a clear formulation;
2. a straight way to a direct solution for the exact buckling value, if possible, or otherwise;
3. a solid basis for numerical computations of the buckling value if an exact analytical solution is no longer possible.

Thus, VAN DE VEN et al. designed a standard tool for this class of problems. In a series of papers they presented solutions for systems reaching from rather simple ones, such as sets of two or more parallel rods [122, 124], or rings [218] to complex ones such as helical or spiral coils [257]. Evident advantages of this method are

1. once a definitive form for the variational principle is chosen, the remaining analysis can be done in an exact way by mere analytical means;
2. whenever the principle is used in an approximate sense, the order of the approximations and the conditions for which they are allowed can be explicitly specified.

In the following sections, we shall present the derivation of this variational principle and its application to a number of magnetoelastic buckling problems.

7.3 Ferromagnetic Systems

7.3.1 Classical Method

In this section, we consider the (in)stability of (systems of) ferromagnetic bodies placed in an external magnetic field. Since a stability problem is always an essentially nonlinear problem, the theory for it must be built upon a nonlinear set of equations for a magnetoelastic model. As seen in the first part of this book, several such models exist and thus one specific model must be chosen first. Once this is done, the general approach to the problem could run as follows:

Consider a ferromagnetic body $\mathscr{B}$ placed in an external magnetic field $\boldsymbol{B}_0$. Due to the action of magnetic forces, $\mathscr{B}$ will deform to a slightly deformed intermediate state G_I. In general, the precise deformation in this state is of minor relevance for the stability considerations. It is the stability of this intermediate state which we want to investigate: we say that $\mathscr{B}$ buckles when G_I becomes unstable. To investigate the stability of G_I we superpose a perturbation (in both the deformations and the magnetic fields) on G_I leading to the final state G for $\mathscr{B}$. By assuming the perturbations to be small, we can linearize the nonlinear set of equations and boundary conditions, referring to state G, with respect to the perturbations. This results in a *homogeneous linear* set of equations, now referring to G_I. Since the deformations in G_I are always very small (and as said, not essential for the stability problem) we may replace *in the linearized perturbed system* the intermediate state by the rigid-body state (or undeformed state). The rigid-body fields have to be calculated first, and once this is done, the perturbed system has to be solved. However, as this is a homogeneous system it will in general only have the trivial zero-solution. Only for a discrete set of real (eigen)values for $B_0 = |\boldsymbol{B}_0|$ the perturbed system will have a non-trivial solution (the eigenvalues are real because the underlying problem is conservative). The lowest of these eigenvalues is the buckling field B_{0cr}. We will refer to this approach as the *classical method*.

To explain this approach in a more formal, mathematical, way, we use the following scheme:

Every equilibrium state of a body $\mathscr{B}$, influenced by an external magnetic field $\boldsymbol{B}_0$, is governed by a set of equations and boundary conditions. Let us denote this set schematically by

$$\mathscr{S}[\boldsymbol{B}(\boldsymbol{x}), \boldsymbol{M}(\boldsymbol{x}), \mathcal{T}(\boldsymbol{x}), \boldsymbol{x}; \boldsymbol{B}_0] = 0 . \tag{7.3.1}$$

The symbol $\mathscr{S}$ encloses various differential operators; some of them act on the boundary.

In the theory of stability two configurations are distinguished, namely the *intermediate* state G_I, satisfying

$$\mathscr{S}^0[\boldsymbol{B}^0(\xi), \boldsymbol{M}^0(\xi), \mathcal{T}^0(\xi), \xi; \boldsymbol{B}_0] = 0 , \tag{7.3.2}$$

where $\xi \in G_I$ is the position vector in the intermediate state, and the *final* (perturbed) state G, which differs only slightly from the intermediate state, is characterized by

$$\mathscr{S}[\boldsymbol{B}^0(\xi) + b(\xi), \boldsymbol{M}^0(\xi) + m(\xi), \boldsymbol{T}^0(\xi) + \boldsymbol{T}^1(\xi), \xi + u(\xi); \boldsymbol{B}_0] = 0 , \quad (7.3.3)$$

where $\xi + \boldsymbol{u}(\xi) = \boldsymbol{x}$, and $\boldsymbol{T}^1$ is the perturbed stress tensor. Here, the perturbations $\{b, m, \boldsymbol{T}^1, u\}$ are supposed to be small. Subtracting (7.3.2) from (7.3.3) and neglecting terms that are of second order in the perturbations, we arrive at a system that is linear and homogeneous with respect to the perturbations. The derivation of this system however is rather complicated because

1. the original set (7.3.1) is nonlinear;
2. the boundary conditions enclosed in $\mathscr{S}$ refer to the deformed boundary ∂G of $\mathscr{B}$;
3. the constitutive equations for the stresses and the magnetization inside the body are different from those outside the body.

In the sequel, the linearized homogeneous problem is denoted by

$$\mathscr{S}^1[b(\xi), m(\xi), \boldsymbol{T}^1(\xi), \boldsymbol{u}(\xi); \boldsymbol{B}_0] = 0 , \qquad (7.3.4)$$

where all operators enclosed in $\mathscr{S}^1$ act on the boundary of the intermediate state.

The set (7.3.4) always has the trivial zero-solution as a solution, but we are only interested in those values of $B_0 = |\boldsymbol{B}_0|$ for which $[b, m, \boldsymbol{T}^1, u] \neq 0$, is a solution of (7.3.4). The thus-posed problem is an eigenvalue problem; the perturbations $[b, m, \boldsymbol{T}^1, u]$ and the magnitude B_0 of the magnetic field play the role of the eigenvector and the eigenvalue, respectively. In the theory of magnetoelastic stability these eigenvalues are called *buckling values*. Of course, we are especially interested in the lowest buckling value.

The eigenvalue problem is linear with respect to the perturbations, but the buckling appears in a nonlinear way, due to the nonlinear dependence of the magnetic forces and stresses on B_0. Although in many cases a simplification in which the intermediate state is replaced by the rigid-body state is allowed, this simplification does not make the nonlinear dependence on B_0 of the eigenvalue problem less complicated. Generally it is impossible to solve the buckling problem directly from (7.3.4). Therefore, it is important to have at our disposal a procedure, intended to obtain a reliable approximation for the buckling value. Practical experience shows that in general variational principles guarantee reliable approximations. We will come back to this subject further on in this chapter.

MOON and PAO [156] applied the classical method to the buckling problem of a cantilevered ferromagnetic beam of narrow rectangular cross-section placed in a transverse magnetic field and they found a buckling field proportional to the $(3/2)$-power of the thickness-to-length ratio. This result was in disagreement with their own experimental results. An explanation for this discrepancy was found by VAN DE VEN [252, 253], who developed an analysis accounting for the finite width of the rectangular cross-section. The main lines and results of this analysis are presented in the following example.

Example: Magnetoelastic buckling of a soft ferromagnetic beam of elliptic cross-section (classical method)

This example is based on [252] and [253]. Consider a slender cantilever beam of length l having an elliptic cross-section with semi-major axis a and semi-minor axis b ($a > b$). This beam consists of soft ferromagnetic elastic material (with magnetic permeability $\mu \gg 1$) and is placed in a uniform external magnetic field $\boldsymbol{B}_0 = B_0 \boldsymbol{e}_y$, directed along the minor axis of the ellipse (the y-direction).

Our analysis will be based on the general formulation for a magnetoelastic body in interaction with an external magnetic field. When applied to a slightly deflected beam, this general formulation is linearized with respect to the pre-buckled state, which here is identified with the rigid-body state (*classical method*). In this way, two problems are obtained: one for the rigid-body magnetic potentials (for an infinitely long straight beam), and a linear one for the perturbed magnetic potentials. For the determination of the buckling value of B_0, we need the potentials inside the beam only, which are here denoted by $\Phi^{(0)}$ for the rigid-body state, and φ, the perturbed potential. For the description of the magnetoelastic interactions the MAXWELL–MINKOWSKI model (Sect. 2.3) is used. For a beam of a narrow rectangular cross-section, $\{-a < x < a, \ -b < y < b, \ b/a \ll 1\}$, a so-called beam-plate, this would yield the following expression for the normal stress on the wider upper plane $y = b$ of the beam-plate (see e.g. [250]; this expression also follows straightforwardly from (compare with eq. $(3.4.25)_1$ in Sect. 3.4 of this book)

$$t_{yy} = \frac{\mu^2}{\mu_0} \frac{\partial \Phi^{(0)}}{\partial y} \frac{\partial \varphi}{\partial y}\bigg|_{y=b} . \tag{7.3.5}$$

Let the deflection of the slender beam in the y-direction be given by $V(z)$, z being the axial coordinate of the beam; then the bending of the beam is governed by the one-dimensional beam equation

$$EI_y V^{\text{iv}}(z) = q(z) , \tag{7.3.6}$$

where EI_y is the bending stiffness and $q(z)$ the normal load per unit of length,

$$q(z) = \int_\Gamma T_y ds , \tag{7.3.7}$$

with Γ the boundary of the elliptic cross-section, and T_y the traction on Γ in the y-direction given by (compare with (7.3.5))

$$T_y = \frac{\mu^2}{\mu_0} \frac{\partial \Phi^{(0)}}{\partial n} \frac{\partial \varphi}{\partial n} n_y , \tag{7.3.8}$$

with n the unit normal vector on Γ.

As demonstrated in [252], the rigid-body magnetic potential can be calculated by use of elliptic coordinates. However, the advantage of an elliptic cross-section is that the magnetic field inside the beam is uniform; calculations in [252] reveal that this internal potential is given by

$$\Phi^{(0)} = -\frac{(1 + \beta)}{\mu} B_0 y , \tag{7.3.9}$$

where $\beta = b/a$ is the thickness-to-width ratio.

In [252], the following boundary value problem for the internal, $\varphi(x, y, z)$, and external, $\psi(x, y, z)$, perturbed potential is derived:

- *in the vacuum outside the beam*

$$\Delta \psi(x, y, z) = 0 , \quad \psi \to 0 , \quad \text{as} \quad |\boldsymbol{x}| \to \infty ; \tag{7.3.10}$$

- *inside the beam*

$$\Delta \varphi(x, y, z) = 0 ; \tag{7.3.11}$$

- *at the boundary Γ*

$$\psi - \varphi = -\mu \frac{\partial \Phi^{(0)}}{\partial n} n_y V(z) ; \tag{7.3.12}$$

and

$$\frac{\partial \psi}{\partial n} - \mu \frac{\partial \varphi}{\partial n} = \frac{\partial}{\partial n} \left(\frac{\partial \Psi^{(0)}}{\partial y} \right) V(z) . \tag{7.3.13}$$

For the solution of this system, we introduce a separation of variables ansatz according to

$$\psi = \Psi(u, v)V(z) , \quad \varphi = \Phi(u, v)V(z) , \tag{7.3.14}$$

where u, v are elliptical coordinates. This separation of variables is, however, only consistent with $\Delta \varphi = 0$ provided that

$$V''(z) + \lambda^2 V(z) = 0 , \tag{7.3.15}$$

subject to $V'(0) = V''(l) = 0$ for the cantilever, yielding

$$\lambda = \frac{\pi}{2l} , \tag{7.3.16}$$

for the eigenvalue λ.

With this separation of variables, the LAPLACE equation for $\varphi(x, y, z)$, (7.3.11), transforms into the HELMHOLTZ equation for $\Phi(u, v)$,

$$\Delta \Phi(u, v) - \lambda^2 \Phi(u, v) = 0 . \tag{7.3.17}$$

The solution of the resulting system is quite complicated and needs the use of MATHIEU functions. Therefore, we do not give the details here, but only refer to [252] for those readers who are interested in the mathematics of the solution procedure.

Using the results of these calculations, we obtain the following expression for the load $q(z)$:

$$q(z) = \frac{2\pi B_0^2}{\mu_0 \kappa \Lambda}(1 + \mathcal{O}(\varepsilon)) , \qquad (7.3.18)$$

where $\varepsilon = \pi a/4l \ll 1$ for slender beams, $\kappa = -\gamma - \ln((1+\beta)/2) - \ln \varepsilon$, $\gamma = 0.5772$, and $\Lambda = 1 + (2\beta \mu \varepsilon^2 \kappa)^{-1}$. Note that although $\varepsilon \ll 1$, $\mu \gg 1$ so that $\mu \varepsilon^2$ can be strictly of order unity.

Substitution of (7.3.15) into the beam equation (7.3.6) yields (with $I_y = \pi a b^3/4$, for the elliptic cross-section)

$$\frac{\pi}{4} E a b^3 V^{\mathrm{iv}}(z) - \frac{2\pi B_0^2}{\mu_0 \kappa \Lambda}V(z) = 0 . \qquad (7.3.19)$$

Taking into account (7.3.15) and (7.3.16), we infer that (7.3.19) has only then a non-trivial (non-zero) solution if B_0 is equal to its critical or *buckling value*

$$\left(\frac{B_0}{\sqrt{\mu_0 E}}\right)_{cr} = \frac{1}{2}\sqrt{\varepsilon \kappa \Lambda}\left(\frac{\pi b}{2l}\right)^{3/2} . \qquad (7.3.20)$$

Let us now consider two limiting cases for the elliptic cross-section, namely

1. the circle: $a = b = R$, ($\Rightarrow \beta = 1$);
2. the very wide narrow ellipse: $b \ll a$, ($\Rightarrow \beta \ll 1$).

For both cases it is assumed that $\varepsilon = \pi a/4l \ll 1$.

For the circular rod this leads to

$$\left(\frac{B_0}{\sqrt{\mu_0 E}}\right)_{cr} = \frac{1}{2}\sqrt{\frac{\kappa}{2} + \frac{1}{\mu}\left(\frac{2l}{\pi R}\right)^2}\left(\frac{\pi R}{2l}\right)^2 , \qquad (7.3.21)$$

where now $\kappa = -\gamma - \ln((1+\beta)/2) - \ln(\pi R/4l)$. This expression is in agreement with the results of MOON [156, Sect. III.3], ALBLAS [9], or VAN DE VEN [251].

For the more interesting case of the very wide ellipse, we note that in the preceding derivations three dimension ratios are used, viz. b/a, a/l, and b/l, though only two are independent. Since we consider here b/l as the basic variable, there remain two possible choices for the second variable. We choose b/a or β as the second independent variable, to obtain

$$\left(\frac{B_0}{\sqrt{\mu_0 E}}\right)_{cr} = \frac{1}{2}\sqrt{\frac{\kappa}{2\beta} + \frac{1}{\mu}\left(\frac{2l}{\pi b}\right)^2}\left(\frac{\pi b}{2l}\right)^2 , \qquad (7.3.22)$$

where $\kappa = -\gamma - \ln((1+\beta)/4\beta) - \ln(\pi b/2l)$.

Assuming that μ is so large that the second term in the square root in (7.3.22) may be neglected, we see that this relation reduces to

$$\left(\frac{B_0}{\sqrt{\mu_0 E}} \right)_{cr} = \frac{1}{2} \sqrt{\frac{\kappa}{2\beta}} \left(\frac{\pi b}{2l} \right)^2 . \qquad (7.3.23)$$

Ignoring the b-influence in κ, we conclude that B_{0cr} at fixed β is proportional to the *second* power of b/l. This result is in conflict with the result of MOON and PAO in [152], and many others who followed the same approach and found a $(3/2)$-dependence of b/l. The basic reason for this difference is that Moon et al. assumed the narrow rectangular cross-section to be *infinitely wide*, whereas we considered here an ellipse of *finite extent*. The crucial point is that the solution of the HELMHOLTZ equation (7.3.17) for the perturbed potential is fundamentally different for a cross-section of finite width from that of an infinite width, *no matter how small the ratio b/a is*. As indicated in [253], the power 2 is also much better in agreement with the experimental results reported in [152] than the power $2/3$ (see also Table 7.1 at the end of the next example).

Our purely analytical method presented in [252] for an elliptic cross-section is not possible for a rectangular cross-section; however, in [253], we presented an approach based on a very reasonable approximation holding for a narrow rectangular cross-section of finite width yielding an analogous result as found in (7.3.23). In a second example below, we shall show how we can use our variational approach to find the buckling value for a rectangular cross-section. Again, this result confirms our result for the elliptical cross-section.

This concludes our first example.

7.3.2 Variational Method for Ferromagnetic Systems

In the magnetoelastic stability theory, as presented in the preceding example, it has been the usual procedure to start from a linear set of equations for the perturbations and to look for a value of the basic field parameter B_0 for which this set has a non-trivial solution. Since an exact 3-dimensional solution is often very difficult, and magnetoelastic buckling problems almost exclusively occur for slender bodies or structures, one starts looking for adequate approximate solutions. This is usually done in the following way:

> For a slender body the 3-dimensional displacement $\boldsymbol{u}$ is approximated by a 1- or 2-dimensional characteristic displacement w; this can be for instance the deflection of the central line of a beam or that of the central plane of a plate. This chosen w has to satisfy the global equilibrium equation, inclusive forces of magnetic origin, depending on B_0. The buckling value is then found as the first eigenvalue for B_0 for which this equation has a non-trivial (non-zero) solution.

However, as the solution obtained by the procedure described above is not an exact solution, but merely a reasonable approximation, the obtained value for B_0 is also an approximation. Therefore, let us introduce the small parameter ϵ ($0 < \epsilon \ll 1$) as a measure for the approximation error in the perturbed displacement field. Then it is evident that, due to the linear character of the perturbed equations, the error in the found eigenvalue for B_0 is also of the first order in ϵ.

In this respect, use of a variational principle clearly has the advantage that in such a procedure the error in B_0 is of *second order* in ϵ. This can be explained best by describing globally the main lines of a variational method. These lines are successively as follows:

1. Starting from a given functional L, which attains its minimum in the final (deformed) state (and which is quadratic in the perturbations) one has
2. to choose a class of trial functions (for the perturbed magnetic and displacement fields) satisfying the constraints of the variational principle;
3. to determine the best members out of this class by equating the first variations of L equal to zero;
4. to calculate the buckling value for B_0 from the equation $J = 0$, where J is the second variation of L; see (7.3.32), further on.

Due to the stationary behaviour of the quadratic functional J the deviation between the exact buckling value and the approximated one calculated as in point 4. above is of the order of the square of the deviation ϵ between the exact and the approximated perturbations. Hence, *the accuracy of the variational method is one order higher than that of the classical method.*

The choice of a class of trial functions (point 2.) is usually based on the choice of a displacement field. In practice, buckling theory always applies to slender bodies or structures, such as beams, rods, plates, rings, or more complex structural elements. For slender bodies, the displacement in buckling can be characterised by one or two global displacement parameters, such as the deflection of the central line of a beam, or the normal displacement of the central plane of a thin plate.

Clearly, it is assumed here that the intermediate fields are known. However, in many practical problems the deformations in the intermediate state are small, and have a negligible effect on the buckling value. In these cases, the intermediate state may be replaced by the so-called rigid-body state, which in general is more simple to determine.

Let us proceed now with a more advanced explanation of the basic idea behind a variational principle for magnetoelastic buckling. Assume that the set $\mathscr{S}$ in (7.3.1) encloses equations and boundary conditions according to $\mathscr{S} = \{S_1, S_2, ...S_k, ...\}$. Take now some of the equations and boundary conditions of (7.3.1) for granted, say

$$S_i\left[\boldsymbol{B}(\boldsymbol{x}), \boldsymbol{M}(\boldsymbol{x}), \mathcal{T}(\boldsymbol{x}), \boldsymbol{x}; B_0\right] = 0, \quad 1 \leq i \leq k , \tag{7.3.24}$$

and analogously for (7.3.2)

$$S_i^0 \, [\boldsymbol{B}^0(\xi), \boldsymbol{M}^0(\xi), \mathcal{T}^0(\xi), \xi; \boldsymbol{B}_0] = 0, \quad 1 \leq i \leq k \,, \qquad (7.3.25)$$

and consider these equations as constraints for the variations of the functionals

$$L \, [\boldsymbol{B}, \boldsymbol{M}, \mathcal{T}; \boldsymbol{B}_0] = \int_{\mathbb{R}_3} \mathscr{L}[\boldsymbol{B}(\boldsymbol{x}), \boldsymbol{M}(\boldsymbol{x}), \mathcal{T}(\boldsymbol{x}), \boldsymbol{x}; \boldsymbol{B}_0] dV \,, \qquad (7.3.26)$$

$$L^0 \, [\boldsymbol{B}^0, \boldsymbol{M}^0, \mathcal{T}^0; \boldsymbol{B}_0] = \int_{\mathbb{R}_3} \mathscr{L}^0[\boldsymbol{B}^0(\xi), \boldsymbol{M}^0(\xi), \mathcal{T}^0(\xi), \xi; \boldsymbol{B}_0] dV \,, \qquad (7.3.27)$$

respectively. The integrands, the so-called LAGRANGEan densities, are connected with the sets of equations and boundary conditions (7.3.1) and (7.3.2) and need to be specified later on.

Evaluation of $S_i - S_i^0 = s_i$ and $L - L^0$ in terms of the perturbations results in

$$s_i \, [\boldsymbol{b}(\xi), \boldsymbol{m}(\xi), \mathcal{T}^1(\xi), \boldsymbol{u}(\xi); \boldsymbol{B}_0] = 0, \quad 1 \leq i \leq k \,, \qquad (7.3.28)$$

and

$$L - L^0 = \delta L + J + O(\epsilon^3) \,, \qquad (7.3.29)$$

where ϵ denotes the order of magnitude of the perturbations, δL is the first variation of L with respect to the intermediate state, which contains only terms of order ϵ, while J is the second variation of L, containing terms of order ε^2 only. If the LAGRANGEan L is chosen in such a way that

$$\delta L = 0 \quad \wedge \quad S_i^0 \, [\boldsymbol{B}^0(\xi), \boldsymbol{M}^0(\xi), \mathcal{T}^0(\xi), \xi; \boldsymbol{B}_0] = 0, \quad 1 \leq i \leq k \,, \qquad (7.3.30)$$

is equivalent to (7.3.2), then it can be proved that

$$\delta J = 0 \quad \wedge \quad s_i \, [\boldsymbol{b}(\xi), \boldsymbol{m}(\xi), \mathcal{T}^1(\xi), \boldsymbol{u}\xi; \boldsymbol{B}_0] = 0, \quad 1 \leq i \leq k \,, \qquad (7.3.31)$$

is equivalent to (7.3.4). For the proof, we refer to [121]. Hence, the eigenvalue problem for $\boldsymbol{B}_0$ is equivalent to finding a non-zero solution of (7.3.31).

Using that J is a homogeneous quadratic functional in the perturbations, we infer the important property

$$\delta J = 0 \quad \Longrightarrow \quad J = 0 \,. \qquad (7.3.32)$$

The latter two results constitute a variational principle that should be used in the following way:

First, a class of trial functions $\{\boldsymbol{b}(\xi), \boldsymbol{m}(\xi), \mathcal{T}^1(\xi), \boldsymbol{u}(\xi)\}$ is constructed, satisfying the constraints (7.3.28). Those trial functions, which approximate the exact perturbations best, are determined from $\delta J = 0$. Subsequently, an approximation of the buckling value $\boldsymbol{B}_0$ is calculated from the equation $J = 0$; see (7.3.32). Let ϵ be the order of magnitude of the deviations between the exact and the approximate perturbations; then, due to the stationary behaviour of J, the deviation between the exact and approximate buckling value is of order ϵ^2.

In the above analysis, the determination of the LAGRANGEan density $\mathscr{L}$ is rather difficult. Therefore, we follow here the reverse way: a LAGRANGEan density accompanied by some constraints is postulated, and by variation of L a set of equations like (7.3.24) is derived. If this set turns out to coincide with one of the generally accepted models presented in Chap. 3 of this book, we conclude that the right choice for $\mathscr{L}$ has been made.

For ferromagnetic elastic bodies the variational principle is based on the following expression for the LAGRANGEan density $\mathscr{L}$:

$$\mathscr{L} = -\frac{1}{2}\mu_0(\boldsymbol{H}, \boldsymbol{H}) - \rho U + \frac{1}{2\mu_0}(\boldsymbol{B}_0, \boldsymbol{B}_0) , \qquad (7.3.33)$$

where $U = U(\mathcal{F}, \boldsymbol{M})$ is the internal (magnetoelastic) energy density, $\mathcal{F}$ (with components $F_{i\alpha}$) is the deformation gradient, $\boldsymbol{M}$ the magnetization, and $\boldsymbol{B}_0$ the uniform external magnetic field at infinity. The pertinent constraints are (here $G^- = G$, G^+ and ∂G are the configuration of the body $\mathscr{B}$, that of the surrounding vacuum, and the boundary of G, respectively, while an upper index $-$ or $+$ stands for a value inside or outside the body, respectively)

$$
\begin{aligned}
B_i^+ &= e_{ijk}A_{k,j}^+ , & M_i^+ &= 0 , & &\boldsymbol{x} \in G^+ , \\
B_i^- &= e_{ijk}A_{k,j}^- , & T_{ij} &= \rho\frac{\partial U}{\partial F_{i\alpha}} , & \rho &= \frac{\rho_0}{J_F} , & \boldsymbol{x} \in G^- , \\
A_i^+ &= A_i^- , & & & &\boldsymbol{x} \in \partial G , \\
B_i^+ &\to B_{0i} , & & & &|\boldsymbol{x}| \to \infty , \quad (7.3.34)
\end{aligned}
$$

where

$$H_i^+ = \frac{1}{\mu_0}B_i^+ , \qquad H_i^- = \frac{1}{\mu_0}B_i^- - \rho M_i^- , \qquad J_F = \det \mathcal{F} . \qquad (7.3.35)$$

To obtain the total LAGRANGEan L the density $\mathscr{L}$ must be integrated over both the (ferromagnetic) body (G^-) and the surrounding vacuum (G^+). The requirement $\delta J = 0$, taking into account the constraints given above, yields the well-known MAXWELL–MINKOWSKI model for magnetoelastic interactions (Model III in Chap. 3).

What we need next is the second variation J of L ($J = \delta^2 L$); since this is a cumbersome derivation, we refer for the resulting expression of J and the underlying derivation to [121]. Since in the equilibrium state, of which the stability we are investigating, the first variation of L is zero, J must be of second order in terms of the perturbations on the equilibrium state. If we neglect higher-order terms, J becomes a homogeneous quadratic function in the perturbations. However, as the final state is again an equilibrium state, also the first variation of J must be zero, but for a homogeneous quadratic function J this implies that then J itself must be zero. Hence, $\delta J = 0 \implies J = 0$, and this relation delivers us directly an explicit expression for the buckling field B_{0cr}. This can be seen as follows: by splitting up J into a magnetic part $\mathcal{K}$, which is proportional to B_0^2 ($K = B_0^2 \mathcal{K}$) and an elastic part W, according to $J = \mathcal{K} - W = B_0^2 K - W$, we see that $J = 0$ yields (K and W are independent of B_0)

$$B_{0cr} = \sqrt{W/K} \ . \qquad (7.3.36)$$

Hence, if we can calculate W or K, either exactly or in a sufficiently accurate numerical approximation, the relation above immediately delivers the looked-for buckling field. LIESHOUT and VAN DE VEN et al. used this method to calculate the magnetic buckling fields for a cantilevered beam of rectangular cross-section (an exact solution was found) and a set of two parallel rods, both systems being soft ferromagnetic, placed in a transverse magnetic field, [122, 123]; the first problem is discussed in the following example, and the second one in Sect. 7.3.1.

Example: Magnetoelastic buckling of a slender soft ferromagnetic beam (variational method)

This example is based on [122] or [123]. Consider again a slender cantilever beam of length l and of arbitrary simply-connected cross-section. This beam is of soft ferromagnetic elastic material (with magnetic permeability $\mu \gg 1$) and is placed in a uniform external magnetic field $\boldsymbol{B}_0 = B_0 \boldsymbol{e}_y$, directed along the y-direction. This is the same problem, at least for an elliptic cross-section, as in the preceding example, but now we solve the problem by means of our variational principle.

For this variational principle we need an expression for $J = B_0^2 K - W$ in terms of the perturbed magnetic potentials ψ and φ and the displacement $\boldsymbol{u}$. Here, W is the elastic energy due to $\boldsymbol{u}$, and K is the scaled magnetic energy (in the perturbed state). To start with the elastic energy, we choose for $\boldsymbol{u}$ a displacement field owing to the bending of a slender beam, according to the BERNOULLI–NAVIER theory,

$$\boldsymbol{u}(\boldsymbol{x}) = V(z)\boldsymbol{e}_2 - yV'(z)\boldsymbol{e}_3 \ , \qquad (7.3.37)$$

where $V(z)$ represents the deflection of the central line of the beam. This choice yields for W the well-known expression for the bending energy of a slender beam

$$W = \frac{1}{2}EI_y \int_0^l (V''(z))^2 dz \ , \qquad (7.3.38)$$

where EI_y is the bending stiffness of the beam.

The derivation of an expression for K is not so straightforward. Therefore, we omit here all details and immediately give the result, for the derivation of which we refer to [122] or [123][1] (in the next section we will present a general derivation for K, but then for the case of a superconducting structure):

---

[1] In [122] and [123], the potential ψ is defined by $b_i^+ = \psi_{,i}$, whereas we use here the commonly accepted definition $h_i^+ = -\psi_{,i}$; this means that the ψ in [122] and [123] must be replaced by $-\mu_0\psi$ in our notation

$$K = -\int_{\partial G} B_i u_i \frac{\partial}{\partial n}\left(\psi - \frac{1}{\mu_0}B_j u_j\right)dS \ . \qquad (7.3.39)$$

This result holds for slender beams (quadratic terms in the slenderness parameter ε are neglected) and for soft ferromagnetic materials (terms of order μ^{-1} are also neglected). In (7.3.39), $\boldsymbol{B}$ is the rigid-body magnetic field and ψ the perturbed magnetic potential both for the external region (the reason that only the external fields appear in this expression is that the internal fields are order μ^{-1} smaller than the external ones). The perturbed potential $\psi = \psi(x, y, z)$ has to satisfy the following constraints (here, G^+ is the vacuum region external to the beam)

$$\begin{aligned} \Delta\psi(x, y, z) &= 0 \ , & \text{for } \ \boldsymbol{x} \in G^+ \ , \\ \psi + B_i u_i &= 0 \ , & \text{for } \ \boldsymbol{x} \in \partial G \ , \\ \psi &\to 0 \ , & \text{for } \ |\boldsymbol{x}| \to \infty \ . \end{aligned} \qquad (7.3.40)$$

The problem for the 3-dimensional potential ψ is reduced to a 2-dimensional problem by the separation of variables

$$\psi(x, y, z) = \Psi(x, y)V(z) \ . \qquad (7.3.41)$$

This separation of variables is only then consistent with $\Delta\psi(x, y, z) = 0$, if $V(z)$ satisfies the relation

$$V''(z) + \lambda^2 V(z) = 0 \ , \qquad (7.3.42)$$

where the real parameter λ, the separation constant, is related to the length l through the support conditions; for a cantilevered beam $\lambda = \pi/2l$; see (7.3.15) and (7.3.16). With all this, (7.3.40) reduces to the 2-dimensional problem for $\Psi(x, y)$ (here, D^+ is the 2-dimensional region of the vacuum in the plane of the cross-section, and Γ is its boundary)

$$\begin{aligned} \Delta\Psi(x, y) - \lambda^2\Psi(x, y) &= 0 \ , & \text{for } \ (x, y) \in D^+, \\ \Psi + B_y &= 0 \ , & \text{for } \ (x, y) \in \Gamma \ , \\ \Psi &\to 0 \ , & \text{for } \ \sqrt{x^2 + y^2} \to \infty \ . \end{aligned} \qquad (7.3.43)$$

Moreover, substitution of (7.3.41) changes the 2-dimensional integral in (7.3.39), with use of (7.3.37) for $\boldsymbol{u}$, into a product of two 1-dimensional integrals according to

$$K = \left(-\int_{\Gamma} B_y \frac{\partial}{\partial n}(\Psi - \frac{1}{\mu_0}B_y)ds\right)\left(\int_0^l V^2(z)dz\right) \ . \qquad (7.3.44)$$

The problem (7.3.43) for Ψ can be solved by use of a conformal mapping: $z = x + iy = h(\zeta)$, which maps D in the xy-plane onto the unit circle in the ζ-plane. The transformed problem can then be solved by use of the theory of complex functions. However, we do not need a full solution for Ψ but rather a calculation of the integral for K in (7.3.44). How this is done in detail can be found in [122]; we restrict ourselves here to giving only the result; it reads

$$K = \frac{\pi}{\mu_0} \frac{1}{\kappa(\varepsilon c)} \,, \tag{7.3.45}$$

with $\kappa = -\gamma - \ln(\varepsilon c)$, and ε is the slenderness parameter, as in the preceding example, while c is a constant that follows from the conformal mapping as $c = \lim_{\zeta \to \infty} h'(\zeta)$. We note here that for the elliptic cross-section: $\varepsilon = \pi a/4l$ and $c = (1 + \beta)/2$, in agreement with the preceding example.

Substituting this into the equation for the buckling value (7.3.36), we arrive at

$$\left(\frac{B_0}{\sqrt{\mu_0 E}} \right)_{cr} = \frac{1}{2} \sqrt{\frac{\kappa}{2\beta}} \left(\frac{\pi b}{2l} \right)^2 \,, \tag{7.3.46}$$

a result identical to (7.3.22) from the first example. So our variational approach confirms the result of the classical method.

With the variational method, we could also solve analytically the buckling problem for a beam having a (narrow) rectangular cross-section. As can be found in [122] this leads to an expression for c in terms of elliptic integrals. We can calculate this value of c for all types of rectangular cross-sections, ranging from a square to very narrow ones. In [122], we compared the results with those for corresponding ellipses, and we found that the difference nowhere became more than 2%. With 'corresponding' we mean here that the elliptic and rectangular cross-sections do have identical thickness-to-width ratio's and moments of inertia I_y.

In (7.3.46), we have expressed B_{0cr} in terms of the two dimension ratios β and b/l; instead of β we could also have used ε, which would yield (with $\beta = b/a = 2\varepsilon(2l/\pi b)$)

$$\left(\frac{B_0}{\sqrt{\mu_0 E}} \right)_{cr} = \frac{1}{2} \sqrt{\kappa \varepsilon} \left(\frac{\pi b}{2l} \right)^{3/2} \,. \tag{7.3.47}$$

Hence, whether B_{0cr} is proportional to b/l to the power 2 or 3/2 depends on the choice of β and b/l or ε and b/l as independent dimension ratios, respectively. However, as we shall show further on, experimental data of a.o. Moon and Pao [152] fit more accurately to a second-power law than to a (3/2)-power law; see Table 7.1.

Some further conclusions that can be drawn from the latter two results, (7.3.46) and (7.3.47), are:

1. At fixed thickness-to-length ratio b/l the buckling value B_{0cr} increases with increasing width-to-length ratio ε. This statement received experimental confirmation from Dalrymple et al.; see [50, Fig. 4]. In this figure the variation of the buckling field with the plate width a (at constant thickness b and length l) is shown. This curve corresponds very well with our result (7.3.47).

2. At fixed thickness-to-length ratio b/l the buckling value B_{0cr} decreases with increasing thickness-to-width ratio β. Again this statement is supported by experimental results from the literature; this time from MIYA et al.; see [144, Fig. 3]. In addition, we note that the results of this paper also support statement 1.

We conclude this example by giving a table in which for fixed values of β and varying (b/l) our theoretical buckling values are compared with the theoretical and experimental results of MOON and PAO [152, Fig. 6]; see Table 7.1. This table shows good agreement between the theoretical values predicted by (7.3.46) and the experimental data from MOON and PAO. This agreement is so evident that we are convinced that this justifies our conclusion that the discrepancy between the theoretical and experimental results of MOON and PAO is substantially due to the inadequacy of the infinite-width assumption. The results of this example evidently reveal that the influence of the finite width of the beam is essential and remains so even for very small thickness-to-width ratios.

We continue this section with one extra example.

Table 7.1. Theoretical and experimental buckling values for a rod of rectangular cross-section

$\beta = 5.6 \times 10^{-2}$ and $\mu = 6 \times 10^4$			
$\frac{b}{l} \times 10^3$	$\left[\left(\frac{B_0}{\sqrt{\mu_0 E}}\right)_{cr}\right]_{(7.3.46)}$	$\left[\left(\frac{B_0}{\sqrt{\mu_0 E}}\right)_{cr}\right]_{[152],\mathrm{Th}}$	$\left[\left(\frac{B_0}{\sqrt{\mu_0 E}}\right)_{cr}\right]_{[152],\mathrm{Exp}}$
3.0	0.96	1.87	0.90-1.00
2.5	0.69	1.42	0.65-0.70
2.0	0.46	1.02	0.45-0.50
$\beta = 2.4 \times 10^{-2}$ and $\mu = 10^4$			
1.80	0.55	0.87	0.51
1.50	0.40	0.66	0.30-0.33
1.15	0.26	0.44	0.24
0.90	0.18	0.31	0.17-0.20

7.3.3 Magnetoelastic Buckling of a Set of Two Soft Ferromagnetic Parallel Rods

Consider a system of two identical parallel slender soft ferromagnetic rods with radii R, length L, $R/L \ll 1$, placed in an external magnetic field $\boldsymbol{B}_0$ acting in a plane normal to the axes of the rods, i.e. the Z-direction. A normal cross-section of the system in the XY-plane is given in Fig. 7.1. The distance between the centres of the rods is $2d$. The uniform external field $\boldsymbol{B}_0$ lies in the

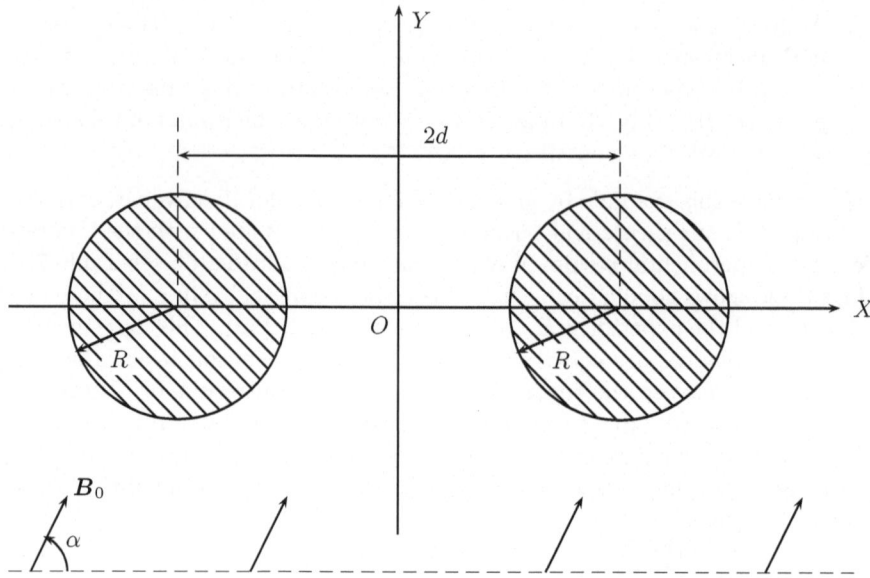

Fig. 7.1. Cross-section of system of two parallel rods

XY-plane and makes an angle α with the positive X-axis. We will present the solution for the buckling problem for this system in two ways: in the classical way and by means of our variational principle. These presentations are based on [255] and [122], respectively.

The classical procedure is almost analogous to that for one single beam. Here, we introduce the displacements of the central lines of the two cantilevered beams by the complex displacement $w^{(m)}(z)$, $m = 1, 2$, $w^{(m)} \in \mathbb{C}$ of the m-th beam by

$$w^{(m)}(z) = u^{(m)}(z) + iv^{(m)}(z) \,, \qquad (7.3.48)$$

where $u^{(m)}$ and $v^{(m)}$ are the displacement in the x and y-directions, respectively, and z is the (real) axial coordinate. With this, the beam equation for the m-th beam becomes

$$EI\frac{d^4 w^{(m)}}{dz^4} = q^{(m)}(z) \,, \qquad (7.3.49)$$

where $I = \pi R^4/4$ and $q^{(m)}$ is the magnetic load per unit of length on the beam. These loads are related to the rigid-body fields and the perturbed fields, and in order to find the $q^{(m)}$'s one has to determine these fields. However, we refrain from doing this here, and refer for the details of the calculations to [255]. It turns out that the $q^{(m)}$'s are of the form

$$q^{(1)}(z) = B_0^2 \left[Q_1 w^{(1)}(z) + Q_2 \bar{w}^{(1)}(z) + Q_3 w^{(2)}(z) + Q_4 \bar{w}^{(2)}(z) \right] , \quad (7.3.50)$$

while $q^{(2)}$ is obtained from $q^{(1)}$ by replacing the upper indices (1,2) by (2,1). In (7.3.50), $\bar{w}^{(m)}$ is the complex conjugate of $w^{(m)}$. However, Q_1 to Q_4 are complex numbers, which are explicitly determined by the solutions of the rigid-body and perturbed fields as can be found in [255]. The analysis presented there makes use of complex function theory and ultimately expresses the perturbed solution in terms of BESSEL functions. We note here that these numbers depend on R, L, d and α, but are *independent* of B_0.

The buckling equation (7.3.49) together with the separation condition (compare with (7.3.15) or (7.3.42))

$$w^{(m)''}(z) + \lambda^2 w^{(m)}(z) = 0 , \quad \lambda = \frac{\pi}{2L} , \quad m = 1, 2 , \quad (7.3.51)$$

yields two homogeneous equations for $w^{(1)}$ and $w^{(2)}$. For a cantilever support at $z = 0$, the boundary conditions are $w^{(m)}(0) = w''^{(m)}(L) = 0$, from which the value for λ actually follows. It turns out that they can be decomposed into two independent equations for $(w^{(1)} + w^{(2)})$ and $(w^{(1)} - w^{(2)})$. Hence, two possible solutions exist:

1. $w^{(1)} = -w^{(2)}$: *symmetrical* buckling mode;
2. $w^{(1)} = +w^{(2)}$: *anti-symmetrical* buckling mode.

The numerical results of [255] reveal that the lowest buckling value for B_0 is associated with the symmetrical buckling mode. The corresponding buckling value reads

$$B_{0cr} = \lambda^2 \sqrt{\frac{EI}{\Omega}} , \quad \Omega = \text{Re}(Q_1 - Q_3) + \sqrt{|Q_2 - Q_4|^2 - (\text{Im}(Q_1 - Q_3))^2} .$$
$$(7.3.52)$$

This buckling value depends on α and d/R; explicit results for B_{0cr}, scaled with respect to the buckling value for one beam $(d/R \to \infty)$, as function of d/R are depicted in Fig. 7.2 for $\alpha = 0$ and $\alpha = \pi/2$.

Discussion. In [255] also the pre-buckling deflections are calculated. Although it turns out that, as expected, these deflections are small, they nevertheless have an effect that can explain the discrepancy between the theoretical and experimental results observed in the left graph of Fig. 7.2. For this, we consider the two cases $\alpha = 0$ and $\alpha = \pi/2$ one by one:

1. In case $\alpha = 0$ both the pre-deflection and the buckling deflection are in the same plane. Hence, if the system is imperfection sensitive the pre-deflection will lower the buckling value. This explains why we experimentally found a somewhat lower buckling value than theoretically predicted.

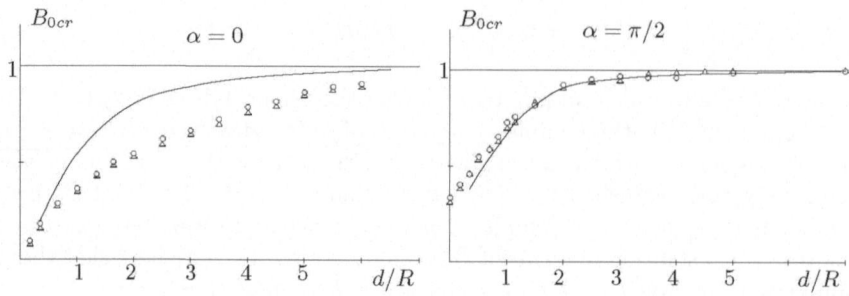

Fig. 7.2. Buckling magnetic field, scaled to the buckling field for one rod, for symmetric buckling mode vs. distance between rods, scaled to radius; *solid lines* are theoretical values and △ and ○ are experimental values. The *left* panel refers to a magnetic field parallel to the plane of the rods ($\alpha = 0$); the *right* to a transverse field ($\alpha = \pi/2$); from [255]

2. In case $\alpha = \pi/2$ the plane in which the pre-deflection takes place is perpendicular to the plane of buckling. Hence, in this case there is no influence of the pre-deflection on the buckling value and, therefore, a much better correspondence between the theoretical and experimental results is observed in the right figure.

We proceed by presenting a solution for the same problem, but now via the *variational principle*. For this, we start with the following approximate assumptions:

1. We neglect the influence of the intermediate deformations (see the **Discussion** above).
2. The magnetic susceptibility $\chi = 1 + \mu$ is so high that all terms of order χ^{-1} are neglected.
3. All terms of order δ^2 are neglected, where $\delta = \lambda R = \pi R/2L$, is the slenderness parameter of the rods.
4. The rods deflect according to BERNOULLI's classical bending theory; the angle between the direction of bending and the X-axis is called θ_1, and the magnitude of the displacement of the center line is $w(z) = w^{(1)}(z) = -w^{(2)}(z)$ (symmetrical buckling mode).
5. The rigid-body (intermediate) fields are independent of z (for long slender rods).
6. For the perturbed magnetic fields a separation of variables with respect to (x, y) and z, and proportional to $w(z)$ can be applied (compare with the example of one rod).
7. Besides $\delta = \lambda R$, also $\lambda d = \delta d/R$ is small.

The variational principle is now applied in the following way: first, $L - L^0$ is developed in the perturbations up to and including the second order, thus yielding J. This expression for J is simplified using the approximations listed above. Next, a class of perturbations $\{b, m, T^1, u\}$ is chosen in such a way that the constraints (7.3.34) and the support conditions for the cantilevered rods are satisfied. We assume a symmetrical buckling mode, meaning that the displacement of the center line of the first rod is given by

$$u = V(z) \cos\theta_1 \; e_x + V(z) \sin\theta_1 \; e_y \;, \tag{7.3.53}$$

where $V(z)$ is the total deflection of the rod in the θ_1-direction. Then the elastic energy W is equal to that given in (7.3.38). For the magnetic functional K we obtain a formula analogous to (7.3.44), namely

$$K = K_1(\theta_1)K_2 \;,$$
$$K_1(\theta_1) = \frac{1}{\mu_0} \int_{\Gamma_1} (B_x \cos\theta_1 + B_y \sin\theta_1) \frac{\partial}{\partial n} \Big(\Psi + B_x \cos\theta_1 + B_y \sin\theta_1 \Big) ds \;,$$
$$K_2 = \int_0^L V^2(z)dz \;, \tag{7.3.54}$$

where $\Psi = \Psi(x, y)$ is the perturbed external magnetic potential after a separation of variables analogous to (7.3.41).

The calculation of the rigid-body fields B_x and B_y and the potential Ψ is done in [122] by using complex function theory together with a conformal mapping of the two circular cross-sections of the rods in the XY-plane onto an annular domain in the complex plane. Using HILBERT theory, we can then derive expressions for B_x, B_y and Ψ. However, it should be emphasized here that for our variational principle we do not need the explicit expressions for these variables, but we only need an explicit expression for the integral for K in (7.3.54). After several mathematical manipulations, for which we refer to [122], the following expression for $J = W - KB_0^2$, which depends on θ_1 only, is obtained (here, we have normalized the integral $\int_0^l V^2(z)dz$ to unity):

$$J(\theta_1) = \frac{\pi}{8} E\delta^4 - \frac{2\pi}{\mu_0} B_0^2 Q(\theta_1) \;, \tag{7.3.55}$$

where

$$Q(\theta_1) = c_0 + c_1 \cos 2\theta_1 + c_2 \sin 2\theta_1 \;, \tag{7.3.56}$$

and c_0, c_1 and c_2 are known functions (calculated in [122, eq. (5.13)]) of the parameters d/R and α. The optimal value $\theta_1 = \hat\theta_1$, corresponding to a minimum value of J, is obtained from

$$\frac{dJ}{d\theta}(\hat\theta_1) = 0, \quad \text{and} \quad \frac{d^2 J}{d\theta^2}(\hat\theta_1) > 0 \;, \tag{7.3.57}$$

yielding

$$\tan 2\hat\theta_1 = \frac{c_2}{c_1} \;. \tag{7.3.58}$$

The buckling value for B_0 then follows from $J(\hat\theta_1) = 0$, and this results in

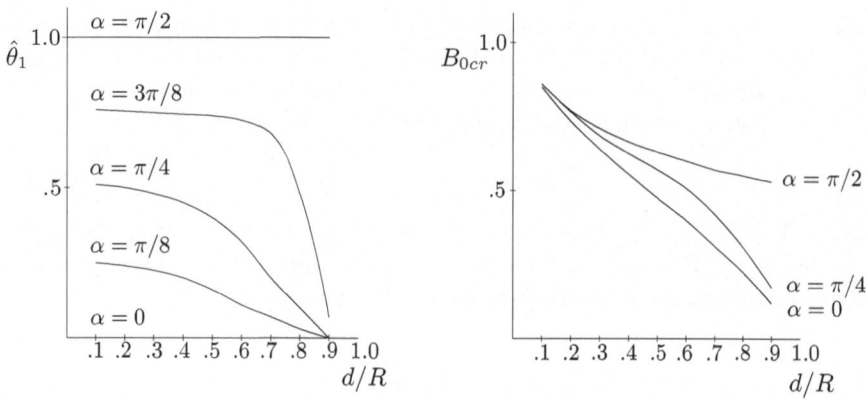

Fig. 7.3. Optimal angle for the deflection $\hat{\theta}_1$ (*left*) and the buckling field B_{0cr}, scaled with respect to that for one rod (*right*) for different values of α; from [122]

$$B_{0cr} = \sqrt{\frac{\mu_0 E}{16Q(\hat{\theta}_1)}} \left(\frac{\pi R}{2L}\right)^2. \tag{7.3.59}$$

Numerical evaluations show that the result (7.3.59) is equal to the previous result (7.3.52), at least for large values of the magnetic susceptibility ($\mu_1 \approx 0$).

Some of the numerical results are depicted in Fig. 7.3: the left panel shows $\hat{\theta}_1$ as a function of R/d for a set of values for α; the right panel displays B_{0cr}, scaled with respect to the buckling value for one rod, as a function of R/d for three values of α. From these graphs, we draw the following conclusions:

1. For fixed values of R/d and α, the buckling value is proportional to $\delta^2 = \left(\frac{\pi R}{2L}\right)^2$.

2. An increase in R/d for fixed values of α results in a decrease of the buckling value, whereas an increase in α for a fixed value of R/d causes an increase of the buckling value.

3. For $0 < \alpha < \pi/2$, the angle of deflection $\hat{\theta}_1$ is a decreasing function of R/d.

4. For $0 < \alpha < \pi/2$, the deflection is NOT in the direction of the field ($\hat{\theta}_1 \neq \alpha$); only for $\alpha = 0$ or $\alpha = \pi/2$ one obtains $\hat{\theta}_1 = \alpha$.

This concludes the section on ferromagnetic bodies; in the ensuing section we shall discuss instabilities in superconducting structures.

7.4 Superconducting Structures

Superconducting coils are extensively used in modern technology, as for instance in magnetic fusion reactors, NMR-scanners, magnetically levitated trains, and many other applications. The current in these devices is often so high that the structures carrying the currents are subjected to strong magnetic forces, which can become so large that the structure collapses (buckles). Hence, in the design of these devices investigation of the stability of the structure is of eminent importance, see e.g. [56, 147]. In reactors and scanners complex structures such as spirals, helical or toroidal coils are in use. The complexity of such structures makes the study of their stability very complicated, see e.g. [280, 284]. These methods often use approximations, of which the degree of accuracy could not always be indicated. To overcome this, we have adapted our variational method for ferromagnetic bodies as presented in the preceding section to one for superconducting structures [218]. In conducting systems, the electromagnetic loading is mainly due to the LORENTZ forces, originating from an interaction of the electrical current with either its own magnetic field or an external field. Whenever these LORENTZ forces become too strong, the structure buckles. The analysis of this (in)stability phenomenon can be performed analogously to the perturbation method presented schematically at the beginning of the preceding section. These methods often use approximations, of which the degree of accuracy could not always be indicated. To overcome this, we have adapted our variational method for ferromagnetic bodies as presented in the preceding section to one for superconducting structures [218].

To make the variational method of Sect. 7.3 suitable for systems of superconducting coils carrying a prescribed current I_0, the LAGRANGEan density must be changed into

$$\mathscr{L} = \frac{1}{2\mu_0}(\boldsymbol{B}, \boldsymbol{B}) - \rho U , \qquad (7.4.1)$$

which is formally a LEGENDRE transformation of (7.3.33) (for non-magnetizable bodies) in which we pass from the variable $\boldsymbol{H}$ to $\boldsymbol{B}$; see [218]. Since the total current I_0 through the coil is prescribed, $\boldsymbol{B}$ must satisfy as a constraint the AMPÈRE law

$$\int_C (\boldsymbol{B}, \boldsymbol{ds}) = \mu_0 I_0 , \qquad (7.4.2)$$

where C is a contour encircling the conductor.

The further evaluation is completely analogous to that in Sect. 7.3 and eventually amounts to (note that here the electromagnetic term is proportional to I_0^2)

$$J = I_0^2 K - W = 0 , \quad \Longrightarrow \quad I_{0cr} = \sqrt{W/K} , \qquad (7.4.3)$$

yielding an explicit expression for the buckling current I_{0cr}.

For not too complicated systems, the values of W and, above all, K can be calculated analytically or numerically, but both exactly, see [122] or [218]. When this is no longer possible, one can choose, on a variational basis, an appropriate (approximate) set of admissible fields to calculate W and K, and thus I_{0cr}; see [257].

A more detailed derivation of this variational principle is presented in the next section.

7.4.1 Formulation of Variational Principle for Superconducting Structures

In this section, we shall start from expression (7.4.1) and derive a formulation for $J = 0$, where J is the second variation of L. We shall do this for a superconducting body (in practice, a slender structure) in vacuum. We denote by G the configuration of the body, and by G^+ that of the vacuum, while ∂G is the boundary of G. We model a superconducting body as a non-magnetizable body, for which the current density $\boldsymbol{J}$ (Ampère/m) is concentrated on the surface of the body, and for which the magnetic fields inside the body vanish $(\boldsymbol{B}^- = \boldsymbol{H}^- = \boldsymbol{0})$. The current density $\boldsymbol{J}$ is related to the boundary value of the external field $\boldsymbol{B}^+$ by

$$\mu_0 \boldsymbol{J} = \boldsymbol{n} \times \boldsymbol{B}^+, \quad \boldsymbol{x} \in \partial G . \tag{7.4.4}$$

We restrict ourselves here to a simply-connected superconducting body in a static situation. For our magnetoelastic model we will use the MAXWELL–MINKOWSKI model (Model III in Chap. 3). Moreover, since the internal fields are zero, we shall simply denote the external fields by $\boldsymbol{B}$ and $\boldsymbol{H}$ (without the superindex $+$).

For the choice of L like in (7.4.1), the magnetoelastic equations for the final state $\boldsymbol{x}$ in the MAXWELL–MINKOWSKI formulation are obtained by variation of L (i.e. $\delta L = 0$) under the following constraints:

$$\boldsymbol{x} \in G , \qquad \rho J_F = \rho_0 , \qquad T_{ij} = \rho \frac{\partial U}{\partial F_{i\alpha}} F_{j\alpha} ,$$
$$\boldsymbol{x} \in G^+ , \qquad e_{ijk} B_{k,j} = 0 , \qquad \boldsymbol{B} \to 0 , \ |\boldsymbol{x}| \to \infty , \tag{7.4.5}$$

plus an extra constraint prescribing the total current I_0 by means of AMPÈRE's law (7.4.2).

To find the first and second variation of L, we develop L in terms of the perturbations with respect to the intermediate state. In doing this, we take for the intermediate state the rigid-body state. The displacement $\boldsymbol{u}$ is then a perturbed field, while we split the total magnetic field as $\boldsymbol{B} = \boldsymbol{B}^0 + \boldsymbol{b}$, where $\boldsymbol{B}^0$ is the field in the rigid-body state, and $\boldsymbol{b}$ is the perturbed field. Developing L up to second-order terms in the perturbations, we obtain $(J = \delta^2 L)$

$$L - L^0 = \delta L + J$$
$$= \frac{1}{\mu_0} \int_{G^+} b_i B_i dV - \frac{1}{2\mu_0} \int_{\partial G} B_k B_k u_i dS + \int_G T_{ij,j} u_i dV$$
$$- \int_{\partial G} T_{ij} N_j u_i dS$$
$$- \frac{1}{\mu_0} \int_{\partial G} \left[b_k B_k u_i + \tfrac{1}{2} B_{k,j} B_k u_i u_j + \tfrac{1}{4} B_k B_k (u_i u_{j,j} - u_j u_{i,j}) \right] N_i dS$$
$$+ \frac{1}{2\mu_0} \int_{G^+} b_i b_i dV - \frac{1}{2} \int_G c_{ijkl} u_{i,k} u_{j,l} dV , \tag{7.4.6}$$

where c_{ijkl} is the classical, HOOKEan, linear elasticity tensor.

Since the constraints (7.4.5) and (7.4.2) have to be satisfied for both the intermediate and the present state, the constraints for the perturbations become

$$\boldsymbol{X} \in G , \qquad t_{ij} = -T_{ij} u_{k,k} + T_{ik} u_{j,k} + c_{ikjl} u_{k,l} ;$$
$$\boldsymbol{X} \in G^+ , \qquad e_{ijk} b_{k,j} = 0 , \qquad \boldsymbol{b} \to 0 , \quad |\boldsymbol{X}| \to \infty ; \tag{7.4.7}$$

and

$$\int_C (\boldsymbol{b} \cdot d\boldsymbol{s}) = 0 . \tag{7.4.8}$$

The constraints $(7.4.7)_2$ and (7.4.8) for $\boldsymbol{b}$ guarantee the existence of a continuous magnetic potential $\psi(\boldsymbol{X})$ for $\boldsymbol{X} \in G^+$, such that

$$b_i = \psi_{,i} , \qquad \boldsymbol{X} \in G^+ . \tag{7.4.9}$$

To dispose of irrelevant constants in ψ, we replace the last constraint of (7.4.7) by

$$\psi \to 0 , \quad |\boldsymbol{X}| \to \infty . \tag{7.4.10}$$

With this we can eliminate $\boldsymbol{b}$ from (7.4.6) in favour of ψ to obtain

$$\delta L = \int_G T_{ij,j} u_i dV + \frac{1}{\mu_0} \int_{G^+} \psi_{,i} B_i dV - \int_{\partial G} \left(T_{ij} N_j u_i + \frac{1}{2\mu_0} B_k B_k N_i \right) u_i dS , \tag{7.4.11}$$

and

$$J = -\frac{1}{2} \int_G c_{ijkl} u_{i,k} u_{j,l} dV + \frac{1}{2\mu_0} \int_{G^+} \psi_{,i} \psi_{,i} dV$$
$$- \frac{1}{\mu_0} \int_{\partial G} \left[\psi_{,k} B_k u_i + \tfrac{1}{2} B_{k,j} B_k u_i u_j + \tfrac{1}{4} B_k B_k (u_i u_{j,j} - u_j u_{i,j}) \right] N_i dS . \tag{7.4.12}$$

By standard procedures, for which we refer to [122], we can prove that $\delta L = 0$ and $\delta J = 0$ yield successively the equations for the intermediate (rigid-body) and the perturbed (linearized) state for the magnetoelastic problem of a superconducting body in vacuum in the MAXWELL–MINKOWSKI formulation. In this respect, we note here that the rigid-body field obtained from $\delta L = 0$ satisfies, besides the constraints $(7.4.5)_2$, also

$$B_{i,i} = 0 , \quad \boldsymbol{X} \in G^+ , \qquad B_i N_i = 0 , \quad \boldsymbol{X} \in \partial G . \tag{7.4.13}$$

For the further evaluation of J we first conjecture that

$$-\frac{1}{2\mu_0} \int_{\partial G} \psi_{,k} B_k u_i N_i dS = -\frac{1}{2\mu_0} \int_{\partial G} (B_{i,j} u_j - B_j u_{i,j}) \psi N_i dS , \tag{7.4.14}$$

a relation which, with the aid of (7.4.13), can be proven as follows

$$\int_{\partial G} (-\psi_{,j} B_j u_i + \psi B_{i,j} u_j - \psi B_j u_{i,j}) N_i dS = \int_{\partial G} e_{ijk}(e_{klm}\psi B_l u_m)_{,j} N_i dS = 0 , \tag{7.4.15}$$

because the integrand in the last integral is a tangential derivative. Secondly, we note that for a pre-stressed linear elasticity problem and with the classical HOOKEan relation for c_{ijkl}, one has

$$c_{ijkl} u_{i,k} u_{j,l} = T_{jk} u_{i,j} u_{i,k} + \frac{E}{1+\nu} \left(\frac{\nu}{1-2\nu} e_{kk} e_{ll} + e_{kl} e_{kl} \right) , \tag{7.4.16}$$

where $2e_{kl} = u_{k,l} + u_{l,k}$. As the third step, we use GAUSS' divergence theorem to evaluate

$$\int_{G^+} \psi_{,i} \psi_{,i} dV = \int_{\partial G} \psi \psi_{,i} N_i dS - \int_{G^+} \psi \psi_{,ii} dV . \tag{7.4.17}$$

Finally, to dispose of the integral over the infinite region G^+ in the result above, we impose the extra constraint

$$\psi_{,ii} = \Delta\psi = 0 , \quad \boldsymbol{X} \in G^+ . \tag{7.4.18}$$

We note that, as the constraints do not prescribe ψ at ∂G, there is still freedom left for the variation of ψ.

Using the four steps above, we obtain J in the form

$$J = \frac{1}{2} \int_G T_{jk} u_{i,jk} u_i dV - \frac{E}{2(1+\nu)} \int_G \left(\frac{\nu}{1-2\nu} e_{kk} e_{ll} + e_{kl} e_{kl} \right) dV$$
$$-\frac{1}{2\mu_0} \int_{\partial G} \Big[(\psi + B_k u_k)_{,j} B_j u_i + \tfrac{1}{2} B_k B_k (u_i u_{j,j} - u_{i,j} u_j - u_{j,i} u_j)$$
$$-\tfrac{1}{2} B_j B_k (u_{j,k} + u_{k,j}) u_i + (\psi_{,i} + B_{i,j} u_j - B_j u_{i,j}) \psi \Big] N_i dS . \tag{7.4.19}$$

Here, the first integral on the right-hand side is the contribution of the pre-stresses T_{jk} to the elastic energy, the second term is the classical (HOOKEan)

linear elastic energy due to the perturbed deformations and the last integral
is the perturbed magnetic interaction energy.

The buckling current follows from $J = 0$; in order to obtain a direct
expression for I_{0cr} we scale the magnetic variables with respect to I_0. For
this we introduce the following dimensionless field variables:

$$\hat{B}_i = \frac{2\pi R}{\mu_0 I_0} B_i , \quad \hat{\psi} = \frac{2\pi}{\mu_0 I_0} \psi , \quad \hat{T}_{ij} = \frac{(2\pi R)^2}{\mu_0 I_0^2} T_{ij} , \quad \hat{u}_i = \frac{1}{R} u_i , \quad (7.4.20)$$

where R is some length parameter that has to be chosen for a specific problem
under consideration. With use of this in (7.4.19), the buckling equation $J = 0$
yields (we subsequently omit the hats)

$$\sqrt{\frac{\mu_0}{E}} \left(\frac{I_0}{2\pi R} \right)_{cr} = \sqrt{\frac{W}{K}} , \qquad (7.4.21)$$

where W is the scaled elastic energy

$$W = \frac{1}{2(1+\nu)} \int_G \left(\frac{\nu}{1-2\nu} e_{kk}e_{ll} + e_{kl}e_{kl} \right) dV , \qquad (7.4.22)$$

and K is

$$K = -\frac{1}{2} \int_{\partial G} [(\psi + B_k u_k)_{,j} B_j u_i + \tfrac{1}{2} B_k B_k (u_i u_{j,j} - u_{i,j} u_j - u_{j,i} u_j)$$
$$- \tfrac{1}{2} B_j B_k (u_{j,k} + u_{k,j}) u_i + (\psi_{,i} + B_{i,j} u_j - B_j u_{i,j})\psi] n_i dS$$
$$+ \tfrac{1}{2} \int_G T_{jk} u_{i,jk} u_i dV . \qquad (7.4.23)$$

The final result (7.4.21)–(7.4.23) immediately gives the buckling current
I_0 once we know ψ and $\boldsymbol{u}$, either exactly or approximately. Assuming that
the buckling deflection is in the x-direction, the external potential Ψ has to
satisfy the constraints:

$$\Delta\psi(x, y, z) = 0 , \quad \text{for } \boldsymbol{x} \in G^+,$$
$$\frac{\partial}{\partial n}(\psi + B_i u_i) = 0 , \quad \text{for } \boldsymbol{x} \in \partial G , \qquad (7.4.24)$$
$$\psi \to 0 , \quad \text{for } |\boldsymbol{x}| \to \infty .$$

We now illustrate the variational method on the example of a set of two
parallel superconducting rods.

Example: Buckling of two parallel superconducting rods (variational method)

We consider a set of two identical parallel infinitely long superconducting
circular rods of radius R. The distance between the centres of the two rods

is $2d$. The rods are periodically supported over a distance L. The currents through the rods have magnitude I_0 and are in the same direction. Axes X, Y, Z are chosen as in Fig. 7.1. The dominant buckling mode turns out to be symmetric and in the X-direction; this means that the deflection of the second rod is equal but opposite to that of the first rod. For the first rod, we choose the displacement field (appropriate for slender rods in bending) as

$$u_1 = V(z) , \qquad u_2 = 0 , \qquad u_3 = -xV'(z) . \qquad (7.4.25)$$

The elastic energy is then, as in (7.3.38),

$$W = \frac{\pi}{8} ER^4 \int_0^L (V''(z))^2 dz . \qquad (7.4.26)$$

With the common separation of variables

$$\psi(x,y,z) = \Psi(x,y)V(z) , \quad \text{and} \quad V''(z) + \lambda^2 V(z) = 0 , \qquad (7.4.27)$$

the constraints (7.4.24) for ψ transform into

$$
\begin{aligned}
\Delta\Psi(x,y) - \lambda^2\Psi(x,y) &= 0 , && \text{for } (x,y) \in D^+, \\
\frac{\partial}{\partial n}(\Psi + B_x) &= 0 , && \text{for } (x,y) \in \Gamma , \\
\Psi &\to 0 , && \text{for } \sqrt{x^2 + y^2} \to \infty ,
\end{aligned}
\qquad (7.4.28)
$$

where now, due to the periodic supports, $\lambda = \pi/L$.

Moreover, with (7.4.25) expression (7.4.23) for K reduces considerably; all terms except the first three are zero (or, better, $O(\delta^2)$). Thus, K for one rod becomes

$$K = -\frac{1}{2}\int_\Gamma \left[(\Psi + B_x)\frac{\partial B_x}{\partial n} + \frac{\lambda^2 x}{2R^2}(B_x^2 + B_y^2)n_x\right] ds \int_0^L V^2(z)dz . \quad (7.4.29)$$

In [122, Sect. 4], the details of the calculation of B_x, B_y and Ψ are presented, and as these analytical calculations are exact, this leads to an exact expression for the buckling current (exact within the concept of the slender beam theory). This eventually results in (see [122, eq. (5.17)])

$$I_{0cr} = \frac{\pi R}{\sqrt{Q}}\left(\frac{\pi R}{L}\right)^2 \sqrt{\frac{E}{\mu_0}} , \qquad (7.4.30)$$

where

$$Q = Q(d/R) = \frac{4}{\beta^2}\sum_{n=1}^\infty na_n^2\left(\frac{1-\alpha^{2n}}{1+\alpha^{2n}}\right)^3 , \qquad (7.4.31)$$

with

$$\beta = \sqrt{\frac{d^2}{R^2} - 1} , \quad a_n = \frac{\alpha^{2n}}{(1+\alpha^{2n-2})(1+\alpha^{2n+2})} , \quad \alpha = \frac{d}{R} - 1 . \quad (7.4.32)$$

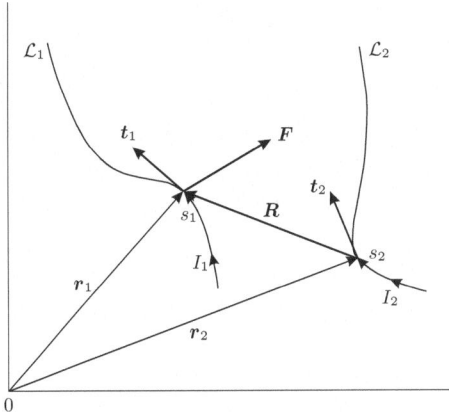

Fig. 7.4. Two curves carrying current; generalization of the law of BIOT and SAVARD

Table 7.2. Values of Q and $1/\sqrt{Q}$ as functions of d/R

d/R	1	1.5	2	3	4	6	8	10
Q	0.311	0.220	0.168	0.0935	0.0568	0.0266	0.0153	0.00985
$1/\sqrt{Q}$	1.79	2.13	2.44	3.27	4.20	6.13	8.09	10.08

Values for Q and $1/\sqrt{Q}$ as functions of d/R are given in Table 7.2. For later reference we note that for large values of d/R the factor $1/\sqrt{Q}$ approaches d/R.

It is of technical interest to compare this exact result with the result of a less accurate, but much simpler solution procedure that is based upon a generalization of the law of BIOT and SAVART. The basic relation for this method is given by MOON in [156, eq. (2-6.4)]. Let $\mathcal{L}_1$ and $\mathcal{L}_2$ be two distinct curves in $\mathcal{R}^3$, carrying the same electric current I_0, see Fig. 7.4. Moreover, let $\mathcal{P}_1$ and $\mathcal{P}_2$ be two points on $\mathcal{L}_1$ and $\mathcal{L}_2$ with position vectors $r_1(s_1)$ and $r_2(s_2)$, respectively, where s_1 and s_2 are the corresponding arc lengths. The force per unit of length in $\mathcal{P}_1$ acting on $\mathcal{L}_1$ is now calculated as the LORENTZ force due to the current through $\mathcal{L}_1$ times the magnetic field created by $\mathcal{L}_2$. The latter follows from a generalization of the law of BIOT and SAVARD as given in [156, eq. (2-6.3)–(2-6.4)]. According to [156, eq. (2-6.4)], this force is then given by

$$\boldsymbol{F}(s_1) = \frac{\mu_0 I_0^2}{4\pi} \int_{\mathcal{L}_2} \frac{(\boldsymbol{t}_1(s_1) \times (\boldsymbol{t}_2(s_2) \times \boldsymbol{R}(s_1, s_2)))}{R^3(s_1, s_2)} \, ds_2 \,, \qquad (7.4.33)$$

where $\boldsymbol{t}_1$ and $\boldsymbol{t}_2$ are unit tangent vectors along $\mathcal{L}_1$ and $\mathcal{L}_2$, respectively, and $\boldsymbol{R}$ is the position vector from $\mathcal{P}_2$ to $\mathcal{P}_1$, i.e.

$$t_1 = \frac{dr_1}{ds_1} , \qquad t_2 = \frac{dr_2}{ds_2} , \qquad R = r_1 - r_2 . \tag{7.4.34}$$

The above formula for F is an approximation in so far as

1. the three-dimensional current-carrying bodies are considered as one-dimensional curves (thus, for instance, the specific shape of the cross-section and the distribution of the current over the cross-section are disregarded);
2. the force due to the self-field of $\mathcal{L}_1$ is neglected.

Both these effects are taken into account by the (exact) variational method. We shall show here that this BIOT–SAVARD method will yield only good agreement with the exact results of the variational method if the two current filaments are sufficiently apart from each other. To this end, we will apply the BIOT–SAVARD method to the example above. In the symmetrical buckling mode the deflections of the two rods are equal but opposite and thus in the deflected state we have

$$
\begin{aligned}
r_1 &= (d + V(s_1)) \, e_x + s_1 \, e_z , \\
r_2 &= (-d - V(s_2)) \, e_x + s_2 \, e_z , \\
R &= (2d + V(s_1) + V(s_2)) \, e_x + (s_1 - s_2) \, e_z , \\
t_1 &= V'(s_1) \, e_x + e_z , \\
t_2 &= -V'(s_2) \, e_x + e_z .
\end{aligned}
\tag{7.4.35}
$$

These formulas enable us to evaluate (7.4.33). In doing so, we must realize that the displacements are small and, hence, a linearization with respect to these displacements is allowed. In this sense, we approximate R^3 by

$$R^3 = R_0^3 \left[1 + \frac{6d}{R_0^2} (V(s_1) + V(s_2)) \right] , \tag{7.4.36}$$

where

$$R_0 = R_0(s_1, s_2) = \sqrt{4d^2 + (s_1 - s_2)^2} \geq 2d . \tag{7.4.37}$$

In the same way linearizing the numerator, we find an expression for F of the form

$$F(s_1) = F^{(0)}(s_1) + f(s_1) , \tag{7.4.38}$$

where $F^{(0)}$ is independent of, and f is linear, in the displacement $V(s_1)$. Hence, $F^{(0)}$ is the force in the pre-buckled state (causing the so-called pre-buckling deflections), which may be discarded for the determination of the buckling current. Therefore, we define $q(s_1)$ as the force per unit of length in the x-direction acting on the deflected rod by

$$q(s_1) = (f(s_1) \cdot e_x) , \tag{7.4.39}$$

and this force governs the deflection of the rod through the beam equation (from here on, we replace s_1 by z, the axial coordinate of the rod)

$$EIV^{\mathrm{iv}}(z) = q(z) \,. \tag{7.4.40}$$

The procedure described above yields the following expression for the load

$$q(z) = \frac{\mu_0 I_0^2}{4\pi} \int_{-\infty}^{\infty} \left[\frac{V(z) + V(\zeta) - (z - \zeta)V'(\zeta)}{R_0^3} - \frac{12d^2(V(z) + V(\zeta))}{R_0^5} \right] d\zeta \,. \tag{7.4.41}$$

After two integrations by parts, in which it is used that $V(z)$ is periodic in z, (7.4.41) becomes

$$q(z) = \frac{\mu_0 I_0^2}{4\pi d^2} \left[V(z) + d^2 \int_{-\infty}^{\infty} \frac{V(\zeta) - V(z)}{R_0^3} \, d\zeta \right] \,. \tag{7.4.42}$$

At this step, we note that the second term on the right-hand side of (7.4.42) is of the order $(d/L)^2$ with respect to the first term, and since we want to restrict ourselves to cases where $d \ll L$ (in fact, d/L is small of the same order as δ), we may neglect this term. Thus, (7.4.40) becomes

$$V^{\mathrm{iv}}(z) - \frac{\mu_0 I_0^2}{4\pi d^2 EI} \, V(z) = 0 \,. \tag{7.4.43}$$

For a simply supported rod, the boundary conditions are

$$V(0) = V(L) = V''(0) = V''(L) = 0 \,, \tag{7.4.44}$$

and the first buckling mode that satisfies these boundary conditions is

$$V(z) = A \sin\left(\frac{\pi z}{L}\right) \,. \tag{7.4.45}$$

After substitution into (7.4.43) this brings us to the following expression for the buckling current

$$I_{0cr} = \pi d \left(\frac{\pi R}{L}\right)^2 \sqrt{\frac{E}{\mu_0}} \,. \tag{7.4.46}$$

Table 7.2 shows that $1/\sqrt{Q} \approx d/R$ for values of $d/R \geq 4$ (the relative difference is then less than 5%) and then (7.4.30) and (7.4.46) yield approximately the same result for I_{0cr}. However, if the filaments are nearer to each other the correspondence becomes worse. In the limit $d/R \to 1$, formula (7.4.46) predicts a buckling current that is about 80% lower than that obtained with (7.4.30). From this we conclude that the so-called *direct* BIOT–SAVARD *method* yields acceptable results for the buckling current only if the current filaments are sufficiently far away from each other.

7.4.2 A Set of Two Concentric Superconducting Rings

Consider two concentric superconducting tori (or rings) both having a circular cross-section of radius R. The central line of the outer torus has radius

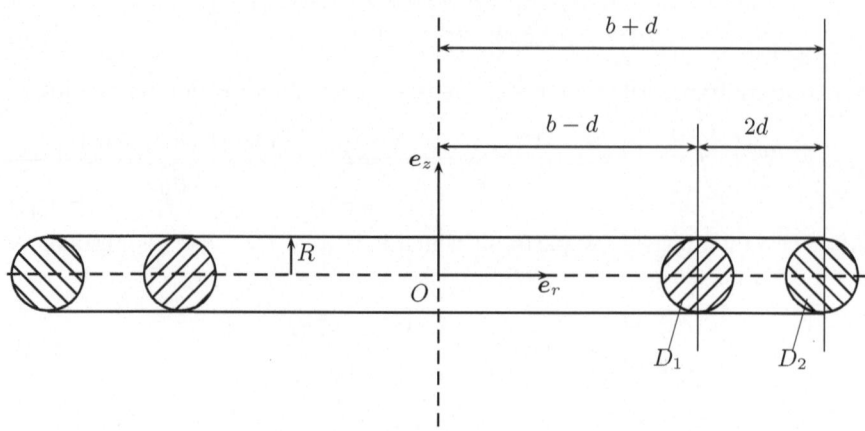

Fig. 7.5. Cross-section of a pair of concentric tori

$b + d$, and that of the inner one $b - d$, where $d > R$. A system of cylindrical coordinates in the center O of the system is introduced as given in Fig. 7.5. We assume that $\varepsilon = R/b \ll 1$, and that d/R is of order one; in view of this, the system of the two tori is called a slender system. The total current through each of the tori is called I_0, and the currents in the tori are equally directed. We consider only in-plane buckling, and we assume that the lowest buckling mode corresponds to an anti-symmetric buckling mode. The deflection of the outer torus ($i = 1$) and the inner ($i = 2$) torus can then be written as $w_i(\phi)e_r + v_i(\phi)e_\phi$, $i = 1, 2$, where $w_2 = -w_1$ for the anti-symmetric buckling mode. The tori are assumed inextensible, which can be shown to be expressible as $v'_i(\phi) + w_i(\phi) = 0$, $i = 1, 2$. In analogy with BERNOULLI's theory for the bending of slender beams, the 3-D displacement field of the i-th torus may be written as, up to $O(\varepsilon^2)$,

$$u_r^{(i)} = w_i(\phi) ,$$

$$u_\phi^{(i)} = v_i(\phi) - \frac{(r - b_i)}{b_i} \left(w'_i(\phi) - v_i(\phi) \right) , \quad i = 1, 2 , \qquad (7.4.47)$$

$$u_z^{(i)} = 0 ,$$

where $b_1 = b + d$, $b_2 = b - d$. The representation (7.4.47) yields the classical expression for the elastic energy of a slender inextensible ring in bending,

$$W_i = \frac{EI_z}{2b^3} \int_0^{2\pi} (w''_i + w_i)^2 d\phi \, (1 + O(\varepsilon)) , \qquad (7.4.48)$$

where we have used that $r = b(1 + O(\varepsilon))$, $b_i = b(1 + O(\varepsilon))$ and where $I_z = \pi R^4/4$.

As in the preceding examples, the perturbed potential $\psi(r, \phi, z)$ is separated according to

$$\psi(r, \phi, z) = \Psi(r, z)\omega(\phi) , \tag{7.4.49}$$

where $\omega(\phi)$ must be periodic in ϕ. For the first buckling mode, we take $\omega(\phi) = \cos(2\phi)$ and, accordingly, choose for the displacements of the central lines

$$w_1(\phi) = w\cos(2\phi) , \quad \text{and} \quad w_2(\phi) = -w\cos(2\phi) . \tag{7.4.50}$$

This yields for the total elastic energy (7.4.48) of the system

$$W = W_1 + W_2 = \frac{9\pi^2}{b^3}EI_z\, w^2 = \frac{9\pi^2}{4}\frac{ER^4}{b^3}\, w^2 . \tag{7.4.51}$$

The electromagnetic interaction term $\mathcal{K} = I_0^2 K$ is found in [218] in a fully analytical way, but we omit the details of this analysis here and refer to [218]. However, what we do want to point out is the following: for the slender system of tori we expect that on a local scale, i.e. for a small line element $bd\phi$, the electromagnetic interaction between two such line elements of two different tori will not differ very much from the corresponding interaction between two parallel rods. This expectation is confirmed by the result for $\mathcal{K}$ found in [218, eq. (4.84)]. In our notation, this result reads ($K = \mathcal{K}/I_0^2$)

$$K = \frac{\mu_0 bQ}{4R^2}\, w^2 , \tag{7.4.52}$$

where $Q = Q(d/R)$ is as given in (7.4.31). This yields for the buckling current

$$I_0 = \frac{3\pi R^3}{b^2}\sqrt{\frac{E}{\mu_0 Q}} . \tag{7.4.53}$$

Note that this result is of the same form as the resulting buckling current for two parallel rods as given in (7.4.30); the only difference is due to the elastic energy. The basic reason for this is that the expression for $\mathcal{K}$ as given by (7.4.29) for slender systems is dominated by its first term. This term takes the same value for all types of slender systems that we will consider in this chapter, at least in a zeroth-order approximation with respect to ε. Since $\mathcal{K}$ is for one part determinant for the buckling value (the other part being determined by the elastic energy W) we may state that the buckling current for any slender pair of parallel curved rods is equal to that of an equivalent pair of parallel straight rods *times the ratio of the elastic energies*. Of course, the concept "slenderness" has to be defined properly in each problem at hand.

As we did already in the example on the buckling of two parallel rods, we shall derive here also an approximate expression for the buckling current based on the so-called *direct* BIOT–SAVARD method as given by MOON in [156, eq. (2-6.4)]. We will do this here not for the system of two concentric tori discussed above, but for a set of two parallel tori, or rings, as depicted in Fig. 7.6. The tori have equally directed currents. In this method, the tori are

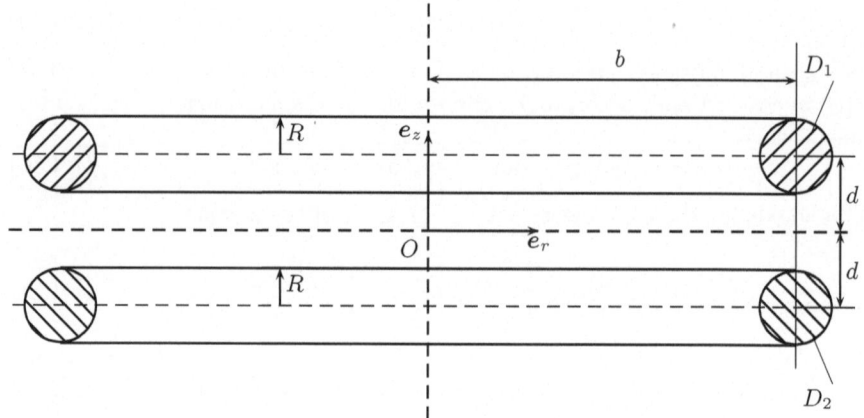

Fig. 7.6. Cross-section of a pair of parallel tori

considered as 1-dimensional rings. We consider here out-of-plane buckling, which consists of out-of-plane bending ($w_i(\phi_i)$) and torsion ($\tau_i(\phi_i)$: the twist per unit of length along the central line). The positions on the two deflected rings ($i = 1, 2$) are then given by

$$
\begin{aligned}
r_1 &= b\, e_{r_1}(\phi_1) + (d + w_1(\phi_1))\, e_z \,, \\
r_2 &= b\, e_{r_2}(\phi_2) + (-d + w_2(\phi_2))\, e_z \,,
\end{aligned}
\tag{7.4.54}
$$

where r_i, ϕ_i, z are cylindrical coordinates connected to the i-th ring. In a way analogous to that used in the derivation of (7.4.43) we can also here derive an expression for the force per unit of length on one ring due to the current in the other ring. In this case, this force is in the z-direction and equal to, for the first ring and when neglecting $O(d/b)$-terms,

$$
f_z(\phi_1) = \frac{\mu_0 I_0^2}{8\pi d^2}\, [w_1(\phi_1) - w_2(\phi_1)] \,.
\tag{7.4.55}
$$

The ring equations for out-of-plane bending and torsion can a.o. be found in [156, eq. 6-7.18]; with $A = EI = (\pi/4)ER^4$ and $C = GI_p = (\pi/2)ER^4/(1+\nu)$ and in our notations they read

$$
-\frac{EI}{b^4}\,[w_1^{\mathrm{iv}}(\phi_1) - b\tau_1''(\phi_1)] + \frac{GI_p}{b^4}\,[w_1''(\phi_1) + b\tau_1''(\phi_1)] + f_z(\phi_1) = 0 \,,
$$
$$
-\frac{EI}{b^2}\,[w_1''(\phi_1) - b\tau_1(\phi_1)] - \frac{GI_p}{b^2}\,[w_1''(\phi_1) + b\tau_1''(\phi_1)] = 0 \,.
\tag{7.4.56}
$$

Putting

$$
w_i(\phi_i) = W_i \cos(2\phi_i) \,, \qquad \tau_i(\phi_i) = T_i \cos(2\phi_i) \,, \qquad i = 1, 2 \,,
\tag{7.4.57}
$$

and using $GI_p = EI/(1 + \nu)$, we obtain from the second relation of (7.4.56) for $i = 1$

$$T_1 = -\frac{4(2+\nu)}{(5+\nu)} \frac{W_1}{b} . \tag{7.4.58}$$

With this result the first equation of (7.4.56) yields

$$\frac{36EI}{(5+\nu)b^4} W_1 \cos(2\phi) = f_z(\phi) = \frac{\mu_0 I_0^2}{8\pi d^2} [W_1 - W_2] \cos(2\phi) . \tag{7.4.59}$$

An analogous relation holds for W_2. It is then easily seen that the lowest buckling current occurs for $W_2 = -W_1$ (anti-symmetric buckling mode) and is equal to (with $I = \pi R^4/4$)

$$I_0 = \frac{6\pi dR^2}{b^2} \sqrt{\frac{E}{\mu_0(5+\nu)}} . \tag{7.4.60}$$

As can be found in [218], the "exact" variational method would yield for the buckling value

$$I_0 = \frac{6\pi R^3}{b^2} \sqrt{\frac{E}{\mu_0(5+\nu)Q}} . \tag{7.4.61}$$

Hence, the result (7.4.60) is only in agreement with (7.4.61), if $1/\sqrt{Q} = d/R$, a result that was already found at the end of Sect. 7.4.1. Thus, we conclude again that the direct BIOT–SAVARD method yields an acceptable approximation (relative error less than 5%) for the buckling current only if the current filaments are far enough away from each other, say $d/R \geq 4$. It seems superfluous to say that an analogous result can also be obtained for the system of two concentric tori; see [218].

7.4.3 How to Use the Law of Biot and Savard in the Variational Principle

In the examples presented in the preceding sections, we used the variational principle in the exact sense. This means that we looked for and found *exact* rigid-body fields B_0 and perturbed fields b, which we used in the variational expression for $\mathcal{K}$, thus obtaining exact values for the buckling current. Exact always means here within the range of slender-beam theory, so for $R/l \ll 1$, where R is the radius of the circular cross-section of the beam and l a characteristic measure of length of the beam. Essential in our model for a superconductor is the assumption that the current runs over the surface of the superconductor only, thus shielding the body from a magnetic field. Hence, there is no magnetic field inside the body. In the preceding section, we also presented a more direct method based on a formula for the Lorentz force on a current carrier due to the current in another conductor derived from the BIOT–SAVARD law. We referred to this method as the *direct* BIOT–SAVARD *method.* The second method is less exact than the variational method, but much easier to work with in practice. A comparison of the two methods showed reasonable agreement as long as the conductors are not too close to one another.

In this respect, it seems reasonable to look for a method that combines the advantages of the two methods, i.e. the greater exactness of the variational method and the convenience in the use pertinent to the direct BIOT–SAVARD method. To this end, we shall use in the variational formulation for $\mathcal{K}$ an admissible magnetic field b obtained on the basis of the law of BIOT and SAVARD. Our expectation that this combined approach will yield a useful approximation for the buckling current is supported by the observed correspondence between the results of the two respective methods. As this method consists of a combination of the variational principle and the BIOT–SAVARD law, we will refer to this method as the *combined method.* Our hope that this method will result in buckling values that are closer to the exact values will be confirmed by the results to be presented in the next section. Moreover, we will show there how we can obtain in a relatively easy way buckling values for systems that we can not solve in an exact way.

In the next section, we shall present the detailed derivation of this method; moreover, we shall give the specific nature of the simplifications and the restrictions under which they are allowable. Applications to such complex systems as sets of helical or spiral conductors will also be presented. The combined method will calculate in a convenient way buckling currents for more or less complex structures, which are sufficiently precise in practical applications.

7.5 Some Results for Superconducting Structures

7.5.1 Review of Specific Structures and Some Results

In this section, we present a series of examples of applications of the variational principle described in the preceding chapter to superconducting systems. These systems always consist of slender rod-like structures, of circular cross-section, radius R, and carrying a total current I_0. These examples concern:

1. A set of two parallel rods, infinitely long, but periodically supported over distances L, and a distance $2d$ apart from each other ($R < d \ll L$); cf. [122]; see also Sect. 7.4.1.
2. A set of two concentric tori (or rings) in one plane; the rings have radii b_1 and b_2, and the distance between them is $2d = b_2 - b_1$ ($R < d \ll b_2$); cf. [218]; see also Sect. 7.4.1.
3. A set of two identical parallel tori (or rings), of radius b and distance $2d$ ($R < d \ll b$); cf. [218].

4. Sets of n , $n > 2$, equidistant parallel rods (like the set in example 1.); cf. [124]; see also Sect. 7.5.2.
5. An infinite helical conductor, periodically supported at every n turns; the radius of the helix is b and the pitch is h, the distance between two turns is $2d = 2\pi h$, and the support length is $2\pi bn$ ($R < \pi h \ll \pi b$), cf. [256] or [257]; see also Sect. 7.5.3.
6. A finite helical conductor (like the one in example 5.) of n turns, simply supported at its end points; cf. [257]; see also Sect. 7.5.3.
7. A flat spiral of n turns; the radius is given as a function of the arc φ, for $\varphi \in [0, 2\pi n]$, by $b(\varphi) = b_0 + h\varphi$, where h is the constant pitch; the distance between two turns is $2d = 2\pi h$ ($R < d \ll b_0$); the spiral is simply supported at its end points; cf. [257]; see also Sect. 7.5.3.

Problems 1, 2, and 3 are solved completely by analytical means, using conformal mapping and complex function theory. These mathematical procedures were necessary to obtain an exact solution for the magnetic fields, both for the rigid-body and the perturbed states (referring to the deformed state of the structure), which, in turn, were needed for the calculation of the electromagnetic interaction integral K, as given in (7.4.23). In this way, an exact value for the buckling current is obtained (exact within the concept of slender-beam theory, i.e. up to $O(R^2/L^2)$-terms). The analytical approach to problem 4 resulted in a set of integral equations for the perturbed magnetic field, which had to be solved numerically. Nevertheless, the buckling value found was exact, in the same sense as in the preceding examples. For problems 5 to 7 no analytical solution was found, but here the variational principle was used to find a good approximation for the buckling current; for this approximation the law of BIOT and SAVARD was used to obtain an admissible magnetic field for the variational method; this method was already explained in Sect. 7.4.3. In the latter three problems a result is also used that states that the value of K for a set of two (or more) rings is equal to the value of K for a set of two straight parallel rods. This result was proved to be true for the set of two concentric rods; see also Sect. 7.4.2. It can also be made plausible by the following reasoning: consider two rings and take a point $\mathcal{A}_1$ on one of the rings; then the interaction of the second ring with $\mathcal{A}_1$ is concentrated at a point $\mathcal{A}_2$ and its direct neighbourhood, where $\mathcal{A}_2$ is that point of the other ring that is closest to $\mathcal{A}_1$. Then, according to the slenderness condition, $L \gg d$, in a d-environment of $\mathcal{A}_2$ the ring may be considered to be locally straight. The errors induced in this way are of $O(d^2/L^2)$. This reasoning also applies to a slender helix or spiral and, therefore, the contribution to K of two interacting turns of a helical or spiral coil can be calculated by replacing them *locally* by a set of two parallel rods, for which the value of K is known.

Results for the buckling current for the 7 examples listed above are presented in Table 7.3. We conclude that for all the problems considered here the buckling current I_{0cr} contains the common factor $\mathcal{J}$, where $\mathcal{J} = (dR^2/L^2)\sqrt{E/\mu_0}$, while the coefficient preceding $\mathcal{J}$, is only a function of

Table 7.3. Buckling currents I_{0cr} for problems 1 to 7; E and ν are YOUNG's and POISSON's moduli, respectively, the common factor $\mathcal{J} = (dR^2/L^2)\sqrt{E/\mu_0}$, where $L = \pi b$ for problems 2, 5, and 6, $L = \pi(b_1+b_2)/2$ for problem 3, and $L = \pi(b_0+n\pi h)$ for problem 7, and $q = R/d\sqrt{Q}$, with $Q = Q(d/R)$ according to (7.4.31); the factors $\alpha_n = \alpha_n(n)$ are given below, and $\kappa = \kappa(d/R)$, $N(n)$, $\lambda(n)$, and $\lambda(n,h)$ will be specified in the next sections

System	1.	2.	3.	4.
I_{0cr}	$\pi^3 q\mathcal{J}$	$3\pi^3 q\mathcal{J}$	$\frac{6\pi^3 q}{\sqrt{5+\nu}}\mathcal{J}$	$\alpha_n\pi^3 q\mathcal{J}$

System	5.		6.	7.
I_{0cr}	$\pi^2\sqrt{\frac{N(n)}{(2n-1)(1+\nu)\kappa}}\mathcal{J}$		$\pi^3\sqrt{\frac{2\lambda(n)}{(1+\nu)\kappa}}\mathcal{J}$	$\pi^3\sqrt{\frac{2\lambda(h,n)}{\kappa}}\mathcal{J}$

R/d and is different for each problem. The factor $\alpha_n(n)$ is for $n = 3, 4, 5$, and ∞ given by $\alpha_3 = \sqrt{2/3}$, $\alpha_4 = 0.753$, $\alpha_5 = 0.723$, and $\alpha_\infty = 2/\pi = 0.637$. Formulas for $\kappa = \kappa(d/R)$, $N(n)$, $\lambda(n)$, and $\lambda(n,h)$ will be derived in the subsequent sections.

In the next section, we shall show how we can use the law of BIOT and SAVARD in the variational method. The basic idea behind this approach is that the magnetic field obtained by the law of BIOT and SAVARD satisfies the constraints of the variational principle and, hence, constitutes an admissible field. In the same way as shown at the end of Sect. 7.4.1, we can directly from the law of BIOT and SAVARD calculate the Lorentz forces on the structures and thus find approximate values for the buckling currents. These values can be obtained from Table 7.3 by substituting $q = 1$ for problems 1 to 4, and $\kappa = 1$ for problems 5 to 7. What we will find with the combined BIOT–SAVARD-variational method are buckling values in between the values from the variational method (exact) and the direct BIOT–SAVARD method. Hence, the proposed *combined method* yields, on the one hand, an improvement of the direct BIOT–SAVARD method, and, on the other hand, an easy way to obtain an admissible magnetic field. The details of the derivation of the combined method will now be presented in the following section.

7.5.2 The Combined (Variational Biot-Savard) Method

Before starting with the derivation of the combined method, we first shortly recapitulate from Sect. 7.4.1 the main lines of the variational principle for a superconducting body in vacuum. In this case, there is only a magnetic field B in the vacuum surrounding the body, which has to satisfy AMPÈRE's law (see (7.4.2))

$$\int_C (B, ds) = \mu_0 I_0 , \tag{7.5.1}$$

where C is a contour entirely in the vacuum, encircling the current carrier and I_0 is the prescribed current through the superconductor.

The variational method is based upon a chosen expression for the LA-GRANGEan for an elastic superconducting body, which, according to (7.4.1), is given by

$$L = \frac{1}{2\mu_0} \int_{G^+} (\boldsymbol{B}, \boldsymbol{B}) dV - \int_{G^-} \rho U dV , \qquad (7.5.2)$$

where the first integral represents the magnetic energy of the vacuum field and the second is the elastic energy of the deformed body. Note that here G^+ and G^- are the configurations of the vacuum and the body, respectively, *in the deformed state*.

Variation of L should satisfy the constraint (7.5.1) and

$$e_{ijk}B_{k,j} = 0 , \quad \boldsymbol{x} \in G^+ , \quad \text{and} \quad \boldsymbol{B} \to 0 , \quad |\boldsymbol{x}| \to \infty . \qquad (7.5.3)$$

The total field $\boldsymbol{B}$ is split into the rigid-body field $\boldsymbol{B}_0$ and the perturbed field $\boldsymbol{b}$, where the latter is due to the displacement $\boldsymbol{u}$ of the elastic body. The perturbed field is expressed in the perturbed magnetic potential ψ by

$$b_i = \psi_{,i} , \qquad (7.5.4)$$

where this potential has to satisfy the extra constraint (see (7.4.18))

$$\psi_{,ii} = 0 , \quad \text{and} \quad \psi(\boldsymbol{x}) \to 0 , \quad |\boldsymbol{x}| \to \infty . \qquad (7.5.5)$$

The second variation of L, called J, can then be determined in terms of $\boldsymbol{B}_0$, ψ, $\boldsymbol{u}$ and the associated linear deformations, or strains, $e_{ij} = (u_{i,j} + u_{j,i})/2$, eventually resulting in the expression (7.4.19).

In the examples presented thus far, we used the exact solutions for $\boldsymbol{B}_0(\boldsymbol{x})$ and $\boldsymbol{b}(\boldsymbol{x})$. However, if the systems become more complex, it becomes increasingly difficult, not to say impossible, to determine these fields exactly. Therefore, we are looking for admissible fields from which we may hope that they are not too far away from the exact ones. Admissible fields are fields $\boldsymbol{B}_0(\boldsymbol{x})$ and $\psi(\boldsymbol{x})$ that satisfy the constraints (7.5.1), (7.5.3) and (7.5.5). We will employ these fields to obtain an approximation for the functional J. Here, we will use the BIOT–SAVARD fields as they can be derived from the law of BIOT and SAVARD (see [156, eq (2-6.3)])

$$\boldsymbol{B}(\boldsymbol{x}) = \frac{\mu_0 I_0}{4\pi} \int_{\mathscr{L}} \frac{(\boldsymbol{t}(s) \times \boldsymbol{R}(\boldsymbol{x}, s))}{R^3(\boldsymbol{x}, s)} ds , \qquad (7.5.6)$$

where $\boldsymbol{t}(s)$ is the unit tangent vector along $\mathscr{L}$ in a point $\mathscr{P}$ on $\mathscr{L}$ having arc length s, and $\boldsymbol{R}(s, \boldsymbol{x})$ is the position vector of the point $x \in G^+$ with respect to $\mathscr{P}$, while $R = |\boldsymbol{R}|$. We will show that these fields are admissible. The already rather good correspondence between the results of the direct BIOT–SAVARD method and the variational method as found in the earlier sections, supports us in our opinion that this choice will lead us to a good approximation of the buckling current.

Before proceeding with the explicit derivation of the admissible fields, we first simplify expression (7.4.19). To this end, we realize that we wish to apply the variational principle to slender, beam-like, bodies. In the bending of such structures the pure deformations are much smaller than the local rotations of the slender body. This motivates us to neglect in the magnetic term in (7.4.19) those terms that contain a factor of order $B^2\epsilon$ (here, the small parameter ϵ is some norm of the deformations e_{ij}). Moreover, since the term containing the prestresses T_{ij} in the elastic energy integral is small of the same order, this term must be neglected too. We like to mention that the neglect of the above terms leads to an error that is small of order R/l and, hence, is justified for systems of slender bodies. Altogether, this leads us to the following simplified expression for J (in which B is used for B^0):

$$
J = -\frac{E}{2(1+\nu)} \int_G \left(\frac{\nu}{1-2\nu} \, e_{kk}e_{ll} + e_{kl}e_{kl} \right) dV
$$

$$
-\frac{1}{2\mu_0} \int_{\partial G} [(\psi + B_k u_k)_{,j} B_j u_i + (\psi_{,i} + B_{i,j}u_j - B_j u_{i,j})\psi] N_i dS
$$

$$
= -W + I_0^2 K \ . \tag{7.5.7}
$$

We have written the last term in (7.5.7) as $I_0^2 K$, because K is independent of I_0 then. This is true because B and ψ are both linear in I_0. With $B_{i,j} = B_{j,i}$, as follows from (7.5.3), we can reduce the integral $I_0^2 K$ still somewhat further to obtain

$$
I_0^2 K = -\frac{1}{2\mu_0} \int_{\partial G} [(\psi + B_k u_k)_{,j}(B_j u_i N_i + \psi N_j)] dS \ , \tag{7.5.8}
$$

where we have neglected again a term proportional to $B^2\epsilon$.

It is possible to derive explicit expressions for admissible $B(x)$ and $\psi(x)$ from (7.5.6) for arbitrarily curved circuits, but we refrain from doing so here. Instead, we use here a result of Sect. 7.5.1 that taught us that for the derivation of the K-integral it suffices to calculate its value for a corresponding system of two parallel rods. We generalize this result by stating that this value for K can also be used for systems of curved beams, not only for a pair of rings (as we showed in Sect. 7.4.2), but also for such structures as helical or spiral conductors, provided the system satisfies the condition of being slender. Hence, in this case we only need to calculate $B(x)$ and $\psi(x)$ for a straight current carrier $\mathscr{L}$ as we shall first do now.

Consider an infinite one-dimensional conductor $\mathscr{L}$, carrying a current I_0. Let $\mathscr{L}$ in its original state be given by a straight line along the e_3-axis, while a point $\mathscr{P}$ on $\mathscr{L}$ then is given by its, undeformed, position vector $\xi = \zeta e_3$. The rigid-body field $B(x)$ for $x \in G^+$, with $x = xe_1 + ye_2 + ze_3$, follows then from (7.5.6) as

$$\boldsymbol{B}(\boldsymbol{x}) = \frac{\mu_0 I_0}{4\pi} \int_{-\infty}^{\infty} \frac{(\boldsymbol{e}_3 \times (\boldsymbol{x} - \boldsymbol{\xi}))}{|\boldsymbol{x} - \boldsymbol{\xi}|^{3/2}} \, d\zeta$$

$$= -\frac{\mu_0 I_0}{4\pi} (y\boldsymbol{e}_1 - x\boldsymbol{e}_2) \int_{-\infty}^{\infty} \frac{d\zeta}{[x^2 + y^2 + (z - \zeta)^2]^{3/2}}$$

$$= -\frac{\mu_0 I_0}{2\pi} \frac{(y\boldsymbol{e}_1 - x\boldsymbol{e}_2)}{(x^2 + y^2)} \ . \tag{7.5.9}$$

Clearly, this field satisfies the constraints (7.5.1) and (7.5.3).

For the calculation of $\psi(\boldsymbol{x})$ we have to consider the deflected beam. For this, we assume that $\mathscr{L}$ has a deflection in the $\boldsymbol{e}_1$-direction: $\boldsymbol{u} = u(\zeta)\boldsymbol{e}_1$, by which

$$\boldsymbol{t}(\zeta) \quad = u'(\zeta)\boldsymbol{e}_1 + \boldsymbol{e}_3 \ ,$$

$$\boldsymbol{R}(\boldsymbol{x}, \zeta) = (x - u(\zeta))\boldsymbol{e}_1 + y\boldsymbol{e}_2 + (z - \zeta)\boldsymbol{e}_3 \ , \tag{7.5.10}$$

$$R^{-3} \quad = R_0^{-3}\left[1 + \frac{3x}{R_0^2} u(\zeta)\right] \ , \qquad R_0 = [x^2 + y^2 + (z - \zeta)^2]^{1/2} \ ,$$

where the last result is only correct up to first order terms in u. With this, (7.5.6) yields for the perturbed field in $\boldsymbol{x} \in G^+$

$$\boldsymbol{b}(\boldsymbol{x}) = \frac{\mu_0 I_0}{4\pi} \int_{-\infty}^{\infty} \frac{1}{R_0^3}\left[\frac{3x}{R_0^2} (x\boldsymbol{e}_2 - y\boldsymbol{e}_1)u(\zeta)\right.$$

$$\left. -[(z - \zeta)\boldsymbol{e}_2 - y\boldsymbol{e}_3]u'(\zeta) - u(\zeta)\boldsymbol{e}_2\right] d\zeta$$

$$= -\frac{\mu_0 I_0}{4\pi}\left[3xy \int_{-\infty}^{\infty} \frac{u(\zeta)}{R_0^5} d\zeta \, \boldsymbol{e}_1\right.$$

$$+ \int_{-\infty}^{\infty} \frac{1}{R_0^3}\left(u(\zeta) + (z - \zeta)u'(\zeta) - \frac{3x^2}{R_0^2} u(\zeta)\right) d\zeta \, \boldsymbol{e}_2 \tag{7.5.11}$$

$$\left. -y \int_{-\infty}^{\infty} \frac{u'(\zeta)}{R_0^3} d\zeta \, \boldsymbol{e}_3\right]$$

$$= -\frac{\mu_0 I_0}{4\pi} \int_{-\infty}^{\infty}\left[\frac{3xy}{R_0^5} \boldsymbol{e}_1 - \left(\frac{1}{R_0^3} - \frac{3y^2}{R_0^5}\right) \boldsymbol{e}_2 + \frac{3y(z - \zeta)}{R_0^5} \boldsymbol{e}_3\right] u(\zeta)d\zeta \ ,$$

where the latter step follows after an integration by parts.

From (7.5.6) it follows that $e_{ijk}b_{k,j} = 0$, and then there exists a magnetic potential $\psi(\boldsymbol{x})$ such that

$$b_j = \psi_{,j} \ . \tag{7.5.12}$$

This potential is equal to

$$\psi(\boldsymbol{x}) = \frac{\mu_0 I_0}{4\pi} \int_{-\infty}^{\infty} \frac{yu(\zeta)}{R_0^3} d\zeta \ . \tag{7.5.13}$$

It is easily checked that (7.5.13) implies $\psi_{,ii} = 0$ and $\psi(\boldsymbol{x}) \to 0$ for $|\boldsymbol{x}| \to \infty$, so this result indeed satisfies the constraints (7.5.5).

We assume that the current carrier $\mathscr{L}$ is periodically supported over distances l. In that case, $u(z) = 0$ in all points $z = kl$, $k \in \mathbb{N}$, and, moreover, $u(z)$ is periodic in z with period $2l$. We thus can write (7.5.13) as

$$
\psi(\boldsymbol{x}) = \frac{\mu_0 I_0}{4\pi} y \left[u(z) \int_{-\infty}^{\infty} \frac{1}{R_0^3} d\zeta + \int_{-\infty}^{\infty} \frac{u(\zeta) - u(z)}{R_0^3} d\zeta \right]
$$

$$
= -\frac{\mu_0 I_0}{4\pi} \frac{2yu(z)}{(x^2 + y^2)} \left[1 + O\left(\frac{R^2}{l^2} \ln \frac{R}{l} \right) \right] . \tag{7.5.14}
$$

In the latter step we have used the mean value theorem and have assumed that for $\boldsymbol{x} \to \partial G$, x and y are such that $(x^2 + y^2)/l^2 = O(R^2/l^2)$; in all our applications the latter condition is fulfilled.

In the next sections, we shall employ the results derived above for the solution of the buckling problem for helical and spiral superconductors. However, before doing so, we first apply the method to our standard example of a set of two parallel straight rods. We do this for three reasons:

1. as an illustration of the method presented in this section;
2. because, and this is the most important reason, one of the results of this example, namely the value of the integral K, can directly be applied to problems more complex than those that will be dealt with in the subsequent sections;
3. in order to show that the combined method leads us to a result that is closer to the (exact) result of the variational method than the direct BIOT–SAVARD method does.

Example: A set of two parallel superconducting rods (combined method)

We consider here exactly the same problem as in the Example of Sect. 7.4.1. Thus two parallel rods with equally directed currents of magnitude I_0, and bending in the $\boldsymbol{e}_1, \boldsymbol{e}_3$-plane. The distance between the rods is $2d$, the rods are periodically supported over distances L, $(l = L)$, and the first buckling mode is the anti-symmetric buckling mode. This brings us to the following choice for the displacements of the two rods:

$$
\boldsymbol{u}^{(1)}(z) = V(z)\boldsymbol{e}_1 = A\sin\frac{\pi z}{L} \boldsymbol{e}_1 ,
$$

$$
\boldsymbol{u}^{(2)}(z) = -V(z)\boldsymbol{e}_1 = -A\sin\frac{\pi z}{L} \boldsymbol{e}_1 . \tag{7.5.15}
$$

From (7.5.9) and (7.5.14) we find the fields $\boldsymbol{B}(\boldsymbol{x})$ and $\psi(\boldsymbol{x})$ in an arbitrary point $\boldsymbol{x}$ in the vacuum by simple superposition. This results in

$$\boldsymbol{B}(\boldsymbol{x}) = -\frac{\mu_0 I_0}{2\pi} \left[\left(\frac{y}{r_1^2} + \frac{y}{r_2^2} \right) \boldsymbol{e}_1 - \left(\frac{x+a}{r_1^2} + \frac{x-a}{r_2^2} \right) \boldsymbol{e}_2 \right] \qquad (7.5.16)$$

and

$$\psi(\boldsymbol{x}) = \frac{\mu_0 I_0}{2\pi} \left[\frac{y}{r_1^2} - \frac{y}{r_2^2} \right] A \sin \frac{\pi z}{L} , \qquad (7.5.17)$$

where

$$r_1^2 = (x+a)^2 + y^2 , \qquad r_2^2 = (x-a)^2 + y^2 . \qquad (7.5.18)$$

We can then calculate $I_0^2 K$ for the two rods and over one full period $-L < z < L$ from (7.5.8) by substituting (7.5.16) and (7.5.17) into this expression. After some lengthy but elementary calculations this results in

$$I_0^2 K = \frac{\mu_0 I_0^2 \kappa}{16\pi d^2} \int_{-L}^{L} [u^{(1)}(z) - u^{(2)}(z)]^2 dz = \frac{\mu_0 I_0^2 \kappa L}{4d^2} A^2 , \qquad (7.5.19)$$

where

$$\kappa = \frac{1 - 4(R/2d)^2 + \frac{3}{2}(R/2d)^4}{[1 - (R/2d)^2]^2} . \qquad (7.5.20)$$

The elastic energy for the two rods over one full period is, like in (7.4.26),

$$W = \frac{\pi}{4} E R^4 \int_{-L}^{L} (V''(z))^2 dz = \frac{\pi^6}{4L^3} E R^4 . \qquad (7.5.21)$$

By combining (7.5.19) and (7.4.32), we obtain the buckling value

$$I_{0cr} = \sqrt{\frac{W}{K}} = \frac{\pi^3 dR^2}{L^2} \sqrt{\frac{E}{\mu_0 \kappa}} . \qquad (7.5.22)$$

We have compared this value with the corresponding results of the Example of Sect. 7.4.1 for the variational method and the direct BIOT–SAVARD method; the result is depicted in Fig. 7.7. In this graph the buckling currents according to the variational method (VM) and to the combined method (CM), both normalized with respect to the buckling current from the direct BIOT–SAVARD method (BS), are displayed as function of d/R. The graph shows that the CM-value lies above the BS-value and below the VM-value. Hence, as could be expected, the combined method yields an improvement of the BS-value in the direction of the VM-value. The differences are significant, but of technical relevance only for values of $d/R < 4$. Moreover, the CM-value is a conservative estimate, meaning that the "real" buckling value is higher. So, using the CM-estimate keeps the designer on the safe side. The reason for the higher VM-value is the following: In the direct BS-method the currents run through the 1-dimensional central lines of the rods, whereas in the VM-approach the current is distributed over the boundary of the cross-section. When the currents in the two rods run in the same direction, the interaction between the currents causes a shift of the current distributions

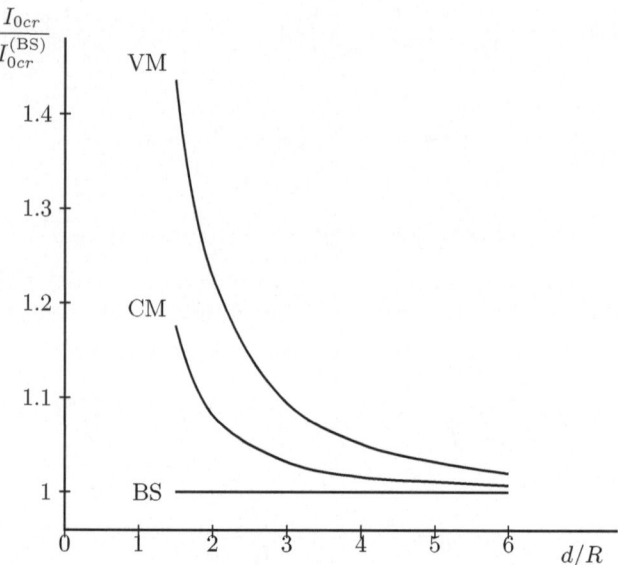

Fig. 7.7. Comparison of the buckling currents for a set of two parallel rods according to the variational method (VM) and the combined method (CM), both normalized with respect to the buckling current from the direct BIOT–SAVARD method (BS); from [257]

to the most opposite sides. If we then would try to imagine that the current runs through a 1-dimensional line, this line would be shifted away from the central line to the outer side. Thus, apparently the distance between the rods has increased and then the set buckles for a higher current. This explains why the VM-value is higher than the CM- or BS-value.

We can easily generalize the above method to sets of n, $n > 2$, parallel rods. Realizing that the integral in (7.5.19) is built up of two integrals (one for the first rod with respect to the second one and vice versa), we see that the generalization to n rods is straightforward, and yields

$$I_0^2 K = \frac{\mu_0 I_0^2 \kappa}{32\pi d^2} \sum_{i=1}^{n} \sum_{\substack{j=1 \\ j \neq i}}^{n} \frac{1}{(j-i)^2} \int_{-L}^{L} [u^{(i)}(z) - u^{(j)}(z)]^2 dz . \qquad (7.5.23)$$

This formula can also be applied to more complex structures such as helical or spiral conductors. Then, n denotes the number of turns of the structure and $2d$ is the distance between two subsequent turns. These cases will be discussed in the following sections.

7.5.3 Helical or Spiral Superconductors

Consider a superconductor in the form of a cylindrical helix; see Fig. 7.8. The radius of the helix is b and the constant pitch is h. For the distance between two turns, we then have $2d = 2\pi h$; for a slender helix $\pi h \ll b$. In the undeformed configuration a point on the central line of the helix is given by its position vector

$$\boldsymbol{X} = \boldsymbol{X}(\varphi) = b\cos\varphi \; \boldsymbol{e}_x + b\sin\varphi \; \boldsymbol{e}_y + h\varphi \; \boldsymbol{e}_z = b \; \boldsymbol{e}_r(\varphi) + h \; \varphi\boldsymbol{e}_z \; . \quad (7.5.24)$$

In this section, we shall consider both finite helices of n turns (in this case $\varphi \in [0, 2\pi n]$) and infinite ones ($\varphi \in \mathbb{R}$). In the first case, the helix is simply supported in its end points, whereas in the second case the helix is periodically supported in the points $\varphi = \varphi_k = 2k\pi n$, $k \in \mathbb{N}$. The total current through the helix is I_0.

Since the pitch angle α is very small ($\alpha \approx b/h \ll 1$) the dominant buckling mode will be a displacement in the axial or z-direction, i.e.

$$\boldsymbol{u} = u(\varphi)\boldsymbol{e}_z \; . \quad (7.5.25)$$

Here, $u(\varphi)$ is the displacement of the central line of the helical conductor, causing, besides bending, also torsion of the helix. We denote the torsion angle by $\beta(\varphi)$. This torsion has no influence on the integral K; it only affects the elastic energy W. By variation of W with respect to β we will derive a relation expressing $\beta(\varphi)$ in terms of $u(\varphi)$.

For the determination of the buckling current, we need expressions for the integrals K and W. The basic idea for K is completely analogous to that leading to (7.5.23). We only have to replace d by πh to obtain

$$K = \frac{\mu_0 \kappa}{32\pi^3 h^2} \sum_{i=1}^{n} \sum_{\substack{j=1 \\ j \neq i}}^{n} \frac{1}{(j-i)^2} \int_{\mathscr{L}_i} [u^{(i)}(\varphi) - u^{(j)}(\varphi)]^2 b\, d\varphi \; , \quad (7.5.26)$$

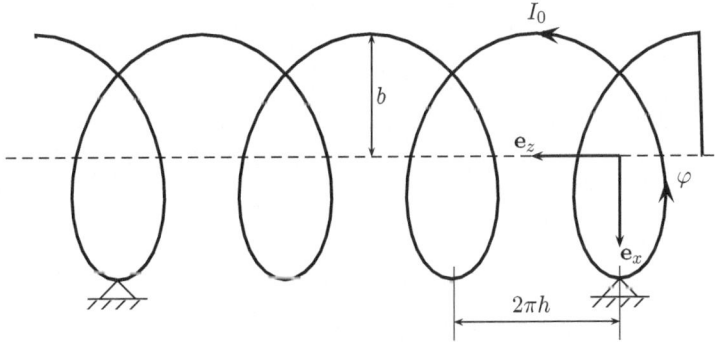

Fig. 7.8. The helical conductor

where $\mathscr{L}_i$ stands for the trajectory

$$\mathscr{L}_i = \{\varphi \mid 2(i-1)\pi \le \varphi \le 2i\pi\} , \tag{7.5.27}$$

while, for $\varphi \in \mathscr{L}_i$

$$u^{(i)}(\varphi) = u(\varphi) , \quad \text{and} \quad u^{(j)}(\varphi) = u(\varphi + 2(j-i)\pi) . \tag{7.5.28}$$

For the elastic energy, we use the classical expression for a slender curved rod

$$
\begin{aligned}
W &= \frac{EI}{2b^4} \int_{\mathscr{L}} (u'' - b\beta)^2 b \, d\varphi + \frac{GI_p}{2b^4} \int_{\mathscr{L}} (u' + b\beta')^2 b \, d\varphi \\
&= \frac{\pi ER^4}{8b^3} \int_{\mathscr{L}} \left[(u'' - b\beta)^2 + \frac{1}{(1+\nu)} (u' + b\beta')^2 \right] d\varphi , \tag{7.5.29}
\end{aligned}
$$

where $\mathscr{L}$ stands for the total length of the helix: $\mathscr{L} = \mathscr{L}_1 \cup \mathscr{L}_2 \cup ....\mathscr{L}_n$.

In the following two subsections, we first consider the infinite and after that the finite helix.

The Infinite Helix

For the infinite helix, periodically supported over n turns, we choose for the lowest buckling mode

$$u(\varphi) = A \sin\left(\frac{\varphi}{2n}\right) . \tag{7.5.30}$$

Due to the periodicity, we may restrict the integral for K to that for one full period, say $\varphi \in [0, 4\pi n)$. According to (7.5.26) we then obtain

$$
\begin{aligned}
K &= \frac{\mu_0 \kappa b}{32\pi^3 h^2} \sum_{i=1}^{2n} \sum_{\substack{j=-\infty \\ j \ne i}}^{\infty} \frac{1}{(j-i)^2} \int_{\mathscr{L}_i} [u(\varphi) - u(\varphi + 2(j-i)\pi)]^2 d\varphi \\
&= \frac{\mu_0 \kappa b}{32\pi^3 h^2} \sum_{k=1}^{\infty} \frac{A^2}{k^2} \int_0^{4\pi n} \left\{ \left[\sin\left(\frac{\varphi}{2n}\right) - \sin\left(\frac{\varphi}{2n} + \frac{\pi k}{n}\right) \right]^2 \right. \\
&\qquad\qquad \left. + \left[\sin\left(\frac{\varphi}{2n}\right) - \sin\left(\frac{\varphi}{2n} - \frac{\pi k}{n}\right) \right]^2 \right\} d\varphi \\
&= \frac{\mu_0 \kappa b n}{4\pi^2 h^2} \sum_{k=1}^{\infty} \frac{A^2}{k^2} \left[1 - \cos\left(\frac{\pi k}{n}\right) \right] = \frac{\mu_0 \kappa b}{8h^2} \left(1 - \frac{1}{2n} \right) A^2 . \tag{7.5.31}
\end{aligned}
$$

In accordance with (7.5.30) we assume the torsion β to have the form

$$\beta(\varphi) = \frac{1}{b} B \sin\left(\frac{\varphi}{2n}\right) . \tag{7.5.32}$$

Substituting (7.5.30) and (7.5.32) into (7.5.29), we obtain for the elastic energy of one full period

$$W = \frac{\pi^2 E R^4}{8b^3} \int_0^{4\pi n} \left[\left(\frac{A}{4n^2} + B \right)^2 \sin^2 \left(\frac{\varphi}{2n} \right) \right.$$

$$\left. + \frac{1}{4(1+\nu)n^2} (A+B)^2 \cos^2 \left(\frac{\varphi}{2n} \right) \right] d\varphi$$

$$= \frac{\pi^2 n E R^4}{8b^3} \left[\left(\frac{A}{4n^2} + B \right)^2 + \frac{1}{4(1+\nu)n^2} (A+B)^2 \right] . \quad (7.5.33)$$

Since the variation of J with respect to β must be zero and K is independent of β, or B, we have $\partial W / \partial B = 0$, yielding

$$B = -\frac{(2+\nu)}{4(1+\nu)n^2 + 1} A , \quad (7.5.34)$$

and with this

$$W = \frac{\pi^2 n E R^4}{4b^3} \frac{(4n^2 - 1)^2 A^2}{16n^4 [4(1+\nu)n^2 + 1]} = \frac{\pi^2 E R^4 A^2}{16(1+\nu)nb^3} [1+O(n^{-2})] , \quad (7.5.35)$$

for $n \gg 1$.

With use of (7.5.31) and (7.5.35) in $J = -W + I_0^2 K$, we find from $J = 0$ the following expression for the buckling current

$$I_{0cr} = \frac{\pi h R^2}{b^2} \sqrt{\frac{N(n)}{(2n-1)\kappa} \frac{E}{(1+\nu)\mu_0}} , \quad (7.5.36)$$

where κ is given by (7.5.20) and $N(n)$ by

$$N(n) = \frac{\left(1 - \frac{1}{4n^2} \right)^2}{\left[1 + \frac{1}{4(1+\nu)n^2} \right]} = 1 + O(n^{-2}) , \quad (7.5.37)$$

for $n \gg 1$.

According to its definition, $N(n)$ depends on ν but only in a very weak sense; for ν running from 0 to 0.5, the value of $N(n)$ changes by less than 2%.

The Finite Helix

Consider a finite helix of n turns, of the same type as described in the preceding section, thus with φ running from $\varphi = 0$ to $2\pi n$. For the evaluation of K and W we have to select a representation for the deflection $u(\varphi)$ and the associated torsion $\beta(\varphi)$. Due to the finiteness of the helix this representation is no longer as simple as it was for the infinite helix. We shall assume a representation in a series of N sine-functions[2] for $u(\varphi)$ according to (we assume the first buckling mode to be symmetric about $\varphi = \pi n$)

$$u(\varphi) = \sum_{k=1}^{N} A_k \, \sin \left[\frac{(2k-1)\varphi}{2n} \right] . \tag{7.5.38}$$

We choose for the torsion the analogous representation

$$\beta(\varphi) = \sum_{k=1}^{N} B_k \, \sin \left[\frac{(2k-1)\varphi}{2n} \right] . \tag{7.5.39}$$

Substituting (7.5.38) and (7.5.39) into expression (7.5.29) for W, we arrive at

$$W = \frac{\pi E R^4}{8(1+\nu)b^3} \int_0^{4\pi n} \left(\left\{ \sum_k \left(\frac{2k-1}{2n} \right)^2 (A_k + B_k) \cos \left[\frac{(2k-1)\varphi}{2n} \right] \right\}^2 \right.$$

$$\left. + (1+\nu) \left\{ \sum_k \left(\frac{(2k-1)^2}{4n^2} A_k + B_k \right) \sin \left[\frac{(2k-1)\varphi}{2n} \right] \right\}^2 \right) d\varphi \tag{7.5.40}$$

$$= \frac{\pi^2 n E R^4}{8b^3} \sum_k \left\{ \frac{(2k-1)^2}{4(1+\nu)n^2} (A_k + B_k)^2 + \left[\frac{(2k-1)^2}{4n^2} A_k + B_k \right]^2 \right\} .$$

From $\partial W / \partial B_k = 0$ we obtain

$$B_k = - \frac{(2+\nu)(2k-1)^2}{[4(1+\nu)n^2 + (2k-1)^2]} A_k , \tag{7.5.41}$$

yielding, after substitution into (7.5.40),

$$W = \frac{\pi E R^4}{8(1+\nu)b^3} \sum_{k=1}^{n} \omega_k A_k^2 , \tag{7.5.42}$$

with

$$\omega_k = \frac{\pi}{4n}(2k-1)^2 \left\{ 1 + \frac{(1+\nu)(2k-1)^2}{4n^2} - \frac{(2+\nu)^2(2k-1)^2}{[4(1+\nu)n^2 + (2k-1)^2]} \right\} . \tag{7.5.43}$$

---

[2] In [257], also a representation in splines was given. However, both the analytical and the subsequent numerical analysis were much more cumbersome and time-consuming than the sine-representation, while the ultimate results were practically the same. Therefore, we present here only the results of the sine-representation.

From (7.5.42) we conclude that in the elastic energy W the A_k-modes are uncoupled.

For the magnetic integral K we obtain from (7.5.26) (with $m = j - i$)

$$
K = \frac{\mu_0 \kappa b}{32\pi^3 h^2} \sum_{i=1}^{n} \sum_{\substack{m=-(i-1) \\ m \neq 0}}^{n-i} \frac{1}{m^2} \int_{\mathcal{L}_i} \left(\sum_k A_k \left\{ \sin\left[\frac{(2k-1)\varphi}{2n} \right] \right. \right.
$$

$$
\left. \left. - \sin\left[\frac{(2k-1)\varphi}{2n} + \frac{(2k-1)\pi m}{n} \right] \right\} \right)^2 d\varphi
$$

$$
= \frac{\mu_0 \kappa b}{16\pi^3 h^2} \sum_{m=1}^{n-1} \frac{1}{m^2} \int_0^{2\pi(n-m)} \left(\sum_k A_k \left\{ \sin\left[\frac{(2k-1)\varphi}{2n} \right] \right. \right.
$$

$$
\left. \left. - \sin\left[\frac{(2k-1)\varphi}{2n} + \frac{(2k-1)\pi m}{n} \right] \right\} \right)^2 d\varphi
$$

$$
= \frac{\mu_0 \kappa b}{16\pi^3 h^2} \sum_{k,l=1}^{N} k_{kl} A_k A_l \,, \tag{7.5.44}
$$

with

$$
k_{kl} = \sum_{m=1}^{n-1} \frac{2n}{m^2} \left\{ \frac{\pi}{n}(n-m) - \frac{1}{(2k-1)} \sin\left[\frac{(2k-1)\pi m}{n} \right] \right\}
$$

$$
\times \sin^2\left[\frac{(2k-1)\pi m}{2n} \right] \,,
$$

$$
\text{if} \quad k = l \,;
$$

$$
\tag{7.5.45}
$$

$$
k_{kl} = -\sum_{m=1}^{n-1} \frac{2n}{m^2} \left\{ \frac{1}{(k+l-1)} \sin\left[\frac{(k+l-1)\pi m}{n} \right] \right.
$$

$$
\left. + \frac{1}{(k-l)} \sin\left[\frac{(k-l)\pi m}{n} \right] \right\} \sin\left[\frac{(2k-1)\pi m}{2n} \right] \sin\left[\frac{(2l-1)\pi m}{2n} \right] \,,
$$

$$
\text{if} \quad k \neq l \,.
$$

With (7.5.42) and (7.5.44) we obtain for $J = -W + I_0^2 K$,

$$
J = -\frac{\pi E R^4}{8(1+\nu)b^3} \left(\sum_{k=1}^{N} \omega_k A_k^2 - \lambda \sum_{k,l-1}^{N} k_{kl} A_k A_l \right) \,, \tag{7.5.46}
$$

where

$$
\lambda = \frac{\kappa b^4}{2\pi^4 h^2 R^4} \frac{(1+\nu)\mu_0}{E} I_0^2 \,. \tag{7.5.47}
$$

Variation of J with respect to A_k yields the linear eigenvalue problem

$$\omega_k A_k - \lambda \sum_{l=1}^{N} k_{kl} A_l = 0 , \qquad k = 1, 2, \ldots, N . \qquad (7.5.48)$$

This problem is solved numerically, and from the lowest eigenvalue λ the buckling current is calculated. This eigenvalue is independent of all system parameters except for the number of turns n (and also, but only in a very weak sense, for ν). Hence $\lambda = \lambda(n)$, and from (7.5.47) it then follows that

$$I_{0cr} = \sqrt{\frac{2\pi^2 \lambda(n)}{(1+\nu)\kappa}} \; \frac{\pi h R^2}{b^2} \sqrt{\frac{E}{\mu_0}} . \qquad (7.5.49)$$

The Flat Spiral

As a final example, consider a flat spiral conductor, lying in the xy-plane; see Fig. 7.9. In the undeformed configuration, a point on the central line of the spiral is given by its position vector

$$\boldsymbol{X} = \boldsymbol{X}(\varphi) = b(\varphi) \cos \varphi \; \boldsymbol{e}_x + b(\varphi) \sin \varphi \; \boldsymbol{e}_y = b(\varphi) \boldsymbol{e}_r(\varphi) , \qquad (7.5.50)$$

with

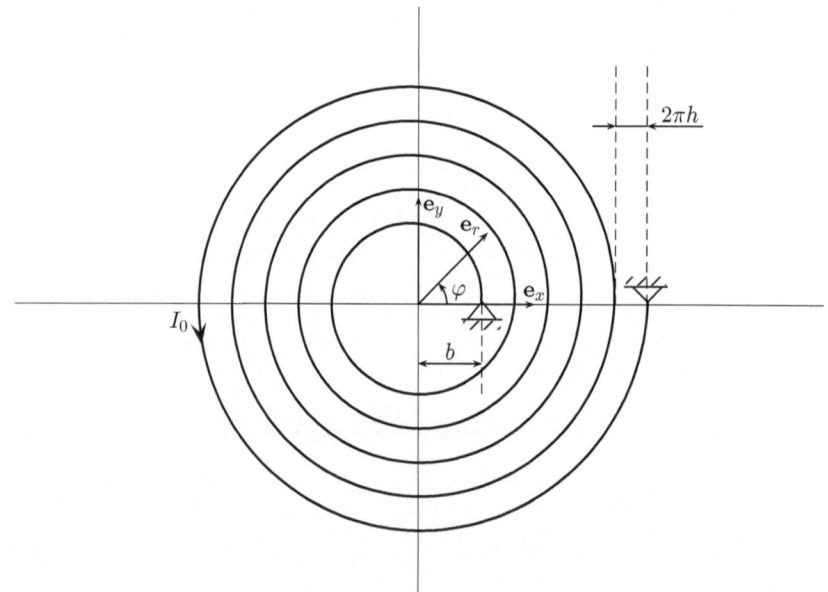

Fig. 7.9. The flat spiral conductor

$$b(\varphi) = b_0 + h\varphi , \quad \text{for} \quad 0 \le \varphi \le 2\pi n , \tag{7.5.51}$$

where b_0 is the radius of the spiral at its starting point, $2\pi h$ is the distance between two adjacent turns $(d = \pi h)$, and n is the number of turns. The cross-section of the conductor is circular, of radius R. The system is called slender if $R/b_0 \ll 1$ (then, $R/b(\varphi) < \pi h/b(\varphi) \ll 1$ for all $\varphi \in [0, 2\pi n]$). The spiral is simply supported in the two end points.

We assume that in the dominant buckling mode the spiral deforms in its plane and that the pertinent displacement of its central line is given by (in polar coordinates)

$$\boldsymbol{u} = \boldsymbol{u}(\varphi) = u(\varphi)\boldsymbol{e}_r + v(\varphi)\boldsymbol{e}_\varphi . \tag{7.5.52}$$

The spiral is taken to be inextensible, implying that

$$u(\varphi) + v'(\varphi) = 0 . \tag{7.5.53}$$

For an inextensible slender curved (almost circular) beam the elastic energy due to in-plane bending is given by

$$W = \frac{\pi}{8} ER^4 \int_0^{2\pi n} \frac{1}{b^3(\varphi)} [u''(\varphi) + u(\varphi)]^2 d\varphi . \tag{7.5.54}$$

Completely analogous to the preceding sections, the expression of K follows again from (7.5.26); we must only realize that now $b = b(\varphi)$.

Just as in the preceding section we use for $u(\varphi)$ a series of N sine-functions; see (7.5.38). Using this representation in W, we arrive at

$$
\begin{aligned}
W &= \frac{\pi}{8} ER^4 \int_0^{2\pi n} \frac{1}{b^3(\varphi)} \left[\sum_k \left(1 - \frac{k^2}{4n^2}\right) A_k \sin\left(\frac{k\varphi}{2n}\right) \right]^2 d\varphi \\
&= \frac{\pi ER^4}{8b_0^3} \sum_{k,l=1}^{N} \left(1 - \frac{k^2}{4n^2}\right) \left(1 - \frac{l^2}{4n^2}\right) I_{kl} A_k A_l ,
\end{aligned} \tag{7.5.55}
$$

where

$$I_{kl} = \int_0^{2\pi n} \frac{1}{\beta^3(\varphi)} \sin\left(\frac{k\varphi}{2n}\right) \sin\left(\frac{l\varphi}{2n}\right) d\varphi , \tag{7.5.56}$$

and

$$\beta(\varphi) = \frac{b(\varphi)}{b_0} = 1 + \frac{h}{b_0}\varphi . \tag{7.5.57}$$

The evaluation of K is analogous to that in the preceding section and leads to

$$K = \frac{\mu_0 \kappa b_0}{16\pi^3 h^2} \sum_{m=1}^{n-1} \frac{1}{m^2} \int_0^{2\pi(n-m)} \beta(\varphi + \pi m)$$

$$\times \left\{ \sum_k A_k \left[\sin\left(\frac{k\varphi}{2n}\right) - \sin\left(\frac{k\varphi}{2n} + \frac{k\pi m}{n}\right) \right] \right\}^2 d\varphi \qquad (7.5.58)$$

$$= \frac{\mu_0 \kappa b_0}{16\pi^3 h^2} \sum_{m=1}^{n-1} \frac{1}{m^2} \sum_{k,l=1}^{N} J_{klm} A_k A_l \, \sin\left(\frac{k\pi m}{2n}\right) \sin\left(\frac{l\pi m}{2n}\right) ,$$

where

$$J_{klm} = 4 \int_{\pi m}^{2\pi(n-m)} \beta(\varphi) \, \cos\left(\frac{k\varphi}{2n}\right) \cos\left(\frac{l\varphi}{2n}\right) d\varphi . \qquad (7.5.59)$$

Introducing λ as (we use here $b_1 = b_0 + n\pi h$, the mean radius of the spiral, instead of b_0 in order to weaken the influence of h on λ)

$$\lambda = \frac{\mu_0 \kappa b_1^4}{2\pi^4 E R^4 h^2} \, I_0^2 , \qquad (7.5.60)$$

substituting (7.5.55) and (7.5.58) into $J = -W + I_0^2 K$ and taking the first variation of J with respect to A_k, we arrive at the linear eigenvalue problem

$$\sum_{l=1}^{N} (\omega_{kl} - \lambda k_{kl}) A_l = 0 , \qquad k = 1, 2, \ldots, N . \qquad (7.5.61)$$

Explicit expressions for ω_{kl} and k_{kl} can easily be read of from (7.5.55) and (7.5.58). By a numerical solution of (7.5.61) the lowest eigenvalue λ is calculated. This λ depends on n and h and is related to the buckling current I_0 through (7.5.60), yielding

$$I_{0cr} = \sqrt{\frac{2\pi^2 \lambda(h, n)}{\kappa}} \, \frac{\pi h R^2}{b^2} \sqrt{\frac{E}{(1 + \nu)\mu_0}} . \qquad (7.5.62)$$

NOTE: The coefficient A_{2n} does neither contribute to W nor to K; this mode represents a rigid-body translation of the spiral. Therefore, in case $N \geq 2n$, we take $A_{2n} = 0$.

In the preceding three sections we have shown how one can find the buckling current for helical or spiral superconductors. We shall present and discuss explicit numerical results in the next section.

7.5.4 Results

We start with the result for the infinite helix as given in (7.5.36). We would like to express this formula in standard form in terms of d and L. Here, the choice for the distance $2d$ is clear, namely $2d = 2\pi h$, but that for L is not so evident. There are two possibilities at our disposal: either $L = \pi b$, half the length of one turn, or $L = 2\pi bn$, the length between two supports. We prefer here the first choice because we believe that this gives the best correspondence with the set of n parallel rods. With this, we can write (7.5.36) in the standard form (like in Table 7.3)

$$I_{0cr} = \pi^2 \sqrt{\frac{N(n)}{(2n-1)(1+\nu)\kappa}} \frac{dR^2}{L^2} \sqrt{\frac{E}{\mu_0}} , \qquad (7.5.63)$$

where κ and $N(n)$ are given by (7.5.20) and (7.5.37), respectively. These definitions show that

$$\kappa = 1 - 2\left(\frac{R}{2d}\right)^2 + O\left(\left(\frac{R}{2d}\right)^4\right) , \qquad \text{for} \qquad \frac{R}{d} \ll 1 , \qquad (7.5.64)$$

while $N(n)$ depends on ν, but only in a very weak sense; for ν running from 0 to 0.5, the value of $N(n)$ changes less than 2%. Moreover,

$$N(n) = 1 + O(n^{-2}) , \qquad \text{for} \qquad n \gg 1 . \qquad (7.5.65)$$

From this, we conclude that I_0 is proportional to $n^{-1/2}$ for large values of n. We recall that the direct BIOT–SAVARD method yields a buckling current that can be found from (7.5.63) by taking $\kappa = 1$.

For the finite helix, we have calculated the eigenvalues λ of the linear eigenvalue problem (7.5.48) numerically for $n = 2, 3, \ldots, 15$. In so doing, we have used $\nu = 0$ for the calculation of λ; this seems permissible because λ depends only very weakly on ν. The results of our calculations are presented in Table 7.4. In these calculations the following peculiarity was observed: the first two coefficients A_k of the eigenvector for any n were dominant, and it seems as if convergence was reached for $N = 3$ or 4 already, but for $N = n$ a sudden jump in λ, small but apparent, occurs. Calculation of the eigenvectors showed that this jump occurred because the coefficients A_n and A_{n+1} were no longer negligible with respect to A_1 or A_2. As a consequence, complete convergence was only found for $N \geq n + 2$. Let us consider as an example the case $n = 8$. For this case we found for the normalized eigenvector

$$A_1 = -0,94 , \quad A_2 = 0.25 , \quad A_3 = 0.02 , \quad A_8 = 0.21 , \quad A_9 = 0.15 , \tag{7.5.66}$$

while all remaining coefficients are less than 10^{-2}. Analogous results were found for other values of n. From this, we conclude that the buckling mode for a finite helix consists of, first, a global part, represented by A_1 and A_2 and, second, a local part on the scale of one winding, represented by A_n and A_{n+1}. The latter part must be due to the direct interaction between two adjacent windings. Although this effect is very striking with regard to the buckling mode, its effect on the buckling value is no more than a few percent.

Table 7.4. Lowest eigenvalues $\lambda \times 10^2$ (second, fifth columns) for the set of numbers n (first, fourth columns); the *values* in the third and sixth columns refer to $\lambda = 1/(4n\pi)(\times 10^2)$, see (7.5.67)

n	$\lambda \times 10^2$	$1/4n\pi \times 10^2$	n	$\lambda \times 10^2$	$1/4n\pi \times 10^2$
2	3.87	*3.98*	9	.866	*.884*
3	2.71	*2.65*	10	.773	*.796*
4	2.05	*1.99*	11	.698	*.723*
5	1.62	*1.59*	12	.636	*.663*
6	1.34	*1.33*	13	.584	*.612*
7	1.13	*1.14*	14	.539	*.568*
8	.983	*.995*	15	.501	*.531*

Scrutiny of Table 7.4 leads us to the following observation: when we compare the calculated values of $\lambda(n)$ in the second row of this table with those in the third row, we notice that the difference is always small; nowhere the relative difference exceeds 6%. Hence, for practical purposes it seems allowed to use for λ the very simple formula

$$\lambda(n) = \frac{1}{4\pi n} \, . \tag{7.5.67}$$

Use of this formula in the expression for the buckling current (7.5.49) yields

$$I_{0cr} = \pi^2 \sqrt{\frac{\pi}{2n(1+\nu)\kappa}} \frac{dR^2}{L^2} \sqrt{\frac{E}{\mu_0}} \, . \tag{7.5.68}$$

Comparing the result (7.5.68) with the analogous result for an infinite helix (7.5.63), we observe that in both cases the buckling current I_{0cr} is proportional to $n^{-1/2}$ for large values of n. However, the buckling current for a finite helix is always a factor $\sqrt{\pi}$ times that for an infinite helix, no matter how large n may be. This result contradicts the expectation, logical at first sight, which assumes that when n is large enough the formula for the infinite helix also governs to a good approximation the buckling of a finite helix of n turns. This expectation was explicitly stated by VAN DE VEN and LIESHOUT in [256]. In the procedure for the infinite helix, one complete period $\mathscr{L}^-$, $\varphi \in [0, 4\pi n]$, was separated from the remaining part of the helix, $\mathscr{L}^+$. The expectation, stated in [256], was motivated by the assumption that the influence of the current in $\mathscr{L}^+$ on the forces in $\mathscr{L}^-$ becomes weaker for increasing n, and was practically restricted to a close neighbourhood of the end points of $\mathscr{L}^-$. If this would be true, the influence of $\mathscr{L}^+$ on the buckling value would diminish and noninterference between $\mathscr{L}^-$ and $\mathscr{L}^+$ would be attained for $n \to \infty$. However, the results of this section reveal that even for exceedingly large values of n the remote parts of the infinite helix persist to

Table 7.5. The lowest eigenvalues $\lambda \times 10^2$ for several values of n and h; the *values* in every second row refer to outcomes of formula (7.5.71)–(7.5.72)

h	$n = 2$	3	4	5	6	7	8	9	10
.050	11.78	6.00	3.82	2.70	2.04	1.61	1.31	1.10	.96
	10.97	*5.97*	*3.88*	*2.77*	*2.11*	*1.67*	*1.37*	*1.15*	*.98*
.075	11.32	5.43	3.36	2.35	1.76	1.39	1.14	.95	.83
	9.45	*5.16*	*3.35*	*2.40*	*1.82*	*1.45*	*1.18*	*.99*	*.85*
.100	10.91	5.00	3.05	2.12	1.59	1.26	1.03	.87	.76
	8.54	*4.65*	*3.02*	*2.16*	*1.64*	*1.30*	*1.07*	*.89*	*.76*
.125	10.54	4.69	2.83	1.97	1.48	1.18	.97	.82	.72
	7.89	*4.29*	*2.79*	*1.99*	*1.52*	*1.20*	*.99*	*.83*	*.71*
.150	10.24	4.45	2.67	1.86	1.40	1.12	.92	.78	.69
	7.38	*4.02*	*2.61*	*1.87*	*1.42*	*1.13*	*.92*	*.77*	*.66*

interfere with the inner part $\mathscr{L}^-$. Hence, the replacement of the finite helix by an infinite one, in order to get an easier problem, is never allowed.

We have also computed the lowest eigenvalues $\lambda = \lambda(h, n)$ of (7.5.61) for n running from 2 to 10 for several values of h; the results are presented in Table 7.5. In performing these computations, the results for the eigenvectors A showed us that the coefficients A_l for l close to $2n$ were dominant. Let us consider as an example the case $h = 0.125$ and $n = 10$. For this case we found for the normalized eigenvector

$$A_{19} = -0,657 , \quad A_{21} = -0.513 , \quad A_{18} = 0.435 , \quad A_{22} = 0.257,$$
$$A_{17} = -0,209 , \quad A_{23} = -0.068 , \quad A_{16} = 0.057 , \tag{7.5.69}$$

while all other coefficients are less than 10^{-2}. Therefore, we changed the order of the summation in (7.5.38) in the following way:

$$u(\varphi) = \sum_{k=1}^{N} \left\{ \hat{A}_{2k-1} \sin \left[\frac{(2n-k)\varphi}{2n} \right] + \hat{A}_{2k} \sin \left[\frac{(2n+k)\varphi}{2n} \right] \right\} . \tag{7.5.70}$$

This reordening improved the rate of convergence substantially; we did obtain a very satisfactory convergence already for $N = 5$.

In [257, Fig. 6] one can find plots of λ as a function of n and h. These graphs are on a doubly logarithmic scale and they give rise to an almost linear behaviour in both h, for $h \in [0.05 \ , \ 0.15]$ and for not too small values of n, i.e. $n \geq 4$. This implies that $\lambda(h, n)$ over this range must be of the form

$$\lambda(h, n) = \Lambda h^\alpha n^\beta \ . \tag{7.5.71}$$

By means of a least square approximation, we found

$$\Lambda = 10.55 \ , \qquad \alpha = -0.36 \ , \qquad \beta = -1.50 \ . \tag{7.5.72}$$

The results of this formula are also presented in Table 7.5 (*as italic numbers*). Everywhere in the range $h \in [0.05 \ , \ 0.15]$, $n \geq 4$, the errors made by using this approximation are less than 5%. Hence, (7.5.71)–(7.5.72) represents a useful empirical formula for the buckling current. The latter then follows from

$$I_{0cr} = \pi^2 \sqrt{\frac{2\pi^2 \lambda}{\kappa}} \ \frac{dR^2}{L^2} \sqrt{\frac{E}{\mu_0}} \ , \tag{7.5.73}$$

where here we have taken $L = \pi b_1 = \pi(b_0 + n\pi h)$.

From the calculated values of the coefficients A_l of the eigenvector we can obtain an impression of the associated buckling mode. For explicit results and graphs of some of these buckling modes, we refer to [257, Fig. 7]; we only mention here that the displacements in the outer windings prevail over those in the inner ones. This could be expected by virtue of the greater mechanical stiffness of the inner windings as compared to the outer ones.

7.6 Magnetoelastic Vibrations of Superconducting Structures

7.6.1 Scope of this Section

Up to here in this chapter, we have only considered static processes, related to magnetoelastic buckling. However, also dynamic processes, such as waves or vibrations, can be affected by electromagnetic elastic interactions. For instance, magnetic fields or electric currents can influence the speed of propagation of waves or the eigenfrequencies or damping characteristics of vibrating magnetoelastic structures, and dynamic electromagnetic fields can generate forced elastic vibrations of such structures. On the other hand, elastic vibrations can change the electromagnetic fields in a magnetizable or electrically conducting body. The best known phenomenon in this is the piezoelectric effect.

Forced vibrations due to dynamic electromagnetic fields can have highly unwanted effects as they generate noise (high acoustic sound pressures) or lead to resonance. Examples are in MRI-scanners, where acoustic noise is generated by the gradient coil; see e.g. YAO et al., [275], or in fusion reactors; see e.g. TANAKA,TANA1. For the analysis of forced vibrations, especially in connection with resonance, it is imperative to know the eigenfrequencies of the system. These eigenfrequencies are influenced by external, or bias, magnetic fields and electric currents, as we shall show in this section.

As said before, electro-elastic vibrations are most well know from, and have found many practical applications in, piezoelectricity. An immens amount of literature is published on vibrations in piezoelectric systems, and it is far beyond the scope of this section to go deeper into this subject. Therefore, here we will not consider piezoelectric vibrations and we will restrict ourselves to mention only one review paper of KARLASH, [107].

Magnetoelastic vibrations of electrically conducting plates and shells have been studied from the 1980's on in the group around AMBARTSUMIAN, and later BAGDASARIAN, in the Yerevan University in Armenia. This research is continued nowadays by HASANYAN and his coworkers, [85, 120, 86], to mention only some recent papers from 2004–2005. These authors looked especially at vibrations of conducting strip-plates placed in an external magnetic field. HASANYAN et al., [85], considered the free vibrations of soft ferromagnetic electrically conducting plates carrying an electric current. Nonlinear bending-stretching vibrations are described by means of the VON-KÁRMÁN theory. For magnetoelastic interactions they used a model that looks like our Model IV. They found that the eigenfrequency ω depends on the bias magnetic field H and current I such that ω increases with H (*magnetic stiffening*) but decreases with I, the latter up to a point $I = I_{cr}$, where ω reduces to zero, implying that the basic state of the plate is unstable for $I > I_{cr}$. Analogous results are found in the examples which we shall present in this section. LIBRESCU et al., [120], considered free bending-stretching vibrations for perfectly conducting plates placed in an external magnetic field H. Again, it is shown that ω increases with increasing H; this effect becomes stronger for thinner plates. In 2005, HASANYAN et al., [85], reported on the effect of finite versus infinite conductivity. For the finite case, the eigenfrequency ω depends, besides on H, also on the conductivity σ. Moreover, special emphasis is given to the effect of σ on the magnetic damping (this effect is zero for perfectly conducting plates).

The bending vibrations of a magnetoelastic beam on a spring foundation are analysed by LIU and CHANG,LIU1. The beam is loaded by a mechanical axial force and a magnetic force. The analysis is based on the classical MOON-approach, [156]. LIU and CHANG found that the magnetic field reduces the deflection of the beam and decreases the eigenfrequency. Moreover, they registered the influence of the magnetic field on the damping of the vibrating beam.

A special paper is that by KUMAR et al., [116], who studied the vibration control of a plate with a magnetostrictive layer by a coil surrounding the plate. The aim was to find the location of the coil that gives the optimal damping; this optimal position depends on the mode shapes of the vibrating plate. The aim of this research was directed to *smart structure design*.

In this section, we present a selection of examples in which eigenfrequencies for specific electromagnetoelastic structures are determined. First, we calculate the eigenfrequencies of thin soft ferromagnetic circular plates placed in a transverse magnetic field for different support conditions of the plate; it is shown how the magnetic field reduces the eigenfrequency until the buckled state is reached. The second example concerns a superconducting ring carrying a prescribed current; here, the current increases the eigenfrequency and, consequently, the ring is always stable. Finally, our variational principle is adapted for magnetoelastic vibrations of superconducting structures and applied to sets of two superconducting rings. By means of a kind of RAYLEIGH quotient the eigenfrequencies of these sets are calculated. It turns out that these eigenfrequencies decrease with increasing current up to a point where the system becomes unstable and buckles. We close this section by presenting a few more results for sets of parallel rods and for an infinite helix.

7.6.2 Magnetoelastic Vibrations of a Thin Soft Ferromagnetic Circular Plate in a Uniform Transverse Magnetic Field

In this section, we consider free flexural vibrations of a thin circular plate placed in a uniform transverse magnetic field B_0. The plate is a soft ferromagnetic elastic body. For the description of the magnetoelastic vibrations of the plate, the MAXWELL–MINKOWSKI model will be used; see Section 3.4 (Model III). However, as far as the electromagnetic part is concerned we only consider *quasi-static* processes. For soft ferromagnetic media (with $\mu \gg 1$, say $\mu > 10^4$), the magnetoelastic stresses according to the MAXWELL–MINKOWSKI model are approximately (for $\mu^{-1} \approx 0$) equal to the purely elastic (HOOKEan) stresses. In this model, also the magnetic body force is negligible ($O(\mu^{-1})$). The only significant magnetoelastic coupling emanates from the boundary conditions. It will turn out that the magnetic fields generate a normal load per unit of area on the upper and lower surfaces of the deflected plate together with a shear force per unit of length at the edge of the plate.

The plate has radius R and thickness $2h$, where $\varepsilon = h/R \ll 1$ for a thin plate. The z-axis of a system of cylindrical coordinates is chosen normal to the plate. The uniform basic field is directed along the z-direction, so $B_0 = B_0 e_z$. For the rigid (undeformed) thin plate, we note that, except for a very small region (of $O(\varepsilon)$) near the edge $r = R$, the magnetic induction field inside the plate is uniform and equal to B_0. In using this approximation, we thus neglect the influences of the edge of the plate on the magnetic fields (we do not neglect the influence of the edge on the elastic stresses as we shall show further on); in other words, we assume for the magnetic problem the plate is infinite in its plane. Our strategy to find the eigenfrequencies of the plate now runs as follows: we split up the problem in a rigid-body part (whose solution we already know) and a perturbed part due to the dynamic deflections of the plates. The latter problem is then linearized with respect to the perturbations. This leads to a set of linearized balance and constitutive equations and boundary conditions; the general form of these relations for the MAXWELL–MINKOWSKI model can be found in Chap. 6 of this book.

We shall neglect here all terms of $O(\varepsilon)$; as one of the consequences, the magnetic body force vanishes and the Maxwell stress tensor becomes equal to the purely elastic tensor. These approximations result in the following set of equations for the magnetic potentials ψ, $|z| > h$, and φ, $|z| < h$, and the equation of motion for the stresses t_{ij} and the (vibrational) displacements $u_i = u_i(\boldsymbol{x}, t)$:

$$
\begin{aligned}
&\Delta\psi = 0, &&|z| > h, &&\psi \to 0, &&\text{for } |z| \to \infty, \\
&\Delta\varphi = 0, &&|z| < h, \\
&t_{ij,j} = \rho\ddot{u}_i, &&|z| < h,
\end{aligned}
\tag{7.6.1}
$$

together with the boundary conditions on the upper and lower surface of the plate (here, $B = B_0$; the stress boundary conditions at the edge $r = R$ will be specified further on)

$$
\begin{aligned}
&\psi_{,r} - \varphi_{,r} = Bw'(r), &&\psi_{,z} - \mu\varphi_{,z} = 0, \\
&t_{xz} = t_{yz} = 0, &&t_{zz} = -\frac{\mu}{\mu_0}B\varphi_{,z}, &&|z| = h,
\end{aligned}
\tag{7.6.2}
$$

where $w(r)$ is the rotationally symmetric normal deflection of the central plane of the plate. To be more precise, $w(r)$ is here the amplitude part of the deflection; the total displacement as well as the perturbed potentials, are assumed to be time harmonic, i.e. proportional to $e^{i\omega t}$, where ω is the real frequency. Hence,

$$
\boldsymbol{u}(\boldsymbol{x}, t) = w(r)e^{i\omega t}\,\boldsymbol{e}_z\,(1 + O(\varepsilon)).
\tag{7.6.3}
$$

We note that the right-hand side of the boundary condition for t_{zz} in (7.6.2) originates from the rotation of the normal vector to the upper and lower surface of the plate; without this rotation these two terms would cancel each other. This boundary condition describes a plate loaded at its upper and lower surface by a normal stress. This is equivalent to the problem of a thin plate under a normal load per unit of area $q(r)$, where

$$
q(r) = t_{zz}\big|_{z=-h} - t_{zz}\big|_{z=h} = -\frac{\mu}{\mu_0}B\varphi_{,z}\bigg|_{z=-h}^{h}.
\tag{7.6.4}
$$

The classical KIRCHHOFF–LOVE theory then leads to the global equation of motion for the flexural vibrations of a thin plate of the form

$$
D\Delta\Delta w(r) = q(r) - 2\rho h\omega^2 w(r),
\tag{7.6.5}
$$

where $D = 2Eh^3/3(1 - \nu^2)$ is the plate constant.

We try to solve equations $(7.6.1)_{1,2}$ for ψ (note that $\psi = \psi(r,z)\, e^{i\omega t}$) and φ by the separation of variables

$$\psi(r,z) = \Psi(z)\Omega(r) , \qquad \varphi(r,z) = \Phi(z)\Omega(r) . \tag{7.6.6}$$

The first boundary condition then yields

$$\Omega'(r) = w'(r) , \qquad \text{or} \qquad w(r) = \Omega(r) + w_0 , \tag{7.6.7}$$

where w_0 represents a rigid-body translation, which we will need further on to satisfy the mechanical (kinematical) boundary conditions at the edge of the plate; one should realize here that w_0 is a *dynamic* rigid-body translation, since the actual displacement is $w_0 e^{i\omega t}$. It is evident that this rigid-body translation does not affect the (quasi-static) magnetic potentials φ and ψ.

The LAPLACE equations in $(7.6.1)_{1,2}$ can now be split up into

$$\frac{d^2}{dr^2}\Omega(r) + \frac{1}{r}\frac{d}{dr}\Omega(r) + \lambda^2\Omega(r) = 0 , \quad \text{and} \quad \frac{d^2}{dz^2}\Phi(z) - \lambda^2\Phi(z) = 0 , \tag{7.6.8}$$

where λ is some real separation constant, which will follow from the mechanical boundary conditions at the edge $r = R$. There, it will turn out that $\lambda R = O(1)$, implying that $\lambda h = O(\varepsilon)$. Analogous equations can be found for $\Phi(z)$.

The equation for $\Omega(r)$ is a BESSEL equation, having the general solution

$$\Omega(r) = W\, J_0(\lambda r) , \tag{7.6.9}$$

while the solutions for $\Psi(z)$, using that $\Psi(z) \to 0$ for $z \to \infty$, and for $\Phi(z)$, accounting for the symmetry in z, read

$$\Psi(z) = C\, e^{-\lambda|z|} , \quad |z| > h , \qquad \Phi(z) = D\, \cosh(\lambda z) . \tag{7.6.10}$$

The constants C and D and an expression for $w(r)$ follow from the boundary conditions $(7.6.2)_1$, which yield, by using $\lambda h = O(\varepsilon) \ll 1$ and neglecting all $O(\varepsilon)$-terms except the term $\mu\lambda h$ (because μ is very large, $\mu\lambda h$ can be of $O(1)$ for $\varepsilon \ll 1$; in practice, we even have $\mu\varepsilon \gg 1$)

$$C = \frac{\mu\lambda h}{1 + \mu\lambda h}\frac{B}{\mu_0} , \qquad D = -\frac{1}{1 + \mu\lambda h}\frac{B}{\mu_0} \tag{7.6.11}$$

and $w'(r) = -\lambda W J_1(\lambda r)$. With these results we find for the load $q(r)$, defined in (7.6.4),

$$q(r) = -2\mu B\Phi'(h)\Omega(r) = \frac{2\mu\lambda^2 h}{\mu_0(1 + \mu\lambda h)}\, B^2 \hat{w} J_0(\lambda r) , \tag{7.6.12}$$

whereupon the equation of motion (7.6.5) becomes

$$D\lambda^4 \hat{w} J_0(\lambda r) = \frac{2\mu\lambda^2 h}{\mu_0(1 + \mu\lambda h)} B^2 \hat{w} J_0(\lambda r) + 2\rho h \omega^2 (\hat{w} J_0(\lambda r) + w_0) . \quad (7.6.13)$$

If $w_0 = 0$, this equation would immediately yield the dispersion relation for ω:

$$\omega^2 = \frac{Eh^2\lambda^4}{3\rho(1 - \nu^2)} \left(1 - \frac{3\mu(1 - \nu^2)}{\mu_0(1 + \mu\lambda h)Eh^2\lambda^2} B^2 \right) . \quad (7.6.14)$$

However, in all of the explicit examples that we shall consider here, we have $w_0 \neq 0$, and then the approach above does not yield a correct dispersion relation. However, we can already draw one general conclusion from this result, that remains valid even for $w_0 \neq 0$. We see that ω^2 decreases with increasing B and becomes zero when the second term between the brackets on the right-hand side of (7.6.14) becomes equal to one. Let us denote by B_{0cr} the associated value for B, and by w_0 the value for ω when $B = 0$. Then (7.6.14) can be written as

$$\omega^2 = \omega_0^2 \left(1 - \frac{B_0^2}{B_{0cr}^2} \right) . \quad (7.6.15)$$

We thus see that if $B > B_{0cr}$ then $\omega^2 < 0$ and ω becomes purely imaginary, implying that the unperturbed solution becomes unstable. Hence, B_{0cr} is indeed the buckling value for B_0. Since the magnetic load $q(r)$ in (7.6.12) and the elastic bending term in the right-hand side of (7.6.13) are not affected by w_0, this result remained valid also for $w_0 \neq 0$. However, w_0 has its effect on the inertia term in (7.6.13) and therefore the value of ω does depend on w_0.

Since the analytical method derived above does not yield results for ω in the way we proposed it, we have to look for an alternative method. For this, we will adapt the variational method as introduced in Sect. 7.4.2, so as to include dynamical vibration problems, and after that we will apply RAYLEIGH's *Principle* to find the eigenfrequencies. To this end, we must calculate also the kinetic energy T and have to replace the LAGRANGEan L by the HAMILTONian H according to $H = T - L$. Because, as far as the electromagnetic part is concerned, we only consider *quasi-static processes*, the expressions found for the perturbed electromagnetic interaction integral $\mathcal{K}$ and the elastic energy W in the preceding sections remain the same here. Moreover, the kinetic energy is only due to the dynamic perturbed displacements and thus T is already the perturbed kinetic energy. Here, T is defined on the perturbed displacement field $\boldsymbol{u} = \boldsymbol{u}(\boldsymbol{x}, t)$ as

$$T = \frac{1}{2} \int_G \rho(\dot{\boldsymbol{u}}, \dot{\boldsymbol{u}}) dV , \quad (7.6.16)$$

in which $(\dot{\boldsymbol{u}}, \dot{\boldsymbol{u}})$ is the inner product of $\boldsymbol{u}$. For steady-state vibrations with (eigen)frequency ω, we can assume the displacements to be harmonic in time according to (7.6.3), whereupon T transforms into

$$T = -\frac{1}{2} \omega^2 \int_G \rho(\hat{u}, \hat{u}) dV \, e^{2i\omega t} = -\omega^2 2\pi \rho h \int_0^R r w^2(r) dr \, e^{2i\omega t}$$
$$= -\omega^2 \hat{T} \, e^{2i\omega t} \, . \tag{7.6.17}$$

Applying RAYLEIGH's Principle, we arrive at the well-known expression for the RAYLEIGH quotient, but now for magnetoelastic systems,

$$\omega^2 = \frac{W - \mathcal{K}}{\hat{T}} = \frac{W}{\hat{T}} \left(1 - \frac{\mathcal{K}}{W} B_0^2 \right) = \omega_0^2 \left(1 - \frac{B_0^2}{B_{0cr}^2} \right) , \tag{7.6.18}$$

where we have used $\mathcal{K} = B_0^2 K$, $W/K = B_{0cr}^2$ and $W/\hat{T} = \omega_0^2$. We point out here that this relation could also directly be derived from the global equation of motion (7.6.5) by multiplying this equation by $w(r)$ and integrating the resulting equation over the plate.

We will apply this method now to the soft ferromagnetic plate. We start with the expression for K according to (7.3.39), which with $B_i = B_{(0)}\delta_{i3}$ and $u_3 \to \Omega(r)$ becomes (we discard the $e^{2i\omega t}$-term from now on)

$$K = -\frac{1}{B^2} \int_{\partial G} B w(r) \frac{\partial}{\partial n} \left(\psi - \frac{B}{\mu_0} \Omega(r) \right) dS$$
$$= -\frac{2\pi}{B} \int_0^R r w(r) \psi_{,z}(r, h) dr \, . \tag{7.6.19}$$

According to (7.6.6), the perturbed potential $\psi(r, z) = \Psi(z)\Omega(r)$ where $\Psi(z)$ is given by (7.6.10)–(7.6.11) and $w(r) = \Omega(r) + w_0$, with $\Omega(r)$ given by (7.6.9). Here it is important to note that (7.3.39) is derived under the assumption that $\mu^{-1} \approx 0$, inclusive $(\mu\lambda h)^{-1} \approx 0$, meaning that (7.6.11) then gives $C \approx B/\mu_0$. This leads us to

$$K = -\frac{2\pi}{B} \int_0^R r w(r) \Phi'(h) \Omega(r) dr = \frac{2\pi\lambda}{\mu_0} \int_0^R r w(r) \Omega(r) dr \, . \tag{7.6.20}$$

For the elastic energy of a circular plate in rotationally symmetric bending we have; see [239, eq. (1.94)],

$$W = \pi D \int_0^R \left(r(\Delta w(r))^2 - 2(1 - \nu) w'(r) w''(r) \right) \, dr \, . \tag{7.6.21}$$

For a clamped plate the contribution of the second term in the integrand for W vanishes, whereupon (7.6.21) reduces to

$$W = \pi D \int_0^R r(\Delta w(r))^2 \, dr \, , \tag{7.6.22}$$

see [239, eq. (1.96)]. With (7.6.7)$_2$ for $w(r)$, expression (7.6.17) for the scaled kinetic energy $\hat{T}$ for the circular plate of thickness $2h$ becomes

$$\widehat{T} = 2\pi\rho h \int_0^R r(w(r))^2 \, dr = 2\pi\rho h \int_0^R r\left(\hat{w}J_0(\lambda r) + w_0\right)^2 dr \,, \quad (7.6.23)$$

an expression that clearly depends on w_0.

We shall now illustrate this method by the following three examples.

1. *Clamped plate*

   In this case, the (kinematic) boundary conditions are

   $$w(R) = w'(R) = 0 \,. \quad (7.6.24)$$

   The second condition is met if

   $$J_1(\lambda R) = J_1(\Lambda) = 0 \,, \quad (7.6.25)$$

where $\Lambda = \lambda R$, so Λ is dimensionless and of $O(1)$. The lowest root of this equation is $\Lambda_1 = 3.832$. Taking this mode as a good approximation for the buckling mode, i.e. taking $\Omega(r) = \hat{w}J_0\left(\Lambda_1 r/R\right)$, we obtain for the deflection $w(r)$

$$w(r) = \hat{w}\left(J_0\left(\Lambda_1\frac{r}{R}\right) - J_0\left(\Lambda_1\right)\right) \,, \quad (7.6.26)$$

where the first condition of (7.6.24) is met by appropriately choosing the coefficient w_0. We note that for the clamped plate w_0 does not affect the value of K or W, but it does affect $\widehat{T}$. Calculating K, W and $\widehat{T}$, we obtain successively

$$K = \frac{2\pi\lambda}{\mu_0}\,\hat{w}^2 \int_0^R rJ_0^2\left(\Lambda_1\frac{r}{R}\right)dr = \frac{\pi\Lambda_1 R}{\mu_0}\,\hat{w}^2 J_0^2\left(\Lambda_1\right)\,, \quad (7.6.27)$$

according to [1, eq.(11.3.34)] and with the use of (7.6.25)

$$W = \pi D\left(\frac{\Lambda_1}{R}\right)^4 \hat{w}^2 \int_0^R rJ_0^2\left(\Lambda_1\frac{r}{R}\right)dr = \frac{\pi D}{2R^2}\,\Lambda_1^4\,\hat{w}^2 J_0^2\left(\Lambda_1\right)\,, \quad (7.6.28)$$

and

$$\widehat{T} = 2\pi\rho h\hat{w}^2\left[\int_0^R rJ_0^2\left(\Lambda_1\frac{r}{R}\right)dr \right.$$

$$\left. +2J_0\left(\Lambda_1\right)\int_0^R rJ_0\left(\Lambda_1\frac{r}{R}\right)dr + \frac{1}{2}R^2 J_0^2\left(\Lambda_1\right)\right]$$

$$= \pi\rho hR^2\hat{w}^2\left[J_0^2\left(\Lambda_1\right) + 0 + J_0^2\left(\Lambda_1\right)\right] = 2\pi\rho hR^2\hat{w}^2 J_0^2\left(\Lambda_1\right) \,. \quad (7.6.29)$$

Substituting these results into (7.6.18), we arrive at

$$\omega^2 = \frac{\pi D\Lambda_1^4}{4\pi\rho hR^4}\left(1 - \frac{2}{\mu_0 D}\left(\frac{R}{\Lambda_1}\right)^3 B_0^2\right) = \omega_0^2\left(1 - \frac{B_0^2}{B_{0cr}^2}\right)\,. \quad (7.6.30)$$

Here,

$$\omega_0 = \frac{\Lambda_1^2}{\sqrt{2}} \frac{1}{R^2} \sqrt{\frac{D}{2\rho h}} = \frac{\Lambda_1^2}{\sqrt{6}} \frac{h}{R^2} \sqrt{\frac{E}{\rho(1-\nu^2)}} , \qquad (7.6.31)$$

is the eigenfrequency when $B = 0$. We find good correspondence with [239, eq.(200)], who found as a first approximation of the coefficient in the first right-hand side of (7.6.31) the value 10.33, whereas we find $\Lambda_1^2/\sqrt{2} = 10.37$ (here one should realize that TIMOSHENKO considers plates of thickness h instead of $2h$ as we do). To get a better correspondence with the more precise value 10.21 (see [239, eq. 201]), we have to take for $w(r)$, or $\Omega(r)$, a FOURIER–BESSEL series according to

$$w(r) = \Omega(r) = \sum_{1}^{N} w_k J_0 \left(\Lambda_k \frac{r}{R} \right) , \qquad (7.6.32)$$

with Λ_k the roots of $J_1(\Lambda_k) = 0$, and were the last coefficient w_N is chosen such that $w(R) = 0$. The best coefficients w_k , $k = 1, 2, \ldots, N-1$ can then be found by equating $\partial \omega^2/\partial w_k = 0$ for $k = 1, 2, \ldots, N$.

We can use the result (7.6.30) also to obtain directly the value for the magnetic field for which the ferromagnetic plate buckles. The value for the buckling field obtained in this way is

$$\frac{B_{0cr}}{\sqrt{\mu_0 E}} = \frac{1}{\sqrt{3(1-\nu^2)}} \left(\frac{\Lambda_1 h}{R} \right)^{3/2} = \frac{4.33}{\sqrt{(1-\nu^2)}} \left(\frac{h}{R} \right)^{3/2} . \qquad (7.6.33)$$

This result is in agreement with the buckling field found by VAN DE VEN, [250, eq. (4.11)].

2. *Simply supported plate*

Here, besides the kinematical boundary condition $w(R) = 0$ also a dynamical condition, expressing that the bending moment along the edge should be zero, holds. However, the latter only holds provided no bending moment of magnetic origin is acting at this edge. This is difficult to prove here, because for this we need the mechanical boundary conditions for the stresses at the edge, and for these we need the magnetic fields there. And this is exactly what we do not know, because at the edge either the magnetic fields become singular or the magnetization becomes saturated. Therefore, we looked for another way. In [250], we showed by using expressions for the total force and the total moment on the plate and requiring that they should be in equilibrium with the total bending moment and shear force along the edge, that indeed the bending moment of magnetic origin is zero along the edge (this in contrast to the shear force of magnetic origin as we shall see in the next case). Thus we obtain the classical boundary condition

$$w''(R) + \frac{\nu}{R} w'(R) = 0 , \qquad (7.6.34)$$

yielding

$$J_0(\lambda R) - \frac{(1-\nu)}{\lambda R} J_1(\lambda R) = 0 , \qquad (7.6.35)$$

of which the lowest root for $\nu = 0.3$ is

$$\Lambda_1 = \lambda_1 R = 2.049 . \qquad (7.6.36)$$

By adding the rigid-body translation w_0, we obtain exactly the same relation for $w(r)$ as in (7.6.26). In an analogous way as we did for the clamped plate we can now calculate the integrals K, W and $\hat{T}$ from the same expressions as found in (7.6.27)–(7.6.29), but the outcomes differ somewhat due to the different boundary condition (7.6.34) (now $J_1(\Lambda_1) = (\Lambda_1/(1-\nu))J_0(\lambda) \neq 0$). Thus we obtain (use [1, eq.(11.3.34)])

$$
\begin{aligned}
K &= \frac{2\pi\Lambda_1}{\mu_0 R} \int_0^R rw(r)\ \hat{w}J_0\left(\lambda_1 r\right) dr \\
&= \frac{2\pi\lambda}{\mu_0}\ \hat{w}^2 \left[\int_0^R rJ_0^2\left(\Lambda_1 \frac{r}{R}\right) dr - J_0(\Lambda_1)\int_0^R rJ_0\left(\Lambda_1 \frac{r}{R}\right) dr \right] \\
&= \frac{\pi\Lambda_1 R}{\mu_0}\ \hat{w}^2 \left[J_0^2\left(\Lambda_1\right) + J_1^2\left(\Lambda_1\right) - \frac{2}{\Lambda_1}J_0(\Lambda_1)J_1(\Lambda_1) \right] \\
&= \frac{\pi\Lambda_1 R}{\mu_0(1-\nu)^2}\ \hat{w}^2 \left[\Lambda_1^2 - (1-\nu^2) \right] J_0^2\left(\Lambda_1\right) , \qquad (7.6.37)
\end{aligned}
$$

where in the last step we have used (7.6.35) to eliminate $J_1(\Lambda_1)$ in favour of $J_0(\Lambda_1)$,

$$
\begin{aligned}
W &= \pi D \left[\frac{\Lambda_1^4}{R^4}\ \hat{w}^2 \int_0^R rJ_0^2\left(\Lambda_1 \frac{r}{R}\right) dr - (1-\nu)(w'(R))^2 \right] \\
&= \frac{\pi D}{2R^2}\ \hat{w}^2 \left[\Lambda_1^4(J_0^2(\Lambda_1) + J_1^2(\Lambda_1)) - 2(1-\nu)\Lambda_1^2 J_1^2(\Lambda_1) \right] \\
&= \frac{\pi D\Lambda_1^4}{2(1-\nu)^2 R^2}\ \hat{w}^2 \left[\Lambda_1^2 - (1-\nu^2) \right] J_0^2(\Lambda_1) , \qquad (7.6.38)
\end{aligned}
$$

and, finally,

$$
\begin{aligned}
\hat{T} &= 2\pi\rho h\hat{w}^2 \left[\int_0^R rJ_0^2\left(\Lambda_1 \frac{r}{R}\right) dr - 2J_0\left(\Lambda_1\right)\int_0^R rJ_0\left(\Lambda_1 \frac{r}{R}\right) dr \right. \\
&\qquad \left. + \frac{1}{2}R^2 J_0^2\left(\Lambda_1\right) \right] \\
&= 2\pi\rho hR^2\hat{w}^2 \left[\frac{1}{2}\left(J_0^2\left(\Lambda_1\right) + J_1^2\left(\Lambda_1\right)\right) - \frac{2}{\Lambda_1}J_0\left(\Lambda_1\right)J_1\left(\Lambda_1\right) + \frac{1}{2}J_0^2\left(\Lambda_1\right) \right] \\
&= \frac{2\pi\rho hR^2}{2(1-\nu)^2}\ \hat{w}^2(\Lambda_1^2 - 2(1-\nu^2))\ J_0^2\left(\Lambda_1\right) . \qquad (7.6.39)
\end{aligned}
$$

Again substituting these results into (7.6.18), we arrive at

$$
\omega^2 = \frac{\pi D \Lambda_1^4}{2\pi \rho h R^4} \frac{\Lambda_1^2 - (1 - \nu^2)}{\Lambda_1^2 - 2(1 - \nu^2)} \left(1 - \frac{2}{\mu_0 D} \left(\frac{R}{\Lambda_1} \right)^3 B_0^2 \right)
$$

$$
= \omega_0^2 \left(1 - \frac{B_0^2}{B_{0cr}^2} \right) , \tag{7.6.40}
$$

where,

$$
\omega_0 = \frac{\Lambda_1^2}{R^2} \sqrt{\frac{\Lambda_1^2 - (1 - \nu^2)}{\Lambda_1^2 - 2(1 - \nu^2)}} \sqrt{\frac{D}{2\rho h}} = \frac{4.936}{R^2} \sqrt{\frac{D}{2\rho h}}
$$

$$
= 2.987 \frac{h}{R^2} \sqrt{\frac{E}{\rho}} , \tag{7.6.41}
$$

is the eigenfrequency when $B = 0$.

For the buckling value, we find the same expression as in (7.6.33), but due to the different value for Λ_1 we obtain here, for $\nu = 0.3$,

$$
\frac{B_{0cr}}{\sqrt{\mu_0 E}} = 1.78 \left(\frac{h}{R} \right)^{3/2} . \tag{7.6.42}
$$

This result is again in agreement with the buckling field found by VAN DE VEN, [250, eq. (4.12)].

We still want to note here that in the calculation of K in (7.6.37), the rigid-body translation $w_0 = \hat{w} J_0 (\Lambda_1)$ affects the outcome for K (i.e. replacing $w(r)$ by $\Omega(r)$ in the first integral would result in a different (wrong) value for K). This may look at first sight somewhat surprising, but this effect is due to the presence of a shear force of magnetic origin at the edge of the plate, as we shall see in the next example. Due to this shear force, the rigid-body translation performs work and thus contributes to the magnetic interaction integral K. This is not actually so here, because in this case the displacement of the edge is zero, and thus the work done by the shear force at the edge is now zero (but this is only so because we included the rigid-body translation w_0 in our choice for the displacement $w(r)$). However, this is no longer true in the next example; we will show there how to incorporate the work done by the shear force in the magnetic interaction integral K.

3. *Free plate*

For the free plate, one would expect as dynamic boundary conditions for this rotationally symmetric problem that the bending moment and the shear force are zero at the edge. However, only the first condition is correct here. For the second one, we calculate the total force due to the magnetic field on the plate, which because of equilibrium should be equal to zero. This force is composed of the force due to the normal load q on the surface of the plate plus the force due to possible shear forces at the edge. Let us denote this shear force by Q, then we have for the total force in the z-direction

$$F_z = 2\pi R Q + 2\pi \int_0^R q(r) r \, dr = 0 \ . \tag{7.6.43}$$

This leaves us with a shear force equal to (for $\mu\lambda h \approx 0$)

$$Q = -\frac{2\lambda}{\mu_0 R} B^2 \hat{w} \int_0^R r J_0(\lambda r) dr = -\frac{2\Lambda}{\mu_0 R^2} B^2 \hat{w} J_1(\Lambda) \ . \tag{7.6.44}$$

The boundary conditions at the free edge then become

$$w''(R) + \frac{\nu}{R} w'(R) = 0 \ , \qquad \left(\frac{d}{dr} \Delta w\right)(R) = -\frac{Q}{D} \ . \tag{7.6.45}$$

Moreover, we now choose the rigid-body translation w_0 such that the total momentum of the plate becomes zero, i.e.

$$\int_0^R r w(r) dr = \hat{w} \int_0^R r J_0(\lambda r) dr - \frac{1}{2} w_0 R^2$$
$$= \hat{w} \frac{R^2}{\Lambda} J_1(\Lambda) - \frac{1}{2} w_0 R^2 = 0 \ , \tag{7.6.46}$$

or

$$w_0 = \frac{2}{\Lambda} J_0(\Lambda) \hat{w} \ . \tag{7.6.47}$$

With this choice, the second boundary condition of (7.6.45) follows immediately from one integration of the equation of motion (7.6.13) over the surface of the plate. We then only have to satisfy the first boundary condition, which, as in the preceding case, is satisfied by choosing $\Lambda = \Lambda_1 = 2.049$, for $\nu = 0.3$.

The choice (7.6.47) for w_0 yields for the deflection of the plate

$$w(r) = \hat{w} \left(J_0(\Lambda_1 \frac{r}{R}) - \frac{2}{\Lambda_1} J_1(\Lambda_1) \right) \ , \tag{7.6.48}$$

and thus $w(R) \neq 0$ now. This implies that the work done by the shear force Q at the edge becomes unequal to zero and thus must be incorporated into K. The easiest way to incorporate this work into K is by replacing $w(r)$ by $w(r) - w(R)$ in the formula for K, (7.6.19), yielding (compare also with (7.6.27))

$$K = \frac{2\pi\lambda}{\mu_0} \hat{w}^2 \int_0^R r J_0 \left(\Lambda_1 \frac{r}{R}\right) \left(J_0 \left(\Lambda_1 \frac{r}{R}\right) - J_0 \left(\Lambda_1\right)\right) dr \ , \tag{7.6.49}$$

the same expression as in the second example. Since also the elastic energy remains the same, this implies that we will here find the same buckling value as in the preceding example, and as in [250, eq. (4.12)].

However, as the kinetic energy depends on the rigid-body translation w_0, the frequency will differ from the value found in the preceding example. Here we obtain

$$
\begin{aligned}
\widehat{T} &= 2\pi\rho h\hat{w}^2 \left[\int_0^R r J_0^2 \left(\Lambda_1 \frac{r}{R}\right) dr - \frac{4 J_1(\Lambda_1)}{\Lambda_1} \int_0^R r J_0 \left(\Lambda_1 \frac{r}{R}\right) dr \right. \\
&\qquad\left. + \frac{2R^2 J_1^2(\Lambda_1)}{\Lambda_1^2} \right] \\
&= 2\pi\rho h R^2 \hat{w}^2 \left[\frac{1}{2} \left(J_0^2(\Lambda_1) + J_1^2(\Lambda_1) \right) - \frac{2}{\Lambda_1^2} J_1^2(\Lambda_1) \right] \\
&= \frac{2\pi\rho h R^2}{2(1-\nu)^2} \, \hat{w}^2 \left(\Lambda_1^2 + (1-\nu)^2 - 4 \right) J_0^2(\Lambda_1) \ .
\end{aligned}
\tag{7.6.50}
$$

Again substituting these results into (7.6.18), we arrive at

$$
\begin{aligned}
\omega^2 &= \frac{\pi D \Lambda_1^4}{2\pi\rho h R^4} \frac{\Lambda_1^2 - (1-\nu^2)}{\Lambda_1^2 + (1-\nu)^2 - 4} \left(1 - \frac{2}{\mu_0 D} \left(\frac{R}{\Lambda_1}\right)^3 B_0^2 \right) \\
&= \omega_0^2 \left(1 - \frac{B_0^2}{B_{0cr}^2} \right) ,
\end{aligned}
\tag{7.6.51}
$$

where,

$$
\begin{aligned}
\omega_0 &= \frac{\Lambda_1^2}{R^2} \sqrt{\frac{\Lambda_1^2 - (1-\nu^2)}{\Lambda_1^2 + (1-\nu)^2 - 4}} \sqrt{\frac{D}{2\rho h}} \\
&= \frac{9.178}{R^2} \sqrt{\frac{D}{2\rho h}} = 5.299 \frac{h}{R^2} \sqrt{\frac{E}{\rho}} ,
\end{aligned}
\tag{7.6.52}
$$

is the eigenfrequency when $B = 0$.

All three examples considered above show the same behaviour: the eigenfrequency decreases with increasing magnetic bias field B_0 up to a value B_{0cr} when ω becomes zero, indicating that the undeflected state of the plate becomes unstable. Moreover, we have seen that a shear force of magnetic origin is acting at the edge of the plate, which can have a dominant effect on the frequency relation. The latter effect is strictly inherent to the MAXWELL–MINKOWSKI model used here as we shall see in the following, where we compare the MAXWELL–MINKOWSKI model with the LORENTZ model (Model V in Chap. 3 of this book).

Comparison between the Maxwell-Minkowski and Lorentz models.
The equations derived above are based on the MAXWELL–MINKOWSKI model
(Model III). We now shall prove that two theories based on different models,
here Model III and the LORENTZ model (Model V) (this model is in the literature also called the AMPÈRE*an current model* by a.o. MOON) yield identical
results. This is of course completely in agreement with the spirit behind this
book.

From Chap. 4, Sects. 4.4–4.5, more specifically, from equations (4.4.2) and
(4.5.4) we can deduce the following relation between the stresses in Models
III and V:

$$^V t_{ij} = {}^{III} t_{ij} + M_i B_j + \tfrac{1}{2}\delta_{ij} M_k (B_k + H_k) \,, \qquad (7.6.53)$$

which for a soft ferromagnetic medium and with the same approximations as
used throughout this section, reduces to

$$^V t_{ij} = t_{ij} - \frac{1}{\mu_0} B_i B_j + \frac{1}{2\mu_0}\delta_{ij} B_k B_k \,, \qquad (7.6.54)$$

where we have used that $^{III} t_{ij} \approx t_{ij}$, the purely elastic stress. Hence, we see
here already one important difference in the two formulations: whereas in the
constitutive equation for $^{III} t_{ij}$ the magnetic part is negligible, for $^V t_{ij}$ this
part is of the same order of magnitude as the elastic part.

After linearization of (7.6.54), we obtain for the perturbed stresses

$$^V t_{ij} = t_{ij} - \mu B (\delta_{i3}\varphi_{,j} + \delta_{j3}\varphi_{,i} - \delta_{ij}\varphi_{,z}) \,. \qquad (7.6.55)$$

Analogously to (7.6.1) and (7.6.2) this leads to the equation of motion and
the boundary conditions in terms of the LORENTZ stresses

$$^V t_{ij,j} = \rho \ddot{u}_i \,, \qquad |z| < h \,, \qquad (7.6.56)$$

and

$$^V t_{i3} = \mu B (-\delta_{i3}\varphi_{,z} + \varphi_{,i}) \,, \qquad |z| = h \,. \qquad (7.6.57)$$

By comparing this with (7.6.1) and (7.6.2), we see that the equation of motion
is the same, but that the boundary conditions are different. This will lead to
a different relation for the shear force as we shall show now.

However, first we show that the relation for the bending moment does not
change. The bending moment is defined as

$$^V M_{rr} = \int_{-h}^{h} z \, ^V t_{zz} dz = M_{rr} - \frac{\mu B}{\mu_0} \int_{-h}^{h} z \varphi_{,z} dz \,, \qquad (7.6.58)$$

where the last step follows after substitution of (7.6.55) and where M_{rr} is the
purely elastic bending moment given by its constitutive equation

$$M_{rr} = -D \left(w''(r) + \frac{\nu}{r} w'(r) \right) \,. \qquad (7.6.59)$$

For the shear force ^VQ we find analogously

$$^VQ = \int_{-h}^{h} z\, ^Vt_{zr}\,dz = Q + \frac{\mu B}{\mu_0} \frac{d}{dr} \int_{-h}^{h} \varphi(r,z)\,dz \;, \tag{7.6.60}$$

with

$$Q = -D\,\frac{d}{dr}\Delta w(r)\;. \tag{7.6.61}$$

If we want to compare the orders of magnitude of the magnetic terms in $^VM_{rr}$ and ^VQ, we must compare $^VM_{rr,r}$ with ^VQ, or

$$\int_{-h}^{h} z\varphi_{,z}\,dz \quad \text{with} \quad \int_{-h}^{h} \varphi\,dz \;.$$

Using (7.6.6) and (7.6.9), we immediately see that

$$\int_{-h}^{h} z\varphi_{,z}\,dz = O((\lambda h)^2)\int_{-h}^{h} \varphi\,dz = O(\varepsilon^2)\int_{-h}^{h} \varphi\,dz \;, \tag{7.6.62}$$

from which we conclude that the magnetic part in $^VM_{rr}$ may be neglected, or

$$^VM_{rr} = M_{rr}\;. \tag{7.6.63}$$

On the other hand, for the shear force this implies that the magnetic term in (7.6.60) may not be neglected as it is of the same order as the elastic term.

In the common way, by integrating the local equations of motion (7.6.56) over the thickness of the plate and using in this process the constitutive equation (7.6.55) and the boundary conditions (7.6.57), we can derive the global equation of motion (compare with (7.6.5))

$$D\Delta\Delta w(r) = q(r) - 2\rho h\omega^2 w(r)\;, \tag{7.6.64}$$

where now

$$q(r) = \frac{\mu B}{\mu_0}\left(\frac{d^2}{dr^2} + \frac{1}{r}\frac{d}{dr}\right)\int_{-h}^{h}\varphi(r,z)\,dz\;, \tag{7.6.65}$$

but, as follows from $\Delta\varphi(r,z) = 0$, this can also be written as

$$q(r) = -\frac{\mu B}{\mu_0}\,\varphi_{,z}\Big|_{z=-h}^{h}\;, \tag{7.6.66}$$

which is identical to (7.6.4).

Finally, we look at the boundary conditions for a free plate. In the same way as we derived (7.6.44) from (7.6.43) we obtain here for the shear force at the edge

$$^VQ = 0\;. \tag{7.6.67}$$

Using (7.6.6) and (7.6.9)–(7.6.11) together with $(\mu\lambda h)^{-1} \ll 1$ and $\lambda h \ll 1$, we derive from (7.6.60)

$$^V Q = \frac{B^2 \hat{w}}{\mu_0 \lambda h} \frac{d}{dr} \left(\int_{-h}^{h} \cosh(\lambda z) dz \, J_0(\lambda r) \right) = \frac{2B^2 \hat{w}}{\mu_0} J_1(\lambda r) \,. \qquad (7.6.68)$$

where Q is the purely elastic shear force. This leads us to the boundary conditions

$$w''(R) + \frac{\nu}{R} \, w'(R) = 0 \,, \qquad - D \left(\frac{d}{dr} \Delta w \right)(R) + \frac{\mu B}{\mu_0} \frac{d}{dr} \int_{-h}^{h} \varphi(r, z) dz = 0 \,, \qquad (7.6.69)$$

which are exactly the same conditions as (7.6.45).

Thus, we have proved that the LORENTZ formulation for this problem is completely equivalent to the MAXWELL–MINKOWSKI formulation. This proof can be extended to the other models presented in Chap. 3 of this book.

7.6.3 Magnetoelastic Vibrations of a Superconducting Ring in its Own Field

In this section, we investigate the in-plane vibrations of a circular ring carrying an electric current, which is confined to the surface of the superconducting ring; see [254]. The investigation is based on a perturbation method: the fields in the final, deformed, state of the ring are decomposed into the fields for the undeformed ring, the rigid-body fields, and the perturbations on these fields. The latter are due to the dynamic deflections of the ring. However, as far as the electromagnetic part is concerned we only consider *quasi-static* processes. First, the rigid-body problem is solved. As a specific result the initial stresses due to the magnetic forces (of LORENTZ type) in the undeformed coil are obtained; it will turn out that these stresses play an essential role in the calculation of the eigenfrequencies of the ring. Next, the linearized perturbed problem is solved, yielding expressions for the load of magnetic origin on the deformed ring. By superposition, the total load on the deformed ring is now known, and from this an equation of motion for the vibrational in-plane motion of the slender ring is derived. The solution of this equation leads to an expression for the eigenfrequencies of the superconducting ring. This expression consists of two terms: one due to the initial stresses and one due to the perturbed ones. The first term increases the frequency with increasing current (and, hence, has a stabilizing effect), whereas the second term causes the frequency to decrease (a destabilizing effect). We will show that the first term dominates the second one, thus implying that the undeformed state of the ring is stable against in-plane vibrations. This result agrees with that of CHATTOPADHYAY [41], who obtained this result by purely numerical means.

The ring that is considered here is an elastic superconducting slender ring of radius b and circular cross-section, radius R; the ring is called slender if

$\varepsilon = R/b \ll 1$. The ring is placed in a vacuum and carries an electric surface current $\boldsymbol{J}$ at its surface, the total current being I_0. The magnetic field inside the ring is zero, and that outside the ring has to satisfy

$$e_{ijk}B_{k,j} = 0 , \quad B_{i,i} = 0 , \quad \boldsymbol{B} \to 0 , \quad |\boldsymbol{x}| \to \infty . \tag{7.6.70}$$

The current density $\boldsymbol{J}$ is related to the boundary value of the external field $\boldsymbol{B}$ through the boundary conditions at the boundary ∂G of the ring

$$\mu_0 J_i = -e_{ijk}B_j n_k , \quad B_i n_i = J_i n_i = 0 , \quad \boldsymbol{x} \in \partial G . \tag{7.6.71}$$

The total current I_0 is prescribed by means of AMPÈRE's law

$$\int_C (\boldsymbol{B} \cdot d\boldsymbol{s}) = \mu_0 I_0 , \tag{7.6.72}$$

where C is a contour entirely encircling a cross-section of the conducting ring.

The CAUCHY stresses T_{ij} have to satisfy the equations of motion

$$T_{ij,j} = \rho \ddot{U}_i , \tag{7.6.73}$$

where $\boldsymbol{U}$ is the total displacement vector.

In the MAXWELL–MINKOWSKI model the current $\boldsymbol{J}$ produces a surface tension at ∂G according to

$$T_{ij}n_j = \frac{1}{\mu_0}\left[B_i B_j - \frac{1}{2}\delta_{ij}B_k B_k\right] = \frac{1}{2}e_{ijk}J_j B_k , \quad \boldsymbol{x} \in \partial G . \tag{7.6.74}$$

The above equations hold on the unknown, deformed state of the ring and are thus nonlinear. To derive a linear system, we first suppose the existence of an intermediate static equilibrium state. Since for our considerations the deformations in this state are not so relevant (this in contrast with the pre-stresses in this state), we approximate this state by the rigid-body state. The fields B_i^0 and T_{ij}^0 in this state satisfy the same equations (7.6.70)–(7.6.74) as above, but now in the known reference configuration. For later use, we especially need the equation for the traction at ∂G

$$T_{ij}^0 n_j = T_i^0 = \tfrac{1}{2}e_{ijk}J_j^0 B_k^0 . \tag{7.6.75}$$

The perturbation on the rigid-body state is characterized by the displacement vector $\boldsymbol{u} = \boldsymbol{u}(\boldsymbol{x}, t)$. The perturbed magnetic fields may be considered as being time-harmonic functions of $\boldsymbol{x}$ and t. These fields are decomposed as $\boldsymbol{B} = \boldsymbol{B}^0(\boldsymbol{x}) + \boldsymbol{b}(\boldsymbol{x})e^{i\omega t}$, $\boldsymbol{J} = \boldsymbol{J}^0(\boldsymbol{x}) + \boldsymbol{j}(\boldsymbol{x})e^{i\omega t}$ and $T_{ij} = T_{ij}^0(\boldsymbol{x}) + t_{ij}(\boldsymbol{x})e^{i\omega t}$ (the harmonic term $e^{i\omega t}$ will be discarded in the quasi-static electromagnetic equations). The linearization procedure now results in the following sets of equations (from here on we omit the the upper ^{0})

inside the ring

$$t_{ij,j} - T_{ij,k}u_{k,j} = \rho\ddot{u}_i \; ; \qquad (7.6.76)$$

in vacuum

$$e_{ijk}b_{k,j} = 0 \; , \qquad b_{i,i} = 0 \; , \qquad \boldsymbol{b} \to 0 \; , \quad |\boldsymbol{x}| \to \infty \; ; \qquad (7.6.77)$$

at the surface ∂G of the ring

$$
\begin{aligned}
&e_{ijk}b_j n_k + e_{ijk}B_{j,l}u_l n_k - e_{ijk}B_j u_{l,k}n_l = -\mu_0 j_i + \mu_0 J_i u_{j,k}n_j n_k \; , \\
&b_i n_i = B_i u_{j,i}n_j - B_{i,j}u_j n_i \; , \\
&j_i n_i = J_i u_{j,i}n_j \; , \\
&t_{ij}n_j = T_i u_{j,k}n_j n_k + T_{ij}u_{k,j}n_k + t_i \; ,
\end{aligned}
\qquad (7.6.78)
$$

where

$$t_i = \tfrac{1}{2}e_{ijk}j_j B_k + \tfrac{1}{2}e_{ijk}J_j(b_k + B_{k,l}u_l) \; , \qquad (7.6.79)$$

and the incremental stresses t_{ij} are given by

$$t_{ij} = -T_{ij}u_{k,k} + T_{jk}u_{i,k} + T_{ik}u_{j,k} + \tau_{ij} \; , \qquad (7.6.80)$$

where τ_{ij} is the incremental elastic stress, which is directly related to the infinitesimal deformations by HOOKE's law. Here, one must realize that all these equations refer to the undeformed reference state of the ring.

Let $\boldsymbol{u} = \boldsymbol{u}(\boldsymbol{x},t) = \hat{\boldsymbol{u}}(r,\phi,z)e^{i\omega t}$ be the time-harmonic displacement of the ring, where $\{r,\phi,z\}$ are cylindrical coordinates in the centre of the ring. Since we consider in-plane vibrations only, the displacement of the central line of the slender ring can be presented by (compare with (7.4.47))

$$\hat{\boldsymbol{u}}(r,\phi,z) = \left(w(\phi)\boldsymbol{e}_r + \left(v(\phi) - \frac{(r-b)}{b}(w'(\phi) - v(\phi))\right)\boldsymbol{e}_\phi\right)(1 + O(\varepsilon^2)) \; . \qquad (7.6.81)$$

For an inextensible ring, the displacements are restricted by the constraint

$$v'(\phi) + w(\phi) = 0 \; . \qquad (7.6.82)$$

In [254], the rigid-body problem and the linear perturbed problem for the magnetic field and the current are solved by introducing a separation of variables in toroidal coordinates. We do not repeat here these mathematical manipulations, which result in expressions for the mechanical body forces and the tractions at ∂G in the rigid-body state and the perturbed state. Integration of the local equations of motion (7.6.76) over the cross-section of the ring, with use of the boundary conditions (7.6.78)$_4$, leads us to the global equation of motion for the in-plane vibrations of the ring. These global equations can also be found in [156, eqs. (6-7.3)-(6-7.4)]; for in-plane vibrations $u_y = N_y = G_x = H = 0$. Moreover, in our notations, $u_x = -w$, $u_z = v$.

After the elimination of N_x and T and with the use of the well-known constitutive equation for the bending moment G_y in an inextensible slender ring,

$$G_y = -\frac{EI}{b^2}(w''(\phi) + w(\phi)) = -\frac{\pi E R^4}{4b^2}(w''(\phi) + w(\phi)) , \qquad (7.6.83)$$

one obtains

$$\frac{\pi E R^4}{4b^2}(w^v(\phi) + 2w'''(\phi) + w'(\phi)) = \Gamma + \pi\rho\omega^2 b^2 R^2(w'(\phi) + v(\phi)) . \quad (7.6.84)$$

Here, Γ is a load parameter, arising from the body force in (7.6.76) and the traction (7.6.78)$_4$, and given by

$$\Gamma(\phi) = bK_r'(\phi) + bK_\phi(\phi) + L''(\phi) + L(\phi) , \qquad (7.6.85)$$

where K_r and K_ϕ are forces per unit of length in r- and ϕ-direction, respectively, and L is a (bending) moment per unit of length about the z-axis.

As is shown in [254], the contribution of the moment L to Γ is of order ε^2 compared to the other terms, and thus negligible ($L(\phi) = L''(\phi) = 0$), while the remaining part can be split up according to $\Gamma = \Gamma_1 + \Gamma_2$, where Γ_1 is the rigid-body contribution and Γ_2 the perturbation part, with

$$\Gamma_1(\phi) = -\frac{3\mu_0 I_0^2}{4\pi} w'(\phi) \left[\log\left(\frac{8}{\varepsilon}\right) - \frac{1}{2}\right](1 + O(\varepsilon \log \varepsilon)) , \qquad (7.6.86)$$

and

$$\Gamma_2(\phi) = \frac{3\mu_0 I_0^2}{4\pi} w'(\phi) \left[\log\left(\frac{8}{\varepsilon}\right) - \frac{17}{6}\right](1 + O(\varepsilon \log \varepsilon)) . \qquad (7.6.87)$$

From these results we conclude that the highest-order contributions of Γ_1 and Γ_2 (of $O(\log \varepsilon)$) are equal but opposite, so they cancel each other. This implies that the remaining part of $\Gamma = \Gamma_1 + \Gamma_2$ is of $O(\varepsilon^0) = O(1)$, and equal to

$$\Gamma(\phi) = -\frac{7\mu_0}{4\pi} I_0^2 w'(\phi)(1 + O(\varepsilon \log \varepsilon)) . \qquad (7.6.88)$$

As in most preceding examples in this chapter, the separation of variables in toroidal coordinates requires the relation

$$w''(\phi) + \lambda^2 w(\phi) = 0 , \qquad (7.6.89)$$

to be satisfied. Taking account of the periodicity in ϕ and excluding rigid-body translations, we see that the lowest eigenmode is represented by $\lambda = 2$, yielding

$$w(\phi) = W \cos 2\phi . \qquad (7.6.90)$$

Then, according to the inextensibility condition (7.6.82),

$$v(\phi) = -\tfrac{1}{2} W \sin 2\phi \ . \tag{7.6.91}$$

Substituting (7.6.90) and (7.6.91) into the global equation of motion (7.6.84), we arrive at the dispersion relation for ω

$$\frac{18\pi E R^4}{4b^2} = -\frac{7\mu_0}{2\pi} I_0^2 + \frac{5}{2}\pi\rho b^2 R^2 \omega^2 \ . \tag{7.6.92}$$

With the definition

$$\omega_0^2 = \frac{R}{b^2}\sqrt{\frac{9E}{5\rho}} \ , \tag{7.6.93}$$

the eigenfrequency of the free ring ($I_0 = 0$), the dispersion relation (7.6.92) yields for the eigenfrequency

$$\omega^2 = \omega_0^2 \left(1 + \frac{7\mu_0 b^2}{9\pi^2 E R^4} I_0^2\right) \ . \tag{7.6.94}$$

Hence, the frequency of the current-carrying ring increases with increasing current. For a conservative problem, as that considered here, instability occurs when ω becomes zero. Since the right-hand side of (7.6.94) remains positive for all values of I_0, we infer that the equilibrium state of the ring (i.e. the pre-stressed rigid-body state) is always stable against in-plane vibrations. In analyzing the effects of Γ_1 and Γ_2 separately, we observe that Γ_1, due to the initial unperturbed field, has a stabilizing effect, whereas Γ_2, due to the perturbed field, has a destabilizing effect. When taken together, the stabilizing effect dominates over the destabilizing one, and therefore the ring remains stable.

The out-of-plane vibrations can be treated in an analogous way, and there is no reason to expect that this will yield essentially different results. This is confirmed by the results of [41].

In order to illustrate the influence of the current I_0 on the eigenfrequency ω quantitatively, we use the following numerical values (these values are from [41])

$$E = 8 \times 10^{10} \ \text{N/m}^2 \ , \qquad I = 2.2 \times 10^{-4} \ \text{m}^4 \ ,$$
$$R = 3.03 \ \text{m} \ , \qquad \mu_0 = 4\pi \times 10^{-7} \ \text{H/m} \ . \tag{7.6.95}$$

For these values, (7.6.94) becomes

$$\omega^2 = \omega_0^2 \left(1 + 4.06 \times 10^{-14} \ I_0^2\right) \ . \tag{7.6.96}$$

This result agrees, at least in order of magnitude, reasonably well with that of CHATTOPADHYAY [41, eq. (42)]. In this aspect it must be mentioned that in [41] it was assumed that the current is uniformly distributed over the cross-section. This correspondence between our results and those of [41] suggests that the specific distribution of the current over the cross-section (i.e.

a uniform distribution in [41] against a distributed surface current in our approach) can at most be of quantitative influence and not qualitative. A different current distribution will change the numerical value of the coefficient of I_0 in the dispersion relation for ω, but it is not to be expected that it can disturb the stability of the ring.

7.6.4 Variational Principle for Magnetoelastic Vibrations of Superconducting Structures

In this section, we will demonstrate how we can generalize our variational method presented in the preceding sections to include also vibrational problems for superconducting structures, and how we can find the eigenfrequencies of these structures from this. The basic idea for this is rather simple: we only replace the LAGRANGEan L introduced in Sect. 7.4 by the HAMILTONian $H = T - L$, where T is the kinetic energy. However, in doing this we shall as far as the electromagnetic part is concerned only consider *quasistatic processes*. This also holds for the perturbed part and, therefore, the expressions found for the perturbed electromagnetic interaction integral $\mathcal{K}$ in the preceding sections remain the same here. Of course, this also holds for the elastic energy. Moreover, the kinetic energy is only due to the perturbed displacement and thus T is already the perturbed kinetic energy.

Therefore, we have to formulate here only the Hamiltonian for the perturbed case (which is in fact the second variation of the full Hamiltonian) and we find for this $H = T - (W - \mathcal{K}) = T - W + \mathcal{K}$. Here, T is defined on the perturbed displacement field $\boldsymbol{u} = \boldsymbol{u}(\boldsymbol{x}, t)$ as

$$T = \frac{1}{2} \int_G \rho(\dot{\boldsymbol{u}}, \dot{\boldsymbol{u}}) dV . \tag{7.6.97}$$

For steady state vibrations with (eigen)frequency ω we can assume the displacements harmonic in time according to

$$\boldsymbol{u}(\boldsymbol{x}, t) = \hat{\boldsymbol{u}}(\boldsymbol{x})\, e^{i\omega t} , \tag{7.6.98}$$

whereupon T transforms into

$$T = -\frac{\omega^2}{2} \int_G \rho(\hat{\boldsymbol{u}}, \hat{\boldsymbol{u}}) dV = -\omega^2 \hat{T} . \tag{7.6.99}$$

The equation of motion and the electromagnetic equations for the perturbed fields follow from equating the first variation of H to zero, but since H is a homogeneous quadratic functional in the perturbations, then also $H = 0$ in the perturbed state (analogous to $J = 0$ in the magnetostatic case). Thus, putting H equal to zero, we arrive at the well-known RAYLEIGH quotient, but now for magnetoelastic superconducting systems, in the form

$$\omega^2 = \frac{W - \mathcal{K}}{\hat{T}} = \omega_0^2 \left(1 - \left(\frac{I_0}{I_{0cr}} \right)^2 \right) , \tag{7.6.100}$$

where in the latter step we have used $\mathcal{K} = I_0^2 K$ and $K/W = (I_{0cr})^2$, and where we have introduced $\omega_0^2 = W/\hat{T}$.

Since we have derived values for I_{0cr} for various systems in the preceding sections, we can, from the relation above, directly read off the dependencies of the eigenfrequencies on the imposed current I_0. We conclude that for all such systems the eigenfrequency decreases from ω_0 (in case $I_0 = 0$) to a value going to zero for $I_0 \rightarrow I_{0cr}$. For $I_0 > I_{0cr}$, ω^2 becomes negative, meaning that the eigenfrequency becomes purely imaginary, implying instability of the system. Hence, this frequency analysis confirms that the system buckles whenever $I_0 > I_{0cr}$.

Example: A set of two concentric superconducting rings

We consider here the same system as described in Sect. 7.4.2; see also Fig. 7.5. To start with the kinetic energy, we describe the displacement as in (7.6.98) with $\hat{\boldsymbol{u}}^{(i)}$, the deflection of the central line of the i-th ring, given by (7.4.47). This yields the following expression for $T^{(i)}$, the kinetic energy of the i-th ring,

$$T^{(i)} = -\frac{\omega^2}{2} \int_{G^{(i)}} \left[w_i^2(\phi) + \left(v_i(\phi) - \frac{r - b_i}{b_i} (w_i'(\phi) - v_i(\phi)) \right)^2 \right] \rho dV$$

$$= -\frac{\pi \omega^2}{2} \rho b_i R^2 \int_0^{2\pi} \left(w_i^2(\phi) + v_i^2(\phi) \right) d\phi \left(1 + O(\varepsilon^2) \right) , \tag{7.6.101}$$

where in the latter step we have used that the integral of $(r - b_i)$ over $G^{(i)}$ is zero, while the integral with $(r - b_i)^2$ yields a term of $O(\varepsilon^2)$ where $\varepsilon = R/b$ and $b = (b_1 + b_2)/2$.

Use of, see (7.4.50), $w_i = \pm w \cos(2\phi)$ and $v_i = \mp \frac{1}{2} w \sin(2\phi)$, such that the inextensibility condition $v_i'(\phi) + w_i(\phi) = 0$ is fulfilled, leads us to

$$T^{(i)} = -\frac{5}{8} \pi \omega^2 \rho b_i R^2 w^2 . \tag{7.6.102}$$

This finally gives for the total kinetic energy T, or better $\hat{T}$,

$$\hat{T} = \frac{5}{4} \pi \rho b R^2 w^2 \left(1 + O(\varepsilon) \right) , \tag{7.6.103}$$

where we have used that $b_i = b (1 + O(\varepsilon))$ for $i = 1, 2$.

With the use of relations (7.4.51) and (7.4.52) for W and K, respectively, in (7.6.100) we then arrive at the relation for the eigenfrequency of a set of two concentric superconducting rings

$$\omega^2 = \frac{W - I_0^2 K}{\hat{T}} = \frac{9ER^2}{5\rho b^4} \left[1 - \frac{\mu_0 Q}{E} \left(\frac{b^2 I_0}{3\pi R^3} \right)^2 \right] . \tag{7.6.104}$$

A few more results

We can apply the method above also to some of the other examples dealt with in the preceding section. We only present the results for the eigenfrequencies here.

1. *A pair of two parallel rods*

$$\omega^2 = \frac{E}{\rho} \left(\frac{\pi^2 R}{2L^2} \right)^2 \left[1 - \frac{\mu_0 Q}{E} \left(\frac{L^2 I_0}{\pi^3 R^3} \right)^2 \right] . \tag{7.6.105}$$

2. *A pair of two parallel rings*

$$\omega^2 = \frac{9E}{5\rho} \left(\frac{R}{b^2} \right)^2 \left[1 - \frac{\mu_0 (5+\nu) Q}{E} \left(\frac{b^2 I_0}{6\pi R^3} \right)^2 \right] . \tag{7.6.106}$$

3. *A set of n parallel rods*

$$\omega^2 = \frac{E}{\rho} \left(\frac{\pi^2 R}{2L^2} \right)^2 \left[1 - \frac{\mu_0 Q}{E} \left(\frac{L^2 I_0}{\alpha_n R^3} \right)^2 \right] . \tag{7.6.107}$$

4. *An infinite helix periodically supported at every n-th turn*

$$\omega^2 = \frac{EN(n)}{8(1+\nu)\rho} \left(\frac{R}{nb^2} \right)^2 \left[1 - \frac{\mu_0 \kappa (1+\nu)(2n-1)}{EN(n)} \left(\frac{b^2 I_0}{\pi h R^2} \right)^2 \right] . \tag{7.6.108}$$

8 Electrorheological Fluids

8.1 Introduction

"Smart" materials can adaptively change or respond to an external stimulus producing a useful effect. Mechanical stresses, temperature, electric or magnetic fields, photon radiation or chemicals are typical examples of stimuli. A useful effect usually means a dramatic change of either one physical property (mechanical, electrical, appearance), the structure or the composition, which can be monitored and used in certain applications. A useful effect may be completely reversed when the stimulus is removed and this important feature permits an easy control through simply changing the environmental conditions. A variety of smart materials exists, which are being researched extensively. These include piezoelectric and thermoelectric materials, magnetorheological and electrorheological fluids, photochromic and thermochromic materials, electroluminiscent, fluorescent and phosphorescent materials and shape memory alloys.

Electrorheological fluids (often abbreviated as ERF) are such intelligent materials that exhibit drastic changes in their rheological properties upon the application of an outer electric field on the order of 1 kV/mm. The ER phenomenon is characterized by full reversibility and a very fast response (often quoted in milliseconds). Upon removal of the field, the corresponding relaxation time is of a comparable scale. The term ER-effect refers to the abrupt change in the apparent viscosity. When the viscosity increases we deal with a positive ER-effect while a decrease in viscosity is called negative ER-effect [82, 207]. Both the positive and negative ER effect can be enhanced by ultraviolet illumination in some ER systems [83]. This phenomenon is called the photo-electrorheological (PER) effect.

Most ERFs are dispersions of polarizable small particles within a nonconducting carrier liquid. The typical range size of the particles entering the structure of an ER-fluid is on the order of 0.10 to 100 μm while the particle volume fraction ranges between 2%–50%. Particles with dimensions below the stated range are liable to execute BROWNian motion while larger particles are more liable to sedimentation and also to draw excessive currents. A wide variety of particulate media have been employed in ER suspensions starting from starch, flour, cellulose, ceramic, glass to complex particles such as polyelectrolytes, composite particles (conducting particles coated with a

K. Hutter et al.: *Electromagnetic Field Matter Interaction in Thermoelastic Solids and Viscous Fluids*, Lect. Notes Phys. **710**, 279–366 (2006)
DOI 10.1007/3-540-37240-7_8

thin non-conducting outer layer, doubly coated particles with dielectric cores of high strength and lower mass (see [228]). The impact of the particle shape on ER performance was recently investigated experimentally in [192] by using microspheres and micro-rods as the component of the solid phase of an ER fluid. The dispersing phase of an ER fluid is an insulating oil or other non-conductive liquid. Currently silicone oil, vegetable oil, mineral oil, paraffin etc. are used. Besides the suspended particles and the carrier fluid, an ER fluid contains also some additives which could be any polar material that can enhance the ER effect or the stability of the whole suspension. ER fluids that contain a small amount of water are normally called hydrous, in contrast to water-free or anhydrous ER fluids in which no detectable water residue exists. It was demonstrated that the addition of water can enhance the ER-effect. Moreover, the influence of water in connecting together the particles has been used as the basis of a theory to explain the ER mechanism (the water bridging mechanism). However, a big disadvantage of ER-fluids with moisture content is the limited range of operating temperatures by the freezing and boiling points of water. Fortunately, it was shown that the operating mechanism does not depend on the presence of water and recently, considerable emphasis has been placed on the development of anhydrous particle suspensions. Extensive reviews centred on the material science aspects of ERFs are available [26, 83] and much work continues to be done in order to find optimal combinations of material properties (see e.g. [197, 228, 276]).

The explanation for the ER effect can be given with the aid of experimental observations at the microscopic level. Under the influence of an external electric field, the initially unordered particles become oriented and attract each other to form particle chains in the fluid along the field lines. The chains then aggregate to form columns. These chain-like and columnar structures cause significant changes in the resistance to the flow, and the material switches in this way from the liquid state to a solid-like state. In 1949, WINSLOW [265] reported an ER-effect for certain suspensions and described for the first time the phenomenon of induced fibration even though earlier observations on electroviscous effects were reported since 1896 (reviewed in [112]). The basic mechanism for this behaviour is thought to be the field-induced particle polarization which is a consequence of the dielectric mismatch between particles and solvent. It should be pointed out that other mechanisms for the field-induced increase in viscosity have been suggested including overlap of the diffuse counter-ion clouds surrounding neighboring particles [112, 113], electrostatic torque preventing particle rotation in the flow field [26], inter-electrode circulation [54] and field induced aggregation due to water bridges between particles [208, 222, 223, 229]. A lot of research has been done to develop theoretical models describing these mechanisms and relating the material properties and microscopic phenomena to the measurable macroscopic properties. Most theories are based on the electrostatic polarization mechanism. For an overview of the fundamental physical mechanisms and strategies in relating the microstructural models to the rheological behaviour we refer the reader to the review papers summarizing the main results in this domain [176, 209].

Besides particulate ER suspensions (heterogeneous ERF), there have been developments of homogeneous physical systems which also show dramatic changes in rheological properties upon application of an external electric field. Oil-in-oil emulsions and liquid-crystal polymer/oil immiscible systems display a relatively strong ER-effect [102, 110, 111, 199, 245, 272]. This is explained by the increase of domain interactions due to the orientation of elongated molecules. In [66, 167, 168] it is shown that also simple dielectric liquids (insulating oils), of which the viscosity hardly changes when being subject to uniform fields, can be ER-active when subjected to non-uniform electric fields (the ER effect is attributed to the electrohydrodynamic convection enhanced by the use of electrodes with flocked fabrics). At last we mention the delicious study of the melt of milk chocolate which also displays ER behaviour [52].

ERFs can be modeled in several different ways. ERFs may be analysed by means of molecular dynamic simulations by using different models (such as the dipole model, conduction model, equivalent plate conduction model) to establish the equations of motion of the particles. [277, 278]. Another possibility consists in the investigation of their microstructure in order to obtain a macroscopic description of the material [90, 178, 199, 210, 258, 261].

A different approach is pursued in the context of continuum mechanics. There are descriptions of ERFs as mixtures of two constituents (the particulate medium and the fluid) [195]. However, many researchers adopted the approach in which ER fluids are treated in a homogenized sense [16, 17, 67, 194, 206]. RAJAGOPAL and RŮŽIČKA in [196] and ECKART in [63] formulated independently governing equations of ER fluids. These formulations have the advantage that they take into account the interactions between the electro-magnetic and the mechanical fields. After assuming the constitutive law characterizing a certain fluid (see Sect. 8.3 of the present chapter) this approach permits mathematical modeling of the ER behaviour [35, 57, 205]. In our study we also assume ER-fluids to be homogeneous and continuous liquids, and we apply a phenomenological modeling approach in order to predict their macroscopic behaviour.

Recently, continuum models were developed which try to reflect (at the macroscopic level) field induced effects of the micro- and mesostructure of ERFs. In [60] thermodynamic continuum modeling is pursued and the influence of the field generated microstructure is described with the aid of an internal variable theory. We mention the works of BRUNN and ABU-JDAYIL [33, 34] who carried out a phenomenological study by considering ERFs as fluids with transverse isotropy. This assumption is based on experimental observations according to which fibers are formed upon the application of an electric field. In their approach the extra stress tensor of an ERF depends on the rate of strain but also on a vector which characterizes the orientation and size of the field induced fibers. In this way it is possible to describe normal stress effects appearing in viscometric flows. In [65] ECKART and SADIKI applied a polar theory to electrorheological fluids in the context of extended thermodynamics. They succeeded in obtaining a model which accounts for different material responses, if the applied electric field (assumed to be constant) is either perpendicular or parallel to the flow direction. This fact was expected but the previous models were not capable to reflect it.

ER-fluids are potentially useful in numerous technical applications. Many of them belong to the automotive industry: shock absorbers, clutches, valves, brakes, dampers, actuators [36, 217]. A good review of the engineering application of ERFs in vibration control can be found in [224]. Non-conventional and advanced actuators may be built using ERFs [75, 139]. Another technological area in which ER fluids offer large promises is virtual reality and telepresence enhanced with haptic (tactile and force) feedback systems [114, 138, 179]. A haptic feedback is a modality for interacting with remote and virtual worlds compared with visual and auditory feedback. ER fluids can be used as smart inks, or to produce photonic crystals or in the polishing industry [83].

Despite the rich research literature about ERFs and the acquired progress in this subject, the application in real-life problems and commercialization of devices based on ERFs have been very limited. The need of high voltage input creates safety problems for the operators especially for the devices that are designed to be in contact with humans. Besides, the problem of their feedback (closed-loop) is difficult to solve because of their complex behaviour. Other obstacles in the development of ERF technologies are related to the composition of ER fluids (e.g. the instabilities caused by the sedimentation tendency of the particles or the limited range of operational temperature). Nevertheless, recent advanced studies led to significant improvements in the fluid formulation.

It can be foreseen that the large interest concerning these materials and the multitude of research studies focused on their potential applications will finally improve the capabilities of ER fluids on the one hand and lead to an optimal design for ER devices on the other hand.

8.1.1 Overview

The approach described here will be formulated within the framework of continuum mechanics of electromechanical interactions of polarizable materials. In particular, electrorheological fluids are considered to be homogeneous single constituent materials. Following this line, Sect. 8.2 presents the derivation of the governing equations of electrorheology (based on the phenomenological approach developed in [63, 64] with the corresponding jump conditions and the exploitation of the entropy inequality once general constitutive equations for the variables of the problem have been assumed.

Section 8.3 treats the constitutive assumptions for the CAUCHY stress tensor in greater detail. A review of the constitutive models used to describe ER fluids in the literature is given and two-dimensional constitutive laws, appropriate for numerical simulations originating from the CASSON-like and power law models are introduced.

The last section deals with applications of the results presented in the previous sections on channel flow of ERF. The boundary value in which the electrodes, flush with the channel, is formulated for both electrically conducting and non-conducting fluids. The simple case of infinite electrodes is analytically solved for the CASSON-like model and for the power-law model. Subsection 8.4.3 treats the case of electrodes with finite length when the electric field is inhomogeneous. The equations are non-dimensionalized for both cases, the alternative CASSON-like and the alternative power law models. The flow is simulated numerically using a software based on the finite element method. We study the behaviour of different fields such as velocity, pressure, generalized viscosity and the second invariant of the strain rate tensor near the electrode edges. Comparison with experimental data is performed that validates the simulations. Then we numerically optimize the configuration of the electrodes to obtain an enhancement of the ER-effect. In the last subsection we present shortly the main experimental results as inferred from the literature and obtained with electrodes of different geometries (restricted to bidimensional geometries) under a direct current.

We, finally, emphasize here that unlike the general theory dealt with in Chaps. 1 to 6, in which the material response was restricted to thermoelastic behaviour of polarizable and magnetizable bodies, or to magnetoelastic solid bodies, the constitutive behaviour of electrorheological materials includes viscous or plastic effects. This makes the material class somewhat larger than in the earlier chapters, but not significantly more complex than before.

8.2 Governing Equations and Constitutive Framework in Electrorheology

We use a continuum mechanical model in which electrorheological fluids are considered as homogeneous single constituent materials. Following [63] and essentially [64], we will summarize in this section the main steps of the phenomenological approach, conducted in order to obtain the governing equations of electrorheology with the corresponding jump conditions.

The starting point in deriving the system of equations that characterize electrorheological fluids consists in recording the balance laws of thermodynamics of fluids in electromagnetic fields. We recall here the balances of mass, momentum, moment of momentum and internal energy (2.2.9) and the entropy inequality (2.3.6)

$$\dot{\rho} + \rho \dot{x}_{i,i} = 0 \ ,$$

$$\rho \ddot{x}_i = t_{ij,j} + \rho F_i^{\mathrm{e}} + \rho F_i^{\mathrm{ext}} \ ,$$

$$\varepsilon_{ijk} t_{kj} = 0 \ ,$$

$$\rho \dot{U} = t_{ij} \dot{x}_{i,j} - q_{i,i} + \rho r^{\mathrm{e}} + \rho r^{\mathrm{ext}} \ ,$$

$$\rho \dot{\eta} + \phi_{i,i} \geq \frac{\rho r^{\mathrm{ext}}}{\Theta} \ ,$$

(8.2.1)

where we assumed that the electromagnetic body couple vanishes. To complete the physical picture we have to add the MAXWELL equations which will be considered here in the MAXWELL–MINKOWSKI formulation (see Chap. 3, Sect. 3.4), namely (see equations (3.4.10))

$$e_{klm} \mathcal{E}_{m,l} = - \overset{*}{B}_k \ , \tag{8.2.2}$$

$$e_{jlm} \mathcal{H}_{m,l} = \overset{*}{D}_j + \mathcal{J}_j \ , \tag{8.2.3}$$

$$D_{i,i} = \mathcal{Q} \ , \tag{8.2.4}$$

$$B_{i,i} = 0 \ . \tag{8.2.5}$$

We recall the definitions of the effective electric and magnetic field strengths (expressed in terms of the MINKOWSKIan electric and magnetic field strengths E_m and H_m) and the conductive current density according to

$$\mathcal{E}_m := E_m + e_{mpq} \dot{x}_p B_q \ , \tag{8.2.6}$$

$$\mathcal{H}_m := H_m - e_{mpq} \dot{x}_p D_q \ , \tag{8.2.7}$$

$$\mathcal{J}_j := J_j - \mathcal{Q} \dot{x}_j \ . \tag{8.2.8}$$

Introducing the electric polarization P_i and the magnetization M_i we have the relations

$$D_i := \varepsilon_0 E_i + P_i \ , \qquad D_i := \varepsilon_0 (\mathcal{E}_i - e_{ijk} \dot{x}_j B_k) + P_i \ , \tag{8.2.9}$$

$$H_i := \mu_0^{-1} B_i - M_i \ , \qquad \mathcal{H}_i := \mu_0^{-1} B_i - \varepsilon_0 e_{ijk} \dot{x}_j E_k - \mathcal{M}_i \ , \tag{8.2.10}$$

where $\mathcal{M}_i$ denotes the effective magnetization

$$\mathcal{M}_i := M_i + e_{ijk} \dot{x}_j P_k \ . \tag{8.2.11}$$

Relations $(8.2.9)_1$, $(8.2.10)_1$ and $(8.2.11)$ are identical with relations $(3.4.3)$ and $(3.4.4)$ given in Chap. 3 except for the notation (the superscript M in the expressions for E_i, H_i, P_i, M_i is now dropped).

8.2.1 The Electromagnetic Momentum Balance

The mathematical expressions for the electromagnetic force and energy supply rate in [63, 64] differ from those treated in Chap. 3. We will now sketch this derivation, but omit details which can be found in [63, 64].

In order to calculate the expression for the electromagnetic force F_i^e the procedure used in [64] is the following: from the MAXWELL–MINKOWSKI equations one may derive an identity which can be interpreted as the electromagnetic momentum balance. Of course, the expression which one may derive in this way contains relativistic terms; since the mechanical balance laws are only non-relativistically valid it follows that one must remove the terms of relativistic order in F_i^e. This reduction can be sought by non-dimensionalising the equations by introducing adequate scales.

In [64] ECKART derives two different forms of the electromagnetic momentum balances, which contain different expressions for the electromagnetic force. Later on he shows how these different expressions for F_i^e influence the mechanical quantities. Of course, the velocity obtained from the total momentum balance must – as an observable quantity – be the same for both variants.

The first variant of the electormagnetic momentum balance is derived by algebraically manipulating the MAXWELL-MINKOWSKI equations. Further manipulation with the emerging balance expression then yields the second form of the electomagnetic momentum balance. In [118], LANDAU and LIFSCHITZ derive the electromagnetic body force from a general expression for the internal energy and deduce an expression for the electromagnetic body force that agrees with the second variant mentioned above. Therefore there are two ways to arrive at these expressions, a formal one from the MAXWELL-MINKOWSKI equations and a more elegant one from an adequate postulate of the internal energy. Both variants differ in their electromagnetic stress tensors and forces, but not in the quantities (appearing in the MAXWELL-MINKOWSKI equations), $\mathcal{E}_j$, D_j, $\mathcal{H}_k$, B_k, $\mathcal{J}_j$ and $\mathcal{Q}$. These are identical in both forms.

In order to derive the electromagnetic momentum balance one starts by postulating the expression for the electromagnetic momentum, viz.,

$$g_i := e_{ijk}\,\varepsilon_0\,\mu_0\,\mathcal{E}_j\,\mathcal{H}_k \ . \tag{8.2.12}$$

From the MAXWELL equations (8.2.2)–(8.2.5), and the definitions (8.2.9)$_2$ and (8.2.10)$_2$ one may then derive the following identity, that can be regarded as the electromagnetic momentum balance

$$\dot{g}_i + \dot{x}_{l,l}\,g_i = {}^{(I)}t_{ij,j}^M - \rho\,{}^{(I)}F_i^e \ , \tag{8.2.13}$$

an identity, in which the electromagnetic stress tensor ${}^{(I)}t_{ij}^M$ and the electromagnetic force ${}^{(I)}F_i^e$ are given by

$$^{(I)}t_{ij}^M := \mathcal{E}_i D_j + \mathcal{H}_i B_j - \tfrac{1}{2}(\mathcal{E}_k D_k + \mathcal{H}_k B_k)\delta_{ij} + \varepsilon_0\mu_0 e_{ilk}\mathcal{E}_l\mathcal{H}_k\dot{x}_j$$
$$- \tfrac{1}{2}(\varepsilon_0 e_{kmn}\dot{x}_m B_n \mathcal{E}_k - \varepsilon_0\mu_0 e_{kmn}\dot{x}_m E_n \mathcal{H}_k)\delta_{ij} , \qquad (8.2.14)$$

$$\rho\,^{(I)}F_i^e := \mathcal{Q}_{el}\mathcal{E}_i + e_{ijk}(\overset{*}{D}_j + \mathcal{J}_j)B_k + \tfrac{1}{2}(P_j\mathcal{E}_{j,i} - P_{j,i}\mathcal{E}_j)$$
$$+ \frac{\mu_0}{2}(M_k\mathcal{H}_{k,i} - M_{k,i}\mathcal{H}_k) + e_{ijk}D_j\overset{*}{B}_k - \varepsilon_0 e_{jmn}\dot{x}_m B_n\mathcal{E}_{j,i}$$
$$+ \varepsilon_0\mu_0 e_{kmn}\dot{x}_m E_n\mathcal{H}_{k,i} - \varepsilon_0\mu_0\frac{\partial}{\partial t}(e_{ijk}\mathcal{E}_j\mathcal{H}_k) . \qquad (8.2.15)$$

This form of the electromagnetic momentum balance is marked with the left superscript "(I)" as an identifier and will subsequently be denoted as the first variant. A second form of the electromagnetic momentum balance, marked with the left superscript "(II)" is obtained by differently writing the divergence of the pressure term in the first two lines of (8.2.14) as

$$-\tfrac{1}{2}\left[\mathcal{E}_k\underbrace{(D_k + \varepsilon_0 e_{kmn}\dot{x}_m B_n)}_{=\,\varepsilon_0\mathcal{E}_k + P_k} + \mathcal{H}_k\underbrace{(B_k - \varepsilon_0\mu_0 e_{kmn}\dot{x}_m E_n)}_{=\,\mu_0(\mathcal{H}_k + M_k)}\right]_{,i} ,$$

and incorporating it in the force term. One then obtains the following electromagnetic momentum balance

$$\dot{g}_i + \dot{x}_{l,l}\, g_i = \,^{(II)}t_{ij,j}^M - \rho\,^{(II)}F_i^e , \qquad (8.2.16)$$

where the electromagnetic stress tensor $^{(II)}t_{ij}^M$ and the electromagnetic force $^{(II)}F_i^e$ are given by

$$^{(II)}t_{ij}^M := \mathcal{E}_i D_j + \mathcal{H}_i B_j + \varepsilon_0\mu_0 e_{ilk}\mathcal{E}_l\mathcal{H}_k\dot{x}_j , \qquad (8.2.17)$$

$$\rho\,^{(II)}F_i^e := \mathcal{Q}_{el}\mathcal{E}_i + e_{ijk}(\overset{*}{D}_j + \mathcal{J}_j)B_k + D_j\mathcal{E}_{i,j} - B_k\mathcal{H}_{k,i}$$
$$- \varepsilon_0\mu_0\frac{\partial}{\partial t}(e_{ijk}\mathcal{E}_j\mathcal{H}_k) . \qquad (8.2.18)$$

In comparison with the first variant the second variant has fewer terms and does not contain the quantities P_j and $\mathcal{M}_j$.

We mention that the two electromagnetic momentum balances (8.2.13) and (8.2.16) differ from the electromagnetic momentum balance defined in $(2.4.7)_1$. However, if we consider

$$g_i\,(2.4.7) = -g_i\,(8.2.13, 8.2.16) , \qquad (8.2.19)$$
$$t_{ij}^M\,(2.4.7) = \,^{(I),(II)}t_{ij}^M\,(8.2.13, 8.2.16) - g_i\,(8.2.13, 8.2.16)\dot{x}_j , \quad (8.2.20)$$

we obtain the same form of the electromagnetic momentum equations in both cases.

8.2.2 The Electromagnetic Energy Balance

In an analogous manner ECKART [64] derives from the MAXWELL–
MINKOWSKI equations another identity which can be viewed as the electromagnetic energy balance. The starting point of the derivation consists in
the choice of the electromagnetic energy. In [64] the expression

$$w := \tfrac{1}{2}(D_j \mathcal{E}_j + B_j \mathcal{H}_j), \qquad (8.2.21)$$

is used. Analogously, $e_{ijk}\mathcal{E}_j\mathcal{H}_k$ is used as the expression for the POYNTING
vector. Using then the same equations (8.2.2)–(8.2.5), $(8.2.9)_2$ and $(8.2.10)_2$
and employing similar manipulations as for the derivation of the electromagnetic momentum balance the first variant of the energy balance, viz.,

$$\dot{w} + \dot{x}_{l,l}\, w + (e_{ijk}\mathcal{E}_j\mathcal{H}_k)_{,i} = {}^{(I)}t_{ij}^M \dot{x}_{i,j} - \rho\,{}^{(I)}r^e, \qquad (8.2.22)$$

is obtained, where the electromagnetic energy supply ${}^{(I)}r^e$ is given by

$$\begin{aligned}
\rho\,{}^{(I)}r^e := {} &\tfrac{1}{2}(\mathcal{E}_j\dot{D}_j - \dot{\mathcal{E}}_j D_j + \dot{B}_j\mathcal{H}_j - B_j\dot{\mathcal{H}}_j) + \mathcal{J}_j\mathcal{E}_j \\
&- \tfrac{1}{2}(\varepsilon_0 e_{kmn}\dot{x}_m B_n \mathcal{E}_k - \varepsilon_0\mu_0 e_{kmn}\dot{x}_m E_n\mathcal{H}_k)\dot{x}_{j,j} \\
&+ \varepsilon_0\mu_0 e_{imn}\mathcal{E}_m\mathcal{H}_n\,\dot{x}_j\dot{x}_{i,j}.
\end{aligned} \qquad (8.2.23)$$

The electromagnetic energy balance 8.2.22 is an identity. Its second variant
is described by the equation

$$\dot{w} + \dot{x}_{l,l}\, w + (e_{ijk}\mathcal{E}_j\mathcal{H}_k)_{,i} = {}^{(II)}t_{ij}^M \dot{x}_{i,j} - \rho\,{}^{(II)}r^e, \qquad (8.2.24)$$

where the electromagnetic energy supply ${}^{(II)}r^e$ is given by

$$\begin{aligned}
\rho\,{}^{(II)}r^e := {} &\tfrac{1}{2}(\mathcal{E}_j\dot{D}_j - \dot{\mathcal{E}}_j D_j + \dot{B}_j\mathcal{H}_j - B_j\dot{\mathcal{H}}_j) + \mathcal{J}_j\mathcal{E}_j \\
&- \tfrac{1}{2}(\mathcal{E}_k D_k + \mathcal{H}_k B_k)\dot{x}_{j,j} + \varepsilon_0\mu_0 e_{imn}\mathcal{E}_m\mathcal{H}_n\,\dot{x}_j\dot{x}_{i,j}.
\end{aligned} \qquad (8.2.25)$$

In order to apply these balances in electrorheology one has to find their
corresponding non-relativistic approximation. This is done in [64] through a
non-dimensionalization process, which we now proceed to explain.

8.2.3 Non-relativistic Approximation

Since equations (8.2.13), (8.2.16) and (8.2.22), (8.2.24) contain some relativistic contributions but equations (8.2.1) are valid only in the non-relativistic
case, the relativistic terms should be removed. In order to identify them one
needs to non-dimensionalize the equations.

One can define the non-dimensional form of most quantities introduced
up to this point, if we choose the dimensionless fields as follows:

$$\begin{aligned}
E_i^+ &= \frac{E_i}{E_0}, \quad \mathcal{E}_i^+ = \frac{\mathcal{E}_i}{E_0}, \quad D_i^+ = \frac{D_i}{\varepsilon_0 E_0}, \\
P_i^+ &= \frac{P_i}{\varepsilon_0 E_0}, \quad \mathcal{J}_i^+ = \frac{\mathcal{J}_i\, T}{\varepsilon_0 E_0},
\end{aligned} \qquad (8.2.26)$$

$$\dot{x}_j^+ = \frac{\dot{x}_j}{V_0}, \quad x_j^+ = \frac{x_j}{L}, \quad t^+ = \frac{t}{T}, \tag{8.2.27}$$

where the following characteristic quantities were introduced: E_0 is a typical field strength of the external electric field, V_0 is a characteristic velocity, L a characteristic length (e.g. the extent of the flow domain) and T is a characteristic time which is in general the inverse value of the electromagnetic frequency. The correct choice of the characteristic quantities of the problem is an important step. This choice determines the domain of validity of the resulting balance laws and consequently of the whole theory.

As is evident, the magnetic quantities B_i, $\mathcal{H}_i$, $\mathcal{M}_i$ and M_i are missing from the above list. No characteristic magnetic quantity is introduced because, by assumption, there is neither an external magnetic field nor a permanent magnetization present.[1] So, it seems to be reasonable to build a new characteristic quantity from the above characteristic quantities, that possesses the dimension of a "magnetic unit."[2] In order to do this let us look at the properties of ERFs. Such fluids possess only dielectric and no magnetic properties. In this case we can use the so-called dielectric assumption (see [80], p. 161 and 175), according to which

$$M_k = -e_{kmn}\dot{x}_m P_n . \tag{8.2.28}$$

In other words: The magnetization, measured in the rest frame, consists only of a part that appears due to the motion of the electrically polarised medium. This assumption is based on the idea that an observer traveling with the fluid particle experiences no magnetization[3]. Using (8.2.11), relation (8.2.28) is tantamount to the statement

$$\mathcal{M}_k = 0 . \tag{8.2.29}$$

So, a co-moving observer does not observe a measurable effective magnetization.

The crucial relation is, however, (8.2.28). We know already how the quantities on the right-side of (8.2.28) have to be non-dimensionalized. From (8.2.26) and (8.2.27) there follows

$$e_{kmn}\dot{x}_m^+ P_n^+ = \frac{e_{kmn}\dot{x}_m P_n}{V_0 \varepsilon_0 E_0} .$$

---

[1] It is assumed here that the electrorheological systems studied are free of a permanent magnet (esp. no permanent magnetic electrodes are present).

[2] In [205] the approach is different. There, a magnetic induction B_0 is first introduced. Later this is considered small in comparison with the electric field strength and it is neglected. This is of course another possibility. But it must be pointed out that there is no a priori knowledge of the magnetic induction. It results as a variable only from the MAXWELL-MINKOWSKI equations.

[3] One must exclude here that a magnetization due to acceleration effects can appear (remarks related to this see also in [219]).

Thus, we may write for the magnetization

$$M_k^+ = \frac{M_k}{V_0 \varepsilon_0 E_0} , \qquad (8.2.30)$$

in which $V_0 \varepsilon_0 E_0$ may be interpreted as the characteristic magnetization in the electrorheological case. Now, we are able to non-dimensionalize the remaining magnetic quantitites[4]. One obtains

$$B_k^+ = \frac{B_k}{V_0 \varepsilon_0 \mu_0 E_0} , \quad M_k^+ = \frac{M_k}{V_0 \varepsilon_0 E_0} ,$$

$$\mathcal{H}_k^+ = \frac{\mathcal{H}_k}{V_0 \varepsilon_0 E_0} , \quad H_k^+ = \frac{H_k}{V_0 \varepsilon_0 E_0} , \qquad (8.2.31)$$

By applying all these transformations the electromagnetic momentum balance in the first variant takes the dimensionless form

$$\frac{\varepsilon_0 E_0^2 V_0^2}{L c^2} \left(\frac{d g_i^+}{d t^+} + \dot{x}_{l,l}^+ g_i^+ \right) = \frac{\varepsilon_0 E_0^2}{L} {}^{(I)} t_{ij,j}^{M,+} - \frac{\varepsilon_0 E_0^2}{L} {}^{(I)} F_i^{e,+} , \qquad (8.2.32)$$

where

$$\frac{d\,()}{dt} = \frac{V_0}{L} \frac{d\,()^+}{dt^+} := \frac{V_0}{L} \left[Str \frac{\partial\,()^+}{\partial t^+} + \dot{x}_l^+ ()_{,l}^+ \right] , \quad Str = \frac{L}{T V_0} , \qquad (8.2.33)$$

isolates in the bracketed factor the non-dimensionalized material derivative; Str represents the (electric) STROUHAL number, and g_i and $^{(I)} t_{ij}^M$ are given by

$$g_i = \frac{\varepsilon_0 E_0^2 V_0}{c^2} g_i^+ := \frac{\varepsilon_0 E_0^2 V_0}{c^2} e_{ijk} \mathcal{E}_j^+ \mathcal{H}_k^+ , \qquad (8.2.34)$$

$$^{(I)} t_{ij}^M = \varepsilon_0 E_0^2 \, {}^{(I)} t_{ij}^{M,+} := \varepsilon_0 E_0^2 \left[\mathcal{E}_i^+ D_j^+ + \frac{V_0^2}{c^2} \mathcal{H}_i^+ B_j^+ \right.$$

$$- \frac{1}{2} \left(\mathcal{E}_k^+ D_k^+ + \frac{V_0^2}{c^2} \mathcal{H}_k^+ B_k^+ \right) \delta_{ij} + \frac{V_0^2}{c^2} e_{ilk} \mathcal{E}_l^+ \mathcal{H}_k^+ \dot{x}_j^+$$

$$\left. - \frac{1}{2} \left(\frac{V_0^2}{c^2} e_{kmn} v_m^+ B_n^+ \mathcal{E}_k^+ - \frac{V_0^2}{c^2} e_{kmn} v_m^+ E_n^+ \mathcal{H}_k^+ \right) \delta_{ij} \right] , \quad (8.2.35)$$

---

[4] More precisely, we require that there is no other magnetic quantity, the value of which clearly exceeds the just introduced quantity. Then (8.2.30) may be interpreted as the highest estimation of the magnetic quantities. If e.g. an electric field is used which varies rapidly in time, one should non-dimensionalize the magnetic induction better with $\varepsilon_0 \mu_0 E_0 L/T$ instead of $\varepsilon_0 \mu_0 E_0 V_0$. When one takes into account the electric current one must use a kind of effective conduction instead of ε_0/T. The choice $\varepsilon_0 \mu_0 E_0 V_0$ is considered an acceptable compromise at the present point.

and, with the use of (8.2.28), $^{(I)}F_i^e$ takes the form

$$
\begin{aligned}
^{(I)}F_i^e = \frac{\varepsilon_0 E_0^2}{\rho L} \, {}^{(I)}F_i^{e,+} :=&\ \frac{\varepsilon_0 E_0^2}{\rho L} \left[\mathcal{Q}^+ \mathcal{E}_i^+ + \frac{V_0^2}{c^2} e_{ijk} \left(\overset{\star}{D_j^+} + Str \, \mathcal{J}_j^+ \right) B_k^+ \right. \\
&+ \frac{V_0^2}{c^2} e_{ijk} D_j^+ \overset{\star}{B_k^+} + \tfrac{1}{2} \left(P_j^+ \mathcal{E}_{j,i}^+ - P_{j,i}^+ \mathcal{E}_j^+ \right) - \frac{LV_0}{Tc^2} \frac{\partial}{\partial t^+} \left(e_{ijk} \mathcal{E}_j^+ \mathcal{H}_k^+ \right) \\
&\left. - \frac{V_0^2}{c^2} e_{jmn} \dot{x}_m^+ B_n^+ \mathcal{E}_{j,i}^+ + \frac{V_0^2}{c^2} e_{kmn} \dot{x}_m^+ E_n^+ \mathcal{H}_{k,i}^+ \right] ,
\end{aligned}
\tag{8.2.36}
$$

in which, the convective derivative of a vector $A_i = A_0 A_i^+$ has been written as

$$
\overset{\star}{A_i} = \frac{A_0 V_0}{L} \overset{\star}{A_i^+} := \frac{A_0 V_0}{L} \left[Str \frac{\partial A_i^+}{\partial t^+} + \dot{x}_j^+ A_{i,j}^+ - \dot{x}_{i,j}^+ A_j^+ + \dot{x}_{j,j}^+ A_i^+ \right] .
\tag{8.2.37}
$$

All bracketed terms in (8.2.33)–(8.2.37) are dimensionless. In the ensuing analysis we wish to impose the non-relativistic approximation and shall drop all terms of $\mathcal{O}(V_0^2/c^2)$, thus assuming $V_0 \ll c$. Beyond this assumption we also request L/T to be of order V_0, so that $(LV_0)/(Tc^2) = \mathcal{O}(V_0^2/c^2)$ as well. This bounds characteristic lengths not to be too large and characteristic times not to be too small[5]. In electrorheological fluids the estimate $L/T = \mathcal{O}(V_0)$ is amply fulfilled, so that all above mentioned terms can indeed be neglected.

If in the momentum equation all terms of order V_0^2/c^2 are dropped, the final form of the non-relativistic electromagnetic momentum balance for electrorheological purposes is given by[6]

$$
\frac{\varepsilon_0 E_0^2}{L} \left[{}^{(1)}t_{ij,j}^{M,+} - {}^{(1)}F_i^{e,+} + \mathcal{O}_i \left(\frac{V_0^2}{c^2}, \frac{V_0 L}{Tc^2} \right) \right] = 0 ,
\tag{8.2.38}
$$

where the non-relativistic electromagnetic stress-tensor, $^{(1)}t_{ij}^{(M),+}$ and the non-relativistic electromagnetic force $^{(1)}F_i^{e,+}$ are given by

$$
^{(1)}t_{ij}^{M,+} := \mathcal{E}_i^+ D_j^+ - \tfrac{1}{2} \mathcal{E}_k^+ D_k^+ \delta_{ij} ,
\tag{8.2.39}
$$

$$
^{(1)}F_i^{e,+} := \mathcal{Q}^+ \mathcal{E}_i^+ + \tfrac{1}{2} \left(P_j^+ \mathcal{E}_{j,i}^+ - P_{j,i}^+ \mathcal{E}_j^+ \right) .
\tag{8.2.40}
$$

For the second variant of the electromagnetic momentum balance the same procedure yields

---

[5] This would not be reasonable in e.g. astrophysics or high frequency physics. In the first case we would have as the characteristic lengths the distances between the planets; in the second case we would have as the characteristic electromagnetic times the inverses to the high frequencies.

[6] To distinguish it from the relativistic balances we use now the left superscript "(1)" for the first variant instead of "(I)" and analogously the left superscript "(2)" for the second variant instead of "(II)". From now on the symbol $\mathcal{O}_i (\dots)$ (and later also $\mathcal{O}(\dots)$) means "terms of order $(\dots)$".

$$\frac{\varepsilon_0 E_0^2}{L}\left[{}^{(2)}t_{ij,j}^{M,+} - {}^{(2)}F_i^{e,+} + \mathcal{O}_i\left(\frac{V_0^2}{c^2}, \frac{V_0 L}{Tc^2}\right)\right] = 0\,, \tag{8.2.41}$$

with the non-relativistic electromagnetic stress-tensor, ${}^{(2)}t_{ij}^{(M),+}$ and the non-relativistic electromagnetic force, ${}^{(2)}F_i^{e,+}$ given by

$$ {}^{(2)}t_{ij}^{M,+} := \mathcal{E}_i^+ D_j^+ \,, \tag{8.2.42}$$

$$ {}^{(2)}F_i^{e,+} := Q^+ \mathcal{E}_i^+ + D_j^+ \mathcal{E}_{i,j}^+\,. \tag{8.2.43}$$

By applying the same procedure to the electromagnetic energy balances (8.2.22) and (8.2.24) the equation

$$\frac{\varepsilon_0 E_0^2 V_0}{L}\left[\frac{d\omega^+}{dt^+} + \dot{x}_{l,l}^+ \omega^+ - {}^{(I)}t_{ij}^{M,+}\dot{x}_{i,j}^+ + \left(e_{ijk}\mathcal{E}_j^+ \mathcal{H}_k^+\right)_{,i} + {}^{(I)}r^{e,+}\right] = 0\,, \tag{8.2.44}$$

is obtained for the first variant, where

$$\omega = \varepsilon_0 E_0^2 \omega^+ := \varepsilon_0 E_0^2 \frac{1}{2}\left[D_k^+ \mathcal{E}_k^+ + \frac{V_0^2}{c^2} B_k^+ \mathcal{H}_k^+\right]\,. \tag{8.2.45}$$

The electromagnetic energy supply ${}^{(I)}r^{e,+}$ becomes

$$ {}^{(I)}r^e = \frac{\varepsilon_0 E_0^2 V_0}{\rho L}{}^{(I)}r^{e,+} := \frac{\varepsilon_0 E_0^2 V_0}{\rho L}\left[Str\,\mathcal{J}_j^+ \mathcal{E}_j^+ - \frac{1}{2}\left(D_j^+ \frac{d\mathcal{E}_j^+}{dt^+}\right.\right.$$
$$\left.- \frac{dD_j^+}{dt^+}\mathcal{E}_j^+ + \frac{V_0^2}{c^2}B_j^+ \frac{d\mathcal{H}_j^+}{dt^+} - \frac{V_0^2}{c^2}\frac{dB_j^+}{dt^+}\mathcal{H}_j^+\right) + \frac{V_0^2}{c^2}e_{imn}\mathcal{E}_m^+ \mathcal{H}_n^+ \dot{x}_j^+ \dot{x}_{i,j}^+$$
$$\left.- \frac{1}{2}\left(\frac{V_0^2}{c^2}e_{kmn}\dot{x}_m^+ B_n^+ \mathcal{E}_k^+ - \frac{V_0^2}{c^2}e_{kmn}\dot{x}_m^+ E_n^+ \mathcal{H}_k^+\right)\dot{x}_{j,j}^+\right]\,. \tag{8.2.46}$$

Neglecting in (8.2.45) and (8.2.46) the terms containing the factors V_0^2/c^2 and $V_0 L/Tc^2$ the following non-relativistic electromagnetic energy balance

$$\frac{\varepsilon_0 E_0^2 V_0}{L}\left[\frac{d\tilde{\omega}^+}{dt^+} + \dot{x}_{l,l}^+ \tilde{\omega}^+ - {}^{(m)}t_{ij}^{M,+}\dot{x}_{i,j}^+ + \left(e_{ijk}\mathcal{E}_j^+ \mathcal{H}_k^+\right)_{,i} + {}^{(m)}r^{e,+}\right.$$
$$\left. + \mathcal{O}\left(\frac{V_0^2}{c^2}, \frac{V_0 L}{Tc^2}\right)\right] = 0\,, \quad (m = 1, 2)\,, \tag{8.2.47}$$

is obtained where the non-relativistic electromagnetic energy, $\tilde{\omega}^+$, and the non-relativistic electromagnetic energy supply, $(\rho^{(m)}r^e)$, $m = 1, 2$, are given by

$$\tilde{\omega} = \varepsilon_0 E_0^2 \tilde{\omega}^+ := \frac{\varepsilon_0 E_0^2}{2}D_k^+ \mathcal{E}_k^+\,, \tag{8.2.48}$$

$$^{(1)}r^e = \frac{\varepsilon_0 E_0^2 V_0}{\rho L} \, ^{(1)}r^{e,+}$$

$$:= \frac{\varepsilon_0 E_0^2 V_0}{\rho L} \left[\text{Str } J_j^+ \mathcal{E}_j^+ - \frac{1}{2} \left(D_j^+ \frac{d\mathcal{E}_j^+}{dt^+} - \frac{dD_j^+}{dt^+} \mathcal{E}_j^+ \right) \right], \qquad (8.2.49)$$

$$^{(2)}r^e = \frac{\varepsilon_0 E_0^2 V_0}{\rho L} \, ^{(2)}r^{e,+}$$

$$:= \frac{\varepsilon_0 E_0^2 V_0}{\rho L} \left[\text{Str } J_j^+ \mathcal{E}_j^+ - \frac{1}{2} \left(D_j^+ \frac{d\mathcal{E}_j^+}{dt^+} - \frac{dD_j^+}{dt^+} \mathcal{E}_j^+ \right) + \frac{\mathcal{E}_k^+ D_k^+ \dot{x}_{j,j}^+}{2} \right]. \qquad (8.2.50)$$

In the above relations (8.2.44)–(8.2.50) the bracketed terms, [·], are dimensionless as before.

From (8.2.6), there follows

$$\mathcal{E}_m^+ := E_m^+ + \frac{V_0^2}{c^2} e_{mpq} \dot{x}_p^+ B_q^+ = E_m^+ + \mathcal{O}_m \left(\frac{V_0^2}{c^2} \right), \qquad (8.2.51)$$

and from relations (8.2.9)$_2$ and (8.2.10)$_2$ one obtains, with the help of (8.2.29),

$$D_i^+ = E_i^+ + P_i^+, \qquad (8.2.52)$$

$$\mathcal{H}_i^+ = B_i^+ - e_{imn} v_m^+ E_n^+. \qquad (8.2.53)$$

Finally, let us quote the dimensionless form of the MAXWELL–MINKOWSKI equations (8.2.2)–(8.2.5). They read

$$\frac{E_0}{L} \left(e_{klm} \mathcal{E}_{m,l}^+ + \frac{V_0^2}{c^2} \overset{\star}{B}_k^+ \right) = 0, \qquad (8.2.54)$$

$$\frac{\varepsilon_0 E_0 V_0}{L} \left(e_{jlm} \mathcal{H}_{m,l}^+ - \overset{\star}{D}_j^+ - \text{Str } J_j^+ \right) = 0, \qquad (8.2.55)$$

$$\frac{\varepsilon_0 E_0}{L} \left(D_{i,i}^+ - \frac{QL}{\varepsilon_0 E_0} \right) = 0, \qquad (8.2.56)$$

$$\frac{E_0 V_0}{Lc^2} B_{i,i}^+ = 0. \qquad (8.2.57)$$

Now, if one neglects all terms with the factors V_0^2/c^2 and $LV_0/(Tc^2)$ one obtains the electrorheological approximation of the MAXWELL–MINKOWSKI equations

$$\frac{E_0}{L} \left[e_{klm} E_{m,l}^+ + \mathcal{O}_k \left(\frac{V_0^2}{c^2}, \frac{V_0 L}{Tc^2} \right) \right] = 0, \qquad (8.2.58)$$

$$\frac{\varepsilon_0 E_0 V_0}{L} \left[e_{jlm} \left(B_m^+ - e_{mpq} \dot{x}_p^+ E_q^+ \right)_{,l} - \overset{\star}{D}_j^+ - \text{Str } J_j^+ \right] = 0, \qquad (8.2.59)$$

$$\frac{\varepsilon_0 E_0}{L} \left[D_{i,i}^+ - \frac{QL}{\varepsilon_0 E_0} \right] = 0, \qquad (8.2.60)$$

$$\frac{E_0 V_0}{Lc^2} B_{i,i}^+ = 0. \qquad (8.2.61)$$

Equation (8.2.58) is crucial. Provided boundary conditions for the electric field at domain boundaries are expressible in the electric field strength alone, it permits us to calculate the electric field decoupled from the magnetic induction and the mechanical fields.

However it is not easy to get rid of the magnetic induction in the electro-rheological approximation in all the MAXWELL-MINKOWSKI equations. The field B_i appears in equations (8.2.54) and (8.2.55) (in $\mathcal{H}_m$) with different factors; this can better be seen in equations (8.2.58) and (8.2.59): the factors in front of the brackets differ through $\varepsilon_0 V_0$. Luckily, the magnetic induction can be calculated from equations (8.2.59) and (8.2.61) if the electric field and the velocity field are known, so it represents a dependent quantity.

In the following subsection we will revert the non-dimensionalization and return to dimensional quantities; i.e. we will write e.g. instead of $V_0 v_j^+$ again v_j and so on. Owing to (8.2.51) in the remainder of this chapter we will write E_j instead of $\mathcal{E}_j$.

8.2.4 The Total Balance Laws of Electrorheology

Having put the non-relativistic approximation on a rational footing by non-dimensionalizing the equations, we may now again return to all approximate equations back in dimensional form. This is often more convenient and certainly physically more transparent.

The mass balance $(8.2.1)_1$ remains unchanged. By using (8.2.38) and (8.2.39) the momentum balance equation $(8.2.1)_2$ in the first variant takes the non-relativistically correct form

$$\rho \ddot{x}_i - \left({}^{(1)}t_{ij} + E_i D_j - \tfrac{1}{2} E_k D_k \delta_{ij} \right)_{,j} - \rho F_i^{ext} = 0 \,. \qquad (8.2.62)$$

Similarly, with the help of (8.2.41) and (8.2.42) its second variant becomes

$$\rho \ddot{x}_i - \left({}^{(2)}t_{ij} + E_i D_j \right)_{,j} - \rho F_i^{ext} = 0 \,. \qquad (8.2.63)$$

Obviously, the Cauchy stress tensor in the first and second variants cannot be identical if the emerging models are requested to be the same. This is the reason why they are distinguished in (8.2.62) and (8.2.63) and in the sequel by the superscripts "(1)" and "(2)".

With the aid of (8.2.49) the energy balance $(8.2.1)_4$ for the first variant becomes

$$\rho^{(1)}\dot{U} + {}^{(1)}q_{j,j} - {}^{(1)}t_{ij}\dot{x}_{i,j} - \rho r^{ext} - \mathcal{J}_j E_j \,,$$
$$+ \tfrac{1}{2} \left(D_j \dot{E}_j - \dot{D}_j E_j \right) = 0 \,, \qquad (8.2.64)$$

whilst for the second variant, on using (8.2.50), it reads

$$\rho^{(2)}\dot{U} +^{(2)}q_{j,j} - {}^{(2)}t_{ij}\dot{x}_{i,j} - \rho r^{ext} - \mathcal{J}_j E_j$$
$$+ \tfrac{1}{2}\left(D_j \dot{E}_j - \dot{D}_j E_j\right) - \tfrac{1}{2}E_k D_k \dot{x}_{j,j} = 0 \,. \tag{8.2.65}$$

Here we took into account that, in the two variants, the internal energies and the heat flux vectors can differ from one another. However the energy balances can be written in a more appropriate form. After a simple transformation, by using (8.2.48), we obtain from (8.2.64)

$$\rho^{(1)}\dot{\tilde{U}} + {}^{(1)}q_{j,j} - {}^{(1)}t_{ij}\dot{x}_{i,j} - \rho r^{ext} - \mathcal{J}_j E_j - E_j \dot{D}_j - \tfrac{1}{2}E_k D_k \dot{x}_{j,j} = 0 \tag{8.2.66}$$

with a modified internal energy[7] ${}^{(1)}\tilde{U}$ according to

$$\rho^{(1)}\tilde{U} := \rho^{(1)}U + \tfrac{1}{2}D_k E_k = \rho^{(1)}U + \tilde{\omega} \,. \tag{8.2.67}$$

An analogous change as in (8.2.67) yields instead of (8.2.65)

$$\rho^{(2)}\dot{\tilde{U}} + {}^{(2)}q_{j,j} - {}^{(2)}t_{ij}\dot{x}_{i,j} - \rho r^{ext} - \mathcal{J}_j E_j - E_j \dot{D}_j - E_k D_k \dot{x}_{j,j} = 0 \,. \tag{8.2.68}$$

For the derivation of the jump conditions, it is advantageous to use yet another form of the energy balance. What is needed is the conservative form of the total energy balance. To obtain it, let us start from the energy balance $(8.2.1)_4$

$$\frac{d}{dt}\left(\rho U + \tfrac{1}{2}\rho\dot{x}_i\dot{x}_i\right) + \left(\rho U + \tfrac{1}{2}\rho\dot{x}_i\dot{x}_i\right)\dot{x}_{j,j} = \left(t_{ij}\dot{x}_i - q_j\right)_{,j}$$
$$+ \dot{x}_i\left(\rho F_i^{ext} + \rho F_i^e\right) + \rho r^{ext} + \rho r^e \,, \tag{8.2.69}$$

Using (8.2.69), (8.2.38) and (8.2.47) in the dimensional form, respectively, we obtain after straightforward calculations the energy balance for the first variant

$$\frac{d}{dt}\left(\rho^{(1)}U + \tfrac{1}{2}\rho\dot{x}_i\dot{x}_i + \tfrac{1}{2}E_k D_k\right) + \left(\rho^{(1)}U + \tfrac{1}{2}\rho\dot{x}_i\dot{x}_i + \tfrac{1}{2}E_k D_k\right)\dot{x}_{j,j}$$
$$+ \left({}^{(1)}q_j + e_{jik}E_i\mathcal{H}_k - {}^{(1)}t_{ij}\dot{x}_i - E_i D_j\dot{x}_i + \tfrac{1}{2}E_k D_k\dot{x}_j\right)_{,j}$$
$$= \dot{x}_i\rho F_i^{ext} + \rho r^{ext} \,, \tag{8.2.70}$$

and for the second variant

$$\frac{d}{dt}\left(\rho^{(2)}U + \tfrac{1}{2}\rho\dot{x}_i\dot{x}_i + \tfrac{1}{2}E_k D_k\right) + \left(\rho^{(2)}U + \tfrac{1}{2}\rho\dot{x}_i\dot{x}_i + \tfrac{1}{2}E_k D_k\right)\dot{x}_{j,j}$$
$$+ \left({}^{(2)}q_j + e_{jik}E_i\mathcal{H}_k - {}^{(2)}t_{ij}\dot{x}_i - E_i D_j\dot{x}_i\right)_{,j} = \dot{x}_i\rho F_i^{ext} + \rho r^{ext} \,. \tag{8.2.71}$$

---

[7] The same modification was used also in [118].

The balances (8.2.70), (8.2.71), now in conservative form, will be used −
as mentioned − exclusively to derive the jump conditions for the energy.
One may remark that in both equations the magnetic quantity $\mathcal{H}_k$ appears.
Unlike the energy balances (8.2.66), (8.2.68) the conductive electric current
$\mathcal{J}_j$ does not appear in (8.2.70), (8.2.71). Thus, effectively one has replaced a
constitutive quantity, namely $\mathcal{J}_j$, by another quantity, namely $\mathcal{H}_k$.

Concerning the entropy inequality we know or we can relatively easily
prove with MÜLLER's entropy principle that for the MAXWELL-MINKOWSKI
model the entropy flux obeys the DUHEM relation

$$^{(1),(2)}\phi_j := \frac{^{(1),(2)}q_j}{\Theta} .$$
(8.2.72)

Furthermore for both variants the density of the HELMHOLTZ free energy
$\widetilde{\Psi}^{(1),(2)}$ is defined in [64] by

$$^{(1),(2)}\widetilde{\Psi} := {}^{(1),(2)}\tilde{U} - {}^{(1),(2)}\eta \Theta - \frac{1}{\rho}D_j E_j ,$$
(8.2.73)

where η denotes the entropy as in $(8.2.1)_5$.

If we solve (8.2.73) for the internal energy $^{(1),(2)}\tilde{U}$ and replace in (8.2.66)
and (8.2.68) $^{(1),(2)}\dot{\tilde{U}}$ by the relation obtained from (8.2.73), then new forms of
energy balances for the two variants are obtained. These relations contain the
external energy supplies $^{(1),(2)}r^{ext}$, which also occur in the entropy imbalance
$(8.2.1)_5$. Eliminating these yields new forms of the entropy inequality in the
two variants, which are, respectively, given by

$$-\rho^{(1)}\dot{\tilde{\Psi}} - \tfrac{1}{2}D_j E_j \dot{x}_{i,i} - \rho^{(1)}\eta \dot{\Theta} - \Theta_{,j}\frac{^{(1)}q_j}{\Theta} - D_j \dot{E}_j + \mathcal{J}_j E_j$$
$$+ {}^{(1)}t^M_{ij}\dot{x}_{i,j} \geq 0 ,$$
(8.2.74)

$$-\rho^{(2)}\dot{\tilde{\Psi}} - \rho^{(2)}\eta \dot{\Theta} - \Theta_{,j}\frac{^{(2)}q_j}{\Theta} - D_j \dot{E}_j + \mathcal{J}_j E_j$$
$$+ {}^{(2)}t^M_{ij}\dot{x}_{i,j} \geq 0 .$$
(8.2.75)

These relations achieve the fundamental goal of this chapter: the electrorheological entropy inequality for both variants. As one can see, the difference
between both inequalities consists in the factor in front of the velocity divergence.

Let us resume now all the total balance equations given so far for both
variants. These are (8.2.62), (8.2.66), (8.2.74) and (8.2.63), (8.2.68), (8.2.75),
respectively. For comparison of both variants the identifications

$$^{(1)}t_{ij} = {}^{(2)}t_{ij} + \tfrac{1}{2}D_k E_k \delta_{ij} ,$$
(8.2.76)

$$^{(1)}q_j = {}^{(2)}q_j , \implies {}^{(1)}\phi_j = {}^{(2)}\phi_j ,$$
(8.2.77)

$$^{(1)}\widetilde{\Psi} = {}^{(2)}\widetilde{\Psi} , \quad {}^{(1)}U = {}^{(2)}U , \implies {}^{(1)}\eta = {}^{(2)}\eta .$$
(8.2.78)

are required if the two variants are to yield identical models. Evidently, the only difference between the two variants appears in the Cauchy stress tensor. One can compare the results from one variant with those of the other if one takes into account relations (8.2.76)–(8.2.78). Recall also that the electromagnetic quantities E_i, D_i, B_i, $\mathcal{H}_i$ and $\mathcal{J}_i$ are the same in both models.

8.2.5 Jump Conditions

In order to complete the electro-mechanical description of electrorheological materials as continuous media, we have to add the jump conditions that must be obeyed by the mechanical and electromagnetic quantities across a discontinuity surface. Let Σ be a smooth surface, not necessarily material (e.g. an infinitely thin wall or a membrane) which separates one part of the body under consideration from another part, and let $w_i n_i$ be its velocity in the positive direction of the unit normal to Σ.

In [64] the easiest case is considered: with the exception of the classical electric quantity $\mathcal{Q}^s$ (electric surface charge density of the free charges) all surface quantities vanish. In general, the jump conditions for volume balances can be given as (see e.g. [24, 72] and also [161])

$$\llbracket \Phi_j + \Xi\,(\dot{x}_j - w_j) \rrbracket n_j = 0\ , \tag{8.2.79}$$

in which the abbreviation $\llbracket A_j \rrbracket := A_j^{(+)} - A_j^{(-)}$ is used where $A_j^{(+)}$, $A_j^{(-)}$ represent the values of A_j on the immediate exterior $(+)$ and interior $(-)$ side of the surface, and $\boldsymbol{n}$ is the unit normal vector pointing into the positive region of Σ. In addition, Φ_j denotes the non-convective flux of the volume quantity Ξ and w_j is the surface velocity. If Σ is a material surface then we have $w_j = \dot{x}_j$. Let us look now at the physical balances in detail.

For the mass balance (8.2.1)$_1$ $\Xi = \rho$ and the non-convective flux, Φ_j, vanishes. So,

$$\llbracket \rho\,(\dot{x}_j - w_j) \rrbracket n_j = 0\ . \tag{8.2.80}$$

For the total momentum balance (8.2.62) of the second variant we have $\Xi_i = \rho\dot{x}_i$, $\Phi_{ij} = -\,^{(2)}t_{ij} - E_i D_j + \tfrac{1}{2} E_k D_k \delta_{ij}$ implying

$$\llbracket \rho\dot{x}_i\,(\dot{x}_j - w_j) - \,^{(2)}t_{ij}^M - E_i D_j + \tfrac{1}{2} E_k D_k \delta_{ij} \rrbracket n_j = 0\ . \tag{8.2.81}$$

The jump condition in the first variant looks very similarly, namely

$$\llbracket \rho\dot{x}_i\,(\dot{x}_j - w_j) - \,^{(1)}t_{ij}^M - E_i D_j \rrbracket n_j = 0\ . \tag{8.2.82}$$

For the energy balance we start from (8.2.70) and (8.2.71), respectively. With obvious identifications of Ξ and Φ one obtains

$$\left[\!\!\left[\left(\rho U + \tfrac{1}{2}\rho \dot{x}_i \dot{x}_i + \tfrac{1}{2} E_k D_k \right) \left(\dot{x}_j - w_j \right) + q_j + e_{jik} E_i \mathcal{H}_k \right.\right.$$

$$\left.\left. - \left({}^{(1)}t_{ij} + E_i D_j - \tfrac{1}{2} E_k D_k \delta_{ij} \right) \right]\!\!\right] n_j = 0 \,, \tag{8.2.83}$$

$$\left[\!\!\left[\left(\rho U + \tfrac{1}{2}\rho \dot{x}_i \dot{x}_i + \tfrac{1}{2} E_k D_k \right) \left(\dot{x}_j - w_j \right) + q_j + e_{jik} E_i \mathcal{H}_k \right.\right.$$

$$\left.\left. - \left({}^{(2)}t_{ij} + E_i D_j \right) \right]\!\!\right] n_j = 0 \,. \tag{8.2.84}$$

in the two variants, in which $\mathcal{H}_k$ can be eliminated by means of (8.2.53).

For the entropy imbalance we start from $(8.2.1)_5$. For both variants the jump condition is

$$\left[\!\!\left[\rho \eta \left(\dot{x}_j - w_j \right) + \frac{q_j}{\Theta} \right]\!\!\right] n_j \geq 0 \,. \tag{8.2.85}$$

Finally, the MAXWELL-MINKOWSKI equations in the electrorheological approximation, (8.2.58)−(8.2.61), imply

$$e_{ijk}\, n_j \left[\!\!\left[E_k \right]\!\!\right] = 0 \,, \tag{8.2.86}$$

$$e_{ijk}\, n_j \left[\!\!\left[\mathcal{H}_k + e_{klm} \left(\dot{x}_l - w_l \right) D_m \right]\!\!\right] = 0 \,, \tag{8.2.87}$$

$$\left[\!\!\left[D_j \right]\!\!\right] n_j = \mathcal{Q}^s \,, \tag{8.2.88}$$

$$\left[\!\!\left[B_j \right]\!\!\right] n_j = 0 \,, \tag{8.2.89}$$

where $\mathcal{Q}^s$ represents the electric surface charge density.

In what follows we shall exclusively use the second variant.

8.2.6 Discussion

The purpose of the first part of this section was to motivate the balance equations for electrorheological fluids as they were derived in [64].

Starting from the general balance laws of rational thermodynamics and electrodynamics the system of total balance equations was derived by a consistent specialization to electrorheological behaviour. The total electromechanical balance laws of mass, momentum and the MAXWELL equations in the MAXWELL–MINKOWSKI formulation were given both in local form and as jump conditions across singular surfaces. The electromagnetic body force and energy supply rates were derived from the MAXWELL equations by deducing from these the electromagnetic momentum and energy balances as identities. The production terms in these expressions were identified as electromagnetic body force and energy supply rate, respectively. However, since the division of the time rates of change of the electromagnetic momentum and energy into flux and production terms are not unique, two variants of electrorheological models were presented, which required connecting relations between energies and stresses, if equivalence of the emerging models are attempted.

The crucial step put forward in this approach was the non-dimensionalization process specific to the electrodynamic equations. The assumptions that there is no external magnetic field and that the magnetic induction is produced exclusively by the motion of the polarized dielectrics, allowed us to neglect the magnetic terms compared with the electric terms in most (but not all) of the electrodynamical equations and, consequently, also the total balances[8].

The essential results are the simplified balance equations of thermomechanics and electrodynamics which will be used in the remainder of this work; this is why we put them here together. These equations are (in due order) the mass balance, the linear momentum balance, the balance of moment of momentum, the energy balance, the entropy inequality as well as the four MAXWELL-MINKOWSKI equations in the electrorheological approximations; explicitly,

$$\dot\rho + \rho\, \dot x_{j,j} = 0 \,, \tag{8.2.1}_1$$

$$\rho\ddot x_i - \left({}^{(2)}t_{ij} + E_i D_j \right)_{,j} - \rho F_i^{ext} = 0 \,, \tag{8.2.63}$$

$$e_{ijl}\, {}^{(2)}t_{lj} = 0 \,, \tag{8.2.1}_3$$

$$\rho\dot U + q_{j,j} - {}^{(2)}t_{ij}\dot x_{i,j} - \rho r^{ext} - \mathcal{J}_j E_j - E_j \dot D_j$$
$$-E_k D_k \dot x_{j,j} = 0 \,, \tag{8.2.68}$$

(ERF) $$-\rho\dot\psi - \eta\dot\Theta - \theta_{,j}\frac{q_j}{\Theta} - D_j \dot E_j + \mathcal{J}_j E_j + {}^{(2)}t_{ij}\dot x_{i,j} \geq 0 \,, \tag{8.2.75}$$

$$e_{klm}\, E_{m,l} = 0 \,, \tag{8.2.90}$$

$$e_{jlm}\left[\mu_0^{-1}B_m + e_{mpq}v_p\left(D_q - \varepsilon_0 E_q\right) \right]_{,l} - \frac{\partial D_j}{\partial t} - \mathcal{J}_j \,,$$
$$- \mathcal{Q}\dot x_j = 0 \,, \tag{8.2.91}$$

$$D_{i,i} - \mathcal{Q} = 0 \,, \tag{8.2.4}$$

$$B_{i,i} = 0 \,. \tag{8.2.5}$$

Henceforth they will be labeled and called ERF equations.

8.2.7 Constitutive Equations

The balance equations in the electrorheological approximation given above are not sufficient to determine all the unknowns of the problem; so additional relations reflecting the specific properties of the studied material must be postulated. The starting point of this procedure consists of the choice of the

---

[8] Even if the magnetic induction would be produced primarily by a time changing electric field or by an electric current, the neglect of the magnetic compared to the electric quantities is under practically relevant circumstances still justified. In general, the situation changes only drastically when an external magnetic field is present.

independent fields from all the physical variables involved in the governing equations. Then we must establish the appropriate constitutive equations for the remaining variables. In [64], by obeying the principle of material frame indifference, the independent variables have been chosen as

$$\rho, \Theta, E_j, d_{ij}, \tag{8.2.92}$$

where d_{ij} denotes the rate of the strain tensor

$$d_{ij} := \tfrac{1}{2}(\dot{x}_{i,j} + \dot{x}_{j,i}), \tag{8.2.93}$$

which is objective.

The presence of this quantity among the independent variables is motivated by the fact that ER-materials behave in general as viscous fluids. That $\Theta_{,i}$ is not among the variables (8.2.92) implies that effects of heat conduction are not considered. Its incorporation would be formally quite obvious and easy, but we omit such a dependence, because in subsequent applications it will not be considered.

It is important to note that the magnetic flux density B_i is not an independent constitutive variable. Such an assumption is reasonable, because B_i is likely very small, since according to (8.2.28) magnetization is induced only by polarization. On this basis B_i can then be computed from (8.2.91) and (8.2.5) if the electric field strength and the velocity are known. It then follows that B_i will not influence the mechanical equations, and it will neither intervene in the electrical problem as will soon be seen.

Although the electric field may also be space-dependent, in a first approximation its gradient will not be considered an independent constitutive variable either.

The dependent constitutive quantities of the problem are

$$\mathbb{C} := \{t_{ij}, D_j, \mathcal{J}_j, U, \tilde{\Psi}, q_j\}, \tag{8.2.94}$$

where we have dropped the left upper index "(2)" from t_{ij}, and they are of the form

$$\mathbb{C} = \mathbb{C}(\rho, \Theta, E_i, d_{ij}). \tag{8.2.95}$$

Henceforth we shall focus attention on formulation "(2)".

Evaluation of the entropy inequality

Evaluation of the entropy inequality will bring the constitutive equations to their final form. Substituting the constitutive function for the HELMHOLTZ free energy given in (8.2.94) and (8.2.95) into (8.2.75) and performing the time differentiation according to the chain rule results in

$$-\rho\left(\eta+\frac{\partial\tilde{\Psi}}{\partial\Theta}\right)\dot{\Theta}-\left(D_j+\rho\frac{\partial\tilde{\Psi}}{\partial E_j}\right)\dot{E}_j-\rho\frac{\partial\tilde{\Psi}}{\partial d_{kl}}\dot{d}_{kl}-\Theta_{,j}\frac{q_j}{\Theta}$$

$$+\left[t_{ij}+\rho^2\frac{\partial\tilde{\Psi}}{\partial\rho}\delta_{ij}\right]d_{ij}+\mathcal{J}_jE_j\geq0\,.\tag{8.2.96}$$

Exploitation of this inequality follows the same procedure outlined earlier in Chap. 3. The inequality is explicitly linear in

$$\dot{\Theta}\,,\dot{E}_j\,,\dot{d}_{kl}\,,\Theta_{,j}\,.\tag{8.2.97}$$

Since in admissible thermodynamic processes all these terms may have any arbitrarily assigned values, it follows that each of the coefficients of these variables in 8.2.96 must be identically zero. This implies the relations

$$\eta=-\frac{\partial\tilde{\Psi}}{\partial\Theta}\,,\tag{8.2.98}$$

$$D_j=-\rho\frac{\partial\tilde{\Psi}}{\partial E_j}\,,\tag{8.2.99}$$

$$0=-\frac{\partial\tilde{\Psi}}{\partial d_{kl}}\,,\tag{8.2.100}$$

$$\frac{q_j}{\Theta}=0\,,\tag{8.2.101}$$

and the residual inequality

$$\left[t_{ij}+\rho^2\frac{\partial\tilde{\Psi}}{\partial\rho}\delta_{ij}\right]d_{ij}+\mathcal{J}_jE_j\geq0\,.\tag{8.2.102}$$

Cross-differentiating (8.2.98)–(8.2.100) we obtain

$$\frac{\partial\eta}{\partial d_{kl}}=0\,,\tag{8.2.103}$$

$$\frac{\partial D_j}{\partial d_{kl}}=0\,,\tag{8.2.104}$$

$$\rho\frac{\partial\eta}{\partial E_j}=\frac{\partial D_j}{\partial\Theta}\,.\tag{8.2.105}$$

From (8.2.103), (8.2.104) it follows that the entropy and the electric displacement cannot depend on the rate of strain tensor. Alternatively, (8.2.105) establishes a connection between the constitutive laws for η and D_j; so not all the constitutive assumptions can be chosen independently.

If we use the definitions of the thermodynamic pressure and if we decompose the Cauchy stress tensor into a spherical tensor containing the dynamic pressure $-p\delta_{ij}$ and an extra stress tensor t_{ij}^e, viz,

$$p := \rho^2 \frac{\partial \tilde{\Psi}}{\partial \rho} , \qquad (8.2.106)$$

$$t_{ij}^e := t_{ij} + p\delta_{ij} , \qquad (8.2.107)$$

the imbalance (8.2.102) can be written as

$$\rho\gamma := t_{ij}^e d_{ij} + \mathcal{J}_j E_j \geq 0 , \qquad (8.2.108)$$

where γ is the density of entropy production. An important step is now the satisfaction of (8.2.108). A necessary condition to fulfill the thermodynamic equilibrium requirement, $\gamma = 0$ (no entropy is produced), is

$$\left.\frac{\partial\gamma}{\partial d_A}\right|_E = 0 , \qquad \left.\frac{\partial\gamma}{\partial E_p}\right|_E = 0 , \qquad (8.2.109)$$

where we denoted by d_A the 6 independent components of the rate of strain tensor

$$(d_A) := (d_{11}, d_{12}, d_{13}, d_{22}, d_{23}, d_{33}) . \qquad (8.2.110)$$

In (8.2.109) and in the sequel the index "E" will denote thermodynamic equilibrium. Further, one has to treat separately the case of electric conducting fluids (namely $\mathcal{J}_j \neq 0$) and the case of non-conducting fluids (namely $\mathcal{J}_j = 0$) because the results of this evaluation will be different in the two cases.

1. Electrically conducting fluids. In equilibrium it follows from (8.2.108) that the independent variables d_{pq} and E_p must vanish,

$$d_{pq}|_E = 0 , \qquad \text{and} \qquad E_p|_E = 0 . \qquad (8.2.111)$$

If we replace in (8.2.109) γ by the formula given on the left-hand side of (8.2.108) and take into account that $t_{ij}^e = t_{ij}^e(\rho, \Theta, E_k, d_{kl})$ as well as $\mathcal{J}_j = \mathcal{J}_j(\rho, \Theta, E_k, d_{kl})$, then (8.2.111) yields

$$\left.\left(\frac{\partial t_{ij}^e}{\partial d_A} d_{ij} + t_{ij}^e \frac{\partial d_{ij}}{\partial d_A} + \frac{\partial \mathcal{J}_j}{\partial d_A} E_j\right)\right|_E = \left.\left(t_{ij}^e \frac{\partial d_{ij}}{\partial d_A}\right)\right|_E = 0 , \quad (8.2.112)$$

and

$$\left.\left(\frac{\partial t_{ij}^e}{\partial E_p} d_{ij} + \frac{\partial \mathcal{J}_j}{\partial E_p} E_j + \mathcal{J}_p\right)\right|_E = \mathcal{J}_p|_E = 0. \qquad (8.2.113)$$

For the conductive current $\mathcal{J}_j$, (8.2.113) implies

$$\text{when} \quad d_{pq} = 0 \quad \text{and} \quad E_p = 0 , \quad \text{then} \quad \mathcal{J}_j = 0 . \qquad (8.2.114)$$

In view of the choice of the independent variables (8.2.92) this condition is always fulfilled.

From (8.2.112), by taking into account the symmetry of the extra stress, it follows that t_{ij}^e cannot be composed of terms containing only ρ and Θ. Consequently, when $d_{pq} = 0$ and $E_p = 0$,   then $t_{ij}^e = 0$.   (8.2.115)

2. Electrically non-conducting fluids. Unlike for conducting fluids the electric field does not need to vanish in thermodynamic equilibrium. From $(8.2.109)_1$ there follows a condition identical to (8.2.112), whilst from $(8.2.109)_2$ one obtains a condition which is always fulfilled if $\partial t_{ij}^e/\partial E_p|_E$ is bounded which we will assume. However, condition (8.2.112) is stronger in this case: t_{ij}^e must vanish whenever d_{mn} is zero and consequently t_{ij}^e cannot contain terms formed only with ρ, Θ and E_l.

A sufficient condition guaranteeing that (8.2.108) is fulfilled at thermodynamic equilibrium is that the matrix of the second-order derivatives (with respect to d_A and E_i) must be positive semidefinite (see [161]). More precisely, this means that all the principal subdeterminants of the matrix

$$
\left. \begin{pmatrix} (3 \times 3) & (3 \times 6) \\ (6 \times 3) & (6 \times 6) \end{pmatrix} \right|_E = \left. \begin{pmatrix} \left(\frac{\partial^2 \gamma}{\partial E_p \, \partial E_r} \right) & \left(\frac{\partial^2 \gamma}{\partial E_p \, \partial d_A} \right) \\ \left(\frac{\partial^2 \gamma}{\partial d_A \, \partial E_r} \right) & \left(\frac{\partial^2 \gamma}{\partial d_A \, \partial d_B} \right) \end{pmatrix} \right|_E \tag{8.2.116}
$$

must be larger or equal to zero. It is difficult to evaluate these conditions in general, but we will do it later for concrete constitutive assumptions.

Constitutive functions for the electrical quantities

Let us discuss now the constitutive laws for the dependent electrical quantities. Starting from (8.2.94) and (8.2.95), ECKART considers in [64] the most general expressions for isotropic polar vector functions (here D_j and $\mathcal{J}_j$) of a polar vector (here E_j) and a symmetric tensor of second order (here d_{jk}). These representations are[9]

$$
D_j = \varepsilon E_j , \tag{8.2.117}
$$
$$
\mathcal{J}_j = \sigma_1 E_j + \sigma_2 \, d_{jk} E_k + \sigma_3 \, d_{j\ell} d_{\ell k} E_k . \tag{8.2.118}
$$

(see e.g. [72] and [220]). The quantity ε denotes the *effective permittivity* of the ER-material at hand and $\sigma_1, \sigma_2, \sigma_3$ represent the *effective conductivities*.

In view of (8.2.104) one can easily see that ε cannot depend on the strain rate tensor d_{ij}; therefore

$$
\varepsilon = \varepsilon \left(\rho, \Theta, E_k E_k \right) . \tag{8.2.119}
$$

Alternatively the effective conductivities can be any functions of the invariants of the independent variables

---

[9] In (8.2.117) two additional terms $\varepsilon_2 d_{jk} E_k$ and $\varepsilon_3 d_{jl} d_{lk} E_k$ are missing because of (8.2.104)

$$\sigma_A = \sigma_A \left(\rho, \Theta, E_k E_k, d_{kk}, d_{jk} d_{kj}, d_{jk} d_{kl} d_{lj}, E_j d_{jk} E_k, E_j d_{jk} d_{kl} E_l \right) ,$$
$$(8.2.120)$$

where $A = 1, 2, 3$. This is, of course very complicated, and in practice one will restrict the number of independent variables to just a few, e.g.

$$\sigma_A = \sigma_A \left(\rho, \Theta, E_k E_k, d_{jk} d_{kj} \right) . \qquad (8.2.121)$$

The electrical model of order zero

The most simple meaningful constitutive assumptions for the electrical quantities are models for which ε and σ_A $(A = 1, 2, 3)$ do not depend on the rate of strain tensor [they are of "zeroth order" in d_{ij}[10] (ε is of course always of zeroth order in d_{ij} according to (8.2.119))]. Moreover, the constitutive parameters must also be independent of the electric field: we consider here linear dielectrics and OHMian conductors. With these restrictions one obtains from (8.2.119) and (8.2.121)

$$\varepsilon = \varepsilon \left(\rho, \Theta \right), \ \sigma_1 = \sigma_{100} \left(\rho, \Theta \right), \ \sigma_2 = \sigma_3 = 0 . \qquad (8.2.122)$$

So, the electrical constitutive parameters are constant for constant density and temperature. All these assumptions agree with the experimental results presented by ECKART [64] who quotes a relevant selection from the measurements done by the Rheology work-group, Department of Fluid Mechanics of the University of Elangen-Nürenberg, see [2, 269, 270]. In our further approach we will adopt the material description given by (8.2.122).

A discussion about the constitutive function of the Cauchy extra stress tensor t_{ij}^e is postponed until the next section where a literature review of the constitutive models for t_{ij}^e used to describe ER fluids is given.

8.3 Constitutive Laws for the Cauchy Stress Tensor

8.3.1 Models Proposed in the Literature

A key step in electrorheology is to relate the theory with practical applications namely with the results from measurements and computations. Usually, the theoretical approaches are very abstract, general and difficult to use in concrete situations whereas the empirical approaches are applicable but often too particular. In the electrorheological field noticeable efforts are made

---

[10] Measurements (see especially [2]) show that the measurable electric current can depend on the shear rate. This dependence is, however, largely influenced by the temperature, frequency of the electric field and mostly by the composition of the ERF. For the ERF Rheobay the shear rate dependence is, however, small; in this work this dependence will be neglected.

from both sides to describe in a better and more accurate way the response of electrorheological fluids to external fields. We will review briefly the most important theoretical models for the expression of the Cauchy stress tensor according to (8.2.94) and (8.2.95) that have so far been proposed in the literature. While these proposals are three-dimensional expressions for the Cauchy stress, in most applications one-dimensional models are used. If one deals with numerical computations involving two or three-dimensional models one can either generalize the one-dimensional models or choose particular forms of the general models taking into account also the experimental characterization of the ERFs.

The most general form of the constitutive function for the Cauchy stress tensor that depends on the objective independent variables ρ, Θ, d_{ij} and E_i (see (8.2.94) and (8.2.95)) is given by (see [220])

$$t_{ij} = -p\delta_{ij} + t_{ij}^e = (-p + \alpha_1)\delta_{ij} + \alpha_2 E_i E_j + \alpha_3 d_{ij} + \alpha_4 d_{ik}d_{kj}$$

$$+\alpha_5(E_i d_{jk}E_k + d_{ik}E_k E_j) + \alpha_6(E_i d_{jk}d_{kl}E_l + d_{ik}d_{kl}E_l E_j)\,, \quad (8.3.1)$$

where α_i, $i = 1, \ldots, 6$ are functions of the invariants

$$\rho\,,\, \Theta\,,\, E_k E_k\,,\, d_{kk}\,,\, d_{jk}d_{kj}\,,\, d_{jk}d_{kl}d_{lj}\,,\, E_j d_{jk}E_k\,,\, E_j d_{jk}d_{kl}E_l\,. \quad (8.3.2)$$

This general constitutive law was first proposed by RAJAGOPAL and WINE-MAN in [194]. However, they treated the electric field as a constant when calculating the velocity field for the flow problems formulated in their paper.

In [196] two special cases of (8.3.1) are discussed. In the first it is assumed that the stress is linear in d_{ij} and quadratic in E_i and hence the material parameters have the form

$$\alpha_1 = \alpha_{11} + \alpha_{12}d_{kk} + \alpha_{13}E_k E_k + \alpha_{14}E_k E_k d_{jj} + \alpha_{15}E_j d_{jk}E_k\,,$$
$$\alpha_2 = \alpha_{21} + \alpha_{22}d_{kk}\,,$$
$$\alpha_3 = \alpha_{31} + \alpha_{32}E_k E_k\,,$$
$$\alpha_4 = 0\,, \quad\quad\quad\quad\quad\quad\quad\quad\quad\quad\quad\quad\quad\quad\quad\quad (8.3.3)$$
$$\alpha_5 = \alpha_{51}\,,$$
$$\alpha_6 = 0\,,$$

where α_{ij} are functions of ρ and Θ only. The subcases of (i) a compressible, (ii) a mechanically incompressible but electrically compressible and (iii) an incompressible fluid are considered and for each of them the restrictions imposed on t_{ij} by the CLAUSIUS–DUHEM inequality are given. The second case pertains to the non-linear model of incompressible ERFs with shear dependent viscosities. First it is assumed that

$$\alpha_4 = \alpha_6 = 0\,. \quad\quad\quad\quad\quad\quad\quad\quad\quad\quad\quad\quad (8.3.4)$$

The choices for the material parameters α_2, α_3 and α_5 reflect a combination of a NEWTONian and power-law like behaviour where the power exponent[11]

---

[11] We adopted a different notation than in [196], where p is used instead of n, to avoid confusion with the pressure p

can be a function of $E_k E_k$. Concretely, it is assumed that

$$
\begin{aligned}
\alpha_2 &= \alpha_{20} + \alpha_{21}(d_{lm}d_{ml})^{(n-1)/2} , \\
\alpha_3 &= \alpha_{30} + \alpha_{31}(d_{lm}d_{ml})^{(n-2)/2} + \alpha_{32}E_k E_k \\
&\quad + \alpha_{33}E_k E_k (d_{lm}d_{ml})^{(n-2)/2} , \\
\alpha_5 &= \alpha_{50} + \alpha_{51}(d_{lm}d_{ml})^{(n-2)/2} ,
\end{aligned}
\tag{8.3.5}
$$

where α_{ij} are functions of Θ. The material function n depends on $E_k E_k$ and satisfies

$$
1 < n_\infty \leq n(E_k E_k) \leq n_0 < \infty , \tag{8.3.6}
$$

where

$$
n_0 = \lim_{E_k E_k \to 0} n(E_k E_k) , \quad n_\infty = \lim_{E_k E_k \to \infty} n(E_k E_k) . \tag{8.3.7}
$$

An alternative model to (8.3.5) is also given; in this model $(d_{lm}d_{ml})^{\beta/2}$ is replaced by

$$
(\xi + d_{lm}d_{ml})^{\beta/2} \quad \text{or} \quad (\xi + (d_{lm}d_{ml})^{1/2})^{\beta} , \tag{8.3.8}
$$

where $\beta = n - 1$ or $n - 2$. The purpose of the addition[12] of ξ in the representation 8.3.8 prevents the model for $n \in (1, 2)$ from developing infinite shear viscosity at zero stretching and a yield-like behaviour as in the first model (8.3.5). Simplified models of (8.3.5) and (8.3.8) corresponding to specific electrorheological fluids ($n \equiv 2$ or $n \not\equiv 2$) are considered and restrictions for the corresponding coefficients α_{ij} are obtained.

Finally, a model which includes all the discussed approximating models except that with $n \in (1, 2)$ and $\alpha_{30} = \alpha_{32} = \alpha_{50} = 0$ is proposed as follows:

$$
\begin{aligned}
t_{ij} &= -p\delta_{ij} + \alpha_{21}((\xi + d_{lm}d_{ml})^{(n-1)/2} - 1)E_i E_j \\
&\quad + (\alpha_{31} + \alpha_{33}E_k E_k)(\xi + d_{lm}d_{ml})^{(n-2)/2}d_{ij} \\
&\quad + \alpha_{51}(\xi + d_{lm}d_{ml})^{(n-2)/2}(E_i d_{jk}E_k + d_{ik}E_k E_j) ,
\end{aligned}
\tag{8.3.9}
$$

where n satisfies (8.3.6). All cases formulated and discussed in [196] are amenable to mathematical analysis. In [205], Růžička studied in detail mathematical issues such as existence, uniqueness and stability of weak and strong solutions for steady flow of incompressible shear dependent electrorheological fluids with the stress given by (8.3.9). In view of the dependence of the material function n on the magnitude of the electric field, this problem is described by an elliptic or parabolic system of partial differential equations (PDE) exhibiting so-called non-standard growth conditions, i.e. the elliptic operator $t_{ij}^e = t_{ij} + p\delta_{ij}$ satisfies the inequalities

---

[12] Here ξ is a quantity having the dimension $1/s^2$ in $(8.3.8)_1$ and $1/s$ in $(8.3.8)_2$ which for the sake of the mathematical treatment is taken to be equal to $1\ s^{-2}$ (s^{-1}). The choice $\xi = 1$ is justified, because the role of the extra constant ξ is to regularize the system of partial differential equations at zero stretching

$$t_{ij}^e(d, E)d_{ij} \geq c_0(1 + E_k E_k)(\xi + d_{lm}d_{ml})^{\frac{n_\infty - 2}{2}} d_{lm}d_{ml} , \quad (8.3.10)$$

$$t_{ij}^e(d, E)\, t_{ij}^e(d, E) \leq c_1(\xi + d_{lm}d_{ml})^{\frac{n_0 - 1}{2}} E_k E_k . \quad (8.3.11)$$

In [205] the case of unsteady flows of shear dependent ERFs is also treated for the constitutive function

$$t_{ij} = -p\delta_{ij} + \alpha_{31}(1 + E_k E_k)(\xi + d_{lm}d_{ml})^{(n-2)/2}d_{ij} , \quad (8.3.12)$$

where $\alpha_{31} > 0$ and $n = n(E_k E_k)$ satisfies (8.3.6) with $n_\infty \geq 2$. The existence of weak and strong solutions global in time for large data under certain restrictions on n_∞ and n_0 and the uniqueness of the strong solution are proved.

The model proposed by ECKART in [63][13] consists of an extension of the model (8.3.5) modified as in (8.3.8)$_1$,

$$t_{ij} = -p\delta_{ij} + [\alpha_{20} + \alpha_{21}Z^{-a+1/2} + \alpha_{22}Z^{-b+1/2}]E_i E_j$$

$$+[\alpha_{30} + \alpha_{31}Z^{-a} + \alpha_{32}Z^{-b}]d_{ij}$$

$$+[\alpha_{40} + \alpha_{41}Z^{-a-1/2} + \alpha_{42}Z^{-b-1/2}]d_{ik}d_{kj}$$

$$+[\alpha_{50} + \alpha_{51}Z^{-a} + \alpha_{52}Z^{-b}](E_i d_{jk}E_k + d_{ik}E_k E_j)$$

$$+[\alpha_{60} + \alpha_{61}Z^{-a-1/2} + \alpha_{62}Z^{-b-1/2}](E_i d_{jk}d_{kl}E_l + d_{ik}d_{kl}E_l E_j) , \quad (8.3.13)$$

where $Z := D_0^2 + d_{mn}d_{nm}$. Here α_{ij} as well as the exponents a and b are material parameters that can depend only on ρ, Θ and $E_k E_k$, while D_0 denotes a constant reference shear rate. The reasons for this choice of the constitutive function are explained in detail in [63] (see also [64]). We mention only some selected arguments. First, this model includes the CASSON model as a special case, to be presented below in (8.3.49)–(8.3.50) and quite successfully used (in one-dimensional form) to fit measured data for one of the ERFs, tested in [2]. Second, by introducing coefficient expansions in Z with two exponents a and b, it is possible to describe the measurements with *constant exponents*.

---

[13] The same author presents in [64] a more detailed approach where he treats also the so-called mechanical model of order 1 (the stress is linear in d_{ij} and quadratic in E_i) which was discussed also in [196] as mentioned above. Unlike RAJAGOPAL and RŮŽIČKA, ECKART considers only incompressible fluids but in addition to the results presented in [196] he gives the consequences of the entropy inequality on this model also for electrically conducting fluids. The particularization of this model on a viscometric flow under uniform electric field perpendicular to the flow direction shows NEWTONian behaviour with a viscosity depending on the electric field. This behaviour is not supported by the experimental data where one observes strongly non-linear dependence of the shear stress on shear rate (for more details see [64]).

[Having only one exponent that does not depend on the electric field, it is in general not possible to fit the data]. This is advantageous for inhomogeneous electric fields (for which E_k depends on x_j) since we then avoid the exponents to be dependent on the coordinate x_j.

The quantity D_0^2, which in [196] has been chosen as $D_0^2 = 1$, exhibits an important meaning. A non-vanishing D_0 plays the same role as ξ in (8.3.8). Specifically, it prevents the model for certain choices of the exponents a and b from developing an infinite viscosity limit at vanishing stretching. For numerical calculations a non-vanishing value $D_0 \neq 0$ avoids singularities that may otherwise cause problems in the computations. For some particular cases of (8.3.13), however, e.g. in viscometric flows (see (8.3.27)–(8.3.30), (8.3.33)–(8.3.36)), the choice $D_0 = 0$ allows calculation of analytical solutions for the velocity (see Subsect. 8.4.2). However, when an inhomogeneous electric field is considered, the flow is no longer viscometric and numerical solutions must be sought by choosing certain positive values for D_0.

The phenomenological approach was continued in [63] (see also [64]) with the investigation of model (8.3.13) in a viscometric flow with a constant electric field perpendicular to the flow direction (e.g. shear flow in a plane channel under an electric field produced by two infinite electrodes placed along the channel walls – see Subsect. 8.4.2), i.e.,

$$E_1 = E_3 = 0, \quad E_2 = -\frac{V}{h}, \tag{8.3.14}$$

$$d_{12} = d_{21} = \tfrac{1}{2}\,\dot{x}_{1,2} =: \tfrac{1}{2}\,\dot{\gamma}, \quad d_{ij} = 0 \quad \text{otherwise}. \tag{8.3.15}$$

Then, it is straightforward to show that

$$Z = D_0^2 + \tfrac{1}{2}\dot{\gamma}^2, \tag{8.3.16}$$

$$\tau := t_{12} = t_{21} = (\bar{\beta}_0 + \bar{\beta}_1 Z^{-a} + \bar{\beta}_2 Z^{-b})\dot{\gamma}, \tag{8.3.17}$$

$$t_{13} = t_{31} = 0, \tag{8.3.18}$$

$$t_{23} = t_{32} = 0, \tag{8.3.19}$$

$$t_{11} = -p + \tfrac{1}{4}(\alpha_{40} + \alpha_{41} Z^{-a-1/2} + \alpha_{42} Z^{-b-1/2})\dot{\gamma}^2, \tag{8.3.20}$$

$$t_{22} = -p + (\alpha_{20} + \alpha_{21} Z^{-a+1/2} + \alpha_{22} Z^{-b+1/2})\frac{V^2}{h^2}$$
$$+ (\bar{\psi}_0 + \bar{\psi}_1 Z^{-a-1/2} + \bar{\psi}_2 Z^{-b-1/2})\dot{\gamma}^2, \tag{8.3.21}$$

$$t_{33} = -p, \tag{8.3.22}$$

$$N_1 := t_{11} - t_{22} = -(\alpha_{20} + \alpha_{21} Z^{-a+1/2} + \alpha_{22} Z^{-b+1/2})\frac{V^2}{h^2}$$
$$-\tfrac{1}{2}(\alpha_{60} + \alpha_{61} Z^{-a-1/2} + \alpha_{62} Z^{-b-1/2})\frac{V^2}{h^2}\,\dot{\gamma}^2, \tag{8.3.23}$$

$$N_2 := t_{22} - t_{33} = (\alpha_{20} + \alpha_{21} Z^{-a+1/2} + \alpha_{22} Z^{-b+1/2})\frac{V^2}{h^2}$$
$$+ (\bar{\psi}_0 + \bar{\psi}_1 Z^{-a-1/2} + \bar{\psi}_2 Z^{-b-1/2})\dot{\gamma}^2. \tag{8.3.24}$$

where

$$\bar{\beta}_i := \tfrac{1}{2}\left(\alpha_{3i} + \alpha_{5i}\,V^2/h^2\right), \ i = 0,1,2\,, \tag{8.3.25}$$

$$\bar{\psi}_i := \tfrac{1}{4}\left(\alpha_{4i} + 2\alpha_{6i}\,V^2/h^2\right), \ i = 0,1,2\,, \tag{8.3.26}$$

and $\alpha_{ij} = \alpha_{ij}(\rho,\Theta,V^2/h^2)$.

Unfortunately there are no measurements available for the normal stresses t_{11}, t_{22}, t_{33}. That is why the parameters α_{4i} and α_{6i}, $i = 0,1,2$, which are primarily responsible for non-vanishing values of N_1 and N_2, will be set equal to zero in the constitutive models used in the numerical computations described below, see (8.3.39), (8.3.65), (8.3.66). We also mention that $N_1 = 0$ when the electric field vanishes which is an unrealistic property, but for an ERF this is acceptable since these fluids exhibit no normal stress effects in the absence of the electric field.

For certain choices of D_0, a, b and $\bar{\beta}_i$ the shear stress formula (8.3.17) agrees with the corresponding formula that may also be derived from model (8.3.8) if that is restricted to the viscometric flow in question. Model (8.3.17) with $D_0 = 0$ includes also the most popular models[14] used in a one-dimensional form in electrorheology: the BINGHAM model, the CASSON model and the power-law model [2, 176, 258, 269, 270]. Let us employ the upper indices B, C and P to denote the quantities corresponding to each of them.

- By choosing $\bar{\beta}_0 = \eta^B$, $a = \tfrac{1}{2}$, $\bar{\beta}_1 = \tfrac{1}{\sqrt{2}}\tau_y^B$ and $\bar{\beta}_2 = 0$ the linear BINGHAM model[15] is found

$$\tau^B = \tau_y^B + \eta^B\dot{\gamma}\,, \qquad \tau^B > \tau_y^B\,, \quad \dot{\gamma} > 0\,, \tag{8.3.27}$$

$$\tau^B = -\tau_y^B + \eta^B\dot{\gamma}\,, \qquad \tau^B < -\tau_y^B\,, \quad \dot{\gamma} < 0\,, \tag{8.3.28}$$

where $\tau_y^B \geq 0$ is the yield stress and $\eta^B > 0$ the viscosity.
- The non-linear CASSON-model possesses also two material parameters: the yield stress $\tau_y^C \geq 0$ and the viscosity $\eta^C > 0$. Its equation reads

$$\tau^C = \tau_y^C + 2\,(\tau_y^C\eta^C\dot{\gamma})^{1/2} + \eta^C\dot{\gamma}\,, \qquad \tau^C > \tau_y^C\,, \quad \dot{\gamma} > 0\,, \tag{8.3.29}$$

$$\tau^C = -\tau_y^C - 2\,(\tau_y^C\eta^C(-\dot{\gamma}))^{1/2} + \eta^C\dot{\gamma}\,, \qquad \tau^C < -\tau_y^C\,, \quad \dot{\gamma} < 0\,, \tag{8.3.30}$$

and is obtained by choosing the parameters in (8.3.17) as follows

$$\bar{\beta}_0 = \eta^C\,, \quad a = \tfrac{1}{2}\,, \quad \bar{\beta}_1 = \tfrac{1}{\sqrt{2}}\tau_y^C\,, \quad b = \tfrac{1}{4}\,, \quad \bar{\beta}_2 = 2^{3/4}(\eta^C\tau_y^C)\,. \tag{8.3.31}$$

---

[14] For models with yield stress one can deduce only the equations describing the liquid-like behaviour.

[15] In [63], equations (8.3.27), (8.3.29), (8.3.33) and (8.3.35) are said to be valid also for $\dot{\gamma} = 0$. We agree with this only in the sense of a subsequent extension by continuation and not as a deduction from the model (8.3.17) since when $a > 0$, Z is not defined for $\dot{\gamma} = 0$ (except when $n > 1$ for the power law model which is valid for $\dot{\gamma} = 0$). Equations (8.3.28), (8.3.30) and (8.3.34) are not given in [63]; these can also be deduced from (8.3.17) and we give them here for completeness.

- The power-law model is also non-linear but it does not possess a yield region and thus continuously connects a zero stretching regime with any such non-zero regime. For its parameterization one must choose

$$\bar{\beta}_0 = 0, \quad a = \tfrac{1}{2}(1-n), \quad \bar{\beta}_1 = 2^{-(1-n)/2}m, \quad \bar{\beta}_2 = 0, \qquad (8.3.32)$$

where $m > 0$ and $n > 0$ are the two model parameters and (8.3.17) becomes

$$\tau^P = m\dot{\gamma}^n, \qquad \dot{\gamma} > 0, \qquad (8.3.33)$$

$$\tau^P = -m(-\dot{\gamma})^n, \quad \dot{\gamma} < 0. \qquad (8.3.34)$$

Then, by analysing the experimental data measured for the electrorheological fluid Rheobay TP AI 3565, the CASSON-like model, introduced by choosing $D_0 = 0$, $a = \tfrac{1}{2}$ and $b = \tfrac{1}{4}$, is found to be a very suitable model (at least for the fluid Rheobay)

$$\tau^{Cl} = (\eta_0 + \beta_0)\dot{\gamma} + 2^{1/2}\beta_1 + 2^{1/4}\beta_2\dot{\gamma}^{1/2},$$
$$\tau^{Cl} > 2^{1/2}\beta_1 \geq 0, \ \dot{\gamma} > 0, \qquad (8.3.35)$$

$$\tau^{Cl} = (\eta_0 + \beta_0)\dot{\gamma} - 2^{1/2}\beta_1 - 2^{1/4}\beta_2(-\dot{\gamma})^{1/2},$$
$$\tau^{Cl} < -2^{1/2}\beta_1 \leq 0, \ \dot{\gamma} < 0, \qquad (8.3.36)$$

where the upper index "Cl" is used to denote the shear stress for the CASSON-like model. Here, $\beta_1 = \bar{\beta}_1$, $\beta_2 = \bar{\beta}_2$ and $\beta_0 = \bar{\beta}_0 - \eta_0$ must obey the relation

$$\beta_i(V = 0) = 0, \quad i = 0, 1, 2, \qquad (8.3.37)$$

where η_0 is the dynamic viscosity in the absence of an electric field. By imposing (8.3.37) it is demanded that the fluid has NEWTONian behaviour at vanishing electric field. Whereas in the CASSON model, the coefficient of $\dot{\gamma}^{1/2}$ is connected to the other material coefficients, in the model (8.3.35)–(8.3.36) it is independent of these. Consequently, the CASSON-like model can be particularized to the BINGHAM model. The values of η_0 and β_1, β_2 may differ for different electric fields. A table with values found by fitting the data from the measurements obtained in a rotational viscometer for the ER-fluid Rheobay at different electric fields is given in Subsect. 8.4.3 (Table 8.1).

Finally, taking into account the experimental results, we now impose additional, new assumptions in the general model (8.3.13) to obtain a simpler constitutive function which is still able both to describe the measurements and to make simpler analytical solutions possible. First, the parameters responsible for the normal stress effects, (see (8.3.23), (8.3.24), are neglected[16],

---

[16] Actually, the parameter α_2 can also be responsible for non-trivial normal stress differences. However, it is almost impossible to describe realistically the normal stress effects only with this parameter.

$$\alpha_{40} = \alpha_{41} = \alpha_{42} = \alpha_{60} = \alpha_{61} = \alpha_{62} = 0 \,, \tag{8.3.38}$$

and, second, the values $a = \frac{1}{2}$ and $b = \frac{1}{4}$ are chosen. This leads to

$$
\begin{aligned}
t_{ij} = &-p\delta_{ij} + [\alpha_{20} + \alpha_{21} + \alpha_{22}(D_0^2 + d_{mn}d_{nm})^{1/4}]E_i E_j + 2\eta_0 d_{ij} \\
&+[\alpha_{30} + \alpha_{31}(D_0^2 + d_{mn}d_{nm})^{-1/2} + \alpha_{32}(D_0^2 + d_{mn}d_{nm})^{-1/4}]d_{ij} \\
&+[\alpha_{50} + \alpha_{51}(D_0^2 + d_{mn}d_{nm})^{-1/2} + \alpha_{52}(D_0^2 + d_{mn}d_{nm})^{-1/4}] \\
&[E_i d_{jk}E_k + d_{ik}E_k E_j] \,.
\end{aligned}
\tag{8.3.39}
$$

In [63] this is called the *extended* CASSON model. When the electric field vanishes, the factors α_{30}, α_{31} and α_{32} must equally vanish to guarantee NEWTONian behaviour for vanishing electric field. The entropy inequality imposes restrictions on the material parameters. The necessary condition (8.2.114) is always fulfilled due to the choice of the independent variables. On the other hand, (8.2.115) is identically fulfilled for conducting fluids while for non-conducting fluids it requests that

$$\alpha_{20} + \alpha_{21} + \alpha_{22}D_0^{1/2} = 0 \,, \tag{8.3.40}$$

The fact that the dependence of the parameters α_{ij} on the electric field is unknown makes evaluation of the sufficient condition (8.2.116) difficult. For non-conducting fluids the following relations can be deduced:

$$\eta_0 \geq 0 \,, \tag{8.3.41}$$

$$
\begin{aligned}
2\eta_0 &+ [\alpha_{30} + \alpha_{31}D_0^{-1} + \alpha_{32}D_0^{-1/2}] \\
&+2[\alpha_{50} + \alpha_{51}D_0^{-1} + \alpha_{52}D_0^{-1/2}]E_1^2 \geq 0 \,,
\end{aligned}
\tag{8.3.42}
$$

$$
\begin{aligned}
2\eta_0 &+ [\alpha_{30} + \alpha_{31}D_0^{-1} + \alpha_{32}D_0^{-1/2}] \\
&+[\alpha_{50} + \alpha_{51}D_0^{-1} + \alpha_{52}D_0^{-1/2}](E_1^2 + E_2^2) \geq 0 \,.
\end{aligned}
\tag{8.3.43}
$$

We regard the two approaches presented in [63] and [196] presently as the most advanced single constituent constitutive proposals for the CAUCHY stress and electric field. They are therefore important and relevant for our treatment of particular problems to be attacked below. Nevertheless, in order to provide a broader view of attempts to describe ER-fluids, we consider it worthwhile to mention also other constitutive models introduced in the specific literature. We continue by mentioning two studies which are interesting, to a greater extent from the mathematical point of view. Both of them treat extensions of the BINGHAM model.

In [67] the authors propose a so-called extension of the BINGHAM model determined in terms of the minimization of the global dissipation energy. The CAUCHY stress (in tensorial form) is given by

$$t = -p\boldsymbol{I} + \gamma\frac{|\boldsymbol{E}|}{|d\boldsymbol{E}|}(d\boldsymbol{E} \otimes \boldsymbol{E} + \boldsymbol{E} \otimes d\boldsymbol{E}) + \eta\,\boldsymbol{d} \,. \tag{8.3.44}$$

Obviously, it is not well defined when $dE = 0$, which characterizes the "rigid zones". It can be proved that the shear stress has to exceed the threshold $\gamma|E|^2$ outside the rigid zones; so this quantity may be viewed as an equivalent of the yield limit of the standard BINGHAM model. However, the stress tensor is not used directly to formulate the boundary value problem to be solved. The velocity field is computed as the solution of a non-smooth minimization problem for the global energy dissipation. This minimization problem is solved numerically by the method of augmented LAGRANGEans combined with an operator-splitting technique. Numerical results are given that illustrate the ER-effect for a pure shear mode (COUETTE flow) and for a more complicated flow structure for an electrorheological clutch.

Another mathematical study of the flow of electrorheological fluids and of their constitutive description of BINGHAM behaviour is given in [35]. First, a boundary value problem for unsteady flow of an electrorheological fluid is formulated. Starting from the most general constitutive function for the stress (8.3.1), the authors assume that t is quadratic in E and affine in d and $d/|d|$ and neglect the term containing $E \otimes E$. The resulting constitutive law has the form

$$t_{ij} = -p\delta_{ij} + \left(\alpha_{30} + \alpha_{31}\frac{1}{|d|} + \alpha_{32}E_kE_k + \alpha_{33}E_kE_k\frac{1}{|d|}\right)d_{ij}$$
$$+\alpha_{50}(E_id_{jk}E_k + d_{ik}E_kE_j)\,, \tag{8.3.45}$$

where the coefficients α_{ij} are constants that have to fulfill

$$\alpha_{30} \geq 0\,,\ \alpha_{31} \geq 0\,,\ \alpha_{32} \geq 0\,,\ \alpha_{33} \geq 0\,,\ \alpha_{32} + \tfrac{4}{3}\alpha_{50} \geq 0\,. \tag{8.3.46}$$

The restrictions (8.3.46) are deduced from the CLAUSIUS–DUHEM inequality; they are essential in the further proofs of existence and uniqueness of solutions. Model (8.3.45) is viewed as a combination of NEWTONian and BINGHAM behaviour. As in [67], (8.3.45) does not make sense if $|d| = 0$, so it cannot be used in the boundary value problem. Instead of this, a variational inequality is formulated for the stress which makes sense. So, the problem is formulated in a variational form. Existence and uniqueness are proved for the solution in the two-dimensional case for any initial data, while in the three-dimensional case global existence of a weak solution is proved for small initial data only. In the end of the paper an interesting result is given concerning the estimation of the time when the fluid stops.

Both articles emphasize mathematical aspects and do not present identification of parameters by experiments.

8.3.2 Constitutive Laws Used in Our (Numerical) Approach

In the majority of the experimental evaluations, the constitutive assumptions for ERFs used in the literature are confined to one-dimensional modeling.

The standard configuration is (steady) plane shear flow for which the constitutive equations take the form of a stress-shearing relation [2, 176, 198, 224, 269, 270]). Most popular in the literature are the BINGHAM, CASSON and power law models, depending on the electrorheological material at hand. As mentioned previously, the CASSON-like model was introduced in [63]. It generalizes the usual CASSON model and includes the BINGHAM model as a particular case. However, when the more realistic case of finite electrodes is considered (see Subsect. 8.4.3), the flow is two-dimensional and bidirectional. So, more generally, we have to deal with two-dimensional constitutive equations. Let us recall here the two-dimensional forms of the aforementioned constitutive equations (the indices take the values 1 and 2).

- The BINGHAM *model* is described by

$$t_{ij}^e = 2\eta_0 d_{ij} + 2^{1/2}\tau_y \frac{d_{ij}}{|d|} \, , \quad |t^e| > 2^{1/2}\tau_y \, , \tag{8.3.47}$$

$$d_{ij} = 0 \, , \qquad\qquad |t^e| \le 2^{1/2}\tau_y \, , \tag{8.3.48}$$

where $|d| := \sqrt{d_{mn}d_{nm}}$ is one form of the second invariant of d.
- The *classical* CASSON *model* is given by

$$t_{ij}^e = 2\eta_0 d_{ij} + 2^{1/2}\tau_y \frac{d_{ij}}{|d|} + 2^{7/4}(\eta_0\tau_y)^{1/2}\frac{d_{ij}}{|d|^{1/2}} \, , \quad |t^e| > 2^{1/2}\tau_y \, , \tag{8.3.49}$$

$$d_{ij} = 0 \, , \qquad\qquad |t^e| \le 2^{1/2}\tau_y \, . \tag{8.3.50}$$

When these equations are used to describe the ERF behaviour, one has to take into account the dependence of the yield stress and, eventually of the viscosity, on the magnitude of the electric field.
- For the *power-law model* we have (see [221])

$$t_{ij}^e = m\dot\gamma^{n-1}2d_{ij} \, , \tag{8.3.51}$$

where $n > 1$ (shear-thickening behaviour). Here $\dot\gamma = 2^{1/2}|d|$ denotes a generalized shear rate. When $n < 1$ (shear-thinning or pseudoplastic behaviour) we can no longer use (8.3.51), since for $\dot\gamma \to 0$ the generalized viscosity $m\dot\gamma^{n-1} \to \infty$. The difficulty is overcome by modifying the model (8.3.51) through the introduction of a new free parameter $\dot\gamma_0$

$$t_{ij}^e = \begin{cases} m\dot\gamma_0^{n-1}2d_{ij} \, , & \dot\gamma \le \dot\gamma_0 \, , \\ m\dot\gamma^{n-1}2d_{ij} \, , & \dot\gamma > \dot\gamma_0 \, . \end{cases} \tag{8.3.52}$$

Here, $\dot\gamma_0$ is a constant value of the generalized shear rate below which NEWTONian behaviour with viscosity $\eta_0 = m\dot\gamma_0^{n-1}$ prevails[17]. For electrorheological fluids the parameters m, n and $\dot\gamma_0$ may depend on the electric field.

---

[17] An alternative to regularize 8.3.51 is to replace $\dot\gamma^{n-1}$ by $(\dot\gamma^{n-1} + \dot\gamma_0^{n-1})$, where $\dot\gamma_0$ is a very small constant.

- Let us also generalize (8.3.35)–(8.3.36) and introduce the two-dimensional form of the CASSON-*like constitutive function* as

$$t_{ij}^e = 2\eta_0 d_{ij} + \beta_1 \frac{2d_{ij}}{|d|} + \beta_2 \frac{2d_{ij}}{|d|^{1/2}}, \quad |t^e| > 2|\beta_1|, \qquad (8.3.53)$$

$$d_{ij} = 0, \qquad\qquad\qquad\qquad |t^e| \leq 2|\beta_1|, \qquad (8.3.54)$$

where the parameters η_0, β_1 and β_2 are positive and may depend on the magnitude of the electric field. The BINGHAM model and the classical CASSON model are obtained from (8.3.53), (8.3.54) if $\beta_1 = 2^{-1/2}\tau_y$, $\beta_2 = 0$ and $\beta_1 = 2^{-1/2}\tau_y$, $\beta_2 = 2^{3/4}(\eta_0\tau_y)^{1/2}$, respectively, are chosen.

All these models may cause serious mathematical difficulties since they are expressed by two-branched functions which are not smooth at the branching point. The presence of the denominator $|d|$ is an obstacle against straightforward numerical modeling of flows in complex geometries because it is difficult to determine a priori where it vanishes. For instance, for models with yield behaviour such as the BINGHAM, CASSON and CASSON-like models, it is not possible to determine a priori the yield surfaces (the interfaces which separate a non-deforming solid from a fluid state region) since they have to be determined as part of the solution. Similarly, it is not possible to determine explicitly in what regions of the problem domain $\dot\gamma = \dot\gamma_0$ for the model (8.3.52). Some attempts to overcome this difficulty have been proposed. Most of them concern the BINGHAM model. In 1999 BARNES published a review [21] on models with yield stress. In a special section of the article, "Problems with yield stress and mathematics", the author mentions important approaches for several complex flow configurations dealing with yield stress. These are mainly based on the modification of the BINGHAM model in such a way that the mathematical problem concerning the yield stress is avoided. The most important approximations of the BINGHAM model mentioned by BARNES are the 'bi-viscosity' model and the PAPANASTASIOU model. In the bi-viscosity model, the rigid-body character (8.3.48) at low stresses is replaced by a NEW-TONian fluid behaviour with very high viscosity ($\eta_N \gg \eta_0$). Then, instead of (8.3.47)–(8.3.48) we have the law

$$t_{ij}^e = 2\eta_0 d_{ij} + 2^{1/2}\tau_y \frac{d_{ij}}{|d|}, \quad |t^e| > 2^{1/2}\tau_N, \qquad (8.3.55)$$

$$t_{ij}^e = 2\eta_N d_{ij}, \qquad\qquad\quad |t^e| < 2^{1/2}\tau_N, \qquad (8.3.56)$$

where the constant τ_N is related with τ_y by

$$\tau_y = \tau_N(1 - \eta_0/\eta_N). \qquad (8.3.57)$$

The law (8.3.55)–(8.3.56) for $\eta_N \to \infty$ becomes (8.3.47)–(8.3.48). In Fig. 8.1 one can see how the shear stress depends on the shear rate (one-dimensional

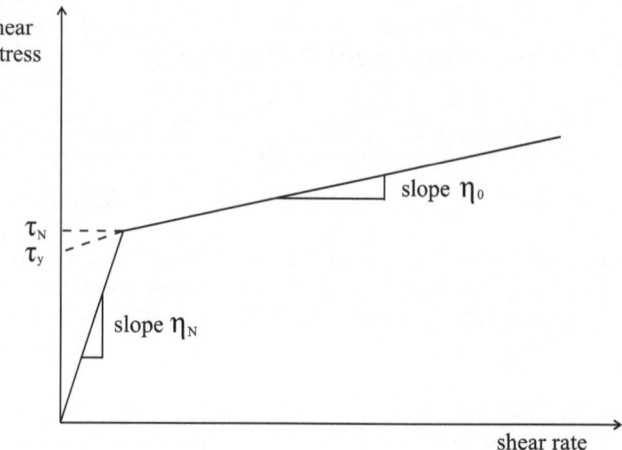

Fig. 8.1. One-dimensional form of the bi-viscosity model (8.3.55)–(8.3.56) (only for positive shear rates)

case) within the bi-viscosity assumption. This model was used in the treatment of squeeze-flow of an electrorheological fluid (see [224]).

PAPANASTASIOU introduced in [173] a modified constitutive equation that smoothes the yield criterion, permitting the numerical treatment of the flow problems based on this model. In our notation his law reads

$$t^e_{ij} = \left(2\eta_0 + 2^{1/2}\tau_y \frac{1 - \exp\left[-n2^{1/2}|\boldsymbol{d}|\right]}{|\boldsymbol{d}|} \right) d_{ij} , \qquad (8.3.58)$$

where the exponent n is a relatively great material parameter that can be determined by experiments. The BINGHAM law in the unyielded region is recovered from equation (8.3.58) for $n \to \infty$. By contrast to (8.3.47), equation (8.3.58) is not singular since

$$\lim_{|\boldsymbol{d}| \to 0} t^e_{ij} = 2(\eta_0 + n\tau_y)d_{ij} , \qquad (8.3.59)$$

and, consequently, it is valid for both the yielded and "unyielded" regions.

We should also mention here the so-called *alternative* BINGHAM *model* introduced by MELLGREN in [141]

$$t^e_{ij} = \left(2\eta_0 + 2^{1/2}\tau_y \frac{1}{(\varepsilon/2 + |\boldsymbol{d}|^2)^{1/2}} \right) d_{ij} , \qquad (8.3.60)$$

where ε is a positive material parameter. For $\varepsilon \to 0$ (8.3.60) reduces to (8.3.47) but since ε is required to be positive, (8.3.60) is defined for all possible values of d_{ij}. As in (8.3.58) the material takes on only a liquid state and therefore does not cause any mathematical difficulties.

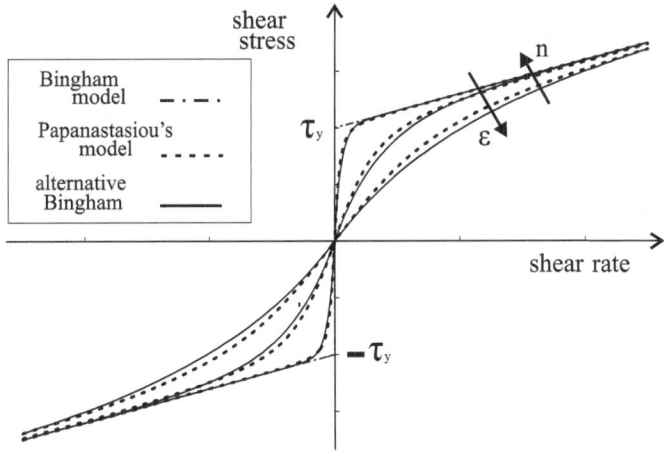

Fig. 8.2. Comparison (one-dimensional form) between the BINGHAM constitutive function (8.3.27), (8.3.28) and the modified BINGHAM models (PAPANASTASIOU's model, alternative BINGHAM model) for different values of the parameters n and ϵ, respectively

In Fig. 8.2 one can see how the models (8.3.58) and (8.3.60) approximate the BINGHAM model in the one-dimensional case. The yield stress τ_y functions only as a material parameter without the significance of a yield stress. Relations (8.3.58) and (8.3.60) may be seen not only as mathematical approximations of the BINGHAM model. In [20] a revolutionary but controversial idea in rheology was introduced according to which the concept of yield stress should be seen only as an idealization since "given accurate measurements, no yield stress exists". So, the viscosity is always finite. This point of view was supported by some experiments done with rheometers which allow stress measurements for very low shear rates. When the experiment was performed for lower shear rates (for the same material), the value of the yield stress was found to be smaller. Consequently, from this point of view, models (8.3.58) and (8.3.60) are closer to reality.

Let us denote by f and g the functions "shear stress vs. shear rate" appearing in the one-dimensional versions of (8.3.58) and (8.3.60),viz.,

$$f(x) = \left(\eta_0 + \tau_y \frac{1}{|x|} \frac{\exp(-n|x|)}{}\right) x \,, \tag{8.3.61}$$

$$g(x) = \left(\eta_0 + \tau_y \frac{1}{(\varepsilon + x^2)^{1/2}}\right) x \,. \tag{8.3.62}$$

If we seek n and ε such that the slopes of the curves are equal in $x = 0$ i.e. $f'(0) = g'(0)$ (where $f'(0) := f'(0^+) = f'(0^-)$) then we obtain $n = 1/(\varepsilon)^{1/2}$. For n and ε satisfying this relation, the curves for f and g are very close (see Fig. 8.2) showing very similar behaviour.

One can apply the ingenious ideas of the models (8.3.58) and (8.3.60) to avoid the singular behaviour of the stress to the CASSON and CASSON-like models, too and even to the power-law model. However, when we use these functions to describe the electrorheological fluids, the material parameters η_0, τ_y and n, ε, respectively, may depend on the electric field. For a space-dependent electric field the constitutive equations modified as in (8.3.58) become too complicated due to the presence of the exponential. Due to its simplicity, (8.3.60) is easier to handle than (8.3.58). Let us show how this same regularization can be applied to the power-law model: for instance, in the one-dimensional case of (8.3.52), one may consider instead of the two-branched function (we use the notation of (8.3.17))

$$\tau = \begin{cases} m|\dot{\gamma}|^{n-1}\dot{\gamma}\,, & |\dot{\gamma}| > \dot{\gamma}_0\,, \\ m\dot{\gamma}_0^{n-1}\dot{\gamma}\,, & |\dot{\gamma}| \le \dot{\gamma}_0\,, \end{cases} \tag{8.3.63}$$

the form

$$\tau = m(\varepsilon + \dot{\gamma}^2)^{(n-1)/2}\dot{\gamma}\,, \tag{8.3.64}$$

which eliminates the branching (for illustration and comparison with the original model (8.3.63) see Fig. 8.3). This law was used by HUTTER (see [97, 98, 99]) to derive the generalized GLEN law in glaciology. As $\dot{\gamma} \to 0$ this law exhibits NEWTONian behaviour. We found this approach very appropriate for our study.

Even though the boundary value problem is quite different from ours, recent work of HILD et al. [89] ought to be mentioned who applied the BINGHAM model to modeling landslides. In their model the viscosity coefficient and the yield stress depend on density which in turn is time- and space-dependent. This makes this approach interesting also for our case. By using variational methods the authors study the blocking property of the flow and describe the rigid zones and the stagnant regions (which are stuck on the boundaries)

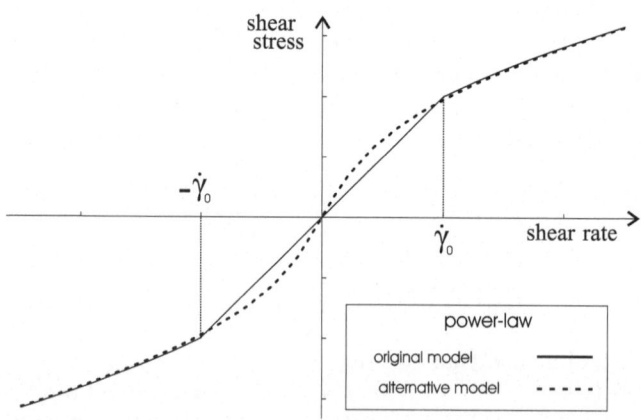

Fig. 8.3. Comparison between models (8.3.63) and (8.3.64)

for certain boundary value problems. However, they explicitly find the yield surfaces only for one-dimensional cases.

Now synthesizing all the previous issues we will introduce two constitutive models for electrorheological fluids which will further be used in the numerical approach: the alternative CASSON-like model (which contains also the alternative forms of the BINGHAM and CASSON models),

$$t_{ij}^e = 2\eta_0 d_{ij} + \beta_1(E)\frac{2d_{ij}}{(\delta + |d|^2)^{1/2}} + \beta_2(E)\frac{2d_{ij}}{(\delta + |d|^2)^{1/4}} \ , \tag{8.3.65}$$

and the alternative power-law model

$$t_{ij}^e = m(E)(\delta + 2|d|^2)^{(n(E)-1)/2}2d_{ij} \ , \tag{8.3.66}$$

where $E = (E_1^2 + E_2^2)^{1/2}$ is the electric field modulus and δ is a small positive material parameter. With these two models it is possible to cover a large class of electrorheological materials. They are suitable to numerical simulations and consistent with the phenomenological approaches presented in the previous section. Namely, the alternative CASSON-like model may be seen as a particularization of (8.3.39) if we take

$$\alpha_{20} = \alpha_{21} = \alpha_{22} = \alpha_{30} = \alpha_{50} = \alpha_{51} = \alpha_{52} = 0 \ , \tag{8.3.67}$$

$$\alpha_{31} = \beta_1 \ , \quad \alpha_{32} = \beta_2 \ , \tag{8.3.68}$$

$$D_0^2 = \delta \ , \tag{8.3.69}$$

while the alternative power-law model is included in (8.3.13) as a particular case; it is similar to a particularization of $(8.3.9)_1$ (for $\alpha_1 = \alpha_2 = \alpha_5 = 0$) with one small deviation: the function m in (8.3.66) depends only on the magnitude of the electric field and its expression has to be determined for each ER-material; the dependence of α_3 (with the coefficients $\alpha_{31} = \alpha_{33} = 0$) on the magnitude of the electric field, takes the particular form $\alpha_{30} + \alpha_{33}E^2$.

For both models the dependence of the coefficients, β_1, β_2 for the CASSON-like model and m, n for the power-law model, on the electric field are established for a certain ERF with the aid of the experimental data. The measurements are usually performed with rotational viscometers based on the COUETTE system (made by two concentric cylinders or plate-plate geometry). These devices provide graphs "shear stress vs. shear rate" for different values of the electric fields. The material parameters are obtained from the measured data by fitting techniques, and they are given in tables for different values of the electric field [2, 63, 269]. By interpolation we can obtain the desired functions.

8.4 Applications: Channel Flow of ERFs under Homogeneous and Inhomogeneous Electric Fields

In general, the working behaviour of devices using ER fluids is classified by three fundamental modes: shear, flow and squeeze [224]. In the flow mode, which is also called Poiseuille flow (the flow occurs under the effect of an axial pressure), it is assumed that the two electrodes are fixed. The study presented in this section is focused on a special case of the flow mode, the steady pressure-driven flow of electrorheological fluids under isothermal conditions ($\Theta = $ const.) in a plane channel under electric fields produced by several kinds of electrodes. A number of ER equipments, including valves, dampers and actuators is based on this configuration.

In the remaining part of this chapter we assume incompressible fluids, i.e.,

$$d_{kk} = v_{k,k} = 0 \, , \tag{8.4.1}$$

where $v_i := \dot{x}_i$. Then the pressure is an unknown of the problem which will be determined as a consequence of the constraint (8.4.1) up to a constant. As mentioned already in Subsect. 8.2.7 we assume (8.2.122) to be valid in the description of ERFs.

The first ensuing three subsections treat the case of the electrodes flush with the channel walls, parallel to the flow as illustrated in Figs. 8.4–8.7. This case was tackled both theoretically and experimentally. The last subsection is concerned with some more complex configurations in which the electrodes protrude into the channel (decreasing the channel height) or retreat from the channel (enlarging the channel height). They can still be parallel to the flow (see Fig. 8.53) or oblique (see Fig. 8.54). Another interesting case deals with crenated electrodes (see Figs. 8.55–8.57). Due to the complexity of these last mentioned cases, no theoretical results are yet available for them. We will present the most important experimental results given in the literature and will make some comments about an eventual theoretical modelling and numerical implementation of such cases.

8.4.1 Formulation of the Problem – Electrodes Flush with the Channel

Let Ox_1x_2 be a Cartesian coordinate system. We consider an infinitely long channel of height $2h$ made by two infinite parallel planes of zero thickness. These planes are situated at $x_2 = -h$, $x_2 = h$, respectively. Along the channel walls, finite electrodes, charged with different potentials are placed. They may be disposed in various configurations as one can see in the examples illustrated in Figs. 8.4–8.7. If the electrodes were of infinite length (as in Fig. 8.8), the electric field would be homogeneous. However, in every realistic application the electrodes are finite and the field is inhomogeneous, especially close to the edges of the electrodes. The electrodes are isolated outside the channel with a dielectric material having constant electric permittivity ε_1. The ER-medium inside the channel has electric permittivity ε_2 which is also a constant according to (8.2.122) and to the fact that the density and temperature are also constant. Since the problem is two-dimensional the indices i and j used henceforth take only the values 1 and 2.

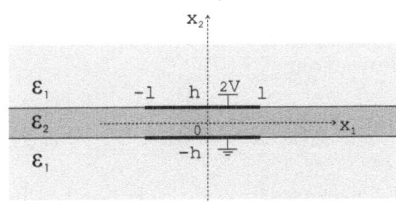

Fig. 8.4. Configuration with two finite electrodes of equal lengths ($n_1 = n_2 = 1$, $V_1 = 2V$, $V_2 = 0$)

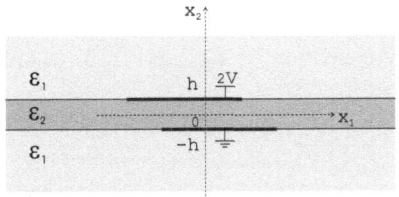

Fig. 8.5. Configuration with two equal but shifted finite electrodes ($n_1 = n_2 = 1$, $V_1 = 2V$, $V_2 = 0$)

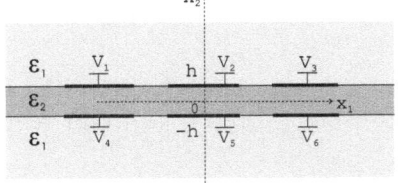

Fig. 8.6. Configuration with finite electrodes interrupted by electrically neutral walls (periodic structure) ($n_1 = n_2 = 3$)

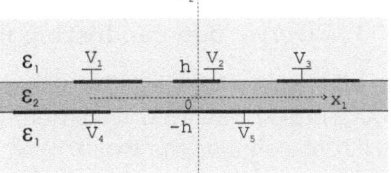

Fig. 8.7. Configuration with finite electrodes interrupted by electrically neutral walls (general case) ($n_1 = 3$, $n_2 = 2$)

Equation (8.2.90) is equivalent with the statement that E_i is the gradient of a scalar function

$$E_i = -\varphi_{,i} \ . \tag{8.4.2}$$

Here, φ is the electric potential. In the further formulation of the boundary value problem (BVP) one has to distinguish between the two cases of conducting and non-conducting ER-fluids. More precisely, the equations from which the electrical unknowns of the problem are determined are different in the part of the domain inside the channel: $-\infty < x_1 < \infty$, $|x_2| < h$. In the domain outside the channel, $-\infty < x_1 < \infty$, $|x_2| > h$, the formulation is the same in both cases. The boundary conditions are formulated identically for the two cases in the whole domain.

1. Electrically conducting fluids

In this case the electrical part of the BVP in the channel domain is given by the divergence of the second MAXWELL–MINKOWSKI equation (8.2.91)

$$\frac{\partial D_{j,j}}{\partial t} + (D_{j,j}v_k)_{,k} + \mathcal{J}_{j,j} = 0 \ , \quad -\infty < x_1 < \infty, \quad |x_2| < h \ , \tag{8.4.3}$$

where the electric charge $\mathcal{Q}$ was eliminated by using (8.2.4). It is worth mentioning that in this case $\mathcal{Q}$ is not a variable that can be given a priori but it has to be determined from (8.2.4) after the calculation of the electric field.

From relations (8.2.117), (8.2.118) and (8.2.122), from the incompressibility condition (8.4.1) and from (8.4.2) it follows that

$$\varepsilon_2 \frac{\partial \varphi_{,jj}}{\partial t} + \varepsilon_2 \varphi_{,jjk} v_k + \sigma_{100} \varphi_{,jj} = 0 , \quad -\infty < x_1 < \infty, \quad |x_2| < h . \quad (8.4.4)$$

Since we treat only the stationary case, the first term on the left-hand side of (8.4.4) vanishes. As one can see from (8.4.4), due to the presence of the velocity in this equation, the electrical problem is explicitly coupled with the mechanical problem.

1. Electric non-conducting fluids

When $\mathcal{J}_j = 0$, then the electric potential is calculated from the third MAXWELL-MINKOWSKI equation (8.2.4). Using (8.2.117), (8.2.122) and (8.4.2) one obtains the POISSON equation

$$\varphi_{,jj} = -\frac{\mathcal{Q}}{\varepsilon_2} . \quad (8.4.5)$$

In general the electric charge vanishes in electrorheological fluids since only uncharged fluids are treated here[18]. Consequently the potential must fulfill the LAPLACE equation in the channel domain

$$\varphi_{,jj} = 0 , \quad -\infty < x_1 < \infty , \quad |x_2| < h . \quad (8.4.6)$$

In the domain outside the channel we assume that the material is electrically non-conducting and that its electric charge can be neglected. Therefore equation (8.4.6) is valid also in the domain $-\infty < x_1 < \infty, \quad |x_2| > h$.

We must specify now the boundary conditions. The following DIRICHLET boundary conditions are given on the part of the boundary where the electrodes are placed

$$\varphi(x_1, h) = V_{i_1} , \quad x_1 \in I_{i_1}^{el} , \quad (8.4.7)$$

$$\varphi(x_1, -h) = V_{i_2} , \quad x_1 \in I_{i_2}^{el} , \quad (8.4.8)$$

where $i_1 = 1, 2, \ldots, n_1$, $i_2 = n_1 + 1, n_1 + 2, \ldots, n_1 + n_2$ and n_1, n_2 are the numbers of electrodes placed on the upper and lower walls of the channel respectively. Moreover, $I_{i_1}^{el}, I_{i_2}^{el}$ are the interval domains of the x_1-coordinate of the electrodes placed on the upper and lower walls of the channel, respectively.

Using (8.2.117) and (8.4.2) in (8.2.88), we see that the jump conditions of the electrode-free boundary parts imply the following equations for the normal derivative of φ:

---

[18] In our continuum mechanical modeling the assumption of charge neutrality is guaranteed. In a description at the mesoscale or microscale level this would not be fulfilled because in ER-fluids free ions can be present.

$$\varepsilon_2 \varphi_{,2}\left(x_1, h^-\right) = \varepsilon_1\, \varphi_{,2}\left(x_1, h^+\right), \qquad x_1 \in (-\infty, \infty) \setminus \bigcup_{i_1} I_{i_1}^{el}, \qquad (8.4.9)$$

$$\varepsilon_2 \varphi_{,2}\left(x_1, -h^+\right) = \varepsilon_1\, \varphi_{,2}\left(x_1, -h^-\right), \qquad x_1 \in (-\infty, \infty) \setminus \bigcup_{i_2} I_{i_2}^{el}, \qquad (8.4.10)$$

while from (8.2.86) the following continuity conditions in the tangential derivative are obtained

$$\varphi_{,1}\left(x_1, h^+\right) = \varphi_{,1}\left(x_1, h^-\right), \qquad x_1 \in (-\infty, \infty) \setminus \bigcup_{i_1} I_{i_1}^{el}, \qquad (8.4.11)$$

$$\varphi_{,1}\left(x_1, -h^+\right) = \varphi_{,1}\left(x_1, -h^-\right), \qquad x_1 \in (-\infty, \infty) \setminus \bigcup_{i_2} I_{i_2}^{el}. \qquad (8.4.12)$$

To specify the jump of the components of the electric field across $x_2 = \pm h$ for $x_1 \le 0$, we use the upper index "$+$" to indicate the limit as x_2 tends to $\pm h$ from positive values of $(x_2 \mp h)$ and the upper index "$-$" to indicate the limit as x_2 tends to $\pm h$ from negative values of $(x_2 \mp h)$. Moreover we may choose

$$\lim_{x_2 \to \pm \infty} \varphi(x_1, x_2) = 0, \quad -\infty < x_1 < \infty. \qquad (8.4.13)$$

It is important to remark that for non-conductive ER-fluids ($\sigma_1 = 0$) we deal with a boundary value problem in which the electric field is completely separated from the mechanical fields[19]. Consequently, the solution for the electric field can be independently attacked and then used in a second step in the mechanical problem. In the remaining part of the chapter we will study only the case of non-conductive electrorheological fluids.

In order to formulate a boundary value problem for the mechanical part we recall the balance of momentum (8.2.63) for a steady flow

$$\rho v_j v_{i,j} - \left(t_{ij} + E_i D_j\right)_{,j} - \rho F_i^{ext} = 0. \qquad (8.4.14)$$

Using (8.2.107) and (8.2.117) in the last equation, we obtain

$$\rho v_j v_{i,j} - \left(-p\delta_{ij} + t_{ij}^e + \varepsilon_2 E_i E_j\right)_{,j} - \rho F_i^{ext} = 0. \qquad (8.4.15)$$

Using (8.4.2) and (8.4.6), we obtain

$$\left(E_i E_j\right)_{,j} = \tfrac{1}{2}\left(E_j E_j\right)_{,i}. \qquad (8.4.16)$$

---

[19] Of course, the fact that the flow does not affect the electric field in this case is equally a consequence of other assumptions such as the neglect of the time derivatives of the strain rate in the constitutive equations (so that (8.2.104) could follow from the entropy inequality which means that the only remaining pure electromagnetic dependent quantity, D_j, cannot depend on the rate of strain tensor d_{ij} that is, apart from ρ and Θ the only independent mechanical quantity), the neglect of the electric charge and the consideration of constant density and isothermal conditions.

If we suppose that the external force F_i^{ext} is conservative then we may incorporate it into the pressure. Doing this and substituting (8.4.16) in the momentum balance, we obtain

$$- p_{,i} + t_{ij,j}^e + \tfrac{1}{2}\varepsilon_2(E_j E_j)_{,i} = \rho v_j v_{i,j} \,. \qquad (8.4.17)$$

Substituting in (8.4.17) a constitutive function for the CAUCHY stress tensor and the solution for the electric field, we obtain for plane flow two equations which together with (8.4.1) may be used to determine the three unknowns: the components of the velocity vector and the pressure. Unlike the electrical problem, the domain for the mechanical problem is restricted to the channel. We assume the no-slip and impermeability conditions on the channel walls

$$v_1(x_1, \pm h) = 0 \,, \quad v_2(x_1, \pm h) = 0 \,, \quad -\infty < x_1 < \infty \,. \qquad (8.4.18)$$

8.4.2 Particular Case – Infinitely Long Electrodes

If we consider in (8.4.7), (8.4.8) that $n_1 = n_2 = 1$, $I_1^{el} = I_2^{el} = (-\infty, \infty)$ and $V_1 = 2V$, $V_2 = 0$ we obtain the configuration sketched in Fig. 8.8. The solution of the electrical problem in this case can be easily derived. It is

$$\varphi(x_1, x_2) = \frac{V}{h} x_2 + V \,, \quad -\infty < x_1 < \infty \,, \quad |x_2| \le h \,. \qquad (8.4.19)$$

Consequently, the electric field has the form

$$E_1 = 0 \,, \quad E_2 = -\frac{V}{h} \,, \qquad (8.4.20)$$

everywhere in the channel. The magnitude of the electric field is now $E = V/h$. The dielectric material outside the channel has no influence on the solution. Another consequence of the fact that the electric field is constant

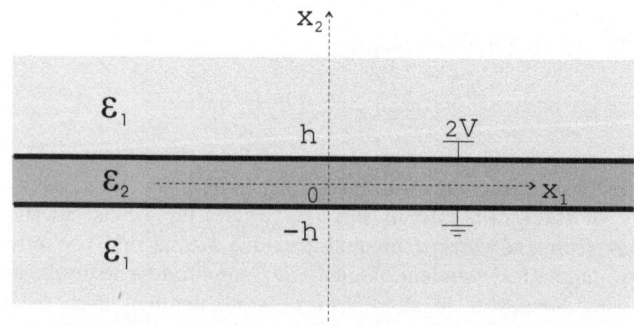

Fig. 8.8. Configuration with infinite electrodes

is that the flow problem will be uni-directional, i.e. the only non-vanishing velocity component varies in the flow direction, provided, of course, that the inlet and outlet boundary conditions agree with this. We may assume first that $v_{1,1} = 0$, $v_{2,1} = 0$. Then, from the continuity equation it follows that $v_{2,2} = 0$. If we take into account the impermeability condition on the walls, we obtain that the x_2-component of the velocity vanishes in the whole domain. So, we have

$$v_1 = v_1(x_2) \,, \tag{8.4.21}$$
$$v_2 = 0 \,. \tag{8.4.22}$$

Consequently, the only non-vanishing component of the rate of strain tensor is

$$D_{12} = D_{21} = \tfrac{1}{2}v_{1,2}(x_2) \,, \tag{8.4.23}$$

and from (8.2.94) and (8.2.95) it follows that all the components of the stress tensor are independent of x_1. By applying all these considerations in the momentum balance (8.4.17) we obtain

$$-p_{,1} = -t^e_{12,2} \,, \tag{8.4.24}$$
$$-p_{,2} + t^e_{22,2} = 0 \,. \tag{8.4.25}$$

From the last equation it follows that the quantity $-p + t^e_{22}$ can be only a function of x_1. However, as just concluded above, $t^e_{22,1} = 0$, hence also $p_{,1}$ is only a function of x_1. Since the right-hand side of (8.4.24) is not a function of x_1, neither is the left-hand side, and, therefore, $p_{,1} = (p - t^e_{22})_{,1}$ should be constant. If we denote this constant by k, it follows from (8.4.24) and (8.4.25) that

$$p = kx_1 + t^e_{22} + c_1 \,, \tag{8.4.26}$$
$$t^e_{12,2} = k \,. \tag{8.4.27}$$

The constant k ($k \leq 0$) is the pressure gradient in the x_1-direction. The constant c_1 can be determined if a boundary condition for p is given.

By assigning the constitutive functions (8.3.65) and (8.3.66) with $\delta = 0$ to t^e_{ij}, one can determine analytically the solution of the problem formulated above. With this choice of the parameter δ, in a viscometric flow, (8.3.65) reduces to the CASSON-like model (8.3.35)–(8.3.36)[20] introduced by ECKART in [63], while (8.3.66) reduces to the classical power-law model (8.3.33)–(8.3.34). The coefficients β_1, β_2, m and n are constant for a certain value of the electric field E, namely of the electric potential V.

---

[20] Unlike in the CASSON-like model we choose as a simplification $\beta_0 = 0$. This is equivalent to saying that $\beta_0 = 0$ is included in the viscosity η_0.

The Casson-Like Model

This case is solved here following ECKART [63] (see also [64]). First, let us recall the constitutive equations (8.3.35)–(8.3.36) (with the notation according to the previous paragraph: $\dot{\gamma}$ is replaced by $v_{1,2}$ and τ^{CL} is replaced by t_{12}^e)

$$t_{12}^e = \eta_0 v_{1,2} + 2^{1/2}\beta_1 + 2^{1/4}\beta_2 v_{1,2}^{1/2} \, ,$$

$$t_{12}^e > 2^{1/2}\beta_1 \geq 0 \, , \quad v_{1,2} > 0 \, , \qquad (8.4.28)$$

$$t_{12}^e = \eta_0 v_{1,2} - 2^{1/2}\beta_1 - 2^{1/4}\beta_2(-v_{1,2})^{1/2} \, ,$$

$$t_{12}^e < -2^{1/2}\beta_1 \leq 0 \, , \quad v_{1,2} < 0 \, . \qquad (8.4.29)$$

Integrating (8.4.27) with respect to x_2 it is found that

$$t_{12}^e = kx_2 + C_2 \, , \qquad |t_{12}^e| > 2^{1/2}\beta_1 \, . \qquad (8.4.30)$$

Since $k < 0$, the shear stress t_{12}^e is a linear decreasing function of x_2. Assuming that the shear stress at the lower wall $\tau_w = -kh + C_2$ is greater than the yield stress $2^{1/2}\beta_1$, one can distinguish three regions in the channel:

- one where the shear stress is greater than or equal to the yield stress: $-h \leq x_2 \leq x^i$, with x^i defining the place where $t_{12}^e = 2^{1/2}\beta_1$; the velocity in this region is increasing with respect to x_2, hence the upper index "i" is used;
- one where the material becomes solid and the velocity denoted by v^p is constant: $x^i \leq x_2 \leq x^d$, with x^d defining the place where $t_{12}^e = -2^{1/2}\beta_1$; this region is called the plug zone, so the upper index "p" is used;
- one where the shear stress is less than or equal to $-2^{1/2}\beta_1$: $x^d \leq x_2 \leq h$; here the velocity is decreasing and the upper index "d" is used.

The places x^i and x^d can readily be found from (8.4.30) as

$$x_2^i = -\frac{C_2}{k} + \frac{2^{1/2}\beta_1}{k} \, , \quad x_2^d = -\frac{C_2}{k} - \frac{2^{1/2}\beta_1}{k} \, . \qquad (8.4.31)$$

The symmetry with respect to the axis $x_2 = 0$ of the boundary value problem implies $x^i = -x^d$. Hence, from (8.4.31) it follows that

$$C_2 = 0 \, . \qquad (8.4.32)$$

By inserting (8.4.28) into (8.4.30) and by taking into account (8.4.32) one arrives at the differential equation for v_1^i

$$v_{1,2}^i + \frac{2^{1/4}\beta_2}{\eta_0}(v_{1,2}^i)^{1/2} + \frac{-kx_2 + 2^{1/2}\beta_1}{\eta_0} = 0 \, . \qquad (8.4.33)$$

Employing the substitution $v_{1,2}^i = (u^i)^2$ in the previous equation, solving the emerging quadratic equation for u^i and using the inverting formula $v_1^i = \int (u^i)^2 dx_2 + C_3^i$, one derives the general solution for v_1^i in the form

$$v_1^i(x_2) = \frac{k}{2\eta_0}x_2^2 + \frac{2^{-1/2}\beta_2^2 - 2^{1/2}\eta_0\beta_1}{\eta_0^2}x_2$$
$$\mp \frac{4}{3}\frac{2^{-3/4}\beta_2}{k}\left(\frac{2^{-3/2}\beta_2^2}{\eta_0^2} + \frac{kx_2 - 2^{1/2}\beta_1}{\eta_0}\right)^{3/2} + C_3^i. \quad (8.4.34)$$

One proceeds analogously for negative shear rates: insertion of (8.4.29) and (8.4.32) into (8.4.30), use of the substitution $v_{1,2}^d = -(u^d)^2$ and then of the inverting formula $v_1^d = -\int(u^d)^2 dx_2 + C_3^d$ yields

$$v_1^d(x_2) = \frac{k}{2\eta_0}x_2^2 - \frac{2^{-1/2}\beta_2^2 - 2^{1/2}\eta_0\beta_1}{\eta_0^2}x_2$$
$$\mp \frac{4}{3}\frac{2^{-3/4}\beta_2}{k}\left(\frac{2^{-3/2}\beta_2^2}{\eta_0^2} - \frac{kx_2 + 2^{1/2}\beta_1}{\eta_0}\right)^{3/2} + C_3^d. \quad (8.4.35)$$

To get rid of the constants $C_3^{i/d}$, the boundary condition (8.4.18) are imposed on (8.4.34), (8.4.35), so

$$v_1^i(x_2) = \frac{k}{2\eta_0}(x_2^2 - h^2) + \frac{2^{-1/2}\beta_2^2 - 2^{1/2}\eta_0\beta_1}{\eta_0^2}(x_2 + h)$$
$$\mp \frac{4}{3}\frac{2^{-3/4}\beta_2}{k}\left(\frac{2^{-3/2}\beta_2^2}{\eta_0^2} + \frac{kx_2 - 2^{1/2}\beta_1}{\eta_0}\right)^{3/2}$$
$$\pm \frac{4}{3}\frac{2^{-3/4}\beta_2}{k}\left(\frac{2^{-3/2}\beta_2^2}{\eta_0^2} - \frac{kh + 2^{1/2}\beta_1}{\eta_0}\right)^{3/2}, \quad (8.4.36)$$

$$v_1^d(x_2) = \frac{k}{(2\eta_0)}(x_2^2 - h^2) + \frac{2^{-1/2}\beta_2^2 - 2^{1/2}\eta_0\beta_1}{\eta_0^2}(h - x_2)$$
$$\mp \frac{4}{3}\frac{2^{-3/4}\beta_2}{k}\left(\frac{2^{-3/2}\beta_2^2}{\eta_0^2} - \frac{kx_2 + 2^{1/2}\beta_1}{\beta_0}\right)^{3/2}$$
$$\pm \frac{4}{3}\frac{2^{-3/4}\beta_2}{k}\left(\frac{2^{-3/2}\beta_2^2}{\eta_0^2} - \frac{kh + 2^{1/2}\beta_1}{\eta_0}\right)^{3/2}. \quad (8.4.37)$$

To deduce the correct signs in (8.4.36), (8.4.37) two additional conditions are needed. Notice that the velocity function should be continuously differentiable for $-h < x_2 < h$ and that $v_{1,2}^p = 0$ in the plug zone. This leads to

$$v_{1,2}^i(x_2 = x_2^i) \stackrel{!}{=} 0, \ v_{1,2}^d(x_2 = x_2^d) \stackrel{!}{=} 0. \quad (8.4.38)$$

Using (8.4.38) one finally determines the following solutions for the velocities

$$v_1^i(x_2) = \frac{k}{2\eta_0}(x_2^2 - h^2) + \frac{2^{-1/2}\beta_2^2 - 2^{1/2}\eta_0\beta_1}{\eta_0^2}(x_2 + h)$$

$$- \frac{4}{3}\frac{2^{-3/4}\beta_2}{k}\left(\frac{2^{-3/2}\beta_2^2}{\eta_0^2} + \frac{kx_2 - 2^{1/2}\beta_1}{\eta_0}\right)^{3/2}$$

$$+ \frac{4}{3}\frac{2^{-3/4}\beta_2}{k}\left(\frac{2^{-3/2}\beta_2^2}{\eta_0^2} - \frac{kh + 2^{1/2}\beta_1}{\eta_0}\right)^{3/2},$$

$$-h \le x_2 \le \frac{2^{1/2}\beta_1}{k}, \tag{8.4.39}$$

$$v_1^p = -\frac{k}{2\eta_0}\left(h + \frac{2^{1/2}\beta_1}{k}\right)^2 - \frac{1}{6}\frac{2^{-3/4}\beta_2}{k}\left(\frac{2^{1/2}\beta_2^2}{\eta_0^2}\right)^{3/2}$$

$$+ \frac{4}{3}\frac{2^{-3/4}\beta_2}{k}\left(\frac{2^{-3/2}\beta_2^2}{\eta_0^2} - \frac{k\,h + 2^{1/2}\beta_1}{\eta_0}\right)^{3/2}$$

$$+ \frac{2^{-1/2}\beta_2^2}{\eta_0^2}\left(h + \frac{2^{1/2}\beta_1}{k}\right), \quad \frac{2^{1/2}\beta_1}{k} \le x_2 \le -\frac{2^{1/2}\beta_1}{k}, \tag{8.4.40}$$

$$v_1^d(x_2) = \frac{k}{(2\eta_0)}(x_2^2 - h^2) + \frac{2^{-1/2}\beta_2^2 - 2^{1/2}\eta_0\beta_1}{\eta_0^2}(h - x_2)$$

$$- \frac{4}{3}\frac{2^{-3/4}\beta_2}{k}\left(\frac{2^{-3/2}\beta_2^2}{\eta_0^2} - \frac{kx_2 + 2^{1/2}\beta_1}{\beta_0}\right)^{3/2}$$

$$+ \frac{4}{3}\frac{2^{-3/4}\beta_2}{k}\left(\frac{2^{-3/2}\beta_2^2}{\eta_0^2} - \frac{kh + 2^{1/2}\beta_1}{\eta_0}\right)^{3/2},$$

$$- \frac{2^{1/2}\beta_1}{k} \le x_2 \le h. \tag{8.4.41}$$

Since the CASSON-like model includes the NEWTONian ($\beta_1 = \beta_2 = 0$), BINGHAM ($\beta_2 = 0$) and CASSON ($\beta_2 = 2\sqrt{\eta_0\beta_1}$) behaviours, the given solution may be particularized for all these types of fluids.

Having found the analytical solution for the velocity field one can calculate the volumetric flow rate

$$Q = \int_0^b \int_{-h}^h v_1(x_2)dx_2dx_3 = b \int_{-h}^h v_1(x_2)dx_2, \tag{8.4.42}$$

where b is the width of the channel. The following formula relates the pressure gradient $k = p_{,1}$ with the volumetric flow rate Q:

$$Q = \frac{b}{120\,\eta_0^4 k^2} \left\{ 40\sqrt{2}\,\eta_0\beta_1\beta_2^4 + 80\sqrt{2}\,\eta_0^3\beta_1^3 - 4\sqrt{2}\,\beta_2^6 - 120\sqrt{2}\,k^2 h^2 \eta_0^3\beta_1 \right.$$

$$-120\sqrt{2}\,\eta_0^2\beta_1^2\beta_2^2 + 60\sqrt{2}\,k^2 h^2 \eta_0^2\beta_2^2 + 80\,k^3 h^3 \eta_0^3 + 2^{11/4} kh\sqrt{Z_1}\,\eta_0\beta_2^3$$

$$-48\cdot 2^{1/4}\,k^2 h^2 \sqrt{Z_1}\,\eta_0^2\beta_2 - 2^{19/4}\,kh\sqrt{Z_1}\,\eta_0^2\beta_1\beta_2 + 2^{9/4}\,\sqrt{Z_1}\,\beta_2^5$$

$$\left. -2^{21/4}\,\sqrt{Z_1}\,\eta_0\beta_1\beta_2^3 + 2^{25/4}\,\sqrt{Z_1}\,\eta_0^2\beta_1^2\beta_2 \right\}\,, \tag{8.4.43}$$

where $Z_1 := \sqrt{2}\,\beta_2^2 - 4\,kh\eta_0 - 4\sqrt{2}\,\eta_0\beta_1$. Equation (8.4.43) yields, of course, the volumetric flow rate of a NEWTONian fluid, a BINGHAM fluid and a CASSON fluid if the corresponding particular values of the coefficients β_i, $i = 1, 2$ are selected.

If we use a procedure in which Q is the input value and we want to calculate the pressure drop as output, we have to invert formula (8.4.43) in order to obtain the pressure gradient for non-vanishing values of the electric fields. Analytically, this is not possible, but using the software MATHEMATICA [268] we may obtain k as a function of the volumetric flow rate Q and of the uniform electric field E.

The pressure gradient for a vanishing electric field (when $\beta_1 = \beta_2 = 0$) is

$$k = -\frac{3Q\eta_0}{2bh^3}\,, \tag{8.4.44}$$

and

$$Q = \tfrac{4}{3}hbv_0\,, \tag{8.4.45}$$

where $v_0 = v_2(0)$ is the maximum velocity in the channel.

The above formulas may be used to evaluate approximately the pressure drop in configurations with long finite electrodes when neglecting end effects. Namely, on the electrode-free part of the channel, the pressure drop is calculated by applying the formula for a vanishing electric field (8.4.44) while on the part where the electrodes are placed, the pressure drop is calculated by applying the inversion of formula (8.4.43) for a constant electric field, where the value of the electric field is chosen to be the value established in the homogeneous region, in the middle of the electrodes, far from the electrode edges. Similar methods were used by ECKART in [64] and by WUNDERLICH in [270] in order to compare the analytical results with the experimental results. The relatively small deviations of the analytical plots from the measured values are due to the neglect of the electric field in the region outside the electrodes and of the inhomogeneity of the electric field in the vicinity of the electrode edges.

The Power-Law Model

If we solve problem (8.4.21)–(8.4.27) for a power-law fluid (8.3.33)–(8.3.34) we obtain (on assuming $k < 0$) the following velocity field

$$v_1(x_2) = \frac{k_1 n}{n+1}\left(h^{(n+1)/n} - |x_2|^{(n+1)/n}\right), \tag{8.4.46}$$

where $k_1 = (-k/m)^{1/n}$. Now, calculating the volumetric flow rate yields

$$Q = \frac{2(n+1)}{2n+1}\, v_0 h b, \tag{8.4.47}$$

where

$$v_0 = v_1(0) = \frac{k_1 n}{n+1}\, h^{(n+1)/n} \tag{8.4.48}$$

is the maximum velocity in the channel. Inverting this formula, we obtain

$$k = -\left(\frac{Q(2+1/n)}{hb}\right)^n \frac{m}{h}. \tag{8.4.49}$$

The model presented in this subsection was so far used to describe the flow of an ERF in a channel (see [63]). When interpreting the experimental results in slit flows, it was routinely assumed that the electric field is constant and that it determines a shear flow in the channel [4, 5, 198, 207, 270]. In other words, the model with infinite electrodes was used as an approximation for the real case with finite electrodes. In our view this is not a realistic approximation in all cases since it neglects the electrode end effects which may considerably affect the flow as we will show in the next subsection.

8.4.3 Electrodes of Finite Length

As we mentioned in the Introduction, an impediment to overcome to make an industrial exploitation of the ER-effect on a large scale possible is the very high voltage requirements necessary to obtain the desired increase in viscosity. There are attempts to increase the electrorheological effect by modifying either the surface or the shape and position of the electrodes relative to the flow geometry in such a way that inhomogeneities in the electric field are introduced [3, 4, 5, 6, 7, 33, 81, 108, 150, 166, 269, 270]. All these experimental investigations demonstrated that application of non-uniform electric fields may lead to more efficient effects on the flow (than with homogeneous electric fields).

The necessity of models which reproduce the ER behaviour for inhomogeneous electric fields was formulated in certain fields of applications [36, 114], the purpose being an accurate description of the experiments performed in order to improve the performance of ER devices. Nevertheless, in most theoretical approaches of ERF flows the electric field is only a constant parameter. In channel flow this is an analytical consequence of the fact that the electrodes are considered to be infinite while in cylindrical COUETTE-type configurations this is an assumption that can be made when the fluid channel is small compared with the radius of the inner cylinder. Exceptions are the theoretical results for a radial configuration obtained by ATKIN et al. [16, 17] where the electric field is slightly inhomogeneous in the radial direction. RAJAGOPAL and RŮŽIČKA in [196] and ECKART in [63] developed a theoretical framework which allows for variable field strength. However, in such a case the electric field has to be determined from the MAXWELL equations.

The aim of the study presented in this subsection is twofold. On the one hand, we consider it important to give a more realistic modeling by taking into account the inhomogeneity effects which appear in the vicinity of electrode edges. Comparing this with the case of plane shear flow exposed to infinite electrodes, where the electric field is simply a constant, the non-uniform electric field will cause here inhomogeneities in the flow too which, as a consequence, will be non-viscometric. Consequently, both components of the velocity are variables of the problem and they depend on both coordinates. On the other hand, our intention is to examine numerically how the ER-effect can be enhanced by a space-dependent electric field.

Let us return to the problem formulated in Subsect. 8.4.1 for electrically non-conducting fluids. It consists of the electrical problem which may be solved independently and the mechanical problem, equations which contain terms based on the electric field components, namely on the solution of the electrical problem. As one can see, the electric field is a key element since its dependence on the space coordinates determines the kinematical character of the flow. In order to gain a better understanding of the end effects of the electrodes on the electric field inhomogeneity in the channel and then on the flow, one needs to investigate first the simplest configuration of electrodes.

In [246] and [247] the distribution of the electric potential around two long electrodes charged with different potentials in a symmetric, an anti-symmetric and a non-symmetric way was investigated. The term "long electrode" denotes in these works either a semi-infinite electrode or a finite electrode of a certain length chosen such that the two far edges of the electrode do not interact. The solutions were found (semi)-analytically and they were constructed with the use of the WIENER–HOPF (WH) technique[21]. With the help of the WH-method it was possible to set the singularities at the tips of the electrodes explicitly in evidence, which in a numerical solution must be approximately accounted for in rather costly mesh refinements. Furthermore, the obtained analytical solution was used to test and validate the numerical solution ob-

---

[21] We mention that the problem is solved in a bounded domain in the x_2-direction, where the upper and lower boundaries consist of two infinite grounded electrodes situated at a distance $H > h$ from the x_1-axis. If H is sufficiently large, this is equivalent with the usual infinity conditions which means vanishing potential at $x_2 = \pm\infty$. For smaller H, this configuration can still be easily realized in practice. We mention that the two grounding electrodes at $x_2 = \pm H$ are needed for technical reasons when solving the WH-problem, for otherwise, i.e., when no grounding electrodes are present no solution could be found (see [247] for a detailed explanation). Physically, this is no restriction because the channel will always be earthed and the system can always be looked at for large H.

tained using the software FEMLAB [47] (which is a powerful tool for solving partial differential equations by applying the finite element method).

Apart from the purely analytical and numerical approach to determine the solution of the electrical problem for the configurations with two long electrodes, it may be more advantageous to use a mixed approach, in which one uses the analytical solution in the numerical program and so reduces the domain of numerical approach. Concretely, we can impose in the numerical program DIRICHLET conditions for the electric problem on the whole channel walls deduced from the analytical solution so that the domain of the electric problem is equally reduced to the channel as for the mechanical problem. Since the mathematical expressions of the analytical solutions for the electric potential on the walls are too complicated and create difficulties when trying to introduce them directly into FEMLAB, we will interpolate them and use polynomials which accurately approximate the exact solution. In this way space memory and computing time can be saved. We call this approach **numerical analytic**. When we treat configurations with two short electrodes or with more than two electrodes then the electric field will be calculated numerically. This is the so-called **completely numerical** approach.

Let us recall the equations of the mechanical problem:

$$-p_{,i} + t^e_{ij,j} + \tfrac{1}{2} \varepsilon_2 (E_j E_j)_{,i} = \rho v_j v_{i,j} , \qquad (8.4.50)$$

$$v_{i,i} = 0 , \qquad (8.4.51)$$

with the boundary conditions

$$v_{1,1}(-L, x_2) = 0, \ v_2(-L, x_2) = 0 , \qquad |x_2| \le h , \qquad (8.4.52)$$

$$p(L, x_2) = 0 , \qquad |x_2| \le h , \qquad (8.4.53)$$

$$v_i(x_1, \pm h) = 0 , \qquad |x_1| \le L . \qquad (8.4.54)$$

The x_1-positions $-L$ and L mark the entrance and the exit of the fluid in the channel, respectively. We assume that L is sufficiently large; specifically the inlet boundary is far enough from the electrode edge so that the electric field is negligible. Consequently, one may impose there the velocity as in a uni-directional channel flow (described in Subsect. 8.4.2). Beyond the channel exit we assume vanishing pressure. For the extra stress t^e_{ij} we will use models (8.3.65) and (8.3.66) and thus may define a generalized viscosity η_{gen} so that $t^e_{ij} = 2\eta_{gen} D_{ij}$. Explicitly,

$$\eta_{gen} = \eta_0 + \frac{\beta_1(E)}{(\delta + |D|^2)^{1/2}} + \frac{\beta_2(E)}{(\delta + |D|^2)^{1/4}} \qquad (8.4.55)$$

for the alternative CASSON-like model and

$$\eta_{gen} = m(E)(\delta + 2|D|^2)^{(n(E)-1)/2} \qquad (8.4.56)$$

for the alternative power-law model.

The Dimensionless Problem
for the Alternative Casson-Like Model

We introduce the non-dimensional quantities $\tilde{x}_i$, $\tilde{v}_i$, $\tilde{E}_i$ and $\tilde{p}$ according to

$$x_i = h\tilde{x}_i \,, \quad v_i = v_0\tilde{v}_i \,, \quad E_i = E_0\tilde{E}_i \,, \quad p = p_0\tilde{p} \,, \qquad (8.4.57)$$

where h, v_0, E_0 and p_0 are characteristic quantities of the problem: h is half the channel height, v_0 is the maximum inlet velocity, E_0 is a typical value for the electric field and $p_0 = \eta_0 v_0/h$. Application of these transformations to (8.4.50), multiplying the resulting expression with h/p_0 and dropping the "tilde", yields

$$-p_{,i} + 2\left(\eta_{gen}\, d_{ij}\right)_{,j} + \frac{1}{2\mathbb{M}a}(E_j E_j)_{,i} = \mathbb{R}e\, v_j v_{i,j} \,, \qquad (8.4.58)$$

where

$$\eta_{gen} = 1 + \frac{h\,\beta_1(E_0 E)}{v_0\eta_0}\,\frac{1}{(n_\delta + |d|^2)^{1/2}} + \frac{\sqrt{h}\,\beta_2(E_0 E)}{\sqrt{v_0}\eta_0}\,\frac{1}{(n_\delta + |d|^2)^{1/4}} \qquad (8.4.59)$$

is the dimensionless generalized viscosity,

$$\mathbb{R}e = \rho h v_0/\eta_0 \,, \qquad (8.4.60)$$

is the REYNOLDS number,

$$\mathbb{M}a = \eta_0 v_0/(h\varepsilon_2 E_0^2) \,, \qquad (8.4.61)$$

is the MASON number and $n_\delta = \delta h^2/v_0^2$. The REYNOLDS number is usually interpreted as the ratio of the inertial force to the viscous force while the MASON number can be interpreted as the ratio of the viscous force to the electrostatic force. We have chosen a non-dimensionalization appropriate to creeping flows for which the REYNOLDS number takes small values. The dimensionless form of equation (8.4.51), after dropping the "tilde", remains unchanged.

Usually, β_1, β_2 vanish when the electric field is zero, so that in this case the fluid response is NEWTONian. Consequently, at the channel entrance, where the electric field is approximately zero, we have before non-dimensionalization

$$v_1(-L, x_2) = v_0(1 - (x_2/h)^2) \,. \qquad (8.4.62)$$

The non-dimensionalized boundary conditions (8.4.52)–(8.4.54) and (8.4.62) become

$$v_1(-L/h, x_2) = 1 - x_2^2 \,, \quad v_2(-L/h, x_2) = 0 \,, \quad |x_2| \leq 1 \,, \qquad (8.4.63)$$
$$p(L/h, x_2) = 0 \,, \qquad\qquad\qquad\qquad\qquad |x_2| \leq 1 \,, \qquad (8.4.64)$$
$$v_i(x_1, \pm 1) = 0 \,, \qquad\qquad\qquad\qquad\quad |x_1| \leq L/h \,. \quad (8.4.65)$$

The BVP is formulated now by equations (8.4.51), (8.4.58), (8.4.59) together with the boundary conditions (8.4.63)–(8.4.65), where E_j is the non-dimensionalised solution of (8.4.2)–(8.4.12). It will be solved numerically for the domain $|x_1| \leq L/h$, $|x_2| \leq 1$ using the commercial software FEMLAB [47].

The Dimensionless Problem for the Alternative Power-Law Model

We apply the same procedure as in the previous case, namely we introduce the non-dimensional quantities as in (8.4.57), but now consider $p_0 = m_0(v_0/h)^{n_0}$, where $m_0 = m(0)$, $n_0 = n(0)$ (which are usually non-zero). The non-dimensional momentum balance for a power-law fluid has the same form as (8.4.58) with the MASON number, the REYNOLDS number and the non-dimensional generalized viscosity given by

$$\mathrm{Ma} = \frac{m_0(v_0/h)^{n_0}}{\varepsilon_2 E_0^2}, \quad \mathrm{Re} = \frac{\rho v_0^2}{m_0(v_0/h)^{n_0}}, \tag{8.4.66}$$

$$\eta_{gen} = \frac{m(E_0 E)(v_0/h)^{n(E_0 E)}}{m_0(v_0/h)^{n_0}} (n_\delta + 2|d|^2)^{(n(E_0 E)-1)/2}. \tag{8.4.67}$$

We remark that, since the dimension of m is Pas^n and since n, which is dimensionless, is space dependent, the quantity m has variable dimension in space. Even though we can fit the values of m (from the experimental data) and we may obtain the function $m(E_0 E)$, this quantity is meaningless from a physical point of view. In order to avoid this problem, we will directly fit the dimensionless function (see (8.4.67))

$$f(E_0 E, v_0/h) = m(E_0 E)(v_0/h)^{n(E_0 E)} \tag{8.4.68}$$

to the data. An example of this operation will be presented later in this subsection.

In order to establish the non-dimensional form of the boundary conditions let us apply formula (8.4.46) for vanishing electric field which gives the dimensional velocity at the entrance of the channel

$$v_1(-L, x_2) = v_0 \left(1 - \left(\frac{|x_2|}{h}\right)^{(n_0+1)/n_0}\right). \tag{8.4.69}$$

Consequently, the non-dimensional inlet boundary condition has the form

$$v_1(-L/h, x_2) = 1 - |x_2|^{(n_0+1)/n_0},$$
$$v_2(-L/h, x_2) = 0. \tag{8.4.70}$$

The non-dimensional boundary conditions at the channel exit and on the channel walls are identical with their correspondents from the CASSON-like case: (8.4.64) and (8.4.65). The problem formulated by equations (8.4.51), (8.4.58), (8.4.67) together with the boundary conditions (8.4.70), (8.4.64) and (8.4.65), where E_j is the non-dimensionalised solution of (8.4.2)–(8.4.12) will be solved numerically for the domain $|x_1| \leq L/h$, $|x_2| \leq 1$ using the commercial software FEMLAB [47].

a) Numerical Results – the Alternative Casson-Like Model

Material and Configuration Properties Used for the Simulations

An ER-fluid chosen for simulations is Rheobay TP AI 3565 which is produced by the Bayer Company (Germany). It is water-free and consists of polyurethan particles in silicone oil, some additional additives and an emulgator [2, 270, 23]. In [63], the CASSON-like model (8.3.35)–(8.3.36) is recommended for this fluid since this model reproduces well the data measured in a rotational viscometer. The fluid shows NEWTONian behaviour in the absence of an electric field i.e. $\beta_1 = 0$ and $\beta_2 = 0$ for $E_0 E = 0$, $(E = (E_1^2 + E_2^2)^{1/2}$ is the modulus of the dimensionless electric field). The value $\eta_0 = 0.037$ Pa·s for the viscosity and the values of the parameters β_1 and β_2 given in Table 8.1 were obtained in [64] using the experimental plots for creeping flow "stress vs. shear rate".

Table 8.1. Values of the parameters β_1 and β_2 of (8.3.35) for different electric field strengths

$E_0 E \, [\frac{kV}{mm}]$	0.5	1	2	3
β_1 [Pa]	19.3	100.32	312.94	582.81
β_2 [Pa·s$^{1/2}$]	0.44	5.84	9.7	32.21

Fitting these data we obtained the dependence of the parameters on the dimensional electric field $E_0 E$:

$$\beta_1(E_0 E) = \alpha_{11}(E_0 E) + \alpha_{12}(E_0 E)^2 \,, \tag{8.4.71}$$
$$\beta_2(E_0 E) = \alpha_{21}(E_0 E) + \alpha_{22}(E_0 E)^2 + \alpha_{23}(E_0 E)^3 \,, \tag{8.4.72}$$

where the coefficients α_{ij} are given in Table 8.2.

The functions β_1 and β_2 are written as series expansions by retaining only the terms up to the second and the third order, respectively. The experimental data are provided for a limited range of the electric field and do not

Table 8.2. Values of the coefficients α_{1i} and α_{2i} from (8.4.71), (8.4.72)

i	1	2	3
α_{1i}	55.322 Pa·mm/kV	46.946 Pa·(mm/kV)^2	–
α_{2i}	7.01 Pa·s$^{1/2}$mm/kV	−5.22 Pa·s$^{1/2}$(mm/kV)^2	2.15 Pa·s$^{1/2}$(mm/kV)^3

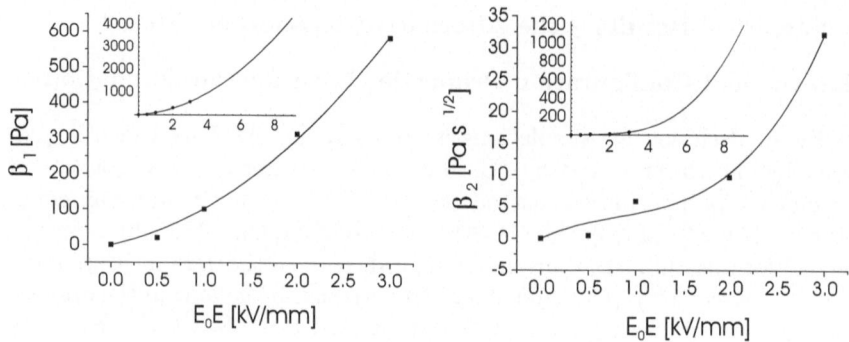

Fig. 8.9. Fitting curves for the dependence of the coefficients β_1 and β_2 on the electric field. The insets show how the fit is extrapolated

cover the range of the electric field used in the simulations. We use formulas (8.4.71) and (8.4.72) also to extrapolate the given data for β_1 and β_2 for larger values of E. Figure 8.9 illustrates how the fitting curves approximate the data and how they are extrapolated. In the subsequent analysis we shall use the alternative bidimensional CASSON-like model (8.3.65). Consequently, the generalized viscosity is expressed by a low-degree polynomial function in E. Substituting (8.4.71) and (8.4.72) in (8.4.59), we obtain

$$\eta_{gen} = 1 + n_{11}\frac{E}{d_\delta} + n_{12}\frac{E^2}{d_\delta} + n_{21}\frac{E}{d_\delta^{1/2}} + n_{22}\frac{E^2}{d_\delta^{1/2}} + n_{23}\frac{E^3}{d_\delta^{1/2}}, \qquad (8.4.73)$$

where we used the short-hand notation

$$d_\delta = (n_\delta + d_{mn}d_{mn})^{1/2}. \qquad (8.4.74)$$

The non-dimensional coefficients n_{ij}, $i = 1, 2$, $j = 1, 2, 3$ are defined as

$$
n_{11} = \frac{h\alpha_{11}E_0}{v_0\eta_0}, \qquad n_{12} = \frac{h\alpha_{12}E_0^2}{v_0\eta_0},
$$
$$
n_{21} = \frac{\sqrt{h}\alpha_{21}E_0}{\sqrt{v_0}\eta_0}, \qquad n_{22} = \frac{\sqrt{h}\alpha_{22}E_0^2}{\sqrt{v_0}\eta_0}, \qquad n_{23} = \frac{\sqrt{h}\alpha_{23}E_0^3}{\sqrt{v_0}\eta_0}. \qquad (8.4.75)
$$

For all graphs that subsequently will be shown we used a geometry characterized by the value $H/h = 10$, where the x_2-coordinates $\pm H$ mark the upper and lower boundaries for the domain of the electrical problem. The value of L/h is always chosen so that the inlet boundary is sufficiently far from the electrode edge in order to insure a negligible electric field at the entrance of the channel. The electric permittivity of Rheobay is $\varepsilon_2 = 1.4 \times 10^{-9}$ A·s/(V·m) and we choose $\varepsilon_1 = 0.02\,\varepsilon_2$. The density of Rheobay is $\rho = 1041\,\text{kg/m}^3$. All simulations except those done to make the comparison with the experimental data from Fig. 8.29 are performed for

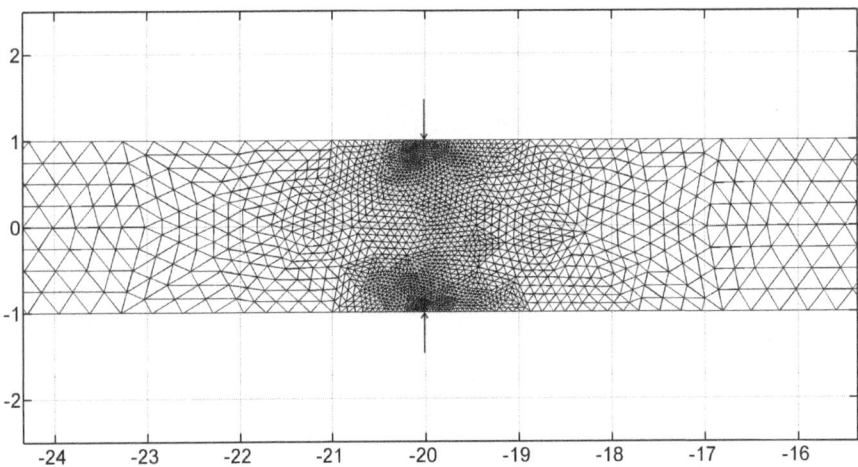

Fig. 8.10. The mesh around the electrode edges. The *arrows* show the position of the electrode edges

$h = 1$ mm, $E_0 = 1$ kV/mm and $v_0 = 0.3$ m/s. So, $\mathbb{R}e = 7.849$, , $\mathrm{Ma} = 0.007$, $n_{11} = 5.417$, $n_{12} = 4.597$, $n_{21} = 11.433$, $n_{22} = -8.515$ and $n_{23} = 3.506$. We considered here $n_\delta = 0.002$. We will study later the influence of this parameter on the solution.

The Flow Near the Electrode Ends (long electrodes)

Let us first illustrate the effect of the electric field inhomogeneity produced by the electrode edges on the flow. To do this we solved the problem (8.4.51), (8.4.58), (8.4.73) together with the boundary conditions (8.4.63)–(8.4.65), where E_j is the non-dimensionalised solution of (8.4.2)–(8.4.12) with two long electrodes charged anti-symmetrically, using the numerical analytic approach. We took $V = 2$ kV.

A triangular mesh consisting of approximately 10000 elements (9984) was used. The default mesh is twice refined (with a regular refinement) so that the obtained mesh is characterized by 8 elements between the channel walls in the regions far from the electrode ends. The maximum element size near the vertices given by the electrode ends is taken to be 0.1. A detail of the mesh in the region close to the electrode edges is plotted in Fig. 8.10.

We choose linear LAGRANGE elements for the electric problem and p2-p1 LAGRANGE elements[22] for the fluid problem. We present and describe here the most relevant fields in an area close to the electrode ends chosen in order to set in evidence the effects ahead, near and after the electrode ends. This is a crossing zone in which the quantities of the studied problem are passing

---

[22] One has quadratic element for the variables v_i, $i = 1, 2$ and linear element for the variable p

from the regime without an electric field to a regime with uniform electric field through a transition zone with strong inhomogeneities. The ranges for the color bars corresponding to Figs. 8.11, 8.19 and 8.20 were chosen so that the transition zone is illustrated in an especially relevant fashion. Consequently, the values from the zones with the darkest red could be greater than the maximum values indicated on the scale. We mention that the maximum values in a given area surrounding the electrode edges are much larger than the maximum value within a similar area with uniform electric field (inside the electrodes, far from the edges). The closest area to the edges is difficult to be described exactly for three reasons. First, the electric field is singular there (see [246]). Second, the material properties (β_1 and β_2) are not provided for high values of the electric field (see Table 8.1). Third, in practice, the electrodes can not be infinitely thin as approximated here, so E is not singular at the edges but has a large value.

Let us show first in Fig. 8.11 the electric field produced by this configuration. The inhomogeneities of the electric field extend over a length of order h. In the vicinity of an order smaller than $h/10$ around the electrode ends, the electric field is very large, its value is exceeding the ranges for which experimental data are provided. In this vicinity, due to the singularity of the electric field, the numerical simulation of the electric field is not stable (since mesh dependence was observed). However, this vicinity is small in comparison with the characteristic length of the experimental configuration.

In Fig. 8.12 we plot the non-dimensional modulus of the velocity, $v(x_1, x_2) = (v_1^2(x_1, x_2) + v_2^2(x_1, x_2))^{1/2}$, in the vicinity of the electrode edges

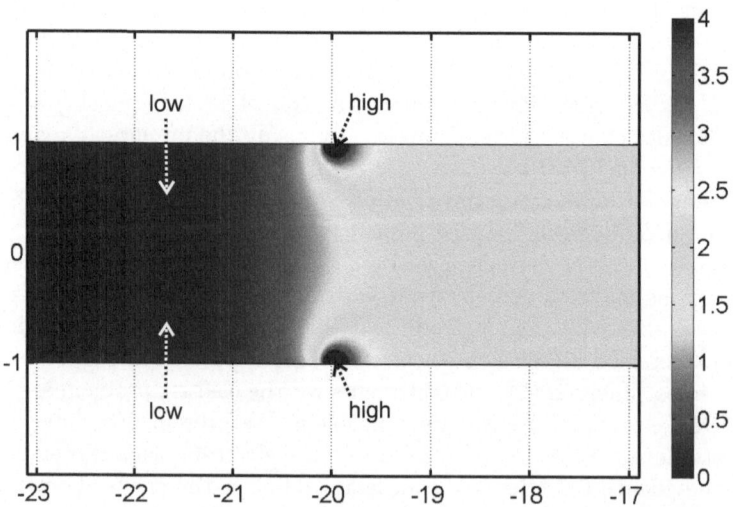

Fig. 8.11. Surface plot of the dimensionless modulus of the electric field

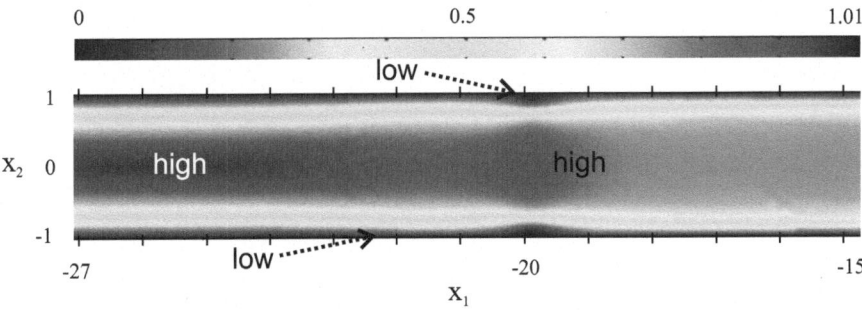

Fig. 8.12. Surface plot of the non-dimensional velocity modulus in the vicinity of the electrodes edges ($x_1 = -20$)

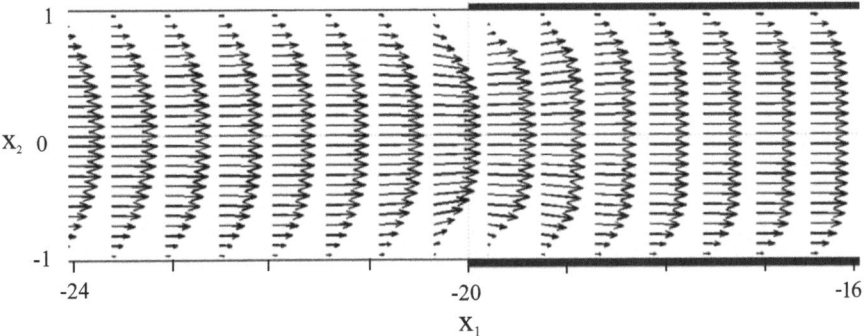

Fig. 8.13. Vector plot of the non-dimensional velocity

($x_1 = -20$). Note the two small domains formed around the electrode ends where the fluid is almost solidified and the difference outside and between the electrodes. To have a better view over the flow in the inhomogeneous area, we present in Fig. 8.13 an arrow plot of the velocity field in a similar, somewhat smaller area. Immediately before and immediately after the electrode ends, v_2 is not negligible; indeed, the transverse velocity influences the profile of the x_1-component as one can see in Figs. 8.14 and 8.15. It is worth mentioning that the two extreme values of v_2 at $x_1 = -20.2$ (first positive and then negative) and $x = -19.5$ (first negative and then positive) indicate that the fluid avoids the electrode tips and flows around them. For x_1 outside this regime, far ahead of the electrode ends where the profile of v_1 is NEWTONian the x_2-component of the velocity almost vanishes (see Figs. 8.16 (left) and 8.17). Between the electrodes, far downstream from the electrode edges the fluid velocity assumes the known one-dimensional CASSON-like profile (see Figs. 8.16 (right) and 8.17).

In Fig. 8.18 we plot the pressure along a part of the channel including the transition area. The distinguished three regions are characterized by different pressure gradients. First one can see a region with a very small gradient

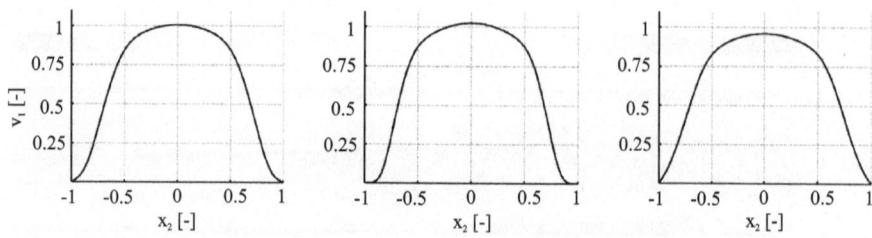

Fig. 8.14. Profiles of v_1 at $x_1 = -20.2, -20, -19.5$, respectively

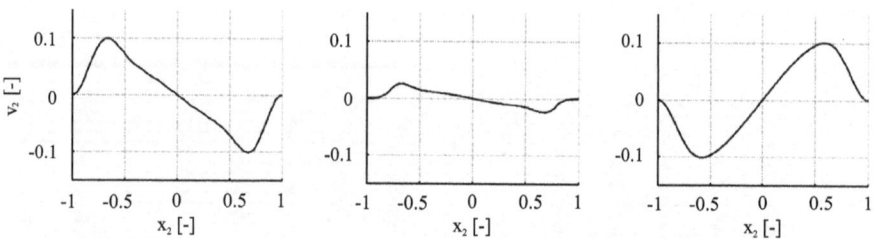

Fig. 8.15. Profiles of v_2 at $x_1 = -20.2, -20, -19.5$, respectively

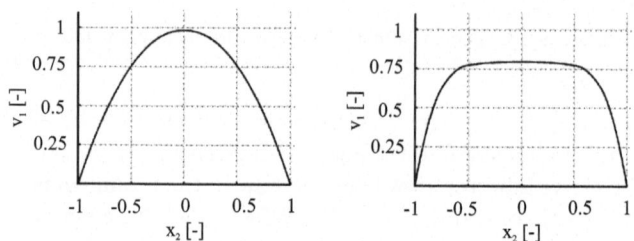

Fig. 8.16. Profiles of v_1 at $x_1 = -23.5, -16.5$, respectively

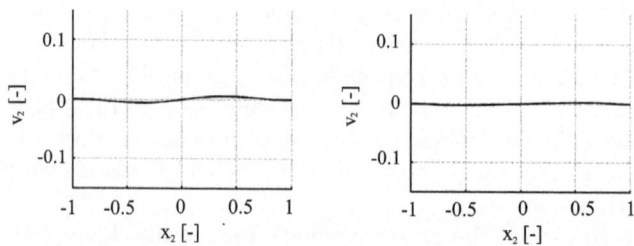

Fig. 8.17. Profiles of v_2 at $x_1 = -23.5, -16.5$, respectively

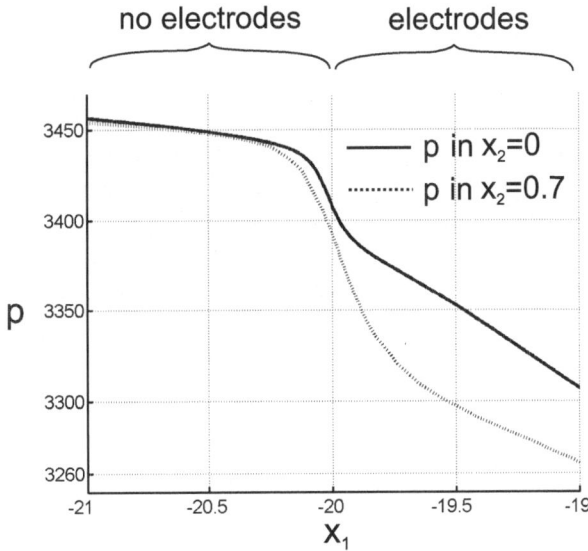

Fig. 8.18. Dimensionless pressure in the transition zone ($x_1 = -20$ marks the electrode ends)

corresponding to the NEWTONian fluid; then a steep region with a high gradient corresponding to the entrance between the electrodes is followed by the third region with the gradient corresponding to the CASSON-like fluid. For $x_1 < -20.3$ and for $x_1 > -19.5$ the gradients may be calculated analytically from the formulas corresponding to the NEWTONian fluid and the CASSON-like fluid, respectively (see Subsect. 8.4.2). The increase in pressure drop produced within the second, so-called transition zone is important and shows that the inhomogeneity produced by the end effects may be used to obtain an enhancement of the ER-effect. This strong inhomogeneity effect can also be seen by the difference of the pressure curves for $x_2 = 0$ and $x_2 = 0.7$, respectively. In Fig. 8.19 the second invariant of the rate of strain tensor d_{ij} is plotted. As expected, the domains with high values of $|d|^2$ are near the walls and in the electrode region, since the velocity is zero on the walls and the velocity gradients are larger there than at electrode-free boudaries, see Fig. 8.13. Exceptions arise in the vicinity of the electrode ends. There one can see zones with reduced shear rate, which suggests the presence of near solid zones. This behaviour can be attributed to the high electric field generated by the electrode edge. In the middle of the channel, $|d|^2$ maintains a small value.

In Fig. 8.20 we display the generalized viscosity. As one can see from (8.4.59) and from Fig. 8.9, η_{gen} is decreasing when $|d|$ is increasing, and it is increasing when E is increasing (since β_1 and β_2 are increasing with E). The electrode edges can again be identified; close to them, both E and $|d|$

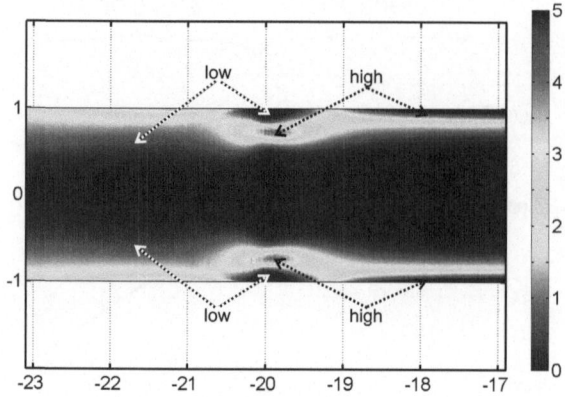

Fig. 8.19. Surface plot of dimensionless $|d|^2$

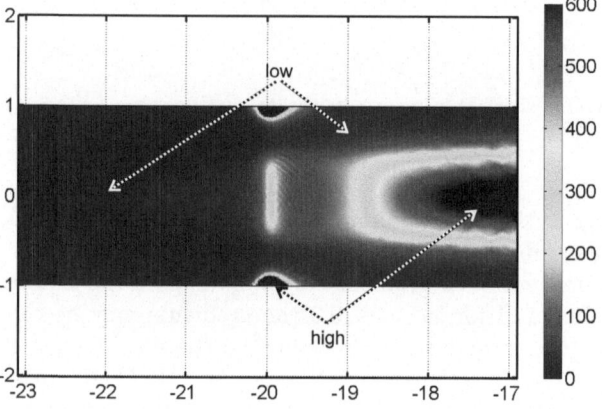

Fig. 8.20. Surface plot of the generalized dimensionless viscosity

give increasing contributions. An interesting shape of the viscosity η_{gen} ought to be noticed: right between the electrode ends a high viscosity "island" is formed after which it quickly decreases; then it increases again and assumed the known shape of the CASSON-like fluid far downstream (with the unyielded region in the middle).

Influence of the Parameters c and δ

Figure 8.21 shows the behaviour of the electric field outside the electrodes for different values of the electric permittivity ratio ($c = \varepsilon_1/\varepsilon_2$). As shown in [246], the electric field is stronger in the area without electrodes when the permittivity of the material outside the electrodes is larger than the permittivity of the ER-fluid. The plot also shows that in the small area close

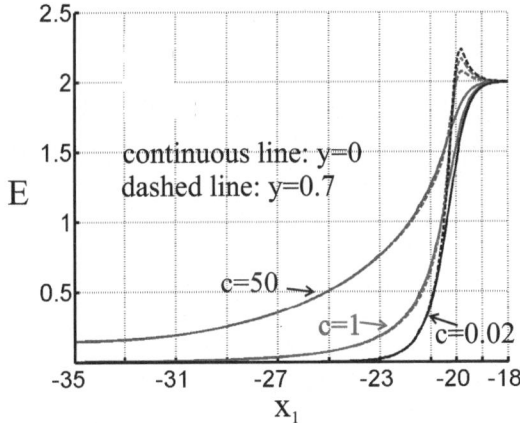

Fig. 8.21. Line plots of $E(x,0)$ (continuous line) and $E(x,0.7)$ (*dashed line*) for different values of c (electrodes end at $x_1 = -20$)

to the electrode edges this effect is reversed. As a consequence, we expect a contribution to the pressure drop from the region outside the electrodes when c is greater than 1. We plot in Fig. 8.22 the pressure in the region where the ERF exits from the electrodes but including also the end of the channel where the pressure takes the reference value $p = 0$. One can see that the pressure drop immediately at the electrode ends is slightly decreasing with the value of c. This effect may be important in configurations with short electrodes. For long electrodes the difference is not relevant because the main contribution to the pressure drop comes from the $E = $ const. region between the electrodes, see Fig. 8.18.

In Fig. 8.23 (left) we plot the pressure drops between $x_1 = 19$ or $x_1 = 21$ and the end of the channel, $x_1 = 35$ (where the outlet boundary condition imposes $p = 0$) versus the electric permittivity ratio (c). These pressures, denoted by p_{19} and p_{21}, respectively, were calculated in the middle of the channel height $x_2 = 0$ and at $x_2 = 0.7$. One can see from the figure that the pressure in the region outside the electrodes does not depend on x_2 while a small dependence of the pressure drop on x_2 can be observed across the electrode edges. For c greater than 50 the change in pressure drop is negligible. Since the permittivity of the ER-fluid is quite small, in practice $c \ll 1$. To see the effects for small values of the ratio c we repeat the same plot in Fig. 8.23 (right) but use now for c a logarithmic scale.

Another key parameter of the model is δ. This can be viewed as a parameter which ought to be chosen as small as possible in order to recover from equation (8.3.65) the CASSON-like model, (equation (8.3.53)). On the other hand, δ can be viewed as a constitutive parameter which can be obtained from equation (8.3.65) by fitting the experimental data.

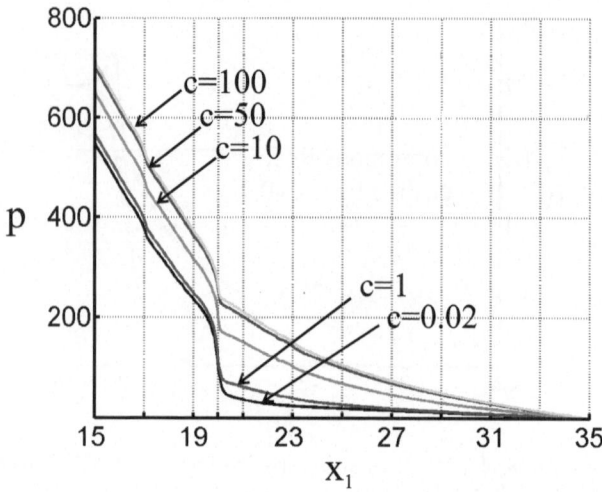

Fig. 8.22. Pressure (at $x_2 = 0$) over the electrode ends for different c

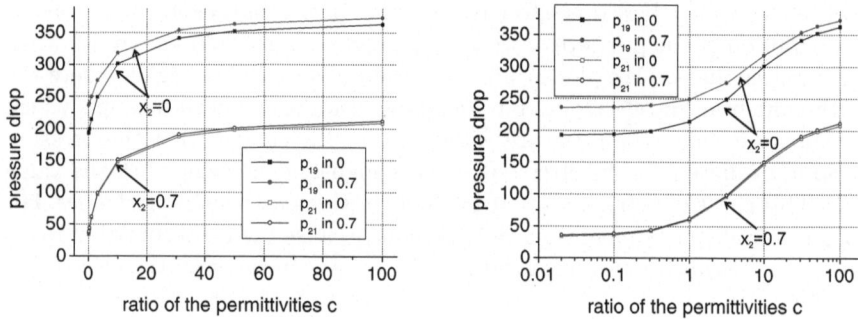

Fig. 8.23. Pressure drop p_{19} and p_{21} vs. ratio of the permittivities c on a linear scale (*left*) and on a semilogarithmic scale (*right*)

In what follows we study the influence of $n_\delta = \delta h^2 / v_0^2$ (the dimensionless form of δ) on the solution. In Figs. 8.24 and 8.25 we plotted $\eta_{gen}(|d|^2, E = 2)$ for different values of n_δ according to formula (8.4.55). The two figures differ in the range of the argument $|d|^2$. One can see from Fig. 8.24 that the choice of n_δ ($n_\delta = 0.002$) implies that the generalized viscosity differs from the classical generalized CASSON-like viscosity (for which $n_\delta = 0$) only for $|d|^2 < 0.01$. Figure 8.25 shows that the parameter δ can still have a strong influence on the generalized viscosity for $|d|^2 < 0.1$.

For determining the yielded/unyielded regions we can introduce a criterion using the quantity $|t^e| = \eta_{gen}|d|$ (we say the material has yielded when $\eta_{gen}|d| > 2|\beta_1(E)|$). As a consequence the yielded regions are influenced by the choice of δ if $|d|^2 < 0.1$; for $|d|^2 > 0.1$, δ does not change significantly the shape of the yielded regions.

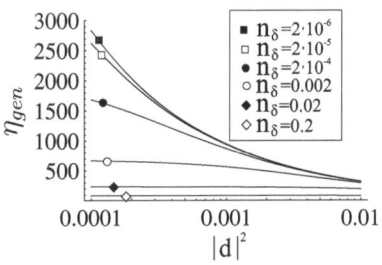

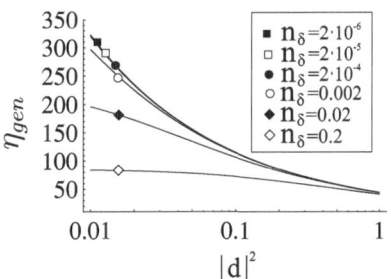

Fig. 8.24. $\eta_{gen}(E = 2, |d|^2)$ for different n_δ in the range $0.0001 < |d|^2 < 0.01$ (CASSON-like model)

Fig. 8.25. $\eta_{gen}(E = 2, |d|^2)$ for different n_δ in the range $0.01 < |d|^2 < 1$ (CASSON-like model)

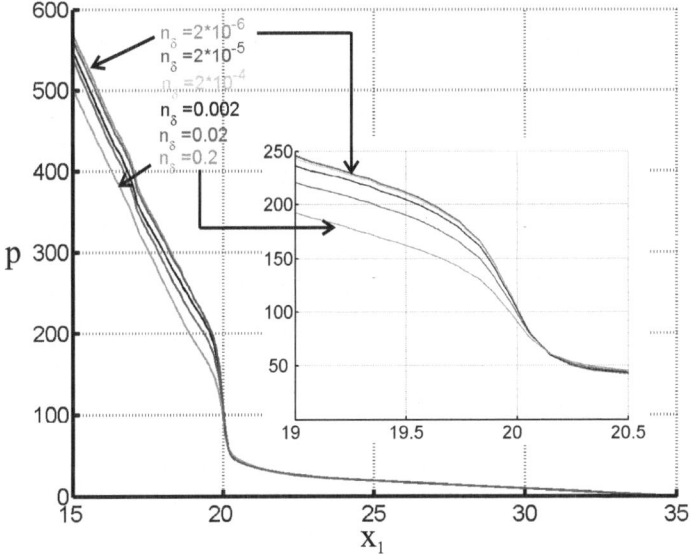

Fig. 8.26. Dimensionless pressure (at $x_2 = 0$) over the electrode ends for different values of n_δ; the inset shows the detail of the pressure near $x_1 = 20$

In Fig. 8.26 we plot the computed pressure evolution in the channel for different values of n_δ. In the region with constant electric field (between the electrodes far from the electrode edge) the pressure drop is not influenced by n_δ; however, in the transition region n_δ produces some differences. For values less than n_δ used in the simulations (0.002) the differences are unnoticeable. In order to see the influence of δ on the flow field we plotted the velocities in the transition area (see Fig. 8.27) and in the middle of the electrodes (see Fig. 8.28) for different values of n_δ. It is seen again that for small values of n_δ ($n_\delta < 0.002$), the profiles are almost congruent. Figures 8.27 and 8.28 permit in principle to identify those fluids which are described by our model con-

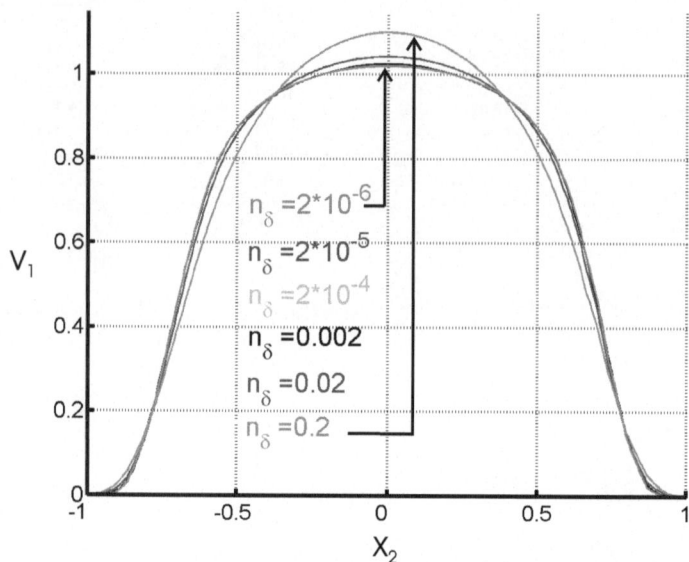

Fig. 8.27. Profiles of v_1 at $x_1 = -20$ (electrode ends) for different values of n_δ

taining a larger n_δ-value. For increasing n_δ, the plateau is curving, indicating a reduced yielded region.

Comparison with Experiment

In order to compare the numerical modeling with the experimental results reported in [64] we considered in the numerical program a configuration consisting of two electrodes, charged non-symmetrically, of length $2l = 40$ mm, with a channel height of $2h = 2$ mm and a length of the channel of $2L = 70$ mm which corresponds to the geometry used in the experiments [64, 269]. Experiments are performed in a channel of width $b = 20$ mm. For computations the input value is the volumetric flow rate in the channel $Q = \frac{4}{3}hbv_0$ and the output is the dimensional pressure drop calculated from $x_1 = -30$ mm on a length of 60 mm. In Fig. 8.29 we compare the obtained results for different values of $E_u = V/h$. The quantity E_u represents the value of the uniform electric field established in the middle of the channel, far from the electrode edges, and it is in fact the value of the magnitude of the electric field if the electrodes were infinitely long (see Sect. 8.4.2). While good agreement is obtained, the remaining inaccuracy of the calculations is given by the uncertainty in fitting values for β_1 and β_2 (see (8.4.71) and (8.4.72)).

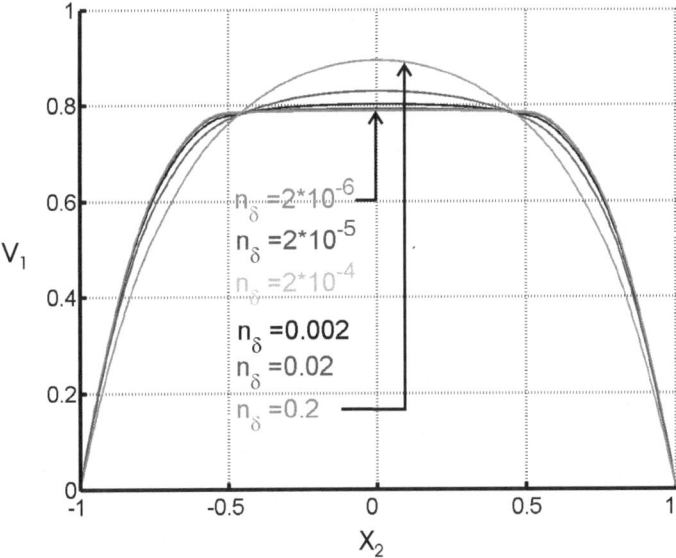

Fig. 8.28. Profiles of v_1 at $x_1 = -16$ (between the electrodes) for different values of n_δ

We considered this result as a validation of our computational model and used it to simulate further configurations in our search for a possible enhancement of the electrorheological effect.

b) Numerical Results – the Alternative Power-Law Model

Material and Configuration Properties Used for the Simulations

To illustrate the power-law constitutive equation we chose the ER-fluid EPS 3301 because, as shown in [269] it is described well by this model. EPS 3301 is produced by the company CONDEA Chemie AG (Germany). It is a particle-free fluid and consists of an acid metal soap which is dissolved homogeneously in a conventional hydraulic basis liquid. It contains also some additional hydraulic oil additives [48, 182, 269]. The fluid was studied experimentally by Wunderlich [269]. In order to better characterize the fluid, it was found that a dilution with 40% (weight percent) white oil (Weissöl) is appropriate, (see [15]).

Unlike the ER-fluid Rheobay 3565, the ER-fluid EPS 3301 shows in the absence of the electric field pseudoplastic behaviour (the shear viscosity decreases with increasing shear rate). Under the application of an electric field, also pseudoplastic behaviour is observed. In [269] the power law model (8.3.33) was found the most appropriate model to describe the data. Fitting

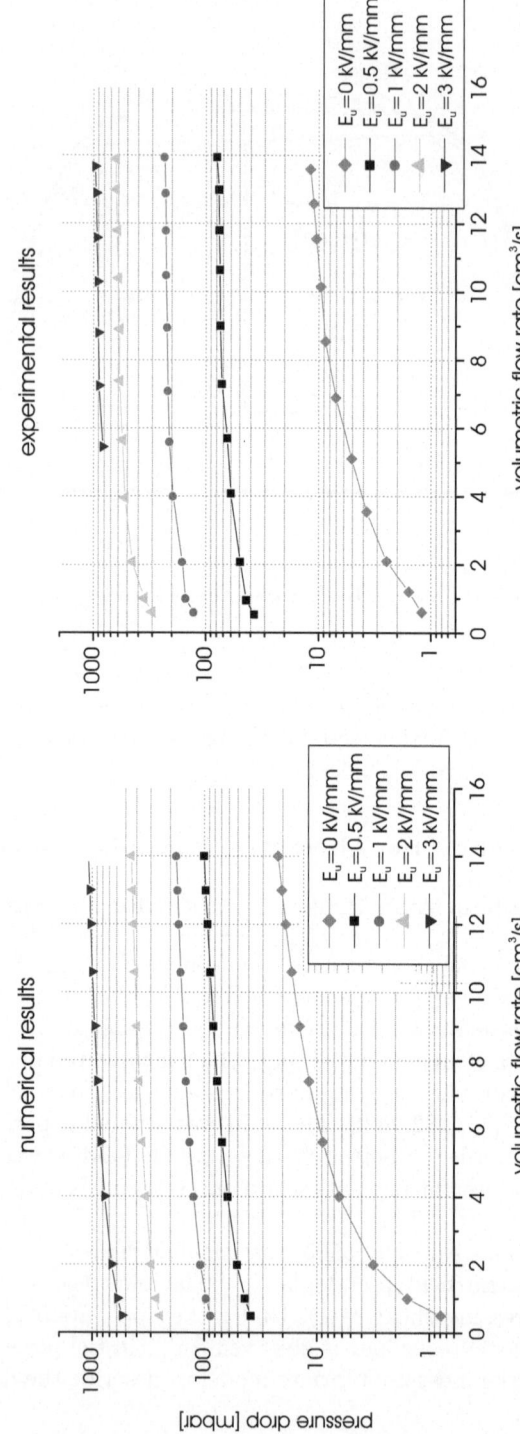

Fig. 8.29. Comparison of the numerical and experimental results (dimensional pressure-drop vs. volumetric flow rate)

Table 8.3. Values of the parameters m and n of (8.3.33) for different electric field strengths

$E_0 E \, [\frac{\text{kV}}{\text{mm}}]$	0	1	2	3	4
$m \, [\text{Pas}^n]$	0.4702	10.8695	23.0439	36.8244	67.0970
$n \, [-]$	0.8697	0.3923	0.3249	0.2921	0.2361

the experimental curves "stress vs. shear rate" (obtained with the rotational viscometer), the values of m and n were calculated (see Table 8.3).

We used these values in our bidimensional model (8.4.67). First, by fitting the values of $f = m(v_0/h)^n$ (see Table 8.3) and n we have to obtain their dependence on the dimensional electric field $E_0 E$:

$$f(E_0 E) = f_1 + f_2(E_0 E) + f_3(E_0 E)^2 \, , \qquad (8.4.76)$$
$$n(E_0 E) = n_1((E_0 E)^2 + n_2)^{n_3} \, . \qquad (8.4.77)$$

The coefficients f_1, f_2 and f_3 depend on the values of v_0 and h. For $v_0 = 0.1647$ m/s and $h = 0.001$ m, we obtained the values given in Table 8.4. The function f is written as a series expansion by retaining only terms up to second order while n is described by a power function. The experimental data (see Table 8.3) are provided for a limited range of the electric field and do not cover the range of the electric field employed in the simulations. We use formulas (8.4.76) and (8.4.77) also to extrapolate the given data for f and n for larger values of E. Figure 8.30 illustrates how the fitted curves approximate the data and how the extrapolations for $v_0/h = 164.7$ s^{-1} look like.

The geometry used for the simulation of EPS 3301 is the same as that used for Rheobay. The electric permittivity of EPS is $\varepsilon_2 = 1.4 \times 10^{-10}$ A·s/(V·m), and we choose $\varepsilon_1 = 0.02 \, \varepsilon_2$. The density of EPS 3301 is 850 kg/m^3. All simulations except those done to conduct the comparison with the experimental data from Fig. 8.44 are performed for $h = 1$ mm, $E_0 = 1$ kV/mm, $V = 2$ kV and $v_0 = 0.1647$ m/s. So, $\mathbb{R}e = 0.578$ and $\mathbb{M}a = 0.284$. We considered here also $n_\delta = 0.002$. We will investigate later the influence of this parameter on the solution.

Table 8.4. Values of the coefficients f_i and n_i from (8.4.76), (8.4.77) for $v_0 = 0.1647$ m/s and $h = 0.001$ m

i	1	2	3
f_i	41.431	33.284 mm/kV	2.958 (mm/kV)^2
n_i	0.398 (mm/kV)$^{2-n_3}$	0.008 (kV/mm)^2	-0.163

The Flow Near the Electrode Ends (long electrodes)

Here a similar analysis as for the alternative Casson-like fluid is carried out. In order to study the influence of the electrode edges on the flow we chose the same configuration of the electrodes and boundary conditions for the electrical problem as for the ER-material Rheobay. Consequently, the different rheological response must be due to the material properties.

One can distinguish three regions: (1) one with small electric field, far upstream of the entrance between the electrodes; (2) the transition zone with strongly inhomogeneous electric field right between the edges and (3) the uniform electric field zone between the electrodes, far downstream the edges.

In Fig. 8.31 the velocity modulus is plotted. In the areas far ahead and far beyond the electrode edges, the x_2-component of the velocity is very small and does practically not influence the x_1 velocity component. The end effect is reduced and the velocity develops smoothly from the region with reduced electric field to the region with constant electric field. Cross-sectional cuts of the surface plot in these regions are presented in Figures 8.34 and 8.35. Figure 8.35 indicates the negligible value of v_2 ($\sim 1/1000$ of v_1). The profile of v_1 from the first zone is close to a NEWTONian profile (reduced pseudoplastic behaviour as it was observed in the experiments for small electric fields in [269]). In the third zone the velocity profile for constant electric field presents the character of the unyielded region in the middle of the channel, as expected in general for an ER-fluid. To quantify the flow in the transition region we choose three points near the electrode ends: $x_1 = -20.2$, $x_1 = -20$ and $x_1 = -19.5$ (see Figs. 8.32 and 8.15). Notice the smooth plateaux in the middle parts of the plots of v_1. The curvature change of v_1 at $x_1 = -20$ is very close to the electrode edge but only slightly visible and much less than in the corresponding plot of the alternative CASSON-like model (middle plot of Fig. 8.14). The x_2 component of the velocity is smaller than 5% relative of the maximum inlet velocity. As for the alternative CASSON-like model, the two extreme values of v_2 in $x_1 = -20.2$ (first positive and then negative) and $x_1 = -19.5$ (first negative and then positive) indicate that the fluid avoids the electrode tips and goes around them.

In order to study the influence of the electric field on the pressure we plot in Fig. 8.36 the pressure along the channel, across the electrode ends (compare also with Fig. 8.18). The transition area is characterized by a higher pressure gradient compared with the pressure gradient in the region with small electric field (outside the electrodes) and within the region with constant electric field (between the electrodes, far from the edges). The difference between these two regions is not significant. The presence of a higher gradient in the transition zone led us to the study of the possibilities to obtain a better ER-effect.

We continue our study of the transition region near the electrode ends with plots of the second invariant of the stretching tensor, $|d|^2$ in Fig. 8.37 and of the generalized viscosity η_{gen} (8.4.67) in Fig. 8.38. Regarding the plot

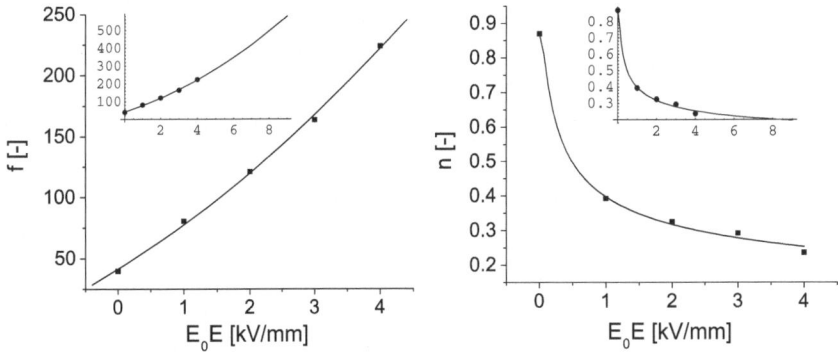

Fig. 8.30. Fitting curves for the dependences of the coefficients f and n on the electric field. The insets show how the fit is extrapolated

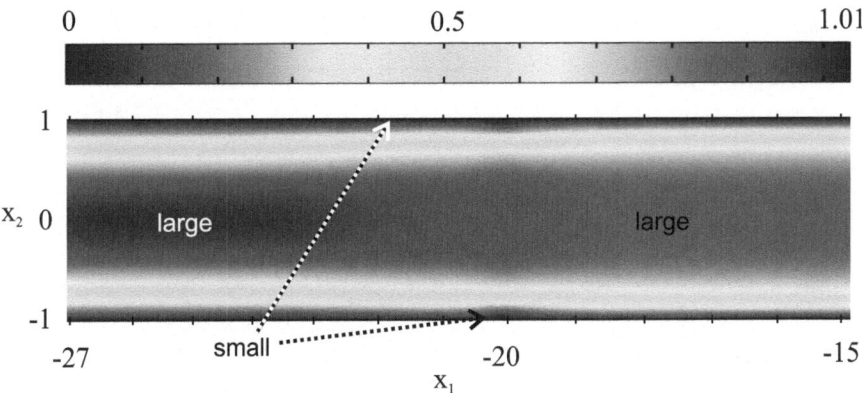

Fig. 8.31. Surface plot of the modulus of the non-dimensional velocity in the vicinity of the electrode edges (which are at $x_1 = -20$)

of $|d|^2$ against x_1, x_2, one can see that the regions with smaller values of $|d|^2$ around the electrode edges are reduced as compared to the alternative CASSON-like model (Fig. 8.19). This can be explained by the fact that the ER-fluid EPS is a material with a reduced ER-effect. However, the plot is qualitatively similar to the corresponding plot for Rheobay (Fig. 8.19).

In Fig. 8.38 as in Fig. 8.20 a transition to a "solid" island between the electrode ends can be observed.

Influence of the Parameter δ

As for the alternative CASSON-like model, δ can be viewed as a parameter which has to be chosen as small as possible to recover the pure power-law model (equations (8.3.51) and (8.3.52)) from equation (8.3.66). On the other hand, δ can be viewed as a constitutive parameter which can be obtained

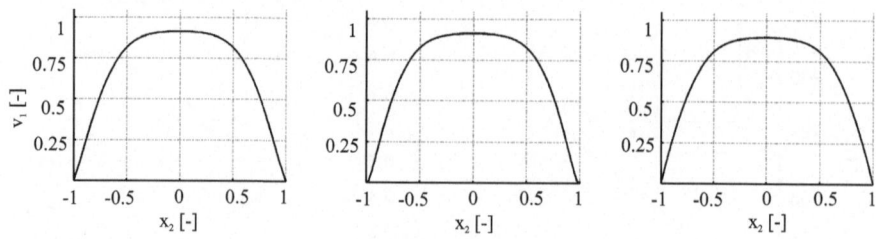

Fig. 8.32. Profiles of v_1 at $x_1 = -20.2, -20, -19.5$, respectively

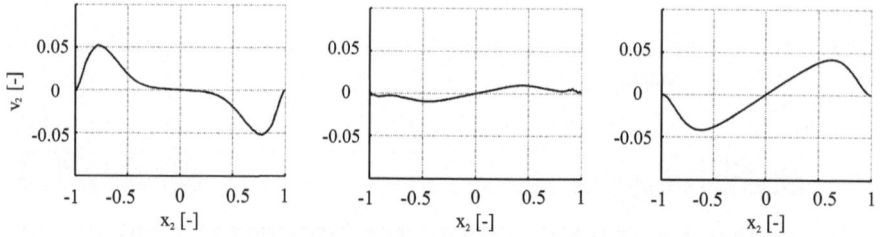

Fig. 8.33. Profiles of v_2 at $x_1 = -20.2, -20, -19.5$, respectively

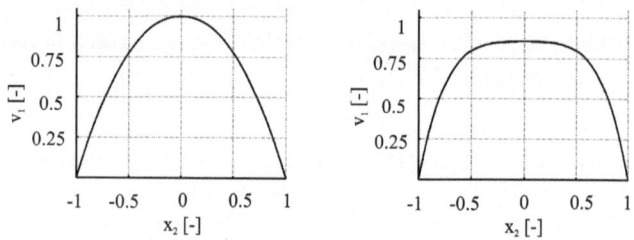

Fig. 8.34. Profiles of v_1 at $x_1 = -23.5, -16.5$, respectively

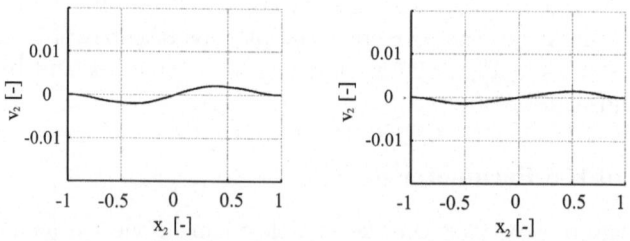

Fig. 8.35. Profiles of v_2 at $x_1 = -23.5, -16.5$, respectively

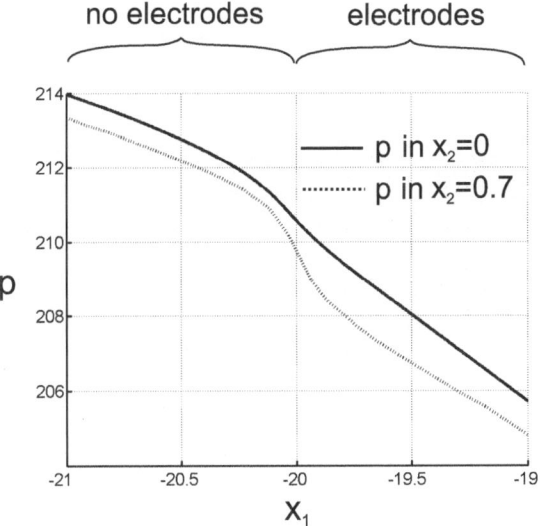

Fig. 8.36. Dimensionless pressure in the transition zone ($x_1 = -20$ marks the electrode ends)

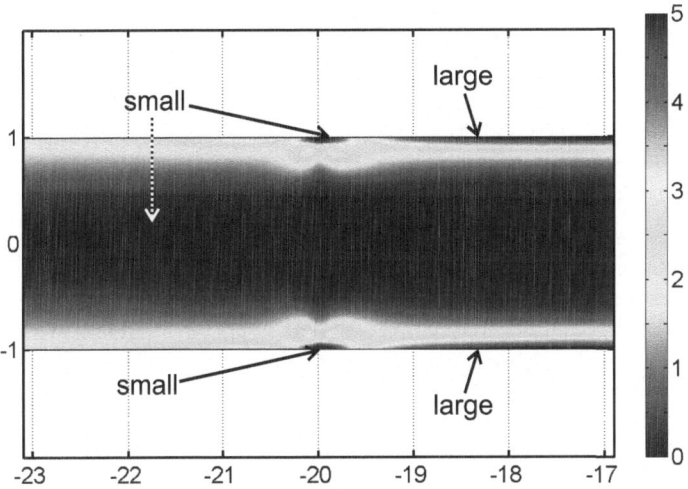

Fig. 8.37. Surface plot of dimensionless $|d|^2$

from the experimental data when fitting equation (8.3.66) with them. Here we treat it as a small parameter.

The number $n_\delta = \delta h^2 / v_0^2$ is the dimensionless form of δ. Figures 8.39 and 8.40 show the influence of n_δ on the dimensionless viscosity $\eta_{gen}(E = 2, |d|^2)$ according to formula (8.4.67) for different ranges of the argument $|d|^2$. One can see that its influence becomes more drastic for small values of $|d|^2$. In

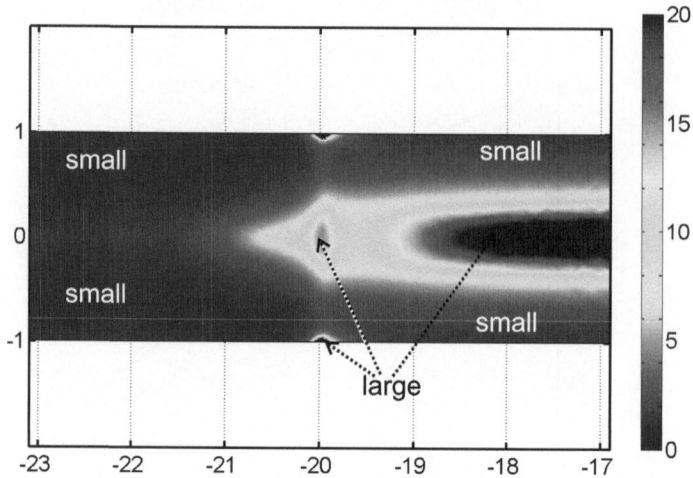

Fig. 8.38. Surface plot of the generalized dimensionless viscosity

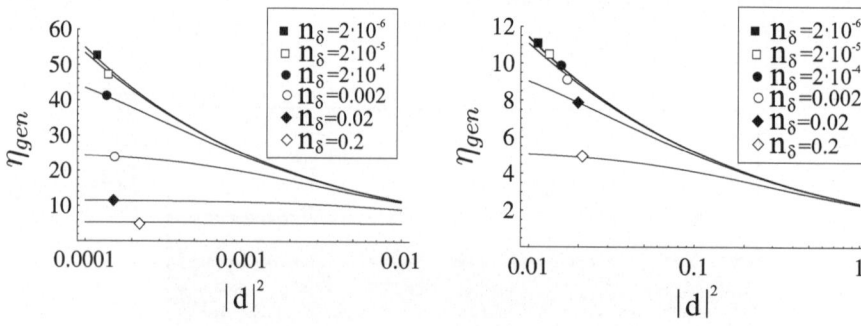

Fig. 8.39. $\eta_{gen}(E = 2, |d|^2)$ for different values of n_δ in the range $10^{-4} < |d|^2 < 0.01$ (alternative power-law model)

Fig. 8.40. $\eta_{gen}(E = 2, |d|^2)$ for different values of n_δ in the range $0.01 < |d|^2 < 1$ (alternative power-law model)

order to study the influence of n_δ on the solution we plot in Fig. 8.41 (compare also Fig. 8.26) the pressure along the channel axis containing also the electrode ends for different values of this parameter. Differences are only marginally visible for $n_\delta < 0.002$ while for larger values of n_δ the curves are practically congruent. Similarly, the plots of the velocity profiles in the uniform electric field and right between the electrode tips are shown in Figs. 8.42 and 8.43, respectively; they indicate a negligible influence of this parameter on the flow.

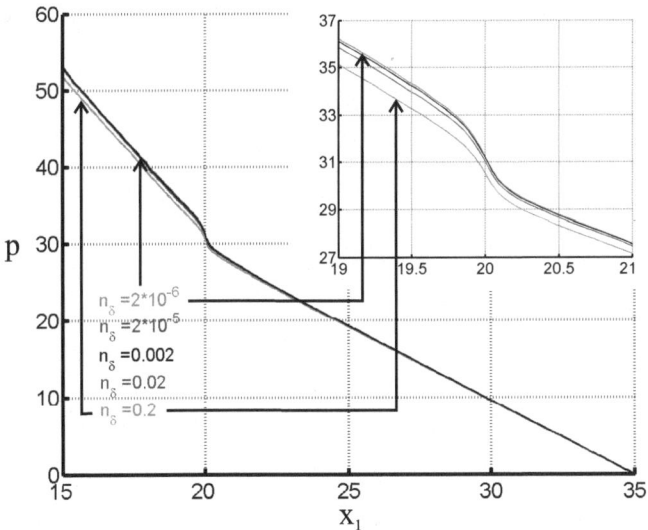

Fig. 8.41. Dimensionless pressure (at $x_2 = 0$) across the electrode end for different values of n_δ; the inset shows the detail of the pressure near $x_1 = 20$

Comparison with Experiment

The experiments given in [269] have been performed for configurations consisting of two parallel electrodes charged non-symmetrically with three different lengths: $2l = 6$ mm, $2l = 20$ mm, and $2l = 40$ mm. The channel height is $2h = 2$ mm, the channel length is $2L = 70$ mm and the channel width is $b = 20$ mm. The potential imposed on the upper electrode is $V = 2$ kV. The input value for computations is the volumetric flow rate in the channel $Q = 2(n+1)\,v_0hb/(2n+1)$. The output is the **ER-effect** which is the dimensionless quantity

$$F_{\Delta p}(Q) = \frac{\Delta p_E(Q)}{\Delta p_{E=0}(Q)} - 1 , \qquad (8.4.78)$$

where Δp is the dimensional pressure drop calculated between $x_1 = -30$ and $x_1 - 30$ mm. In Fig. 8.44, the experimental results are compared with the numerical results for the three lengths of the electrodes. We found qualitative agreement: the ER-effect is decreasing with increasing volumetric flow rate and as expected, the ER-effect is larger for larger electrode lengths l.

The FEMLAB results are quantitatively close to the approximate analytic computations[23]. This means that the effect of the electrode ends is reduced, as is expected for a long electrode.

---

[23] The electrorheological effect (8.4.78) measured from $x_1 = -30$ on a distance of 60 mm for the configuration used in the experiments may be calculated analytically (approximately) by using the procedure described in Subsect. 8.4.2

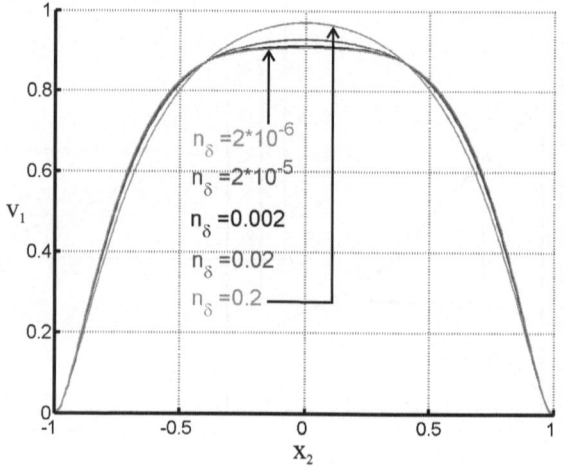

Fig. 8.42. Profiles of v_1 at $x_1 = -20$ (electrode ends) for different values of n_δ

c) Enhancing the ER-Effect

In technical applications there is a strong interest in obtaining an enhanced ER-effect (8.4.78). Many experimental works (see e.g. [2, 269]) have been performed with a focus in this direction. Most attempts are based on the modification of the geometry in such a way that inhomogeneities in the electric field are introduced. These investigations show that changes in the geometrical configurations may lead to a reduction or an enhancement of the ER-effect. However, it is difficult to distinguish by what measure this enhancement is produced, i.e. by the electric field inhomogeneities or by the inhomogeneities in the flow which are due to the modification of the geometry (which appear also when the electric field is switched off).

The configurations studied in this section are based on inhomogeneities caused by the end effects of the electrodes. This means that instead of modifying the geometry, the electric field inhomogeneities are introduced here by changing the boundary conditions in the electrical problem. Since by each such change, the formulation of the problem for vanishing electric field remains unchanged, the increase/decrease in $F_{\Delta p}$ (see (8.4.78)) is equivalent to an increase/decrease in Δp. Consequently, our study will focus on the examination of the possibilities to obtain larger values of the pressure drop at the same value of the volumetric flow rate by using the inhomogeneities produced by the electrode ends.

(by considering a uniform electric field in the part of the channel between the electrodes and a vanishing electric field in the remainder of the channel). One obtains $F_{\Delta p} = 2l(k - k_0)/(0.06k_0)$ where $k_0 = (Q(2 + 1/n_0)/(hb))^{n_0} m_0/h$ is the pressure gradient at a vanishing electric field (see 8.4.49).

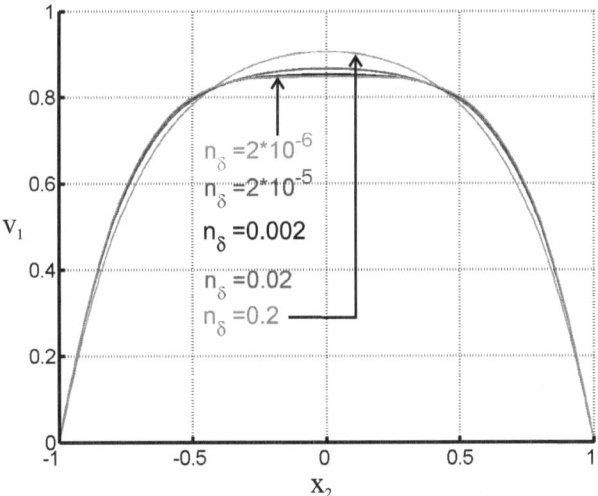

Fig. 8.43. Profiles of v_1 at $x_1 = -16$ (between the electrodes) for different values of n_δ

As Figs. 8.18 and 8.36 show, the pressure decreases more rapidly around the electrode edges than elsewhere. In order to analyse how one can use this effect in an optimal way the simplest elements are investigated: short electrodes (which imply also an interaction between the left and right electrode edges), an electrode interrupted by a "hole" and a short electrode between two "holes". The study is performed for both ER-materials Rheobay and EPS, which are modeled with the alternative CASSON-like and power-law equations, respectively, as described previously. Since we treat here (almost) only cases with short electrodes, all approaches are numerical. The data introduced in FEMLAB are those given above (all lengths in this section are given relative to the channel height).

Let us investigate first the influence of the electrode length. To obtain the plot of Fig. 8.45, different lengths $2l$ of the electrodes were considered, and the pressure drop $\Delta p = p_{x_1=-30} - p_{x_1=0}$ vs. l was computed. For $l < 1.5$ for the CASSON-like fluid and for $l < 2$ for the power-law fluid, the curves (square points interpolation with continuous lines) deviate from the linear behaviour (solid straight lines) which is a consequence of the coupling of the electrode ends in the electrical problem. For very short electrodes (stronger inhomogeneities) the pressure drop is significantly decreased. For the CASSON-like fluid, a better electrorheological effect is obtained when the electrode edges do not interact while for the power-law fluid the largest value of the ER-effect is obtained for short electrodes of length about 0.5 where the electrode ends are strongly interacting. Note that 2 is the dimensionless height of the channel. The dotted lines of the same figures represent the pressure drop calculated

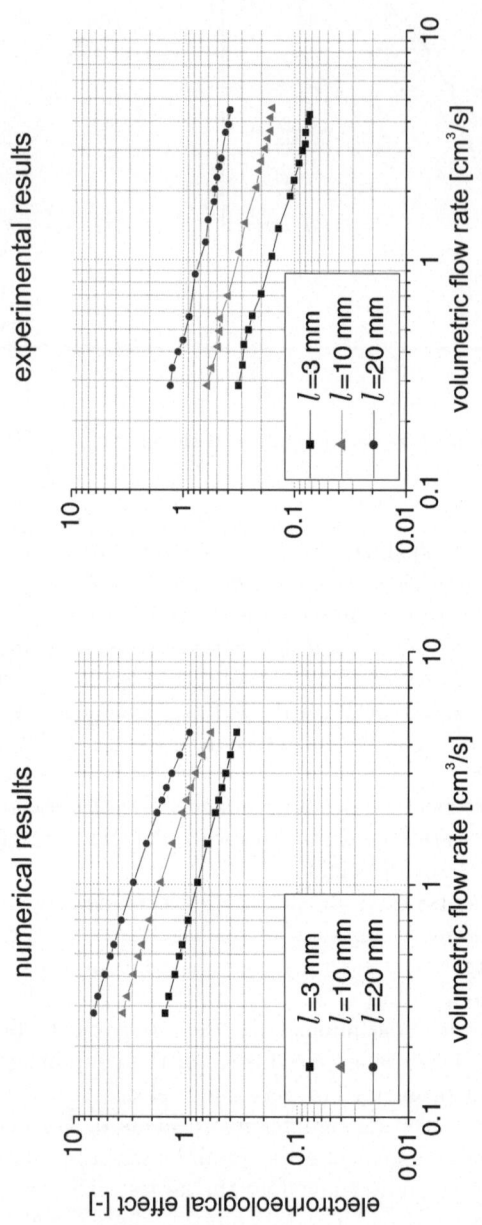

Fig. 8.44. Comparison of the numerical and experimental results (electrorheological effect *vs.* volumetric flow rate)

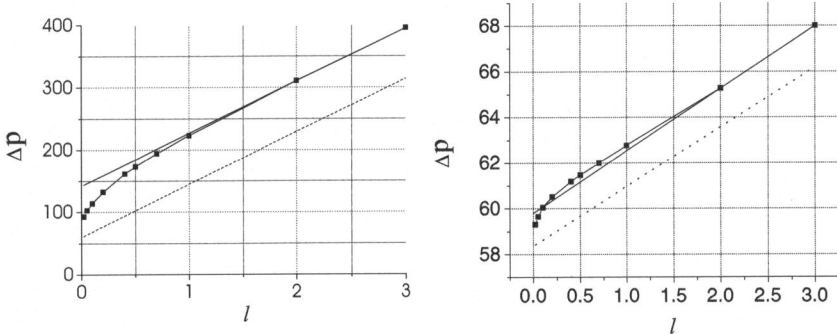

Fig. 8.45. Pressure drop from $x_1 = -30$ to $x_1 = 0$ vs. the half length of the electrodes (CASSON-like fluid (*left*) and power-law fluid (*right*)): *dotted line* – analytical evaluation with 1D model; square points interpolated with *continuous line* – numerical evaluation with 2D model; continuous *straight line* – fit for the behaviour of the pressure drop for long electrodes

analytically by neglecting the end effects. The difference between this curve and the curves obtained numerically is a measure of the values of additive pressure drops due to the end effects.

In order to see how the value of the pressure drop can be increased, we will investigate further what happens when the electrodes are interrupted by "holes" of length d (see Fig. 8.46). We study the effect in two cases of charging the 4 electrodes: for $V_1 = V_2 = V$ and for $V_1 = -V_2 = V$. We call these situations normal and inverse polarity of the electrodes, respectively. For inverse polarity, when the "hole" is sufficiently short, there will appear a stronger electric field that influences the flow more significantly. To illustrate this effect we plot in Fig. 8.47 the velocity surface in the channel for the CASSON-like fluid. The behaviour for the power-law model is similar. The inhomogeneities in the flow are stronger near the electrode edges bounding the "hole" than in the vicinity of the exterior edges. Of course, one should avoid too short "holes" which can lead to short circuits.

We computed the average pressure gradient $k_1 = [p(x_1 = -(l + d)/2) - p(x_1 = (l + d)/2)]/(l + d)$ for $l = 10$ and for variable d. Since the electrodes are sufficiently long, the electric field becomes homogeneous and the flow will be of POISEUILLE type around $x_1 = \pm(l+d)/2$, the middles of the electrodes. By plotting k_1 for different lengths of the "hole" one can compare the "hole effect" on the pressure drop. From Fig. 8.48 one can infer pretty similar qualitative behaviours for both fluids.

For $d > 1$ and normal polarity and for $d > 2$ and inverse polarity, the pressure gradient is smaller than that calculated for electrodes without any "hole", $k_1 = 85.84$ (for Rheobay) and $k_1 = 4.464$ (for EPS) (corresponding to the case $d = 0$ for normal polarity). The behaviour for short "holes" is different for the two polarity cases due to the different interactions of

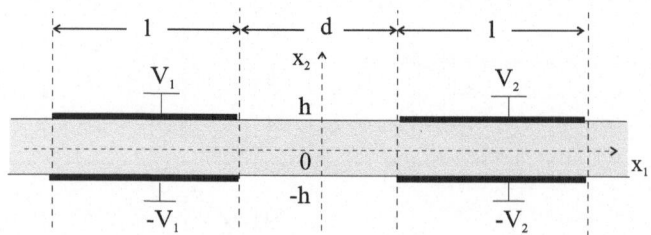

Fig. 8.46. Configuration with electrodes interrupted by one "hole"

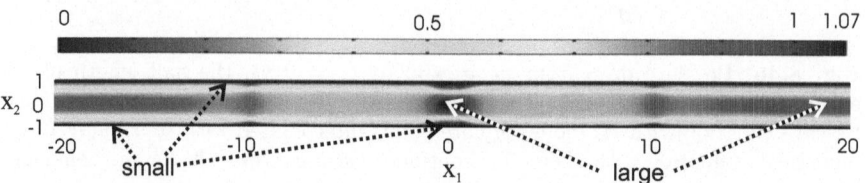

Fig. 8.47. Surface plot of the velocity modulus – long electrodes ($2l + d = 20$) with "holes" (of length $d = 1$) in the middle (inverse polarity) (CASSON-like fluid)

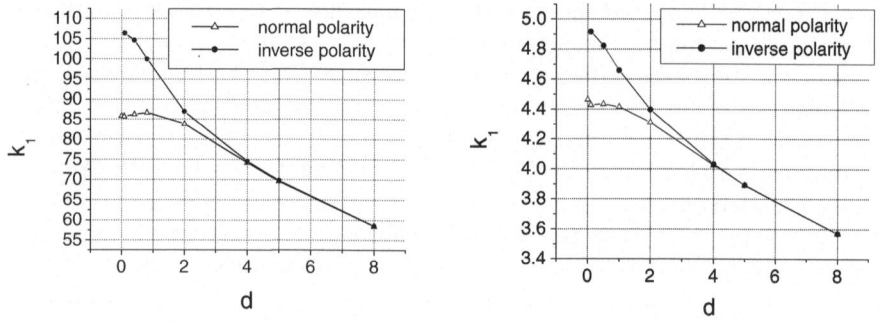

Fig. 8.48. Average pressure gradient vs. "hole" length: normal and inverse polarity (CASSON-like fluid (*left*) and power-law fluid (*right*))

the electrode edges in the "hole": smaller values of the pressure gradient for normal polarity due to a reduced electric field in the "hole" and higher pressure gradient for inverse polarity since a stronger electric field occurs in this case in the "hole". The optimal length of the "hole" is the shortest possible one without electric short circuit.

If we introduce more "holes", the next step is to study, for inverse polarity, how close the "holes" should be placed. To answer this question we study the case of one variable electrode of length l_1 placed between 2 "holes" of constant length d (see Fig. 8.49). The electrodes on the left and right sides have length l_2. In Fig. 8.50 we plotted the average pressure gradient $k_2 = [p(x_1 = -(l_1 + l_2 + 2d)/2) - p(x_1 = (l_1 + l_2 + 2d)/2)]/(l_1 + l_2 + 2d)$ vs. l_1

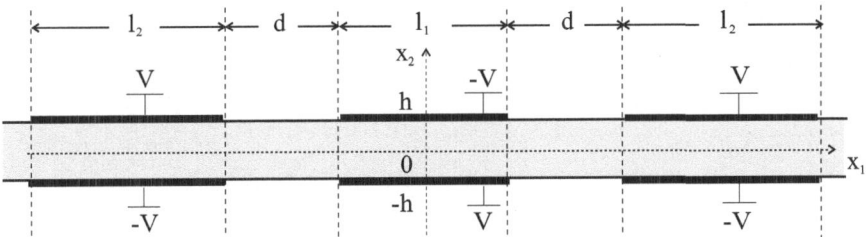

Fig. 8.49. Configuration with electrodes between "holes" (inverse polarity)

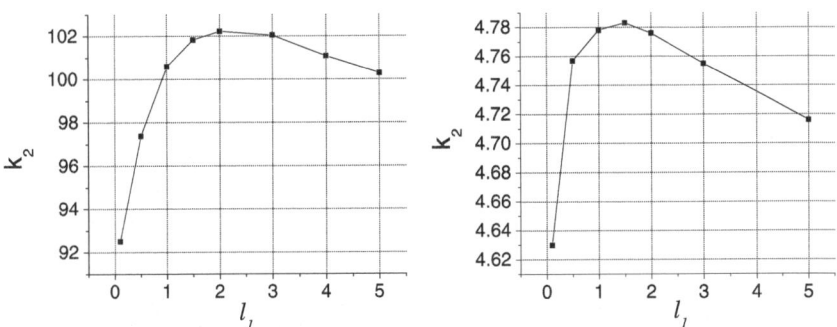

Fig. 8.50. Average pressure gradient vs. length of the middle electrode (one electrode placed between two "holes", alternating polarity) (CASSON-like fluid (*left*) and power-law fluid (*right*))

for $l_2 = 10$ and $d = 1$. The optimal length of an electrode between "holes" of length $d = 1$ is around $l_1 = 2$ in the CASSON-like case while in the power-law case it is around $l_1 = 1.5$.

Having the optimal parameters, one can now design optimal series of electrodes. As an example we consider a series of five electrode pairs and we compare the pressure over an interrupted electrode pair with the pressure over an uninterrupted electrode pair (see Fig. 8.51). One can see that up to 30% enhancement of the pressure drop can be obtained when combined optimized series of "holes" and electrodes are used. To get a better insight we display in Fig. 8.52 the generalized viscosity. For the Rheobay fluid two regions with increased viscosity can be discerned: one along the walls and another one in the middle of the channel. The first one is due to the high electric field around the edges of the electrodes. If we had uninterrupted electrodes, we would have a continuous yielded region characterized by a high viscosity in the middle of the channel. But in the case presented here this region is split into islands of high viscosity. One observes a similar behaviour for the EPS fluid but to a lower extent.

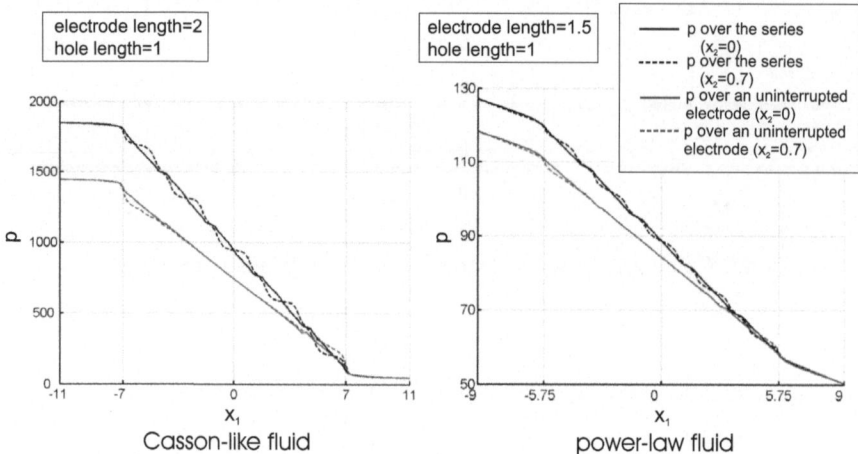

Fig. 8.51. Comparison between the pressure over an uninterrupted electrode pair and the pressure over an optimized series of five electrode pairs (inverse polarity)

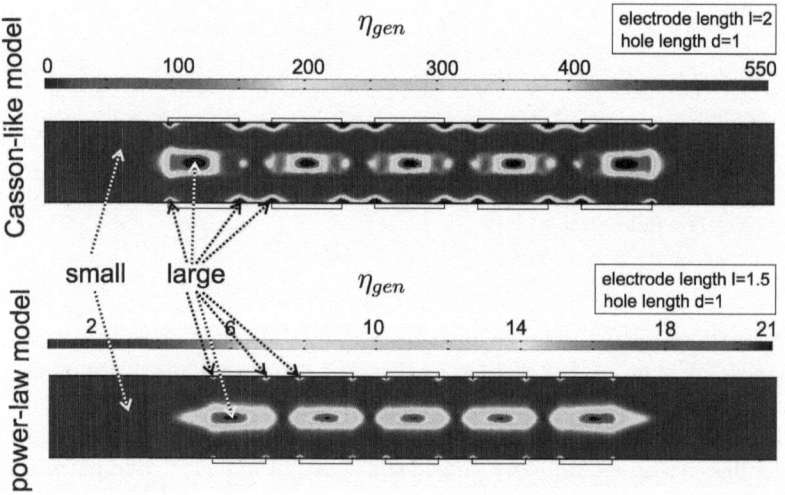

Fig. 8.52. Surface plot of the generalized viscosity for an optimized series of five electrode pairs (inverse polarity)

Conclusions

In this subsection we focused on pressure drop simulations due to its importance for applications. Effects of the "holes" and of the lengths of the electrodes have been computed that provide direct insight into the building blocks of typical experimental set-ups. The analysis has been carried out by taking into account the polarity of the electrodes which critically influences the distribution of the electric field in the channel. A better effect is obtained by introducing "holes" in the electrodes and by alternating the polarity of the electrode cuts as depicted in Fig. 8.49 than taking continuous electrodes. We found optimal lengths for the electrodes and "holes" relative to the height of the channel.

8.4.4 Electrodes with Modified Shape and Position Relative to the Flow – Experimental Results and Discussion

The inhomogeneities of the electric field studied in Subsect. 8.4.3 were introduced by changing the boundary conditions in such a way that the upper and lower electrodes are parallel and placed along the walls of the channel. The geometric configuration of the flow remains unchanged. Another way to create non-uniform electric fields is to modify the shape or the position of the electrodes.

Several experiments with different ER-fluids were performed in order to study the influence of the geometry of the electrodes on the ER response. KATSIKOPOULOS and ZUKOSKI [108] obtained an increase of the stress by a factor of 2 by using electrodes with grooves and ridges perpendicular to the direction of the flow when compared with the case having smooth electrodes. On the other hand and quite contrary, the ER-response was the same as for smooth electrodes when the grooves of the electrodes were disposed in the direction of flow. OTSUBO's experiments [166] show that the non-uniformity of the electric field obtained by using electrodes of different types – one coated with a striped pattern of conductive lines and another one with a honeycomb pattern of conductive lines – is responsible for an enhanced ER-effect not achievable with smooth electrodes. These initiating studies and results were followed by further systematic works of ABU-JDAYIL [2], ABU-JDAYIL and BRUNN [4, 5] and WUNDERLICH [269], whose results we briefly describe in this subsection. All three authors try to find how the ER-effect can be increased by altering the geometry of the smooth electrodes.

We focus in this presentation only on those experimental configurations which are connected with the theoretical modelling presented in this section, namely those performed on slit flows in a stationary electric field that can be reduced to two-dimensional models.

The ER-fluid used in [2, 4, 5] for the slit flow, P 1723, is supplied by the company Robert Bosch GmbH (Germany). It is a transparent fluid consisting of paraffin oil and silica gel with particle diameters from 5 to 30 nm, and a dispersed phase volume fraction of 30%. In the experiments performed by WUNDERLICH in [269] under a direct current (DC) the ER-fluid EPS 3301 was used which was already described in Subsect. 8.4.3.

We mention that in the experiments performed in [2, 4, 5] the length of the electrodes (irrespective of their shape and position), defined as the distance between the electrode ends, is $l = 200$ mm. The measurements were performed with fixed Δp, where Δp is the pressure drop determined in the middle of the electrodes pair over a length of $l_{\Delta p} = 150$ mm. The volumetric flow rate Q was obtained by integrating the measured velocity profiles as a function of Δp. The quantitative measure of the ER-effect is given by the efficiency factor (F) defined as the reduction in volumetric flow rate at constant pressure drop

$$F = \frac{Q|_{E=0} - Q|_E}{Q|_{E=0}}\bigg|_{\Delta p}. \tag{8.4.79}$$

In order to quantify the influence of the modified geometry of the electrodes on the ER-effect in [2] the quantity

$$f_1 = \frac{F}{F_u}\bigg|_{\Delta p} \tag{8.4.80}$$

is formed, where F_u is the efficiency factor determined for smooth electrodes flush with the boundary.

In all experiments reported in [269] and mentioned here the length of the electrodes is $l = 40$ mm. The pressure drop over the electrodes, Δp was measured over a distance of $l_{\Delta p} = 60$ mm containing the electrodes, as a function of the volumetric flow rate at certain constant values of the applied electric field[24]. Since $l_{\Delta p} > l$ one has to take into account also the distance between the channel plates which is $h = 2$ mm. In order to represent the variation of the ER-effect and of the electrical power due to the modified geometry of the electrodes in [269] the quantities

$$f_2 = \frac{F_{\Delta p}}{F_{\Delta p, h=2\,\mathrm{mm}}}\bigg|_Q, \tag{8.4.81}$$

and

$$f_3 = \frac{P_{el}}{P_{el, h=2\,\mathrm{mm}}}\bigg|_Q, \tag{8.4.82}$$

respectively, are chosen. $F_{\Delta p}$ is defined in (8.4.78), while P_{el} represent the electrical power dissipated to maintain the constant electric potential on the electrodes. The reference values $F_{\Delta p, h=2\,\mathrm{mm}}$ and $P_{el, h=2\,\mathrm{mm}}$ (in the denominators of (8.4.81) and (8.4.82), respectively) are evaluated for the configuration with parallel smooth electrodes placed along the channel walls with $Q = $ constant[25].

---

[24] Although the electric field is inhomogeneous the authors calculate for each configuration a mean value of the electric field in order to be able to compare the results obtained with different configurations at the same value of the electric field. It would be perhaps more appropriate to compare the results at the same value of the applied voltage on the electrodes which is a constant.

[25] Although the author does not mention it, we understand that the denominators as well as the corresponding numerators are evaluated for the same value of the electric field. We make an alternative proposal at the end of this subsection.

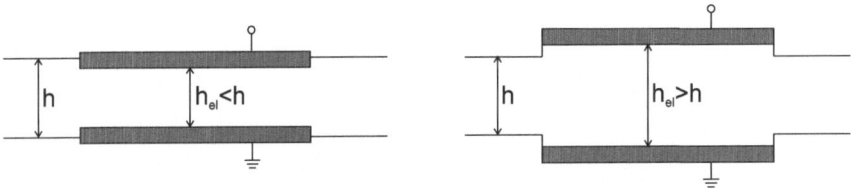

Fig. 8.53. Configurations with electrodes separated by a gap smaller (*left*) and larger (*right*) than the channel height

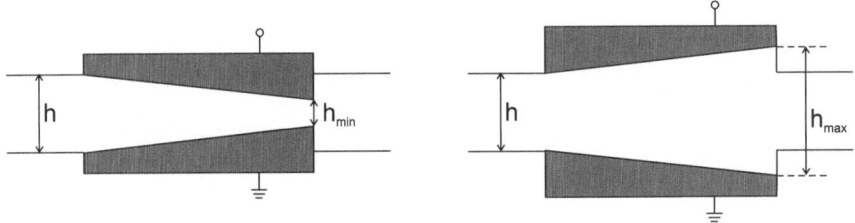

Fig. 8.54. Configurations with convergent (*left*) and divergent (*right*) oblique electrodes

Let us make an important remark: the geometry of the electrodes may directly influence the flow field and beyond the inhomogeneity of the electric field. In other words the flow is non-viscometric due to the geometry even in the absence of the electric field. In order to study what influence such a non-viscometric flow may have on the ER-effect, two configurations with parallel smooth electrodes with different distances (see Fig. 8.53) were investigated in [269]. The flow becomes non-viscometric at least in the vicinity of the edges and corners near the electrode ends. From the measurements with protruding electrodes for $h_{el} = 1$ mm and $h_{el} = 0.5$ mm, $f_2 < 1$ and $f_3 < 1$ were found. From the measurements with retreated electrodes for $h_{el} = 3$ mm instead $f_2 > 1$ and $f_3 > 1$ was obtained.

Another investigated configuration consists of oblique electrodes (see Fig. 8.54). The idea behind this arrangement is to avoid as far as possible the presence of edges and corners and to examine the effects when the distance between the electrodes varies which leads in this case to convergence/divergence of the stream lines. There are no measurements with this configuration conducted for DC in [269]. The configuration with convergent electrodes was studied in [2, 4, 5] for $h = 3$ mm and $h_{min} = 2$ mm. When evaluating f_1 the author compared F with F_u, where F_u is the efficiency factor for a configuration with smooth parallel electrodes separated by a gap

Fig. 8.55. Configuration with corrugated electrodes (asymmetric arrangement)

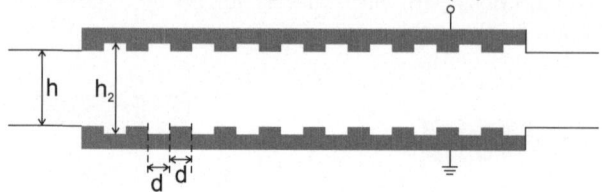

Fig. 8.56. Configuration with corrugated electrodes (symmetric arrangement)

of $h_s = 2.5$ mm[26]. The result was $f_1 = 1$ for $U < 1.7$ kV and $f_1 < 1$ for $U > 1.7$ kV.

The configurations presented now are based on electrodes with modified surfaces. The inhomogeneities of the electric field and the non-viscometric effects of the flow are intensified in this way.

In order to study the influence of the surface of the electrodes on the ER-effect grooved electrodes were used (see Figs. 8.55 and 8.56). This shape can be viewed as a series of protruding and retreating electrodes. For $d = 2$ mm, $h_1 = 2.15$ mm and $h_2 = 2.3$ mm in [269] an increased ER-effect $f_2 > 1$ was obtained at approximately the same electric power $f_3 \approx 1$ for both symmetric and asymmetric arrangements. In [2, 4, 5] the dimensions characterizing the applied grooved electrodes are $d = 10$ mm, $h_1 = 2.5$ mm and $h_2 = 3$ mm. The efficiency factor F was compared with F_u taking the gap between the smooth parallel electrodes, $h_s = 2.3$ mm. The results obtained were $f_1 > 1$ for $U < 1.7$ kV and $f_1 < 1$ for $U > 1.7$ kV. For $U > 1.7$ kV the efficiency factor for the asymmetric configuration is slightly less than for the symmetric configuration.

In the above-described configurations the electric field lines are approximately perpendicular on the stream lines. Wunderlich considered in [269] a configuration (see Fig. 8.57) in which this behaviour is altered: the electrodes are shaped such that the electric field lines are in some regions essentially parallel to the stream lines. Moreover, the flow is obstructed by the beamed extrusions of the electrodes and the stream lines take a snake-like shape (the flow direction will be alternatively horizontal and vertical). The fluid flows

---

[26] This value is equal to the mean distance between the oblique electrodes. In this way, at a certain applied voltage U, the electric field established between smooth electrodes, $E_s = U/h_s$ is the same as the mean value of the electric field between the oblique electrodes, $E_{mean} = U/h_{mean}$.

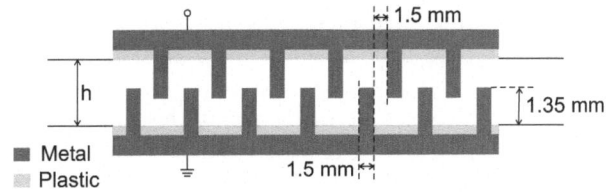

Fig. 8.57. Configurations with beamed electrodes

alternatively through regions with and without an electric field. In spite of all these strong inhomogeneities the obtained ER-effect is smaller than for the smooth electrodes: $f_2 < 1$ and $f_3 < 1$.

The conclusion of all these experimental works is that an enhanced ER-effect can be obtained when the inhomogeneities in the electric field are introduced such that the flow field is perturbed as little as possible (i.e. remains essentially viscometric). The inhomogeneity in the flow field lowers the ER-effect or at least cancels the gained ER-effect due to the non-uniform electric field so that the effects compensate each other.

In their approach the authors differentiate between two types of electrodes. In a configuration based on the first type the flow becomes non-viscometric due to both the shape of the electrodes (presence of corners and edges) and the inhomogeneity of the electric field. This means that the flow is non-viscometric also when the electric field vanishes. Configurations with protruding electrodes, retreated electrodes, beamed electrodes belong to this category. The second types of electrodes have shapes which are supposed to be perturbed only by a small measure of the viscometric flow. Consequently, in this case, the non-viscometry of the flow is due approximately to the non-uniform electric field and in the absence of the electric field the flow is approximately viscometric. Oblique electrodes and corrugated electrodes are considered of this second type. However, since the mechanical problem is coupled with the electrical problem in a non-linear way, this separation can be used only as a guiding hint. A more rigorous approach would be accomplished by complementing the experimental studies by theoretical modeling.

Concerning the theoretical approach to be applied to these complex configurations some remarks should be made. Thanks to its generality, the formulation of the problem given in Subsect. 8.4.1 permits its application and numerical implementation also for modified geometries of the channel (e.g. with oblique electrodes, grooved electrodes or with electrodes separated by a gap smaller/larger than the channel height), by simply adapting the respective boundary conditions. However this formulation is restricted to electrodes of zero thickness. Luckily, one can approximate each of the presented configurations with a corresponding one having electrodes of zero thickness by imposing the applied electric potential only on the interior boundaries of the electrodes. The main difficulty which arises in the numerical implementation is related to the choice of the mesh. A sufficiently refined mesh around the edges and corners requires huge memory space and excessively large CPU times to ensure convergence since at sharp electrode corners the electric field is singular (see [246]).

An important remark must be made about the empirical evaluation of the quantity E, which appears in (8.4.79)–(8.4.82), and it is approximated by the mean value of the electric field. Nevertheless, the electric field is inhomogeneous (at least in certain regions) in each configuration investigated in these works so it is not reasonable to characterize it by a single mean value for the whole domain. Since the aim of the experimental and theoretical investigations is to obtain enhanced effects at the same power consumption by modifying the morphology of the electrodes, perhaps it would be more appropriate to compare the obtained ER-effects for the initial and modified configurations at the same electric power instead of at the same mean electric field. So, in order to conduct a correct comparison between the ER-effect with normal electrodes and the ER-effect with modified electrodes we suggest to choose for each modified configuration an electric voltage *such that the electric power is the same as the power dissipated for the configuration with smooth electrodes.* This means for WUNDERLICH's approach that

$$\frac{P_{el}}{P_{el,h=2\,\mathrm{mm}}}\bigg|_Q = 1\,. \tag{8.4.83}$$

9 Appendix

9.1 Appendix A: On Objectivity

In this Appendix we briefly state how one can show what transformation properties the various electromagnetic field variables introduced in the main body of this book enjoy. To this end, we shall use three and four-dimensional notation. Let x^A be the (contravariant) four-vector (x_i, t), consisting of the position vector x_i $(i = 1, 2, 3)$ of a particle and time t, and let $\mathrm{x}^{\star A} = \mathrm{x}^{\star A}(\mathrm{x}^B)$ be any C^1-transformation $(x_i, t) \rightarrow (x_i^\star, t^\star)$. A covariant four-tensor ψ_{AB} and a contravariant four-tensor ψ^{AB} are then quantities, which under such mappings transform according to

$$\psi^\star_{AB} = \frac{\partial \mathrm{x}^C}{\partial \mathrm{x}^{\star A}} \frac{\partial \mathrm{x}^D}{\partial \mathrm{x}^{\star B}} \psi_{CD} , \qquad \psi^{\star AB} = \frac{\partial \mathrm{x}^{\star A}}{\partial \mathrm{x}^C} \frac{\partial \mathrm{x}^{\star B}}{\partial \mathrm{x}^D} \psi^{CD} . \tag{9.1.1}$$

Likewise, a contravariant four-vector transforms under general transformations $(x_i, t) \rightarrow (x_i^\star, t^\star)$ according to

$$\sigma^{\star A} = \frac{\partial \mathrm{x}^{\star A}}{\partial \mathrm{x}^B} \sigma^B . \tag{9.1.2}$$

Of special interest are EUCLIDian transformations given by

$$x_i^\star = O_{ij}(t)x_j + c_i(t) , \qquad x_i = O_{ji}(t)(x_j^\star - c_j(t)) , \tag{9.1.3}$$

$$t^\star = t , \qquad\qquad\qquad t = t^\star .$$

In what follows we would like to explore some consequences implied by them.

(i) Let

$$\sigma^A = (J_i, \mathcal{Q})$$

be a contravariant vector. Then a routine calculation shows that under EUCLIDian transformations

$$\sigma^{\star A} - (J_i^\star, \mathcal{Q}^\star) = (O_{ij}(J_j \quad \dot{x}_j \mathcal{Q}) + \mathcal{Q}\dot{x}_i^\star, \mathcal{Q}) , \tag{9.1.4}$$

or

K. Hutter et al.: *Electromagnetic Field Matter Interaction in Thermoelastic Solids and Viscous Fluids*, Lect. Notes Phys. **710**, 367–374 (2006)
DOI 10.1007/3-540-37240-7_9   © Springer-Verlag Berlin Heidelberg 2006

$$(J_i - Q\dot{x}_i)^\star = O_{ij}(t)(J_i - Q\dot{x}_j)\,,$$
$$Q^\star = Q\,. \tag{9.1.5}$$

In other words, Q and $(J_i - Q\dot{x}_i) = J_i$ transform under EUCLIDian transformations as an objective scalar and an objective vector, respectively.

(ii) Let ψ_{AB} be a skew-symmetric *covariant* four-tensor with the components

$$\psi_{AB} = \begin{pmatrix} 0 & b_3 & -b_2 & e_1 \\ & 0 & b_1 & e_2 \\ (-) & & 0 & e_3 \\ & & & 0 \end{pmatrix} \tag{9.1.6}$$

(we choose to name these components b_i and e_i for suggestive reasons lateron). If we perform a EUCLIDian transformation, (9.1.3) shows that the three-vectors $b_i : (b_1, b_2, b_3)$ and $e_i := (e_1, e_2, e_3)$ transform as follows:

$$b_i^\star = \det(O)O_{ik}b_k\,,$$
$$e_i^\star + e_{ijk}\dot{x}_j^\star b_k^\star = O_{ij}(t)(e_j + e_{jkl}\dot{x}_k b_l)\,. \tag{9.1.7}$$

Otherwise stated, b_i *is an objective axial vector and* $(e_i + e_{ijk}\dot{x}_j b_k)$ *an objective polar vector under the* EUCLID*ian transformation group*. To prove $(9.1.7)_1$ for instance, note that

$$\psi_{ij}^\star = O_{ik}O_{jl}\psi_{kl}\,, \quad \text{where} \quad \psi_{kl} = e_{klm}b_m\,,$$

where e_{ijk} is the usual three-dimensional permutation tensor. In much the same, though more complicated way, one can also prove that

$$e_i^\star = O_{il}\{e_l + e_{lmn}[\dot{O}_{jm}(x_j^\star - c_j) - O_{jm}\dot{c}_j]b_n\}\,,$$

which, with the aid of the identity

$$\dot{O}_{jm}(x_j^\star - c_j) - O_{jm}\dot{c}_j = \dot{x}_m - O_{jm}\dot{x}_j^\star\,,$$

immediately implies (9.1.7).

A special application of (9.1.6) is the tensor whose components are $b_i := m_i$, $e_i = 0$. Then m_i must be an objective axial tensor under the EUCLIDian transformation group.

(iii) Another covariant skew-symmetric tensor of importance is

$$\psi_{AB} = \begin{pmatrix} 0 & m_3 & -m_2 & (\boldsymbol{m} \times \dot{\boldsymbol{x}})_1 \\ & 0 & m_1 & (\boldsymbol{m} \times \dot{\boldsymbol{x}})_2 \\ (-) & & 0 & (\boldsymbol{m} \times \dot{\boldsymbol{x}})_3 \\ & & & 0 \end{pmatrix}\,. \tag{9.1.8}$$

It can be shown by a straightforward calculation, that the three-vector $m_i := (m_1, m_2, m_3)$ *is an objective axial vector under* EUCLID*ian transformations*. (This is just a special case of (ii).)

(iv) Let ψ^{AB} be a skew-symmetric *contravariant* four-tensor with the components

$$\psi^{AB} = \begin{pmatrix} 0 & h_3 & -h_2 & -d_1 \\ & 0 & h_1 & -d_2 \\ (-) & & 0 & -d_3 \\ & & & 0 \end{pmatrix} . \tag{9.1.9}$$

A calculation identical to that performed above shows that the vectors $d_1 := (d_1, d_2, d_3)$ and $h_1 := (h_1, h_2, h_3)$ transform under the EUCLIDian group as

$$\begin{aligned} d_i^\star &= O_{ij} d_j , \\ h_i^\star - e_{ijk}\dot{x}_j^\star d_k^\star &= \det{(\boldsymbol{O})} O_{ij}(h_j - e_{jkl}\dot{x}_k d_l) . \end{aligned} \tag{9.1.10}$$

Hence d_i and $(h_i - e_{ijk}\dot{x}_j d_k)$ are an objective vector and an objective axial vector under EUCLIDian transformations.

A special situation is again the case for which

$$h_i = m_i \qquad \text{and} \qquad d_i = 0$$

which immediately shows that m_i must be an objective axial vector.

(v) As a last example, consider the contravariant skew-symmetric tensor

$$\psi^{AB} = \begin{pmatrix} 0 & (\boldsymbol{p} \times \dot{\boldsymbol{x}})_3 & -(\boldsymbol{p} \times \dot{\boldsymbol{x}})_2 & p_1 \\ & 0 & (\boldsymbol{p} \times \dot{\boldsymbol{x}})_1 & p_2 \\ & & 0 & p_3 \\ (-) & & & 0 \end{pmatrix} . \tag{9.1.11}$$

Its transformation properties are most easily found in two steps. Firstly, we write

$$(\psi^\star)^{k4} = \frac{\partial x^{\star k}}{\partial x^l} \frac{\partial x^{\star 4}}{\partial x^4} \psi^{l4} ,$$

and obtain with the aid of (9.1.11) and (9.1.3)$_1$,

$$p_i^\star = O_{ij} p_j , \tag{9.1.12}$$

proving that p_i is an objective vector. On the other hand

$$(\psi^\star)^{ij} = p_i^\star \dot{x}_j^\star - p_j^\star \dot{x}_i^\star = \frac{\partial x^{\star i}}{\partial x^k} \frac{\partial x^{\star j}}{\partial x^l} \psi^{kl} + \frac{\partial x^{\star i}}{\partial x^4} \frac{\partial x^{\star j}}{\partial x^l} \psi^{4l} + \frac{\partial x^{\star i}}{\partial x^k} \frac{\partial x^{\star j}}{\partial x^4} \psi^{k4} ,$$

and it is now an easy matter, using (9.1.11) and (9.1.3), to show that the expression on the far right and in the middle of this chain are the same if p_i is assumed to obey (9.1.12).

It is shown in theoretical electrodynamics that the MAXWELL *equations* of deformable continua can be written in the form

$$e^{ABCD}\frac{\partial\varphi_{CD}}{\partial x^B} = 0 \quad \text{and} \quad \frac{\partial\eta^{AB}}{\partial x^B} = \sigma^A, \qquad (9.1.13)$$

where φ_{CD} and η^{CD} are skew-symmetric covariant and contravariant four-tensors, respectively, and where σ^A is a contravariant vector. Furthermore, e_{ABCD} is the four-dimensional permutation tensor, which is anti-symmetric with respect to any interchange of two indices and vanishes if any two indices are the same. Moreover, $e_{1234} = 1$, and lowering and rising of indices is achieved by the use of the metric tensor $g_{AB} = g^{AB}$, whose matrix is given by

$$g_{AB} = \begin{pmatrix} 1 & 0 & 0 & 0 \\ 0 & 1 & 0 & 0 \\ 0 & 0 & 1 & 0 \\ 0 & 0 & 0 & -1 \end{pmatrix}. \qquad (9.1.14)$$

Hence $e^{1234} = -1$, since $e_{1234} = +1$.

The equations (9.1.13) are general and hold in vacuo as well as in matter, but the contribution of matter is usually separated from that of vacuo; this separation is achieved by writing

$$\varphi_{AB} = \Phi_{AB} - \mu_{AB}, \qquad \eta^{AB} = H^{AB} - \pi^{AB}, \qquad (9.1.15)$$

where μ_{AB} and π^{AB} are a covariant and a contravariant skew-symmetric four-tensor, respectively, which vanish in vacuo. Thus Φ_{AB} and H^{AB} are the vacuum fields. Note that in view of the transformation properties explained under (i)–(v) for general skew-symmetric tensors, there will be no need to derive such properties for φ_{AB}, Φ_{AB}, μ_{AB}, η^{AB}, H^{AB} and π^{AB} anew. Before we list these tensors in the various descriptions, recall that the vacuum-fields Φ_{AB} and H^{AB} are related to each other through the equation

$$\Phi_{AB} = \hat{\Phi}_{AB}(H^{CD}), \qquad (9.1.16)$$

a relation which is sometimes called the MAXWELL–LORENTZ-**aether relation**. We shall see that it is *not* invariant under the general transformation $(x_i, t) \rightarrow (x_i^*, t^*)$.

We now list the various formulations and give the invariance properties which their variables enjoy.

(a) MINKOWSKI **formulation.** In this formulation one chooses $\mu_{AB} = 0$ and does not separate η^{AB} into two parts. Thus

$$\varphi_{AB} = \begin{pmatrix} 0 & B_3 & -B_2 & E_1 \\ & 0 & B_1 & E_2 \\ (-) & & 0 & E_3 \\ & & & 0 \end{pmatrix}, \quad \eta^{AB} = \begin{pmatrix} 0 & H_3 & -H_2 & -D_1 \\ & 0 & H_1 & -D_2 \\ (-) & & 0 & -D_3 \\ & & & 0 \end{pmatrix}. \quad (9.1.17)$$

Hence, because of the properties (ii) and (iv) of skew-symmetric tensors, we have under the EUCLIDIAN transformation group

$$\mathcal{D}_i \ , \ \mathcal{E}_i := E_i + e_{ijk}\dot{x}_j^* B_k \ , \qquad \text{transform as objective polar vectors} \ ,$$

$$\mathcal{B}_i \ , \ \mathcal{H}_i := H_i - e_{ijk}\dot{x}_j^* D_k \ , \qquad \text{transform as objective axial vectors} \ .$$

It is also a routine matter to show that (9.1.13) agrees with (3.3.36). Finally, the MAXWELL–LORENTZ–aether relation is formally introduced by writing $\eta^{AB} = H^{AB} - \pi^{AB}$ with

$$H_{AB} = \begin{pmatrix} 0 & H_3^{\mathrm{a}} & -H_2^{\mathrm{a}} & -D_1^{\mathrm{a}} \\ & 0 & H_1^{\mathrm{a}} & -D_2^{\mathrm{a}} \\ (-) & & 0 & -D_3^{\mathrm{a}} \\ & & & 0 \end{pmatrix}, \quad \pi^{AB} = \begin{pmatrix} 0 & -M_3 & M_2 & P_1 \\ & 0 & -M_1 & P_2 \\ (-) & & 0 & P_3 \\ & & & 0 \end{pmatrix},$$

$$(9.1.18)$$

where H_i^{a} and D_i^{a} are auxiliary fields and M_i and P_i are the MINKOWSKIan magnetization and polarization. Again, in view of items (ii) and (iv) above

$$D_i^{\mathrm{a}}, P_i, \qquad \text{transform as objective polar vectors} \ ,$$

$$\left.\begin{array}{l} \mathcal{M}_i := M_i - e_{ijk}\dot{x}_j P_k \\ \mathcal{H}_i := H_i^{\mathrm{a}} - e_{ijk}\dot{x}_j D_k \end{array}\right\} \quad \text{transform as objective axial vectors} \qquad (9.1.19)$$

under the EUCLIDian transformation group. Now, the MAXWELL LORENTZ–aether relation (9.1.16) is given by

$$H_i^{\mathrm{a}} = \frac{1}{\mu_0} B_i \quad \text{and} \quad D_i^{\mathrm{a}} = \varepsilon_0 E_i \ , \qquad (9.1.20)$$

and it is trivial to show that these are *not* invariant under EUCLIDian transformations.

(b) **Statistical Model.** Except for notation this model is identical with the MINKOWSKI model. Hence, under EUCLIDian transformations

$$\begin{array}{ll} D_i^{\mathrm{a}}, \mathcal{E}_i, P_i \ , & \text{transform as objective polar vectors} \ , \\ \mathcal{M}_i, B_i, \mathcal{H}_i, & \text{transform as objective axial vectors} \ . \end{array} \qquad (9.1.21)$$

(c) LORENTZ **formulation.** In this formulation one sets $\mu_{AB} = 0$, as was done in the previous formulations. Furthermore, H^{AB} is given as in (9.1.18), but

$$\eta^{AB} = \begin{pmatrix} 0 & -M_3^L + (\boldsymbol{P}^L \times \dot{\boldsymbol{x}})_3 & M_2^L - (\boldsymbol{P}^L \times \dot{\boldsymbol{x}})_2 & P_1^L \\ & 0 & -M_1^L + (\boldsymbol{P}^L \times \dot{\boldsymbol{x}})_1 & P_2^L \\ & & 0 & P_3^L \\ (-) & & & 0 \end{pmatrix} . \qquad (9.1.22)$$

This tensor can easily be written as the sum of two tensors, one containing the polarization, the other containing the magnetization only. From iv) and v) and the previous results it then follows that

$$D_i^{\mathrm{a}}, \mathcal{E}_i, P_i^L = P_i \,, \qquad \text{transform as objective polar vectors ,}$$
$$M_i^L = \mathcal{M}_i, B_i, \mathcal{H}_i^{\mathrm{a}}, \qquad \text{transform as objective axial vectors .}$$

under the EUCLIDian transformation group.

(d) **Two–Dipole Model** (CHU formulation). This model is the only one with nonvanishing μ_{AB}. Indeed,

$$\Phi_{AB} = \begin{pmatrix} 0 & B_3^{\mathrm{a}} & -B_2^{\mathrm{a}} & E_1^C \\ & 0 & B_1^{\mathrm{a}} & E_2^C \\ & & 0 & E_3^C \\ (-) & & & 0 \end{pmatrix}, \; \mu_{AB} = \begin{pmatrix} 0 & -M_3^C & M_2^C & (\boldsymbol{M}^C \times \dot{\boldsymbol{x}})_1 \\ & 0 & -M_1^C & (\boldsymbol{M}^C \times \dot{\boldsymbol{x}})_2 \\ (-) & & 0 & (\boldsymbol{M}^C \times \dot{\boldsymbol{x}})_3 \\ & & & 0 \end{pmatrix}$$

$$\eta^{AB} = \begin{pmatrix} 0 & H_3^C & -H_2^C & -D_1^{\mathrm{a}} \\ & 0 & H_1^C & -D_2^{\mathrm{a}} \\ (-) & & 0 & -D_3^{\mathrm{a}} \\ & & & 0 \end{pmatrix}, \; \pi^{AB} = \begin{pmatrix} 0 & (\boldsymbol{P}^C \times \dot{\boldsymbol{x}})_3 & -(\boldsymbol{P}^C \times \dot{\boldsymbol{x}})_2 & P_1^C \\ & 0 & (\boldsymbol{P}^C \times \dot{\boldsymbol{x}})_1 & P_2^C \\ & & 0 & P_3^C \\ (-) & & & 0 \end{pmatrix}.$$

$$(9.1.23)$$

Here, all variables are the so-called CHU-variables and

$$B_i^{\mathrm{a}} = \mu_0 H_i^C \quad \text{and} \quad D_i^{\mathrm{a}} = \varepsilon_0 E_i^C$$

are the MAXWELL–LORENTZ–aether relations. It follows from (ii)–(v) above that under EUCLIDian transformations

$$D_i^{\mathrm{a}}, \mathcal{E}_i := E_i + e_{ijk}\dot{x}_j B_k^{\mathrm{a}}, P_i^C \,, \qquad \text{transform as objective polar vectors ,}$$
$$B_i^{\mathrm{a}}, \mathcal{H}_i := H_i^C - e_{ijk}\dot{x}_j D_k^{\mathrm{a}}, M_i^C \,, \qquad \text{transform as objective axial vectors .}$$

It is not difficult to show for each set of four-tensors, introduced above that the three-dimensional MAXWELL equations in the respective formulations are obtained. Moreover, as can be clearly seen from the above derivation, E_i (or E_i^C) and H_i (or H_i^{a} or H_i^C) are not objective vectors under the EUCLIDian transformation group but that $\varepsilon_0 E_i$, $\varepsilon_0 E_i^C$, $\mu_0 H_i$, $\mu_0 H_i^{\mathrm{a}}$ and $\mu_0 H_i^C$ are, as can easily be seen by invoking the MAXWELL–LORENTZ–aether relations in the expressions for $D_i^{\mathrm{a}}, B_i^{\mathrm{a}}$. Note also, that M_i^L and M_i^C are both objective. Of such properties we have freely made use of in the main body of this book.

Finally, we mention once more that it is through the MAXWELL–LORENTZ–aether relations that the MAXWELL equations are not invariant under general transformations $(x_i, t) \rightarrow (x_i^\star, t^\star)$. Their form is such that the MAXWELL equations in vacuo (in which the MAXWELL–LORENTZ–aether relations are substituted) are invariant only under a very restricted transformation group, the LORENTZ *group*. This should not be confused with the basic fact that φ_{AB}, Φ_{AB}, μ_{AB}, η^{AB}, H^{AB} and π^{AB} are four-tensors, which must obey (9.1.1). Consequently, the transformation properties under the EUCLIDian group hold irrespective of the invariance properties of the MAXWELL equations.

Of course, EUCLIDian transformations are one group only, for which the transformations (9.1.1) and (9.1.2) hold. In principle, any other transformation group can be investigated and of special interest are Lorentz transformations, because they leave the MAXWELL equations *including* the MAXWELL–LORENTZ–aether relations invariant. These transformations are well-known and so we do not elaborate on them.

9.2 Appendix B: Some Detailed Calculations of the Maxwell–Minkowski Model

In this Appendix we present a motivation for the choice of ρT and R, (3.4.18), and a derivation of equation (3.4.20).

We start by transforming the Poynting-vector

$$\boldsymbol{E}^M \times \boldsymbol{H}^M.$$

With the aid of (3.4.1) we show that

$$
\begin{aligned}
e_{ijk} E_j^M H_j^M &= e_{ijk}(\mathcal{E}_j - e_{jlm}\dot{x}_l B_m)(\mathcal{H}_k + e_{kpq}\dot{x}_p D_q) \\
&= e_{ijk}\mathcal{E}_j\mathcal{H}_k + (\mathcal{E}_j D_j + \mathcal{H}_j B_j)\dot{x}_i - (\mathcal{E}_j D_i + \mathcal{H}_j B_i)\dot{x}_j \quad (9.2.1) \\
&\quad + e_{jkl} P_k B_l \dot{x}_j \dot{x}_i + \mathcal{O}(V^2/c^2).
\end{aligned}
$$

Next, using the MAXWELL equations (3.4.10) and (3.4.13), we derive

$$
\begin{aligned}
(e_{ijk}\mathcal{E}_j\mathcal{H}_k)_{,i} &= -\mathcal{J}_i\mathcal{E}_i - \overset{*}{D}_i\,\mathcal{E}_i - \overset{*}{B}_i\,\mathcal{H}_i \\
&= -\mathcal{J}_i\mathcal{E}_i - (\varepsilon_0\mathcal{E}_i\dot{\mathcal{E}}_i + \mu_0\mathcal{H}_i\dot{\mathcal{H}}_i) - \mathcal{E}_i\dot{P}_i - \mu_0\mathcal{H}_i\dot{\mathcal{M}}_i \\
&\quad - (\mathcal{E}_j D_j - \mathcal{H}_j B_j)\dot{x}_{i,i} + (\mathcal{E}_i D_j + \mathcal{H}_i B_j)\dot{x}_{i,j} \\
&= -\mathcal{J}_i\mathcal{E}_i - \frac{d}{dt}\left[\tfrac{1}{2}(\varepsilon_0\mathcal{E}_i\mathcal{E}_i + \mu_0\mathcal{H}_i\mathcal{H}_i)\right] - \mathcal{E}_i\dot{P}_i - \mu_0\mathcal{H}_i\dot{\mathcal{M}}_i \\
&\quad - (\mathcal{E}_i P_j + \mu_0\mathcal{H}_i\mathcal{M}_j)\dot{x}_{i,j} - (\varepsilon_0\mathcal{E}_j\mathcal{E}_j + \mu_0\mathcal{H}_j\mathcal{H}_j)\dot{x}_{i,i} \\
&\quad + (\mathcal{E}_i D_j + \mathcal{H}_i B_j)\dot{x}_{i,j}.
\end{aligned}
$$
$$(9.2.2)$$

Substitution of (9.2.1) and (9.2.2) into (3.4.17) yields

$$
\begin{aligned}
&(\dot{\rho} + \rho\dot{x}_{i,i})(U + \tfrac{1}{2}\dot{x}_i\dot{x}_i + T) + \rho\dot{T} - \rho\dot{U} + q_{i,i} - \rho r^{\text{ext}} \\
&- \mathcal{J}_i\mathcal{E}_i - \mathcal{E}_i\,\overset{*}{P}_i - \mu_0\mathcal{H}_i\,\overset{*}{\mathcal{M}}_i - [R_i - (P_j\mathcal{E}_j + \mu_0\mathcal{H}_j\mathcal{M}_j)\dot{x}_i \\
&+ e_{jkl}P_k B_l\dot{x}_j\dot{x}_i]_{,i} - [t_{ij} + \mathcal{E}_i P_j + \mu_0\mathcal{H}_i\mathcal{M}_j]\dot{x}_{i,j} \\
&+ [\rho\ddot{x}_i - \rho f_i^{\text{ext}} - t_{ij,j} - (\mathcal{E}_i D_j + \mathcal{H}_i B_j)_{,j} \\
&+ \tfrac{1}{2}(\varepsilon_0\mathcal{E}_k\mathcal{E}_k + \mu_0\mathcal{H}_k\mathcal{H}_k)_{,i}]\dot{x}_i = 0.
\end{aligned}
$$
$$(9.2.3)$$

For this relation to be invariant under rigid-body translations, the term proportional to $\dot{x}_i\dot{x}_j$, i.e.:

$$\mathrm{e}_{jkl}P_kB_l\dot{x}_i\dot{x}_j \ ,$$

must be compensated by R_i. Moreover, $\boldsymbol{R}$ must vanish with vanishing $\boldsymbol{P}$ and $\boldsymbol{M}$. Both requirements are satisfied by assuming $(3.4.18)_2$ which reads

$$R_i = (P_j\mathcal{E}_j + \mu_0\mathcal{M}_j\mathcal{H}_j)\dot{x}_i + \mathrm{e}_{jkl}P_kB_l\dot{x}_j\dot{x}_i \ . \tag{9.2.4}$$

Furthermore, if T would not be an objective scalar under rigid-body translations (with velocity $\boldsymbol{b}(t)$) the term ρT would lead to a term proportional to $\dot{\boldsymbol{b}}(t)$. Since this would be the only term of this kind, the above assumption leads to a contradiction and, hence, ρT must be objective. In that case, there is no distinction possible between T and U, and so T may be absorbed by U, or in other words, we may take

$$\rho T = 0 \ , \tag{9.2.5}$$

as was done in $(3.4.18)_1$.

We now substitute (9.2.4) and (9.2.5) into (9.2.3) and use the relation

$$
\begin{aligned}
(\mathcal{E}_iD_j + \mathcal{H}_iB_j)_{,j} - \tfrac{1}{2}(\varepsilon_0\mathcal{E}_k\mathcal{E}_k + \mu_0\mathcal{H}_k\mathcal{H}_k)_{,i} \\
= \mathcal{Q}\mathcal{E}_i + \mathrm{e}_{ijk}\mathcal{J}_jB_k + P_j\mathcal{E}_{j,i} + \mu_0\mathcal{M}_j\mathcal{H}_{j,i} + \mathrm{e}_{ijk}(D_j \overset{\star}{B}_k + \overset{\star}{D}_j B_k) \ ,
\end{aligned}
\tag{9.2.6}
$$

which follows from (3.4.10) and (3.4.13). This then leads to (3.4.20).

References

1. M. Abramowitz and I.A. Stegun (eds.): Handbook of Mathematical Functions. New York, Dover Publications, Inc., 1046 pp., (1965)
2. B. Abu-Jdayil: Electrorheological Fluids in Rotational Couette Flow, Slit Flow and Torsional Flow (Clutch). Aachen, Shaker Verlag, 137 pp., (1996)
3. B. Abu-Jdayil and P.O. Brunn: Effects of nonuniform electric field on slit flow of an electrorheological fluid. *J. Rheol.* **39**, 1327–1341, (1995)
4. B. Abu-Jdayil and P.O. Brunn: Effects of electrode morphology on the slit flow of an electrorheological fluid. *J. Non-Newtonian Fluid Mech.* **63**, 45–61, (1996)
5. B. Abu-Jdayil and P.O. Brunn: Study of the flow behaviour of electrorheological fluids at shear- and flow-mode. *Chem. Eng. and Process.* **36**, 281–289, (1997)
6. B. Abu-Jdayil and P.O. Brunn: Effects of coating on the behaviour of electrorheological fluids in torsional flow. *Smart Mater. and Struct.* **6**(5), 509–520, (1997)
7. B. Abu-Jdayil and P.O. Brunn: Effect of electrode morphology on the behaviour of electrorheological fluids in torsional flow. *J. Intel. Mat. Syst. Struct.* **13**(1), 3–11, (2002)
8. J.B. Alblas: Continuum Mechanics of Media with Internal Structure. *Instituto Nazionale di Alta Matematica, Symposia Mathematica* Vol. I, 229–251, London, Academic Press, Oderisi, Gubbio, (1969)
9. J.B. Alblas: A general theory of magnetoelastic stability, in *Vekua's Anniversary Volume*, Akad. Nauk. S.S.S.R., Moscow, 22–39, (1978)
10. J.B. Alblas: General Theory of Electro- and Magneto-Elasticity, in *Electromagnetic Interactions in Elastic Solids*, ed. by H. Parkus, Springer, Wien, (1978)
11. G.A. Alers and P.A. Fleary: Modification of the Velocity of Sound in Metals by Magnetic Fields. *Phys. Rev.* **129**, 2425, (1963)
12. S.A. Ambartsumian: Magneto-elasticity of thin plates and shells. *Appl. Mech. Rev.* **35**, 1–5, (1982)
13. S.A. Ambartsumian and M.V. Belubekian: Vibrations and stability of current-carrying plates with the account of transverse shear deformations, in *Mechanical Modelling of New Electromagnetic Materials*, ed. R.K.T. Hsieh, Elsevier, Amsterdam, 321–328, (1990)
14. S.A. Ambartsumian, M.V. Belubekian and M.M. Minassian: The problem of vibration of current-carrying plates. *Int. J. Appl. Electromagnetics in Materials* **3**, 65–72, (1992)

15. H. Asoud: Messungen von elektrorheologischen Flüssigkeiten in einem Flachkanal mit glatten, schrägen und geriffelten Elektroden. Diplomarbeit, Univ. Erlangen-Nürenberg, Erlangen, (1999)

16. R.J. Atkin, X. Shi and W.A. Bullough: Solutions of the constitutive equations for the flow of an electrorheological fluid in radial configurations. *J. Rheol.* **35**(7), 1441–1461, (1991)

17. R.J. Atkin, X. Shi and W.A. Bullough: Effect of non-uniform field distribution on steady flows of an electro-rheological fluid. *J. Non-Newtonian Fluid Mech.* **86**, 119–132, (1999)

18. G.Y. Bagdasarian, E.H. Danoyan and G.T. Piliposyan: Vibrations and stability of two-layered magnetostrictive plates, in *Proceedings of the SPIE*, Int. Soc. Opt. Eng, USA, **2442**, 532–543, (1995)

19. G.Y. Bagdasarian and G.T. Piliposyan: Mathematical modelling and numerical investigation of magnetoelastic stability of superconducting plates, in *Proceedings of the SPIE*, Int. Soc. Opt. Eng, USA, **3039**, 715–725, (1997)

20. H.A. Barnes and K. Walters: The yield stress myth? *Rheol. Acta*, **24**(4), 323–326, (1985)

21. H.A. Barnes: The yield stress – a review or '$\pi\alpha\nu\tau\alpha$ $\rho\varepsilon\iota$' – everything flows? *J. Non-Newtonian Fluid Mech.*, **81**, 133–178, (1999)

22. J.C. Baumhauer and H.F. Tiersten: Nonlinear Electroelastic Equations for Small Fields Superposed on a Bias. *J. Acoust. Soc. Am.* **54**, 1017–1025, (1973)

23. A.G. Bayer: Sicherheitsdatenblatt. Leverkusen, **6**, (1994)

24. E. Becker and W. Bürger: Kontinuumsmechanik. Teubner Verlag, (1975)

25. R. Benach and I. Müller: Thermodynamics and the Description of Magnetizable Dielectric Mixtures of Fluids. *Arch. Rat. Mech. Anal.* **53**, 312–346, (1974)

26. H. Block and J.P. Kelly: Electro-rheology. *J. Phys. D: Appl. Phys.* **21**, 1661–1677, (1988)

27. R. Borghesani and A. Morro: Thermodynamics and Isotropy in Thermal and Electrical Conduction. *Meccanica*, **9**, 63–69, (1974)

28. R. Borghesani and A. Morro: Thermodynamic Restrictions on Thermoelastic, Thermomagnetic and Galvanomagnetic Coefficients. *Meccanica*, **9**, 157–161, (1974)

29. M. Born and K. Huang: Dynamical Theory of Crystal Lattices, Oxford Press, (1954)

30. P. Boulanger, G. Mayné: Étude Théorique de 'Interaction d'un Champ electromagnétique et d'un Continu non Conducteur Polarisable et Magnétisable. *C.R. Acad. Sc.* Paris, Série A, **274**, 591–594, (1972)

31. P. Boulanger, G. Mayne, and R. van Geen: Magnetooptical, Electrooptical and Photoelastic Effects in an Elastic Polarizable and Magnetizable Isotropic Continuum. *Int. J. Solids Structures*, **9**, 1439–1464, (1973)

32. Jr.W.F. Brown: Magnetoelastic Interactions *Springer Tracts in Natural Philosophy* Vol. 9, Springer-Verlag, Berlin, (1966)

33. P.O. Brunn and B. Abu-Jdayil: Fluids with transverse isotropy as models for electrorheological fluids. *Z. angew. Math. Mech.* **78**(2), 97–107, (1998)

34. P.O. Brunn and B. Abu-Jdayil: A phenomenological model of electrorheological fluids. *Rheol. Acta* **43**, 62–67, (2004)

35. V. Busuioc and D. Cioranescu: On the flow of a Bingham fluid passing through an electric field. *Int. J. Non-Linear Mech.*, **38**, 287–304, (2003)

36. T. Butz and O. von Stryk: Modeling and Simulation of Electro- and magnetorheological Fluid Dampers. *Z. angew. Math. Mech.*, **82**(1), 3–20, (2002)

37. C. Carathéodory: Untersuchungen über die Grundlagen der Thermodynamik. *Math. Anal.* **67** 355–386, (1909)

38. G.F. Carrier, M. Krook and C.E. Pearson: Functions of a Complex Variable. New York, McGraw-Hill, 438 pp. (1966)

39. P. Chadwick: Continuum Mechanics, Concise Theory and Problems, Dover Publications, INC, Mineola, N.Y., 187 p. (1999)

40. S. Chattopadhyay and F.C. Moon: Magnetoelastic buckling and vibration of a rod carrying electric current. *J. Appl. Mech.* **42**, 809–914, (1975)

41. S. Chattopadhyay: Magnetoelastic instability of structures carrying electric current. *Int. J. Solids and Struct.* **15**, 467–477, (1979)

42. K.L. Chowdhury and P.G. Glockner: Constitutive Equations for Elastic Dielectrics. *Int. J. Non-Linear Mech.* **11**, 315–24, (1976)

43. R.V. Churchill: Complex Variables and Applications. New York, McGraw-Hill, 297 pp. (1960)

44. B.D. Coleman, W. Noll: The Thermodynamics of Elastic Materials with Heat Conduction and Viscosity. *Arch. Rat. Mech. Anal.* **13**, 167–178, (1963)

45. B. Collet and G.A. Maugin: Part I, Sur l'Electrodynamique des Milieux Continus avec Interactions; Part II, Thermodynamique des Milieux Continus avec Interactions. *C.R. Acad. Sc. Paris*, Série B, **279**, 379–382, 439–442, (1974)

46. B. Collet and G.A. Maugin: Couplage Magnetoelastique de Surface dans les Materiaux Ferromagnétiques. *C.R. Acad. Sc. Paris*, Série A, **280**, 1641–1644, (1975)

47. A.B. Comsol: Femlab. Version 2.3.0.148 (2002)

48. CONDEA Chemie GmbH, Sicherheitsdatenblatt, Hamburg, **12**, (1996)

49. A. Dalamangas: Magnetoelastic stability and vibration of ferromagnetic thin plates in a transverse magnetic field. *Mech. Res. Communications*, **10**, 279–286, (1983)

50. J.M. Dalrymple, M.O. Peach and G.L. Vliegelahn: Magnetoelastic Buckling of Beams and Thin Plates of Magnetically Soft Material. *J. Appl. Mech.* **39**, 451–455, (1972)

51. J.M. Dalrymple, M.O. Peach and G.L. Vliegelahn: Magnetoelastic Buckling of Thin Magnetically Soft Plates in Cylindrical Mode. *J. Appl. Mech.* **41**, 145–150, (1974)

52. C.R. Daubert, J.F. Steffe and A.K. Srivastava: Predicting the electrorheological behaviour of milk chocolate. *J. Food Process Eng.*, **21**(3), 249–261, (1998)

53. S.R. De Groot, L.G. Suttorp: Foundations of Electrodynamics. *North-Holland Publishing Co.*, Amsterdam, (1972)

54. Y.F. Deinega and G V. Vinogradov: Electric-fields in the rheology of disperse systems. *Rheol. Acta*, **23**(6), 636–651, (1984)

55. H.G. De Lorenzi and H.F. Tiersten: On the Interaction of the Electromagnetic Field with Heat Conducting Deformable Semiconductors. *J. Math. Phys.* **16**, 938–957, (1975)

56. K. Demachi, Y. Yoshida and K. Miya: Numerical analysis of magnetoelastic buckling of fusion reactor components. *Fusion Engineering and Design* **27**, 490–498, (1995)

57. L. Diening: Maximal function on generaized Lebesgue spaces L-p(center dot). *Math. Inequal. Appl.*, **7(2)**, 245–253, (2004)

58. H.J. Ding and W.Q. Chen: Three-Dimensional Problems in Piezoelasticty. *Nova Science Pub.*, New York (2001)

59. R.C. Dixon and A.C. Eringen: A Dynamical Theory of Polar Elastic Dielectrics. *Int. J. Eng. Sc.* **3**, 359–398, (1965)

60. R. Drouot, G. Napoli and G. Racineux: Continuum modeling of electrorheological fluids. *Int. J. of Modern Phys. B*, **16**(17&18), 2649–2654, (2002)

61. R. Drouot and G. Racineux: A continuum modelling for the reversible behaviour of electrorheological media. *Int. J. of Applied Electromagnetics and Mechanics*, **22**, 177–187, (2005)

62. J.W. Dunkin and A.C. Eringen: Propagation of Waves in an Electromagnetic Elastic Solid. *Int. J. Eng. Sc.* **1**, 461–495, (1963)

63. W. Eckart: Phenomenological modeling of electrorheological fluids with an extended CASSON-Model. *Continuum Mech. Thermodyn.*, **12(5)**, 341–362, (2000)

64. W. Eckart: Theoretische Untersuchungen von elektrorheologischen Flüssigkeiten bei homogenen und inhomogenen elektrischen Feldern. Aachen, Shaker Verlag, 162 pp., (2000)

65. W. Eckart and A. Sadiki: Polar theory for electrorheological fluids based on extended thermodynamics. *Int. J. of Applied Mechanics and Engineering*, **6**(4), 969–998, (2001)

66. K. Edamura and Y. Otsubo: Electrorheology of dielectric liquids. *Rheol. Acta*, **43**, 180–183, (2004)

67. B. Engelmann, R. Hiptmair, R.H.W. Hoppe and G. Mazurkevitch: Numerical simulation of electrorheological fluids based on an extended Bingham model. *Comput. Visual Sci.*, **2**, 211–219, (2000)

68. A.C. Eringen and E.S. Suhubi: Elastodynamics, Vol. I, Academic Press, New York and London, (1974)

69. A.C. Eringen: Theory of electromagnetic elastic plates. *Int. J. Engng. Sci.* **27**, 363–375, (1989)

70. A.C. Eringen and G.A. Maugin: Electrodynamics of Continua I. Foundations and Solid Media, Springer Verlag, New York (1989)

71. A.C. Eringen and G.A. Maugin: Electrodynamics of Continua II. Fluids and Complex Media, Springer Verlag, New York (1989)

72. A.C. Eringen and G.A. Maugin: Electrodynamics of Continua I + II, Springer Verlag, (1990/91)

73. R.M. Fano, L.C. Chu and R.B. Adler: Electromagnetic Fields, Energy and Forces, John Wiley & Sons, Inc., New York, (1960) (Reprinted by the M.I.T. Press)

74. R.P. Feynman, R.B. Leighton and M. Sands: The Feynman Lectures on Physics, Vol. 2, Addison-Wesley, Reading, Massachusettes, (1964)

75. D. Garg and G. Anderson: Structural Damping and Vibration Control via Smart Sensors and Actuators. *J. Vib. Control*, **9**, 1421–1452, (2003)

76. C. Goudjo and G.A. Maugin: On the static and dynamic stability of soft ferromagnetic elastic plates. *J. Méca. Théor. Math.* **21**, 947–975, (1983)

77. A.E. Green and R.S. Rivlin: Multipolar Continuum Mechanics. *Arch. Rat. Mech. Anal.* **17**, 113–147, (1964)

78. R.A. Grot and A.C. Eringen: Relativistic Continuum Mechanics, Part I, Mechanics and Thermodynamics; Part II, Electromagnetic Interactions with Matter. *Int. J. Eng. Sc.* **4**, 611–638, 639–670, (1966)

79. R.A. Grot: Relativistic Continuum Theory for the Interaction of Electromagnetic Fields with Deformable Bodies. *J. Math. Phys.* **11**, 109–113, (1970)

80. R.A. Grot: Relativistic Continuum Physics: Electromagnetic Interactions, in *Continuum Physics*, ed. A. C. Eringen, Academic Press, (1976)

81. R. Hanaoka, M. Murakumo, H. Anzai and K. Sakurai: Effects of electrode surface morphology on electrical response of electrorheological fluids. *IEEE Trans. Dielect. Electr. In.* **9**(1), 10–16, (2002)

82. T. Hao, A. Kawai and F. Ikazaki: Dielectric Criteria for the Electrorheological Effect. *Langmuir*, **15**, 918–921, (1999)

83. T. Hao: Electrorheological fluids. *Adv. Mater.*, **13**(24), 1847–1857, (2001)

84. K. Hara and F.C. Moon: Internal buckling and vibration of solenoid magnets for high fields, in *Electromechanical Interactions in Deformable Solids and Strutures*, ed. Y. Yamamoto and K. Miya, Elsevier, Amsterdam, 69–74, (1987)

85. D.J. Hasanyan and L. Librescu and D.R. Ambur: A few results on the foundation of the theory and behavior of nonlinear magnetoelastic plates carrying an electric current. *Int. J. Engin. Science*, **42**, 1547–1572, (2004)

86. D.J. Hasanyan, L. Librescu, Z. Qin and D.R. Ambur: Nonlinear vibration of finitely-electroconductive plate strips in an axial magnetic field. *Computers and Structures*, **83**, 1205–1216, (2005)

87. D.J. Hasanyan and G.T. Piliposian: Stress-strain state of a piecewise homogeneous ferromagnetic body with an interfacial crack. *Int. J. of Fracture*, **133**, 183–196, (2005)

88. L.A.Z. Hefni, A.F. Ghaleb and G.A. Maugin: Surface waves in a nonlinear magnetoelastic conductor of finite electric conductivity. *Int. J. of Engineering Science*, **33**, 2085–2102, (1995)

89. P. Hild, I.R. Ionescu, Th. Lachand-Robert and I. Rosca: The blocking of an inhomogeneous Bingham fluid. Applications to landslides. *M2AN*, **36**(6), 1013–1026, (2002)

90. Z. Huang and J.H. Spurk: Der elektroviskose Effekt als Folge elektrostatischer Kraft. *Rheol. Acta*, **29**, 475–481, (1990)

91. K. Hutter and Y.H. Pao: A Dynamic Theory for Magnetizable Elastic Solids with Thermal and Electrical Conduction. *J. of Elasticity*, **4**, 89–114, (1974)

92. K. Hutter: On Thermodynamics and Thermostatics of Viscous Thermoelastic Solids in the Electromagnetic Fields. A Lagrangian Formulation *Arch. Rat. Mech. Anal.* **58**, 339–368, (1975)

93. K. Hutter: Wave Propagation and Attenuation in Paramagnetic and Soft Ferromagnetic Materials. *Int. J. Eng. Sc.* **13**, 1067–1084, (1975)

94. K. Hutter: Wave Propagation and Attenuation in Paramagnetic and Soft Ferromagnetic Materials, Part II. *Int. J. Eng. Sc.* **14**, 883–894, (1976)

95. K. Hutter: A Thermodynamic Theory of Fluids and Solids in the Electromagnetic Fields. *Arch. Rat. Mech. Anal.* **64**, 269–298, (1977)

96. K. Hutter: Thermodynamic Aspects in Field-Matter Interactions, in: *Electromagnetic Interactions in Elastic Solids*, ed. by H. Parkus, Springer, Wien, (1978)

97. K. Hutter: Time-dependent surface elevation of an ice slope. *J. Glaciology*, **25**, 247–266, (1980)

98. K. Hutter: The effect of longitudinal strain on the shear stress of an ice sheet. In defence of using stretched coordinates. *J. Glaciology*, **27**, 39–66, (1981)

99. K. Hutter, F. Legerer and U. Spring: First order stresses and deformations in glaciers and ice sheets. *J. Glaciology*, **27**, 227–270, (1981)

380   References

100. K. Hutter and K. Jöhnk: Continuum Methods of Physical Modeling, Springer Verlag, Berlin etc., 635 p. (2004)
101. T. Ikeda: Fundamentals of Piezoelectricity, Oxford University Press, Oxford (1990)
102. A. Inoue and S. Maniwa: Electrorheological effect of liquid crystalline polymers. *J. Appl. Polym. Sci.* **55**, 113–118, (1995)
103. J.S. Jang and H.G. Zhou: An interface wave in piezoelectromagnetic materials. *Int. J. of Applied Electromagnetics and Mechanics*, **21**, 63–68, (2005)
104. N.F. Jordan and C.A. Eringen: On the Static Nonlinear Theory of Electromagnetic Thermoelastic Solids. *Int. J. Eng. Sc.* **2**, 59–114, (1964)
105. S. Kaliski and J. Petykiewicz: Dynamical Equations of Motion and Solving Functions for Elastic and Inelastic Anisotropic Bodies in the Magnetic Field. *Proc. of Vibr. Probl.* **1**, 17–35, (1959/1960)
106. M. Kamlah: Ferroelectric and ferroelastic piezoceramics – Modeling of electromechanical hysteresis phenomena. *Continuum Mechanics and Thermodynamics*, **13**, 219–268, (2001)
107. V.L. Karlash: Resonant electromechanical vibrations of piezoelectric plates. *Int. Appl. Mech.*, **41**, 709–747, (2005)
108. P. Katsikopoulos and C. Zukoski: Effects of electrode morphology on the electrorheological response, in *Proceedings of the 4th Conference on Electrorheological Fluids*, ed. R. Tao, World Scientific, Singapore, (1994)
109. K.B. Kazarian and R.A. Kazarian: Stability of a metal elastic shell with a stationary azimuthal-periodic current, in *Mechanical Modelling of New Electromagnetic Materials*, ed. R.K.T. Hsieh, Elsevier, Amsterdam, 337–342, (1990)
110. H. Kimura, K. Aikawa, Y. Masubuchi, J. Takimoto, K. Koyama and K. Minagawa: Phase structure change and ER-effect in liquid crystalline polymer/dimethylsiloxane blends. *Rheol. Acta* **37**, 54–60, (1998)
111. H. Kimura, K. Aikawa, Y. Masubuchi, J. Takimoto, K. Koyama and T. Uemura: "Positive" and "negative" electro-rheological effect of liquid blends. *J. Non-Newtonian Fluid Mech.* **76**, 199–211, (1998)
112. D. Klass and Th. Martinek: Electroviscous fluids. I. Rheological Properties. *J. Appl. Phys.* **38**(1), 67–74, (1967)
113. D. Klass and Th. Martinek: Electroviscous fluids. II. Electrical Properties. *J. Appl. Phys.* **38**(1), 75–80, (1967)
114. D. Klein, D. Rensink, H. Freimuth, G.J. Monkman, S. Egersdörfer, H. Böse and M. Baumann: Modeling the response of a tactile array using electrorheological fluids. *J. Phys. D: Appl. Phys.* **37**, 794–803, (2004)
115. H.-J. Ko and G.S. Dulikravich: A fully non-linear theory of electro-magnetohydrodynamics. *Int. J. of Non-Linear Mechanics*, **35**, 709–719, (2000)
116. J.S. Kumar, N. Ganesan, S. Swarnamani and C. Padmanabhan: Active control of simply supported plates with a magnetostrictive layer. *Smart Mater. Struct.*, **13**, 487–492, (2004)
117. G. Lancioni and G. Tomassetti: Flexure waves in electroelastic plates. *Wave Motion*, **35**, 257–269, (2002)
118. L.D. Landau and E.M. Lifschitz: Theoretische Physik Bd. VIII: Elektrodynamik der Kontinua. Akademie Verlag, (1985)
119. X.-F. Li and J. Yang: Electromagnetoelastic behavior induced by a crack under antiplane mechanical and inplane electric impacts. *Int. J. of Fracture*, **132**, 49–64, (2005)

120. L. Librescu, D. Hasanyan and D.R. Ambur: Electromagneticall conducting elastic plates in a magnetic field: modelling and dynamic implications. *Int. J. Non-Linear Mechanics*, **39**, 723–739, (2004)

121. P.H. van Lieshout, P.M.J. Rongen and A.A.F. Van de Ven: A variational principle for magnetoelastic buckling. *J. Eng. Math.* **21**, 227–252, (1987)

122. P.H. van Lieshout, P.M.J. Rongen and A.A.F. Van de Ven: A variational approach to magnetoelastic buckling problems for systems of ferromagnetic or superconducting beams. *J. Eng. Math.* **22**, 143–176, (1988)

123. P.H. van Lieshout and A.A.F. Van de Ven: A variational principle for magnetoelastic buckling, in *Electromechanical Interactions in Deformable Solids and Structures*, ed. Y.Yamamoto and K. Miya, Elsevier (North-Holland), Amsterdam, 15–20, (1987)

124. P.H. van Lieshout and A.A.F. Van de Ven: A variational approach to the magnetoelastic buckling of an arbitrary number of superconducting rods. *J. Eng. Math.* **25**, 353–374, (1991)

125. S. Lin, F. Narita and Y. Shindo: Electroelastic analysis of a piezoelectric cylindrical fiber with a penny-shaped crack embedded in matrix. *Int. J. Solids and Structures*, **40**, 5157–5174, (2003)

126. M.-F. Liu and T.-P. Chang: Vibration analysis of a magneto-elastic beam with general boundary conditions subjected to axial load and external force. *J. Sound and Vibration*, **288**, 399–411, (2005)

127. I.S. Liu and I. Müller: On the Thermodynamics and Thermostatics of Fluids in Electromagnetic Fields. *Arch. Rat. Mech. Anal.* **46**, 149–176, (1972)

128. H. Ma and B. Wang: The scattering of electroelastic waves by an ellipsoidal inclusion in piezoelectric medium. *Int. J. of Solids and Structures*, **42**, 4541–4554, (2005)

129. C.W. Maranville and J.M. Ginder: Small-strain dynamic mechanical behaviour of magnetorheological fluids entrained in foams. *Int. J. of Applied Electromagnetics and Mechanics*, **22**, 25–38, (2005)

130. H. Matsue, K. Demachi and K. Miya: Numerical analysis of the superconducting magnet outer vessel of a Maglev train by a structural and electromagnetic coupling method. *Physica C: Superconductivity*, **357-360**, 874–877, (2001)

131. G.A. Maugin and C.A. Eringen: Deformable Magnetizable Saturated Media, Part I, Field Equations; Part II, Constitutive Theory. *J. Math. Phys.* **13**, 143–155, 1334–1347, (1972)

132. G.A. Maugin and C.A. Eringen: Polarized Elastic Materials with Electronic Spin- A Relativistic Approach. *J. Math. Phys.* **13**, 1777–1788, (1972)

133. G.A. Maugin: On the Spin Relaxation in Deformable Ferromagnets. *Physica* **A81**, 454–468, (1975)

134. G.A. Maugin: A Continuum Theory of Deformable Ferrimagnetic Bodies, Part I, Field Equations; Part II, Thermodynamics, Constitutive Theory. *J. Math. Phys.* **17**, 1727–1738, 1739–1751, (1976)

135. G.A. Maugin: Deformable Dielectrics, Field Equations for a Dielectric Made of Several Molecular Species. *Arch. of Mech.* **28**, 679–692, (1976)

136. G.A. Maugin and C.A. Eringen: On the Equations of the Electrodynamics of Deformable Bodies of Finite Extent. *J. de Mécanique* **16**, 101–147, (1977)

137. G.A. Maugin, J. Pouget, R. Drouot and B. Collet, Nonlinear Electromechanical Couplings, John Wiley & Sons, New York (1992)

138. C. Mavroidis, Y. Bar-Cohen and M. Bouzit: Haptic Interfaces Using Electrorheologica Fluids, in *Electroactive Polymer Actuators as Artificial Muscles: reality, potentials and challenges*, SPIE Optical Engineering Press, 567–594, (2001)

139. C. Mavroidis: Development of Advanced Actuators Using Shape Memory Alloys and Electrorheological Fluids. *Res. Nondestr. Eval.* **14**, 1–32, (2002)

140. M.F. McCarthy: The Propagation and Growth of Plane Acceleration Waves in a Perfectly Electrically Conducting Elastic Material in a Magnetic Field. *Int. J. Eng. Sc.* **4**, 361–381 (1966)

141. N. Mellgren: A combined viscoelastic plastic material in an oscillating pressure-driven plane channel flow. Diplomarbeit, Technische Universität Darmstadt, (2002)

142. R.D. Mindlin: Polarization Gradient in Elastic Dielectrics. *Int. J. Solids Struct.* **4**, 637–642, (1968)

143. K. Miya, K. Hara and K. Someya: Experimental and theoretical study on the magnetoelastic buckling of a ferromagnetic cantilevered beam-plate. *J. Appl. Mech.* **45**, 355–360, (1978)

144. K. Miya, T. Takagi and Y. Ando: Finite-element analysis of magnetoelastic buckling of a ferromagnetic beam-plate. *J. Appl. Mech.* **47**, 377–382, (1980)

145. K. Miya, T. Rizawa, K. Someya, A. Minato and T. Tone: Analysis of the magnetomechanical behavior of a ferromagnetic beam-plate. *Fusion Engineering and Design* **5**, 167–180, (1987)

146. K. Miya, T. Takagi, M. Uesaka and K. Someya: Finite element analysis of magnetoelastic buckling of eight-coil superconducting full torus. *J. Appl. Mech.* **49**, 180–186, (1982)

147. K. Miya, T. Takagi and T. Takagi: Magnetosolid mechanics in fusion reactor technology. *Fusion Engineering and Design* **7**, 281–292, (1989)

148. L.V. Mol'chenko and P.V. Dikii: Two-dimensional Magnetoelastic Solutions for an Annular Plate. *Int. Appl. Mech.* **39**, 1328–1334, (2003)

149. L.V. Mol'chenko: A Method for Solving Two-dimensional Nonlinear Boundary-Value Problems of Magnetoelasticity for Thin Shells. *Int. J. of Appl. Mech.* **41**, 490–495, (2005)

150. G.J. Monkman: Addition of solid structures to electrorheological fluids. *J. Rheol.* **35**, 1385–1392, (1991)

151. F.C. Moon: Magneto-solid mechanics, John Wiley & Sons, New York (1984)

152. F.C. Moon and Y.H. Pao: Magnetoelastic Buckling of a Thin Plate. *J.A.M.* **35**, 53–58, (1968)

153. F.C. Moon and S. Chattopadhyay: Elastic Stability of a Thermonuclear Reactor Coil, Proc. 5th Symposium on Engineering Problems of Fusion Research, November 1973, Princeton, N.J. *IEEE Nuclear and Plasma* Sci. Soc., N.Y., Publ. No.**73**, CH 0843-3-NPS, 544–578, (April 1974)

154. F.C. Moon and S. Chattopadhyay: Magnetically Induced Stress Waves in a Conducting Solid-Theory and Experiment. *J.A.M.* **41**, 641–645, (1974)

155. F.C. Moon: Problems in Magneto-Solid Mechanics. *Mechanics Today*, ed. S. Nemat-Nasser, Pergamon Press Inc., New York, Vol. **4**, 307–390, (1978)

156. F.C. Moon: Magneto-solid mechanics, John Wiley & Sons, New York, (1984)

157. Moon, F.C. and Y.H. Pao: Magnetoelastic buckling of a thin plate. *Int. J. of Appl. Mech.* **35**, 53–58, (1968)

158. C. Møller: The Theory of Relativity, Oxford University Press, London, (1972)

159. I. Müller: On the Frame Dependence of Stress and Heat Flux. *Arch. Rat. Mech. Anal.* **45**, 241–250, (1972)

160. I. Müller: Thermodynamik. *Bertelsmann Universitätsverlag*, Düsseldorf, Germany, (1973)

161. I. Müller: Thermodynamics. Pitman Publishing, 521 pp. (1985)

162. M. Nemoto, Y. Kawamoto, K. Inoue, K. Ioki, M. Hashimoto, and K. Miya: Experimental study of impulsive electromagnetic buckling of cylindrical shells, in *Electromagnetic Forces and Applications*, ed. J. Tani and T. Takagi, Elsevier, Amsterdam, 453–456, (1992)

163. B. Noble: Methods Based on the Wiener-Hopf Technique for the Solution of Partial Differential Equations. Oxford, Pergamon Press, 246 pp., (1958)

164. J.P. Nowacki: Vibrating string in a magnetic field. *Int. J. Appl. Electromagnetics in Materials* **1**, 127–133, (1990)

165. J.F. Nye: Physical Properties of Crystals, Oxford University Press, Oxford, England, (1957)

166. Y. Otsubo: Effect of Electrode Pattern on the Column Structure and Yield Stress of Electrorheological Fluids. *J. Colloid Interface Sci.* **190**, 466–471, (1997)

167. Y. Otsubo and K. Edamura: Viscoelasticity of a dielectric fluid in nonuniform electric fields generated by electrodes with flocked fabrics. *Rheol. Acta*, **37**, 500–507, (1998)

168. Y. Otsubo and K. Edamura: Electric effect on the rehology of insulating oils in electrodes with flocked fabrics. *Rheol. Acta*, **38**, 137–144, (1999)

169. R.N. Ovakimian: On stability of cylindrical current-carrying shell, in *Mechanical Modelling of New Electromagnetic Materials*, ed. R.K.T. Hsieh, Elsevier, Amsterdam, 343–348, (1990)

170. Y.H. Pao and C.S. Yeh: A Linear Theory of Soft Ferromagnetic Elastic Solids. *Int. J. Eng. Sc.* **11**, 415–436 (1973)

171. Y.H. Pao and K. Hutter: Electrodynamics of Moving Elastic Solids and Viscous Fluids. *Proc. I.E.E.E.* **63**, 1011–1021, (1975)

172. Y.H. Pao: Electromagnetic Forces in Deformable Media. *Mechanics Today* Vol. 4, ed. S. Nemat-Nasser, Pergamon Press Inc., New York, (1978)

173. T.C. Papanastasiou: Flows of Materials with Yield. *J. Rheol.* **31**(5), 385–404, (1987)

174. G. Paria: Magneto-Elasticity and Magneto-Thermoelasticity. *Adv. in Appl. Mech.*, **10**, 73–112, (1967)

175. H. Parkus: Variational Principles in Thermo- and Magneto-Elasticity. *CISM, Courses and Lectures-No. 58*, Udine (1970), Springer-Verlag, Wien, (1972)

176. M. Parthasarathy and D.J. Klingenberg: Electrorheology: Mechanisms and Models. *Mater. Sci. Eng.* **17**, 57–103, (1996)

177. P. Penfield and H.A. Haus: Electrodynamics of Moving Media, The M.I.T. Press, Cambridge, Massachusettes, (1967)

178. J. Perlak and B. Vernescu: Constitutive equations for electrorheological fluids. *Rev. Roumaine Math. Pures Appl.* **45**, 287–297, (2000)

179. Ch. Pfeiffer, C. Mavroidis, Y. Bar-Cohen and B. Dolgin: Electrorheological Fluid Based Force Feedback Device, in *Proceedings of the 1999 SPIE Telemanipulator and Telepresence technologies VI Conference* SPIE Proc. **3840**, 19–21, (1999)

180. A.S. Pipkin and R.S. Rivlin: Electrical Conduction in Deformed Isotropic Materials. *J. Math. Phys.* **1**, 127–130, (1960)

181. A.S. Pipkin and R.S. Rivlin: Galvanomagnetic and Thermomagnetic Effects in Isotropic Materials. *J. Math. Phys.* **1**, 542–546, (1960)

182. D. Pirck: Homogene elektroviskose Flüssigkeiten. Deutsches Patent, DE 41 39 065 A1 (1993)

183. Yu.N. Podil'chuk and I. Yu. Podil'chuk: Stress State of a Transversely Isotropic Ferromagnetic Body with an Elliptic Crack in a Homogeneous Magnetic Field. *Int. Applied Mechanics* **41**, 32–41, (2005)

184. C.H. Popelar: Postbuckling Analysis of a Magnetoelastic Beam. *J.A.M.* **39**, 207–211, (1972)

185. A. Prechtl: On the Electrodynamics of Deformable Media. *Acta Mechanica* **28**, 255–294, (1977)

186. *The Mechanical Behaviour of Electromagnetic Solid Continua.* Proceedings of IUTAM-Symposium, Paris, France, 4–7 July, 1983. Ed: G.A. Maugin, North-Holland, Amsterdam (1984)

187. *Electromagnetomechanical Interactions in Deformable Solids and Structures.* Proceedings of IUTAM-Symposium, Tokyo, Japan, 12–17 October, 1986. Ed: Y. Yamamoto and K. Miya, North-Holland, Amsterdam (1987)

188. *Mechanical Modellings of New Electromagnetic Materials.* Proceedings of IUTAM-Symposium, Stockholm, Sweden, 2–6 April, 1990. Ed: R.K.T. Hsieh, Elsevier, Amsterdam (1990)

189. *Electromagnetic Forces and Applications.* Proceedings of the Third ISEM Symposium, Sendai, Japan, 28–30 January (1991). Ed: J. Tani and T. Takagi, Elsevier, Amsterdam (1992)

190. *Simulation and Design of Applied Electromagnetic Systems.* Proceedings of Fourth ISEM Symposium, Sapporo, Japan 26–30 January, 1993. Ed: T. Honma, Elsevier, Amsterdam (1994)

191. Proceedings of the eleventh International Symposium on Applied Electromagnetics and Mechanics ISEM 03, Versailles. *Special Issue International Journal of Applied Electromagnetics and Mechanics*, **19**, (2004)

192. Y. Qi and W. Wen: Influence of geometry of particles on electrorheological fluids. *J. Phys. D: Appl. Phys.*, **35**, 2231–2235, (2002)

193. Q.-H. Qin: Fracture Mechanics of Piezoelectric Materials, WIT Press, Southampton, Boston (2001)

194. K.R. Rajagopal and A.S. Wineman: Flow of electro-rheological materials. *Acta Mechanica* **91**, 57–75, (1992)

195. K.R. Rajagopal, R.C. Yalamanchili and A.S. Wineman: Modeling electro-rheological materials through mixture theory. *Int. J. Engng. Sci.*, **32**(3), 481–500, (1994)

196. K.R. Rajagopal and M. Růžička: Mathematical modeling of electrorheological materials. *Continuum Mech. Thermodyn.*, **13**(1), 59–78, (2001)

197. L. Rejon: Electrorheological characterisation of suspensions of surface-modified ceramic hollow-sphere particles. *Oral presentation at Eurorheo 2002 Stuttgart.*

198. E.J. Rhee, M.K. Park, R. Yamane and S. Oshima: A study on the relation between flow characteristics and cluster formation of electrorheological fluid using visualization. *Exp. in Fluids* **34**, 316–323, (2003)

199. P. Riha, H. Kimura, K. Aikawa, Y. Masubuchi, J. Takimoto and K. Koyama: The shear-flow properties of electro-rheological liquid polymeric blends. *J. Non-Newtonian Fluid Mech.* **85**, 249–256, (1999)

200. A. Romano: A Macroscopic Nonlinear Theory of Magnetothermoelastic Continua. *Arch. Rat. Mech. Anal* **65**, 1–24, (1977)

201. M. Romeo: Electromagnetoelastic waves at piezoelectric interfaces. *Int. J. of Engineering Science*, **42**, 753–768, (2004)

202. M. Romeo: Electromagnetoelastic waves at piezoelectric interfaces. *Int. J. of Engineering Science*, **44**, 14–25, (2006)

203. B.W. Roos: Analytic Functions and Distributions in Physics and Engineering. New York, Wiley, 521 pp., (1969)

204. L. Rosenfeldt: Theory of electrons, North-Holland, Amsterdam, (1951)

205. M. Růžička: Electrorheological Fluids: Modeling and Mathematical Theory. Berlin, Springer, 176 pp., (2000)

206. A. Sadiki and C. Balan: Rate-type Model for electro-rheological Material Behaviour consistent with Extended Thermodynamics: Application to a steady viscometric flow. *Proc. Appl. Math. Mech.* **2**, 174–175, (2003)

207. U. Schindler, J. Schindler, R. Steger and P.O. Brunn: Optical studies (LDA) of an electrorheological fluid in slit flow. *Rheol. Acta* **34**, 80–85, (1995)

208. H. See, H. Tamura and M. Doi: The role of water capillary forces in electrorheological fluids. *J. Phys. D: Appl. Phys.* **26**, 746–752, (1993)

209. H. See: Advances in modeling the mechanisms and rheology of electrorheological fluids. *Korea-Australia Rheology J.* **11(3)**, 169–195, (1999)

210. H. See: Constitutive equation for electrorheological fluids based on the chain model. *J. Phys. D: Appl. Phys.* **33**, 1625–1633, (2000)

211. Y. Shindo: Plane-strain problem of a crack in a ferromagnetic elastic layer. *Theoret. Appl. Mech.* **30**, 203–214, (1981)

212. Y. Shindo, E. Ozawa and J.P. Nowacki: Singular stress and electric fields of a cracked piezoelectric strip. *Int. J. of Applied Electromagnetics in Materials* **1**, 77–87, (1990)

213. Y. Shindo, K. Horiguchi and A.A.F. Van de Ven: Bending of a magnetically saturated plate with a crack in a uniform magnetic field. *Int. J. of Applied Electromagnetics in Materials* **1**, 135–146, (1990)

214. Y. Shindo, D. Sekiya, F. Narita and K. Hohiguchi: Tensile testing and analysis of ferromagnetic elastic strip with a central crack in a uniform magnetic field. *Acta Materialia*, **52**, 4677–4684, (2004)

215. Shu-Ang Zhou: Electrodynamic Theory of Superconductors, Peter Peregrinus Ltd., London (1991)

216. Shu-Ang Zhou: Electrodynamics of Solids and Microwave Superconductivity, John Wiley & Sons, New York (1999)

217. N.D. Sims, R. Stanway, D.J. Peel, W.A. Bullough and A.R. Johnson: Controllable viscous damping: an experimental study of an electrorheological long-stroke damper under proportional feedback control. *Smart Mater. Struct.* **8**, 601–615, (1999)

218. P.R.J.M. Smits, P.H. van Lieshout and A.A.F. Van de Ven: A variational approach to magnetoelastic buckling problems for systems of superconducting tori. *J. Eng. Math.* **23**, 157–186, (1989)

219. A. Sommerfeld: Vorlesungen über theoretische Physik, Band III: Elektrodynamik. Harri Deutsch, (1988)

220. A.J.M. Spencer: Theory of Invariants, in A.C. Eringen: Continuum Physics Vol. **1**: Mathematics, Academic Press, (1971)

221. J.H. Spurk: Fluid Mechanics. Berlin, Heidelberg, Springer Verlag, 513 pp., (1997)

222. J.E. Stangroom: Electrorheological fluids. *Phys. Technol.* **14**, 290–296, (1983)

223. J.E. Stangroom: Basic considerations in flowing electrorheological fluids. *J. Stat. Phys.* **64**(5-6), 1059–1072, (1991)

224. R. Stanway, J.L. Spronston and A.K. El-Wahed: Applications of electro-rheological fluids in vibration control: a survey. *Smart Mater. Struct.* **5**, 464–482, (1996)

225. E.S. Suhubi: Elastic Dielectrics with Polarization Gradient. *Int. J. Eng. Sc.* **7**, 993–997, (1969)

226. T. Takagi, K. Miya, H. Yamada and T. Takagi: Theoretical and experimental study on the magnetomechanical behaviour of superconducting helical coils for a fusion reactor. *Nucl. Eng. Design/Fusion* **1**, 61–71, (1984)

227. T. Takagi, J. Tani, S. Kawamura and K. Miya: Coupling effect between mag-netic field and deflection in thin structure. *Fusion Engn. and Design*, **18**, 425–433, (1991)

228. W.Y. Tam, G. Yi, W. Wen, H. Ma, M.M.T. Loy and P. Sheng: New Elec-trorheological Fluid: Theory and Experiment. *Phys. Rev. Lett.* **78**, 2987–2990, (1998)

229. H. Tamura, H. See and M. Doi: Model of porous particles containing water in electrorheological fluids. *J. Phys. D: Appl. Phys.* **26**(8), 1181–1187, (1993)

230. Y. Tanaka, T. Horie and T. Niho: Simplified analysis method for vibration of fusion reactor components with magnetic damping. *Fusion Engineering and Design*, **51-52**, 263–271, (2000)

231. J. Tani and K. Otomo: Interaction of two nearby ferromagnetic panels on the magnetoelastic buckling, in *The Mechanical Behaviour of Electromagnetic Solid Continua*, ed. G.A. Maugin, Elsevier (North-Holland), Amsterdam, 385–390, (1984)

232. G.W. Taylor: Piezoelectricity, Gordon and Beach, Amsterdam, (1992)

233. H.F. Tiersten: Coupled Magnetomechanical Equations for Magnetically Sat-urated Insulators. *J. Math. Phys.* **5**, 1298–1318, (1964)

234. H.F. Tiersten: Thickness Vibrations of Saturated Magnetoelastic Plates. *J.A.P.* **36**, 2250–2259, (1965)

235. H.F. Tiersten: Linear Piezoelectric Plate Vibrations, Plenum Press, New York, (1969)

236. H.F. Tiersten: On the Nonlinear Equations of Thermoelectroelasticity. *Int. J. Eng. Sc.* **9**, 587–603, (1971)

237. H.F. Tiersten and C.F. Tsai: On the Interaction of the Electromagnetic Field with Heat Conducting Deformable Insulators. *J. Math. Phys.* **13**, 361–378, (1972)

238. H.F. Tiersten: A Development of the Equations of Electromagnetism in Ma-terial Continua. *Springer Tracts in Natural Philosophy* **36** Springer, Berlin, (1990)

239. S. Timoshenko and D.H. Young: Vibration Problems in Engineering, (Third Edition) D. van Nostrant Company, New York, (1954)

240. R.A. Toupin: The Elastic Dielectric. *J. Rational Mechanics and Analysis* **5**, 850–915, (1956)

241. R.A. Toupin: A Dynamical Theory of Elastic Dielectrics. *Int. J. Eng. Sc.* **1**, 101–126, (1963)

242. C. Trimarco: Stresses and moments in electromagnetic materials. *Mechanics Research Communications* **29**, 485–492, (2002)

243. C. Truesdell and W. Noll: The Nonlinear Filed Theories of Mechanics. *Handbuch der Physik* Vol. III/3, ed. S. Flügge, Springer-Verlag, Berlin (1960)
244. C. Truesdell and R.A. Toupin: The Classical field Theories. *Encyclopedia of Physics* Vol. III/1, ed. S. Flügge, Springer-Verlag, Berlin, (1960)
245. T. Tsukiji and S. Tanabe: ER effect of liquid crystal flowing between two parallel-plate electrodes. *Int. J. Mod. Phys. B* **16**, 2569–2575, (2002)
246. A. Ursescu, W. Eckart, H. Marschall and K. Hutter: Inhomogeneous electric field generated by two long electrodes placed along parallel infinite walls separating different dielectric media. *J. Engrg. Math.* **49**, 57–75, (2004)
247. A. Ursescu: Channel Flow of Electrorheological Fluids under an Inhomogeneous Electric Field. Technische Universität Darmstadt, 139 pp., (2005) (http://elib.tu-darmstadt.de/diss/000556)
248. A.A.F. Van de Ven: On the Vibrations of a Nonconducting Magnetically Saturated Cylinder in a Magnetic Field. *Proc. of Vibr. Probl.* **11**, 89–102, (1970)
249. A.A.F. Van de Ven : Interaction of Electromagnetic and Elastic Fields in Solids. *Dr. of Science Thesis* University of Technology Eindhoven, the Netherlands, (1975)
250. A.A.F. Van de Ven: Magnetoelastic Buckling of Thin Plates in a Uniform Transverse Magnetic Field. *J. of Elasticity* **8**, 297–312, (1978)
251. A.A.F. Van de Ven: Magnetoelastic buckling of magnetically saturated bodies. *Acta Mech.* **47**, 229–244, (1983)
252. A.A.F. Van de Ven: Magnetoelastic buckling of a beam of elliptic cross-section. *Acta Mech.* **51**, 119–138, (1984)
253. A.A.F. Van de Ven: The influence of finite specimen dimensions on the magnetoelastic buckling of a cantilever, in *The Mechanical Behaviour of Electromagnetic Solid Continua*, ed. G.A. Maugin, Elsevier (North-Holland), Amsterdam, 421–426, (1984)
254. A.A.F. Van de Ven and M.J.H. Couwenberg: Magnetoelastic stability of a superconducting ring in its own field. *J. Eng. Math.* **20**, 251–270, (1986)
255. A.A.F. Van de Ven, J. Tani, K. Otomo and Y. Shindo: Magnetoelastic buckling of two nearby ferromagnetic rods in a magnetic field. *Acta Mech.* **75**, 191–209, (1988)
256. A.A.F. Van de Ven and P.H. van Lieshout: Buckling of superconducting structures under prescribed current, in *Proceedings of IUTAM- Symposium on the Mechanical Modellings of New Electromagnetic Materials*, ed. R.K.T. Hsieh, Elsevier, Amsterdam, 25–42, (1991)
257. A.A.F. Van de Ven and L.G.F.C. van Bree: Buckling of superconducting structures. A variational approach using the law of Biot and Savart. *Int. J. Appl. Electromagnetics in Materials* **3**, 111–137, (1992)
258. B. Vernescu: Multiscale Analysis of Electrorheological Fluids. *Int. J. Mod. Phys. B* **16**(17&18), 2643–2648, (2002)
259. K.B. Vlasov and B.Kh. Ishmukhametov: Equations of Motion and State for Magnetoelastic Media. *Soviet Phys. JETP* **19**, 142–148, (1964)
260. D.V. Wallerstein and M.O. Peach, Magnetoelastic Buckling of Beams and Thin Plates of Magnetically Soft Material. *J.A.M* **39**, 451–455, (1972)
261. B. Wang and Z. Xiao: A general constitutive equation of an ER suspension based on the internal variable theory. *Acta Mechanica* **163**, 99–120, (2003)
262. X.D. Wang and L.Y. Jiang: Coupled behaviour of interacting dielectric cracks in piezoelctric materials. *Int. J. of Fracture* **132**, 115–133, (2005)

263. G.H. Wannier: Statistical Physics, John Wiley & Sons, Inc., New York, London, Sydney, (1966)

264. H.F. Weinberger: A First Course in Partial Differential Equations. New York, Dover Publications, Inc., 446 pp., (1995)

265. W. Winslow: Induced fibration of suspensions. *J. Appl. Phys.* **20**, 1137–1140, (1949)

266. P. Wolfe: Equilibrium state of an elastic conductor in a magnetic field: a paradigm of bifurcation theory. *TAMS* **278**, 377–387, (1983)

267. P. Wolfe: Bifurcation theory of an elastic conducting wire subject to magnetic forces. *J. Elast.* **23**, 201–217, (1990)

268. Wolfram Research, Inc., *Mathematica.* Version 4.1.2.0 (2000)

269. Th. Wunderlich: Der Einfluss der Elektrodenoberfläche und der Strömungsform auf den elektrorheologischen Effekt. PhD thesis, Universität Erlangen-Nürnberg, 160 pp., (2000)

270. Th. Wunderlich and P.O. Brunn: Pressure drop measurements inside a flat channel – with flush mounted and protruding electrodes of variable length – using an electrorheological fluid. *Exp. in Fluids* **28**, 455–461, (2000)

271. H. Yabuno: Buckling of a beam subjected to electromagnetic force and its stabilization by controlling the perturbation of the bifurcation. *Nonlinear Dynamics* **10**, 271–285, (1996)

272. I.K. Yang and A.D. Shine: Electrorheology of a nematic poly(n-hexyl isocyanate)solution. *J. Rheol.* **36**(6), 1079–1104, (1992)

273. J.S. Yang: Buckling of a piezoelectric plate. *Int. J. of Applied Electromagnetics and Mechanics* **9**, 399–408, (1998)

274. J.S. Yang: Equations for the extension and flexure of a piezoelectric beam with rectangular cross section and applications. *Int. J. Applied Electromagnetics and Mechanics* **9**, 409–420, (1998)

275. G.Z. Yao, C.K. Mechefske and B.K. Rutt: Vibration analysis and measurement of a gradient coil insert in a 4T MRI. *J. Sound and Vibration* **285**, 743–758, (2005)

276. H. Zhao, Z. Liu and Y. Liu: Mechanical properties of a new electrorheological fluid. *Solid State Commun.* **116**, 321–325, (2000)

277. X. Zhao and D. Gao: Structure evolution in Poiseuille flow of electrorheological fluids. *J. Phys. D: Apl. Phys.* **34**, 2926–2931, (2001)

278. X. Zhao, X.Y. Gao and D.J. Gao: Evolution of chain structure of electrorheological fluids in flow model. *Int. J. Mod. Phys. B* **16**(17&18), 2697–2703, (2002)

279. J.Z. Zheng, Y.-H. Zhou, X.P. Yang and K. Miya: Analysis on dynamic stability of superconducting coils in a three-coil partial torus. *Fusion Engineering and Design* **54**, 31–39, (2001)

280. Y.H. Zhou, X.J. Zheng and K. Miya: Magnetoelastic bending and buckling of three-coil superconducting partial torus. *Fusion Engineering and Design* **30**, 275–289, (1995)

281. Y.H. Zhou, X.J. Zheng and K. Miya: Magnetoelastic bending and buckling of three-coil superconducting partial torus. *Fusion Engn. and Design* **30**, 275–289, (1995)

282. Y.H. Zhou, X.J. Zheng and K. Miya: Magnetoelastic bending and snapping of ferromagnetic plates in oblique magnetic fields. *Fusion Engn. and Design* **30**, 325–337, (1995)

283. Y.H. Zhou and X.J. Zheng: A general expression of magnetic force for soft ferromagnetic plates in complex magnetic fields. *Int. J. of Engng. Sci.* **35**, 1405–1417, (1997)

284. Y.H. Zhou and K. Miya: Mechanical behaviors of magnetoelastic interaction for superconducting helical magnets. *Fusion Engineering and Design* **38**, 283–293, (1998)

285. Y.H. Zhou and K. Miya: A theoretical prediction of increase of natural frequency to ferromagnetic plates under in-plate magnetic fields. *J. Sound and Vibration* **222**, 49–64, (1999)

Name Index

Subject Index

Lecture Notes in Physics

For information about earlier volumes
please contact your bookseller or Springer
LNP Online archive: springerlink.com